Praise for
Food, Agriculture, and Environmental Law

"This book is an essential guide for changing the direction and dynamics of a food production system in deep peril. The good news is that we already have a complex structure of policies, laws, and regulations in place that, if properly applied, could help us right the course of American agriculture. *Food, Agriculture, and Environmental Law* may be just the book to inspire a new generation of policymakers, activists, and lawyers to rebuild a food system around the principles of environmental sustainability, social equity, and rural vitality."

—Daniel Imhoff, Distinguished Author (*Food Fight: The Citizen's Guide to the Next Food and Farm Bill* and *Farming with the Wild: Enhancing Biodiversity on Farms and Ranches*) and Director at Watershed Media

"This groundbreaking book has arrived in the nick of time to provide a carefully crafted blueprint for what must be done to reform our food and agricultural systems through existing laws and policies. A must read for anyone who enjoys healthy food produced in an ecologically sustainable manner, this book provides a ray of hope in a darkening landscape."

—Patrick A. Parenteau, Professor of Law and Senior Counsel to the Environmental and Natural Resources Law Clinic, Vermont Law School

Praise for
Food, Agriculture and Environmental Law

"This book is an essential guide for changing the direction and dynamics of a food production system in deep peril. The good news is that we already have a complex structure of policies, laws, and regulations in place that, if properly applied, could help us right the course of American agriculture. Food, Agriculture, and Environmental Law may be just the book to inspire a new generation of policymakers, activists, and lawyers to rebuild a food system around the principles of environmental sustainability, social equity, and rural vitality."

—Daniel Imhoff, Distinguished Author of Food Fight: The Citizen's Guide to the Next Food and Farm Bill and Farming with the Wild, Publisher at Watershed Media, or Farming and Rancher and Director at Watershed Media

"This groundbreaking book has arrived in the nick of time to provide a carefully crafted blueprint for what must be done to reform our food and agricultural systems through existing laws and policies. A must read for anyone who cares to have food produced in an ecologically sustainable manner, this book provides a ray of hope in a darkening landscape."

—Patrick A. Parenteau, Professor of Law and Senior Counsel to the Environmental and Natural Resources Law Clinic, Vermont Law School

FOOD, AGRICULTURE, AND ENVIRONMENTAL LAW

by
Mary Jane Angelo, Jason J. Czarnezki,
and William S. Eubanks II

ENVIRONMENTAL LAW INSTITUTE

Washington, D.C.

Copyright © 2013 Environmental Law Institute
2000 L Street NW, Washington, DC 20036

Published April 2013.

Printed in the United States of America
ISBN 978-1-58576-160-9

Table of Contents

About the Editors xiii

Contributing Authors xv

Other Contributing Authors xvii

Chapter 1—A Brief History of U.S. Agricultural Policy and the Farm Bill 1

    A. *How Our Agricultural Policy Has Evolved: A Nation's Time Line Through an Agricultural Prism* 1

        1. From Jeffersonian Agrarianism to the Great Depression 1

        2. The "Temporary" Safety Net: The Creation of the Farm Bill 2

        3. The Post-World War II Boom: Maximizing Yields and New Technology ... 4

        4. The 1980s Farm Crisis and Congress' Response 7

        5. Recent History: Decoupling and Beyond 8

    B. *How Agricultural Policy Has Shaped the Modern Farm Economy, and How Historical Trends Can Help Us Predict the Policy Future* 10

    Conclusion 12

Chapter 2—An Overview of the Modern Farm Bill 13

    A. *The Food, Conservation, and Energy Act of 2008: Key Provisions and Programs Impacting the Environment* 14

        1. Title I: Commodities 14

        2. Title II: Conservation 21

        3. Title IX: Energy 25

        4. Title X: Horticulture and Organic Agriculture 29

    B. *Environmental Implications of Farm Bill Programs* 30

    C. *Looking Toward the Future: Trends and Issues for Upcoming Farm Bill Legislation* 31

        1. Budgetary Considerations 31

        2. Trade Considerations 32

        3. Policy Considerations 33

    Conclusion 33

Chapter 3—The Environmental Impacts of Industrial Fertilizers and Pesticides 35

    A. *History of Pesticide and Fertilizer Use* 35

    B. *Environmental Risks of Pesticides* 39

    C. *Environmental Risks of Fertilizers* 44

    D. *Other Considerations Prompted by Pesticide and Fertilizer Use* 48

        1. Carbon Footprint/Energy Usage 48

        2. Groundwater Contamination 49

　　　　3.　Sustainable Pest Management and Fertilization 49
　　Conclusion 50

Chapter 4—Agricultural Irrigation 51
　　A.　Irrigation in Practice 52
　　　　1.　Background 52
　　　　2.　Field Drainage: Purposes 54
　　　　3.　Field Drainage: Techniques 55
　　B.　The Potential Environmental Effects of Agricultural Irrigation 57
　　　　1.　Runoff From Farm Fields 58
　　　　2.　Groundwater Pumping 58
　　　　3.　Responding to Environmental Impacts: Law and Policy 59
　　C.　Other Factors Affecting the Impact of Irrigation 60
　　　　1.　Shift to Large, Integrated Farm Operations 60
　　　　2.　Conversion of Rangeland to Irrigation 60
　　　　3.　Conversion of Wetlands to Irrigated Farmland 60
　　　　4.　Climate Change and Variability 61
　　D.　Irrigation's Challenges and Opportunities 61
　　　　1.　Reducing Polluted Return Flows 61
　　　　2.　Water Conservation 62
　　Conclusion 62

Chapter 5—The Industrialization of Animal Agriculture: Connecting a Model With Its Impacts on the Environment 65
　　A.　Growth and Consolidation of the Livestock and Poultry Industry 66
　　B.　Livestock Production Models 68
　　　　1.　Production Area 68
　　　　2.　Land Application 72
　　C.　Environmental Impacts of Industrial Animal Production 72
　　　　1.　Water Pollution and Water Scarcity 73
　　　　2.　Air Pollution and Climate Change 82
　　　　3.　Land Degradation 89
　　　　4.　Loss of Biodiversity 90
　　　　5.　Environmental Injustice 91
　　Conclusion 91

Chapter 6—Genetically Modified Organisms and the Environment 93
　　A.　Genetically Modified Plants 93
　　　　1.　What Is a GM Plant? 94
　　　　2.　Benefits of GM Plants 95

3. Risks and Concerns 96
4. Environmental Risks 97
5. Regulation of GM Plants 99
6. Case Study: Alfalfa 101

B. *GM Animals* 103
1. Background 103
2. Benefits 103
3. Risks and Concerns 104
4. Regulation of GM Animals in the United States 105
5. Case Study: AquaBounty Salmon 107
6. Labeling for GM Salmon? 110

Conclusion 111

Chapter 7—Environmental and Climate Impacts of Food Production, Processing, Packaging, and Distribution 113

A. *Food, Agriculture, and the Environment* 114
B. *Industrialized Food Production and Cultivation* 116
1. Mechanized Cultivation and Irrigation 117
2. Pesticides and Fertilizers 119
3. Monoculture 122

C. *Food Processing* 123
1. High Fructose Corn Syrup, Commodity Crops, and the Farm Bill 123
2. Greenhouse Gas Emissions and Air Pollution 124
3. Wastewater Pollution 124

D. *Food Packaging* 125
E. *Food Distribution* 127
1. Food Miles 127
2. Importing Food 127

Conclusion 128

Chapter 8—The Federal Insecticide, Fungicide, and Rodenticide Act 129

A. *History and Provisions of the Federal Insecticide, Fungicide, and Rodenticide Act* 130
1. History of the Act 130
2. Registration 130
3. Data Requirements 131
4. Regulation of Pesticide Use 132
5. Restricted-Use Pesticides 133

		6. Other Approval Mechanisms	133
		7. Continuing Duties of Registrants	134
		8. Cancellation and Suspension	135
	B.	*FIFRA's Strengths and Limitations*	136
		1. Strengths	137
		2. Weaknesses	138
	C.	*Recent Legal Developments*	141
	D.	*Encouraging Reduced-Risk Pesticides*	144
		Conclusion	145

Chapter 9—Agriculture and the Clean Water Act 147

	A.	*A Brief Overview*	147
	B.	*The Clean Water Act*	149
		1. The National Pollutant Discharge Elimination System Permitting Program	149
		2. Standards for Effluent Limitations and Water Quality	155
		3. Animal Feeding Operations and Concentrated Animal Feeding Operations	156
		4. Aquaculture	158
		5. Wetlands Regulation Under CWA §404	158
		Conclusion	161

Chapter 10—Agriculture and the Clean Air Act 163

	A.	*Clean Air Act*	164
		1. National Ambient Air Quality Standards and State Implementation Plans	164
		2. New Source Review and Title V Permits	166
		3. NSPS Program	167
		4. Hazardous Air Pollutant Program	168
	B.	*Reporting Requirements Under EPCRA and CERCLA*	169
	C.	*Enforcement Issues*	171
		1. Initial Enforcement Actions	171
		2. EPA's Consent Agreement	173
	D.	*State Efforts*	176
		1. State Authority	176
		2. State Permitting	177
		3. State Air Quality Standards	179
		4. Other State Requirements	181
	E.	*Greenhouse Gas Controls*	181
		1. PSD and Title V Permitting	182

		2. NSPS and Mobile Source Rules 182
		3. State and Regional Measures 183

    Conclusion 183

Chapter 11—Agriculture and the Endangered Species Act 185

    A. *The ESA: Statutory and Regulatory Framework* 185

        1. Background 185

        2. Section 4—The Listing Process 186

        3. Section 9—The Take Prohibition 187

        4. Sections 7 and 10—Interagency Consultation and Incidental Take Permits 188

        5. ESA Litigation 190

    B. *Application of the ESA to Crop-Based Agricultural Inputs* 191

        1. Crop Seeds 192

        2. Fertilizers 194

        3. Irrigation 198

        4. Pesticides 201

    C. *Application of the ESA to Animal Agriculture and Nonplant GE Foods* 203

        1. Concentrated Animal Feeding Operations 203

        2. Nonplant GE Foods 204

    Conclusion 205

Chapter 12—Agriculture, Food, and the National Environmental Policy Act 207

    A. *NEPA's Statutory and Regulatory Framework* 207

        1. Background 207

        2. An Environmental Impact Statement or an Environmental Assessment? 208

        3. Federal Agency Obligations in an EIS 209

        4. NEPA Litigation 210

    B. *NEPA as Applied to Agriculture and Food* 211

        1. Programmatic NEPA Review of the Farm Bill 212

        2. NEPA Review of Independent Statutory Processes 217

    Conclusion 222

Chapter 13—The Food Statutes 223

    A. *Public Health and Safety* 223

        1. The Federal Food, Drug, and Cosmetic Act and the Food Quality Protection Act 224

        2. The Food Safety Modernization Act of 2010 226

B. Organic Food and Labeling 228
   1. The OFPA 228
   2. Country of Origin Labeling Provisions of the 2002 Farm Bill 232
C. The National School Lunch Program 233
   1. Origins of the National School Lunch Program 234
   2. 2010 Reauthorization of the Child Nutrition Act 235
   3. Role of the USDA in the NSLP 236
   4. Challenges to the NSLP 237
   5. Successful Reform Programs 239

Conclusion 240

Chapter 14—Agriculture and Ecosystem Services: Paying Farmers to Do the *New* Right Thing 241

A. *Promoting Farm Multifunctionality* 244
   1. The Ecology and Economics of Farm Multifunctionality 244
   2. Conceiving Alternative Futures for Agricultural Lands 247
   3. Policy Instruments 248
B. *Designing PES and TDR Programs for Agricultural Ecosystem Services* 251
   1. General Design Features 252
   2. Designing Agricultural PES Programs 253
   3. Designing Agricultural TDRs for Ecosystem Service Enhancement 254
   4. Matching PES and TDR Programs With Context 258

Conclusion 259

Chapter 15—Achieving a Sustainable Farm Bill 263

A. *Seeking a Truly "Green" Revolution: Large-Scale Reform for Widespread Problems* 263
   1. Why a Fundamental Shift Will Work: Sustainable Agriculture Already Exists on a Small Scale 266
   2. Scaling Up Sustainable Agriculture With Significant Reform of Farm Bill Commodity Subsidies 269
B. *Breathing New Life Into the Farm Bill: Life by a Thousand Cuts* 270
   1. Eliminating or Limiting Commodity Payments and Crop Insurance Payments 271
   2. Putting the Flexible Back in Planting Flexibility 272
   3. Reestablishing Conservation Compliance Conditions on Federal Crop and Revenue Insurance Payments 273
   4. Ensuring Adequate Funding for the Conservation Stewardship Program and Eliminating Barriers to Enrollment in the Program 275

            5. Prioritizing Organic Agriculture Through Funding, Research, and Targeted Set-Asides 276
            6. Bolstering Local and Regional Food Systems 277
    Conclusion 279

Chapter 16—Regulating Transgenic Crops Pursuant to the Plant Protection Act............281
    A. *Agricultural Biotechnology* 282
    B. *The Impacts of Transgenic Crops* 284
    C. *USDA Oversight of Transgenic Crops* 286
    D. *Applying the Plant Protection Act to Transgenic Crops* 290
            1. Applying the PPA's Plant Pest Authority 291
            2. Transgenic Contamination and Economic Impacts Under the Noxious Weed Authority 292
            3. The Noxious Weed Authority and HR Weeds 292
            4. Following EPA's Example 293
            5. Integrated Resistance Management 294
            6. Updating USDA's Scope of Authority 295
            7. Including Public Health Assessment 295
            8. Implementing USDA's Authority to Promulgate Partial Deregulations and Continuing Post-Market Oversight 296
    Conclusion 298

Chapter 17—The Future of Food Eco-Labeling: A Comparative Analysis 301
    A. *Food and the Environment* 303
            1. Agricultural Practices 303
            2. Livestock and Fishing Industries 304
            3. Food Processing and Distribution Systems 305
    B. *Environmental Labeling Regimes for Food in the United States and Europe* 306
            1. Organic Labeling 306
            2. Carbon Footprint Labeling 308
            3. Country of Origin Labeling and Other Food Labels 309
    C. *The Swedish Experiment* 310
            1. Swedish Dietary Guidelines 310
            2. Klimatmärkning för Mat (Climate Labeling for Food) 311
    D. *Environmental Federalism in the United States and Europe* 314
            1. The Merits of Federal Legislation 315
            2. A State-Sponsored Eco-Label in the United States 316

  3. Environmental Life-Cycle Analysis 319
  4. Implementing an Eco-Labeling Program 321
 Conclusion 323

Chapter 18—Into the Future: Building a Sustainable and Resilient Agricultural System for a Changing Global Environment 325
 A. *The Link Between Agriculture and Climate Change* 325
 B. *Agriculture's Contribution to Climate Change* 326
 C. *Climate Change Impacts on Agriculture* 327
 D. *Adapting to Climate Change* 328
  1. Ecological Resilience 329
  2. Building a Sustainable and Resilient Agro-Ecosystem 330
 E. *Policy Solutions* 331
 Conclusion 332

Index 333

About the Editors

Mary Jane Angelo
Mary Jane Angelo is a Professor of Law, Director of the Environmental and Land Use Law Program, and University of Florida Research Foundation Professor at the University of Florida Levin College of Law. She is also Affiliate Faculty in both the University of Florida School of Natural Resources and Water Institute. Mary Jane has published extensively on a variety of environmental law topics including pesticide law, endangered species law, water and wetlands law, sustainable agriculture, the regulation of genetically modified organisms, and the relationship between law and science. Her articles have been published in the *Texas Law Review*, the *Wake Forest Law Review*, the *George Mason Law Review*, the *Harvard Environmental Law Review, Ecology Law Quarterly*, and *Environmental Law*. Her forthcoming book, *The Law and Ecology of Pesticides and Pest Management*, will be published by Ashgate Publishing in 2013. Mary Jane serves on two National Academy of Sciences, National Research Council Committees: The Committee on Independent Scientific Review of Everglades Restoration Progress; and the Committee on Ecological Risk Assessment under FIFRA and the ESA. Mary Jane is also a member of the Vermont Law School summer faculty and has taught and lectured throughout the United States and other parts of the world, including Belize, Brazil, Costa Rica, Poland, and Uruguay. She is also a Member-Scholar with the Center for Progressive Reform in Washington, D.C. Prior to joining academia, May Jane practiced as an environmental lawyer for many years. She served in the U.S. Environmental Protection Agency Office of the Administrator and Office of General Counsel in Washington, D.C., and as Senior Assistant General Counsel for the St. Johns River Water Management District in Florida. Her substantial environmental law practice has included water law, wetlands law, endangered species law, pesticides law, biotechnology law, and hazardous and toxic substances law. Mary Jane received her B.S., with High Honors, in biological sciences from Rutgers University, and both her M.S., in Entomology, and J.D., with Honors, from the University of Florida.

Jason J. Czarnezki
Jason Czarnezki is, as of the 2013-2014 academic year, the Gilbert & Sarah Kerlin Distinguished Professor of Environmental Law at Pace Law School. Prior to joining the Pace Law faculty, he was Professor of Law in the Environmental Law Center at Vermont Law School and faculty director of the U.S.-China Partnership for Environmental Law. He also has held academic appointments at Marquette University Law School and the DePaul University College of Law. Jason also served as a guest researcher at Uppsala University in Sweden in 2011 and spent the 2009-2010 academic year as a J. William Fulbright Scholar at Sun Yat-Sen University in Guangzhou, China. He has presented his work on environmentalism, natural resources law, food policy, and global climate policy at universities, public interest organizations, government institutions, and conferences throughout the United States, Europe, and Asia. Previously, he served as a law clerk to the Hon. D. Brock Hornby of the U.S. District Court for the District of Maine and as a law clerk for the Bureau of Legal Services at the Wisconsin Department of Natural Resources. His articles have been published in the law journals of Boston College, Boston University, Stanford University, the University of Chicago, the University of Colorado, the University of Maryland, and the University of Virginia, and he is the author of *Everyday Environmentalism: Law, Nature and Individual Behavior* (ELI 2011). Jason received his undergraduate and law degrees from the University of Chicago.

William S. Eubanks II
Bill Eubanks is a partner at one of the nation's leading public interest environmental law firms, Meyer Glitzenstein & Crystal, where he litigates complex federal environmental cases on behalf of conservation organizations under the Endangered Species Act, National Environmental Policy Act, Clean Water Act, National Park Service Organic Act, and other statutes. Cases on which he has worked include challenging *Deepwater Horizon* oil spill response strategies harming sea turtles, garnering protections for endangered Indiana bats from an industrial wind energy project, obtaining agency records regarding federal financing of coal-fired power facilities, forcing a reconsideration of critical habitat for the California tiger salamander,

reducing off-road vehicle use in Florida's Big Cypress National Preserve, and co-authoring several amicus briefs in the U.S. Supreme Court on standing and remedies in environmental cases involving climate change, genetically modified crops, and naval sonar use. Among other topics, Bill has also written and lectured extensively about the environmental and public health impacts of agricultural policy and examined various proposals to create more sustainable and resilient food systems. Bill serves as an Adjunct Professor of Law at Vermont Law School and an Adjunct Associate Professor of Law at American University's Washington College of Law, where he teaches courses on environmental law, food systems, and agricultural policy. Bill received his undergraduate degree from the University of North Carolina at Chapel Hill; his law degree, magna cum laude, from North Carolina Central University School of Law; and his LL.M. in Environmental Law, summa cum laude, from Vermont Law School.

Contributing Authors

Teresa Clemmer
Teresa Clemmer is an attorney with extensive experience in environmental and natural resources law, including air quality, water quality, wetlands, oil spill prevention and response, contaminated site cleanup, solid and hazardous waste management, public lands, natural resource management, and endangered species. She is presently a member of the law firm of Bessenyey & Van Tuyn, LLC, in Anchorage, Alaska, representing conservation-minded clients in a wide range of matters and projects. Before joining the firm, Teresa spent four years as a law professor at Vermont Law School, helping to train the next generation of environmental lawyers. She taught courses on air pollution and environmental law, served as acting director of the Environmental and Natural Resources Law Clinic, and published scholarly work relating to climate change. Before her adventures in academia, Teresa practiced environmental law with the law firms of Cooper, White & Cooper LLP and Perkins Coie, and later as a staff attorney with the nonprofit Trustees for Alaska. Teresa's achievements include successful litigation prompting EPA to update its outdated national air pollution standards for nitric acid plants; successful litigation preventing the construction of a liquefied natural gas terminal on Passamaquoddy tribal lands in Maine, which posed a threat to both Passamaquoddy culture and endangered Northern right whales; and successful advocacy before the U.S. Army Corps of Engineers to protect communities and ecologically sensitive areas of Puerto Rico from the harmful effects of a proposed natural gas pipeline. Teresa received her law degree from Georgetown University and her undergraduate degree from Princeton University.

Hannah Connor
Hannah M. M. Connor is an attorney in the Animal Protection Litigation section of the Humane Society of the United States. Her principal practice areas include environmental, administrative, and animal law. Ms. Connor is committed to employing laws, including the Clean Water Act, the Clean Air Act, the Emergency Planning and Community Right-to-Know Act, and the National Environmental Policy Act, to further the rights of communities, animals, and the natural environment by addressing and preventing the harms caused by the industrialization of animal agriculture and encouraging healthy ecosystems by supporting sustainable food systems. She has written, presented, and litigated on a range of issues relating to the industrialization of animal agriculture. Prior to joining the Humane Society in 2011, Ms. Connor was an attorney with the Waterkeeper Alliance. She received her undergraduate degree from Boston College and her law degree from Vermont Law School.

John H. Davidson
John Davidson is President of the Northern Prairies Land Trust and a law professor at the University of South Dakota School of Law, a position he has held since 1972. He has authored numerous articles, law casebooks, and treatises in the fields of agricultural law, water and irrigation law, and environmental law. In 1995, President William J. Clinton appointed John to the Western Water Policy Review Commission. He has long held an association with the conservation movement, including several terms on the Board of the South Dakota Association of Conservation Districts, and his involvement in pro bono environmental litigation has been continuous since the early 1970s. He has taught the course titled Agriculture and the Environment at the Vermont Law School. He received his undergraduate degree from Wake Forest University, his law degree from the University of Pittsburgh School of Law, and his LL.M. in Natural Resources from the George Washington University Law School.

George Kimbrell
George Kimbrell is Senior Attorney for the Center for Food Safety (CFS), where he practices environmental and administrative law. George's litigation and policy work spans a broad range of CFS program areas, including: genetically engineered (GE) foods; transgenic plants, trees and animals; food labeling; food safety and contamination; organic standards; factory farming pollution; aquaculture; pesticides; agri-

cultural patent law; and nanotechnology. One of his cases, *Monsanto v Geertson Seed Farms* (2010), was the first U.S. Supreme Court case on the oversight of agricultural biotechnology. George also serves as an Adjunct Professor of Law at Lewis and Clark Law School, where he teaches food and agriculture law. He has written and presented extensively on a range of issues related to industrial agriculture's impacts on the environment. George joined CFS upon completing a clerkship with the Honorable Ronald M. Gould, U.S. Court of Appeals for the Ninth Circuit. He received his undergraduate degree, cum laude, from the College of William and Mary, and his law degree, magna cum laude, from Lewis and Clark Law School.

J.B. Ruhl

J.B. Ruhl is the David Daniels Allen Distinguished Chair in Law at Vanderbilt University Law School, where he teaches courses in environmental law, natural resources law, and property. Before he joined Vanderbilt's law faculty in 2011, he was the Matthews & Hawkins Professor of Property at the Florida State University College of Law, where he had taught since 1999. His influential scholarly articles on environmental law relating to climate change, the Endangered Species Act, ecosystems, federal public lands, and other land use and environmental issues have appeared in the *California Law Review*, the *Georgetown Law Review*, the *Stanford Law Review*, the *Duke Law Review*, the *Environmental Law Reporter*, the *Vanderbilt Law Review*, and the specialty environment journals at several top law schools, among other journals. His works have been selected among the best law review articles in the field of environmental law seven times from 1989 to 2012. Over the course of his career, he has been a visiting professor at Harvard Law School, Vermont Law School, George Washington University Law School, the University of Texas Law School, and Lewis & Clark College of Law. He began his academic career at the Southern Illinois University School of Law, where he taught from 1994-1999 and earned his Ph.D. in geography. Before entering the academy, he was a partner with Fulbright & Jaworski in Austin, Texas, where he also taught on the adjunct faculty of the University of Texas Law School.

Other Contributing Authors

James Choate
James Choate holds an LL.M. in Environmental and Land Use Law from the University of Florida, a J.D. from Stetson University College of Law, and a B.S. in Wildlife Ecology & Conservation from the University of Florida. He currently works as an attorney for the U.S. Army Corps of Engineers in Charleston, South Carolina.

Seth Hennes
Seth Hennes, originally from Florida, currently resides in Washington, D.C., where he has worked on a wide variety of environmental issues. He graduated from Wake Forest University with a B.A. in Political Science in 2003. Seth earned his J.D. from Tulane Law School in 2006 and his LL.M. in Environmental and Land Use Law from the University of Florida Levin College of Law in 2011.

Elena Mihaly
Elena Mihaly graduated with a degree in Environmental Science from Colorado College and earned a J.D. and master's degree in Environmental Law and Policy from Vermont Law School in 2013. She has researched and written in the field of food and agriculture policy, including a publication titled, "A Farmer's Handbook to Energy Self-Reliance." Beginning in September 2013, she will be working on sustainable agriculture policy at the Conservation Law Foundation.

Emily Montgomery
Emily Montgomery currently works for the state of Vermont and holds an LL.M. in Environmental and Natural Resources Law from the S.J. Quinney College of Law at the University of Utah, a J.D. from Vermont Law School, and a B.A. in Environmental Studies from Gettysburg College.

Elisa Prescott
Elisa Prescott, originally from Vermont, received her B.A. in Environmental Sociology from St. Lawrence University in 2008 and a master's degree in Environmental Law and Policy from Vermont Law School in 2011. Elisa currently lives in Bozeman, Montana, where she continues to work on agriculture and environmental-related issues in rural communities.

Joanna Reilly-Brown
Joanna Reilly-Brown holds a J.D. and Certificate in Environmental and Land Use Law from the University of Florida Levin College of Law, an M.A. in Environmental Anthropology, and a B.A. in Anthropology from the University of Florida. She currently resides in Florida, where she continues to research and write about agriculture and environmental issues.

Chapter 1
A Brief History of U.S. Agricultural Policy and the Farm Bill
William S. Eubanks II

A historical perspective is helpful in understanding past events as they relate to current circumstances and in divining where that history, in conjunction with innovation and common sense, will lead us in the future. Nowhere is this concept more salient than in our nation's agricultural framework, which has been embedded into the founding principles of our country since its beginning. The rich history of our agricultural system—which has at different times served vastly divergent goals—continues to conjure iconic images in the minds of most Americans and influences the public's perception of agriculture in the 21st century. Therefore, because there are few facets of American society where William Shakespeare's admonition that "what's past is prologue" is more pertinent, this chapter provides a brief historical overview of agricultural policy as it has adapted to satisfy shifting demands and new technologies, and its role in shaping not only the current farming system and the rural economy, but also the value which we ascribe to our natural resources relative to agricultural production.

A. How Our Agricultural Policy Has Evolved: A Nation's Time Line Through an Agricultural Prism

In the more than two centuries since U.S. independence, our nation's deep agrarian roots have had a profound influence on our increasingly complex federal farming and food policies that have attempted to adapt over time to meet the prevailing needs of farmers, consumers, and ultimately the national economy. However, these policies have often aimed to address short-term objectives rather than long-term stability and sustainability of our food supply, and thus the underlying and sometimes contradictory policies have led to confusion among farmers and consumers alike. The following discussion provides a backdrop of watershed moments in our agricultural policy history in order to provide a richer context to what otherwise might appear to be little more than an enigmatic set of policies devoid of a coherent purpose. To the contrary, as this history demonstrates, our policies have in fact been carefully yet flexibly crafted to address market conditions and other economic and social concerns existing at a given time, but too often have succumbed to budgetary short-sightedness and the lack of an integrated approach taking into full account many important noneconomic concerns such as natural resource protection, nutrient-rich food production, and consumer demand.

1. From Jeffersonian Agrarianism to the Great Depression

Our nation's roots are inextricably intertwined with farming. Soon after the British colonies declared independence from England in 1776, Thomas Jefferson and other political leaders encouraged—indeed, advocated—a "national agrarian identity" for the new nation.[1] Jefferson envisioned the United States as

---

This chapter is, with permission, an updated and adapted version of an article previously published as William S. Eubanks II, *A Rotten System: Subsidizing Environmental Degradation and Poor Public Health With Our Nation's Tax Dollars*, 28 STAN. ENVTL. L.J. 213 (2009). Copyright © 2009 by the Board of Trustees of the Leland Stanford Junior University.

1. DENNIS KEENEY, INST. FOR AGRIC. & TRADE POL'Y & LONI KEMP, THE MINNESOTA PROJECT, A NEW AGRICULTURAL POLICY FOR THE UNITED STATES 6 (2003) [hereinafter KEENEY & KEMP, A NEW AGRICULTURAL POLICY], *available at* http://www.mnproject.org/pub-sustainableag.html.

a democracy comprised of yeomen farmers whose impeccable virtues would propel the young nation to stability.[2] In a letter to U.S. Secretary of Foreign Affairs John Jay, Jefferson wrote:

> Cultivators of the earth are the most valuable citizens. They are the most vigorous, the most independent, the most virtuous, and they are tied to their country and wedded to its liberty and interests by the most lasting bonds. As long, therefore, as they can find employment in this line, I would not convert them into mariners, artisans, or anything else.[3]

When Jefferson became president in 1801, more than 95% of the nation's population worked in the farming sector, in fields that produced a diverse mix of crops to sustain local communities. Over the next century, that percentage would rapidly change due to industrialization and new technology, decreasing to 41% by 1900.[4] However, even as the percentage of Americans involved in the farming economy declined over time, Jeffersonian agrarianism continued to leave an indelible imprint on American culture and "remains to this day an important component of our national rural identity and is embedded in farm politics and policies."[5]

Despite early federal policies such as the Homestead Act in 1862, which aimed to encourage small-scale family agriculture, the beginnings of what is now considered commercial agriculture expanded significantly across the 19th century.[6] This shift began early in the century with the invention and implementation of more efficient agricultural tools such as the cotton gin, the steel plow, the reaper, the grain drill, and the harvester.[7] The shift accelerated after the Civil War as the South sought a new agricultural identity without the use of slave labor.[8] As the scope of commercial crop production expanded and farm size began to gradually increase, the number of subsistence farmers declined rapidly.[9] Further, the heightened commercialization of agriculture created a more complex economy both domestically and abroad, in which farmers came to rely more heavily on capital, banking, mechanization, and soil inputs to increase yields.[10] In hindsight, this gradual trend toward commercialization of agriculture had two unintended consequences by the early 1900s that have lasting impacts on farm policy today: (1) control of the agricultural industry began to fall to either large processing companies that consolidated their markets through economic pressure, or to farmers with the most capital who could outcompete smaller farms; and (2) for the first time in our nation's history, the rural yeoman farmer idealized by Jefferson found it more difficult to earn a livelihood on the family farm than previous generations.[11]

Not surprisingly, by the early decades of the 20th century, the commercialization of agriculture coupled with the multitude of employment options in America's industrial economy led to a smaller proportion of Americans in the agricultural sector. In just over one century, from 1801 to around 1910, the percentage of our nation's citizens that farmed full-time had dropped from 95% to approximately 40%.[12] And within a few decades those remaining farmers would be tested severely by the economic woes of the Great Depression and the record droughts that devastated crop yields.

2. The "Temporary" Safety Net: The Creation of the Farm Bill

During the early to mid-1930s, nearly 40% of the nation's population, including a large portion of the farming population, was grinding out an impoverished subsistence as bank closures, home foreclosures, and economic downturn caused difficult times in the United States, and particularly in the nation's heart-

---

2. *See id.*
3. Letter from Thomas Jefferson, U.S Minister to France, to John Jay, U.S. Sec'y of Foreign Aff. (Aug. 23, 1785), *quoted at* http://www.fireandknowledge.org/archives/2007/02/09/cultivators-of-the-earth-are-the-most-valuable-citizens-jefferson/.
4. U.S. Dep't. Agric, Econ. Research Serv., *The 20th Century Transformation of U.S. Agriculture and Farm Policy* at 2, 9 (2005), *available* at http://www.ers.usda.gov/publications/eib-economic-information-bulletin/eib3.aspx.
5. Keeney & Kemp, A New Agricultural Policy, *supra* note 1.
6. *Id.* at 9-11.
7. Benedict A. Leerburger, *Agricultural Machines*, in Science Encyclopedia, http://science.jrank.org/pages/128/Agricultural-Machines.html.
8. Keeney & Kemp, A New Agricultural Policy, *supra* note 1.
9. *Id.*
10. *Id.*
11. *See id.*
12. U.S. Dep't. Agric., Econ. Research Serv., *20th Century Transformation*, *supra* note 4, at 2.

land.[13] At that time, one in four Americans still lived on a farm.[14] Although poverty affected all sectors of society, some scholars contend that the farming economy was the most impacted—and the most visible—because of the convergence of bank foreclosures, drought, dust storms, and floods.[15] These woes were merely the visible causes of what was termed the "farm crisis," but the primary cause for the crisis escaped scrutiny because it was obscured from public view by the prominence of those more tangible problems.[16]

Simply stated, the farm crisis of the 1930s was "triggered not by too little food, but by too much."[17] Innovative advances in both mechanization and soil inputs during the 1920s led to vast overproduction of most staple crops such as corn, wheat, and soybeans despite voluntary attempts to limit production.[18] This immense surplus benefited "distributors, processors, and monopolists who were increasingly dominating the food system," but seriously curtailed the profits of farmers as domestic and global crop prices fell dramatically due to the market glut of these products.[19] As the crop prices fell below their respective costs of production, farmers could no longer stay afloat: net farm income dropped by two-thirds between 1929 and 1932, 60% of farms were mortgaged in hopes of surviving, and by 1933 the price of corn registered at zero and grain elevators refused to buy any surplus corn.[20]

Recognizing the importance of farmers in preserving our nation's food supply throughout the Great Depression, the U.S. Congress acted quickly to temporarily protect small family farms. The Agricultural Adjustment Act of 1933[21] was a "comprehensive and monumentally ambitious" program to address myriad social, cultural, environmental, and economic issues facing the country.[22] As part of President Franklin D. Roosevelt's New Deal agenda, the 1933 farm bill sought to do many things: bring crop prices back to stability by weaning the nation from its affinity for agricultural overproduction through rigid supply controls, utilize surplus crops productively to combat widespread hunger and provide nutritional assistance to children in the form of school lunch programs, implement strategies to prevent further erosion and soil loss from poor land conservation policies and weather events on marginal farmlands, provide crop insurance and credit assurances for subsistence farmers, and build community infrastructure for rural farming towns.[23] In essence, the 1933 farm bill was designed to save small farming in America and, to some, it signaled a return to the Jeffersonian ideal of an agrarian democracy.

Despite the great potential of the initial farm bill, the dream of Jeffersonian agrarianism was quickly grounded as debate set in. However positive the goals espoused by President Roosevelt, the farm bill's agricultural policies were controversial from the outset.[24] Farmers criticized sacrificing any surplus crops or livestock for hunger relief as "shameful charity" and "a threat to free markets."[25] Members of the malnourished public criticized government-induced surplus dumping—the elimination of excess crops by burning,

---

13. *Id.* at 9.
14. *Id.* at 3.
15. U.S. Dep't. Agric. Econ. Research Serv., *History of Agricultural Price-Support and Adjustment Programs 1933-84*, at 1 (1985), *available at* http://naldc.nal.usda.gov/download/CAT10842840/PDF ("The unprecedented economic crisis which paralyzed the Nation by 1933 struck first and hardest at the economy's farm sector. For agriculture and rural America, it was the worst economic-social-political wrenching in history."); *see also* PETER TEMIN, LESSONS FROM THE GREAT DEPRESSION 54-56 (1991) (explaining that "[f]armers suffered, while the rest of the [U.S.] economy gained . . . [because] the prices of agricultural products and raw materials had [already] been falling in the 1920s . . . [and] [a]t about the same time as the stock-market crash, the prices of raw materials and agricultural goods . . . began to fall precipitously").
16. U.S. Dep't. Agric., Econ. Research Serv., *History of Agricultural Price-Support and Adjustment Programs 1933-84*, *supra* note 15, at 1 ("The unprecedented economic crisis which paralyzed the Nation by 1933 struck first and hardest at the economy's farm sector. For agriculture and rural America, it was the worst economic-social-political wrenching in history. Farm foreclosures were the order of the day.").
17. DANIEL IMHOFF, FOOD FIGHT: THE CITIZEN'S GUIDE TO A FOOD AND FARM BILL 34 (2007).
18. U.S. Dep't. Agric., Econ. Research Serv., *History of Agricultural Price-Support and Adjustment Programs 1933-84*, *supra* note 15, at 1 ("Farm journals and farm organizations had, since the 1920s, been advising farmers to control production on a voluntary basis. Attempts were made in some areas to organize crop withholding movements on the theory that speculative manipulation caused price declines. When these attempts proved to be unsuccessful, farmers turned to the more formal organization of cooperative marketing for staple crops. After voluntary organizations of wheat and livestock producers collapsed, farmers began campaigns for Government assistance in solving the farm problem.").
19. *Id.* at 1-3.
20. *See id.* at 1-10.
21. Agricultural Adjustment Act of 1933, Pub. L. No. 73-10, 48 Stat. 31 (1933), *available at* http://www.nationalaglawcenter.org/assets/farmbills/1933.pdf.
22. Jim Chen, *Get Green or Get Out: Decoupling Environmental From Economic Objectives in Agricultural Regulation*, 48 OKLA. L. REV. 333 (1995).
23. *Cf.* U.S. Dep't. Agric., Econ. Research Serv., *History of Agricultural Price-Support and Adjustment Programs 1933-84*, *supra* note 15, at 1-6; *see also* Agricultural Adjustment Act of 1933, *supra* note 21.
24. IMHOFF, *supra* note 17, at 36.
25. *Id.*

dumping into rivers, or selling overseas[26]; in fact, polls showed that the public was "horrified with [the government's] policy of forced scarcity," which proposed to dump many critical food surpluses that could have fed hungry Americans.[27] Further, farmers and nonfarmers alike claimed that the farm bill's supply control mandate to cultivate less than 25% of all cotton fields precluded the manufacture of textile products that could have generated profits for cotton farmers in the domestic marketplace and provided clothing and blankets for those in need—ultimately leading to a U.S. Supreme Court decision that invalidated the farm bill's initial acreage quotas for cotton.[28]

Over the course of several years, the controversy surrounding the first farm bill eventually calmed as the positive results of the bill were universally hailed as New Deal successes: crop prices gradually resorted to stable levels as supply was reined in, rising incomes enabled many farming families to keep their farms out of foreclosure, the school meal program introduced by the farm bill laid the groundwork for today's national school lunch program, and every government dollar spent on agricultural policy under the New Deal had a significant multiplier effect in the American economy that led to stimulation of all facets of the domestic marketplace.[29] Indeed, farmers were delighted when "[g]ross farm income increased by 50%" within three years of the farm bill's enactment.[30]

This increase, however, did not come without a price: many of the farm income increases were artificial market supports in the form of government subsidies designed to meet the shortfall between the price a farmer could get for full production in the free market as compared to the price the farmer could get for what he was allowed to grow, pursuant to the supply controls imposed by the farm bill quotas.[31] Focused solely on day-to-day survival after struggling through the Great Depression, small farmers—then still the backbone of the American agrarian system—arguably failed to grasp the unintended consequences of the initial farm bill's introduction of artificial market supports via commodity subsidies. Although pre-World War II farm bill subsidies provided income supports for over 100 crops,[32] rapid changes were on the horizon because the power to determine which crops to subsidize, and at what levels, would be held by those with political and economic leverage in Washington.

3. The Post-World War II Boom: Maximizing Yields and New Technology

In the three decades after the end of World War II, the American agricultural system saw arguably the most drastic short-term shift that it had ever, and will ever have, seen. Rather than discarding what had been intended to serve as a temporary fix to an emergency farm crisis a decade earlier to return to a free market farm economy, Congress instead memorialized its long-term commitment to providing agricultural market supports by enacting a revised farm bill, the 1948 Agricultural Act.[33] At that time, despite some level of farm commercialization, five million farms remained in the United States, and these farms were "generally small, diversified operations selling primarily to domestic markets behind high tariff walls."[34] In fact, over 100 unique crops received partial price supports in the way of federal subsidies, which gave farmers choices regarding what and how to cultivate.[35]

---

26. ALAN BRINKLEY, AMERICAN HISTORY: A SURVEY 877 (10th ed. 1999) (noting that in 1933 alone, six million piglets and 220,000 pregnant cows were slaughtered in an effort by the government to raise prices); *see also* IMHOFF, *supra* note 17, at 36 ("[M]illions of young hogs purchased by the government to restrict supply (bump up prices) and feed the hungry never reached their intended beneficiaries. Instead, they were slaughtered and dumped into the Missouri River. Likewise, millions of gallons of milk were poured into the streets rather than nourishing famished and distended bellies.").
27. BARRY CUSHMAN, RETHINKING THE NEW DEAL COURT 35 (1998) (noting that a *Washington Post* Gallup Poll revealed that a majority of the American public opposed the 1933 farm bill because of forced scarcity).
28. United States v. Butler, 297 U.S. 1, 77-78 (1936); *see also* BRINKLEY, *supra* note 26.
29. J. PATRICK RAINES & CHARLES G. LEATHERS, DEBT, INNOVATIONS, & DEFLATION 124 (2008).
30. BRINKLEY, *supra* note 26, at 404.
31. *Id.*
32. U.S. Dep't. Agric. Econ. Research Serv., *History of Agricultural Price-Support and Adjustment Programs 1933-84*, *supra* note 15, at 16 ("Under the provisions of this legislation, the supports for both basic and nonbasic commodities continued for 2 years after the declaration of the end of hostilities. In all, by the mid-1940s, well over 100 commodities were being supported.").
33. Agricultural Act of 1948, *available at* http://www.nationalaglawcenter.org/assets/farmbills/1948.pdf.
34. U.S. Dep't. Agric., Econ. Research Serv., *20th Century Transformation*, *supra* note 4, at 9; *see also* JIM MONKE, CONG. RESEARCH SERV., FARM COMMODITY PROGRAMS IN THE 2008 FARM BILL 2 (2008), *available at* http://www.nationalaglawcenter.org/assets/crs/RL34594.pdf [hereinafter CRS, FARM COMMODITY PROGRAMS] (explaining that "in the 1930s, most of the 6 million farms in the United States were small and diversified").
35. U.S. Dep't. Agric. Econ. Research Serv., *History of Agricultural Price-Support and Adjustment Programs 1933-84*, *supra* note 15, at 16.

Fundamental change loomed, however, as the Green Revolution[36] led to new plant breeding and hybridization techniques, and military technology developed during World War II led to new pesticides, herbicides, synthetic fertilizers, and agricultural mechanization that streamlined farming.[37] These modern advances increased yields consistently, but also encouraged monocultures (i.e., the planting of a single bumper crop as opposed to a diverse mix of crops) and more efficient economies of scale, which by the late 1950's and early 1960's once again resulted in gross overproduction and depressed domestic crop prices, reminiscent of the farm crisis during the Great Depression.[38] Unlike the earlier farm crisis, however, the government did not swoop in to protect the small farmer, but rather tweaked existing market support programs to take advantage of the increased efficiency of the new system. These tweaks benefited larger farms that had the ability to stay afloat despite lower crop prices because they could capitalize on this changing farm economy by purchasing foreclosed farms at below-market rates to increase in size and thus crop yield, and, importantly, created an opportunity for consolidation with other large farms and food processors to create modern agribusiness, accompanied by a powerful lobby to advocate in the political arena.[39]

Accordingly, not only were there key scientific and technological advances during the post-war era, but the farm economy itself, and the politics driving it, also began fundamentally charting a new path distinct from that which had served as the basis for New Deal farm policy. For example, "[a]griculture in the 1950's and 1960's moved rapidly to the business (industrial) model, with an emphasis on specialization, simplification, and less [crop] diversity."[40] Particularly during this period, "agriculture policy was formulated to reward large producers, although it was an unintended consequence of commodity subsidy policy."[41] As agroeconomic researchers explain, this new agricultural economic theory built around commodities that Congress incorporated in the early 1960s stemmed from the fact that "[c]ommodity prices in general offer only marginal returns to the producer," and thus to achieve long-term profitability, a farmer primarily producing such commodities "must be producing at low cost" and at high output to result in a net profit.[42]

This new, more efficient agricultural economy of the post-war era reached its height by the early 1970s, when President Richard Nixon appointed Earl Butz to serve as his Secretary of Agriculture.[43] Partly motivated by the Cold War dynamics of the era, Secretary Butz had revolutionary and transformative views for creating a highly managed and heavily centralized food system that maximized yields and efficiency, with surpluses seen as a boon for foreign trade—views that broke not only with the original intent of 1933 farm bill but also with many of those to hold Secretary Butz's office before him in the early decades of farm bill implementation. Indeed, Secretary Butz proclaimed that the changes he planned to implement "represented an historic turning point in the philosophy of farm programs in the United States."[44]

For example, in one of his well-known edicts, Butz called on all American farmers to "get big or get out," which served as a plea to concentrate the food market into an economy of scale.[45] As part of this policy, Butz proclaimed that farming "is now a big business" and that the family farm "must adapt or die" by expanding into large operations reliant on industrial pesticides, herbicides, fertilizers, and mecha-

---

36. In the early 1960s, the Green Revolution led to a tripling in grain yields (namely of the wheat, rice, and corn that prove to be the most heavily subsidized crops today) due to scientific advances in crop hybridization. Although the crop yields increased substantially, some writers argue that the consequences outweighed the benefits. *See, e.g.*, Richard Manning, *The Oil We Eat*, HARPER'S, Feb. 2004, at 37, *available at* http://harpers.org/archive/2004/02/the-oil-we-eat/ ("The accepted term for this strange turn of events is the green revolution, though it would be more properly labeled the amber revolution, because it applied exclusively to grain—wheat, rice, and corn. Plant breeders tinkered with the architecture of these three grains so that they could be hypercharged with irrigation water and chemical fertilizers, especially nitrogen. This innovation meshed nicely with the increased 'efficiency' of the industrialized factory-farm system. . . . [However,] it disrupted long-standing patterns of rural life worldwide, moving a lot of no-longer-needed people off the land and into the world's most severe poverty.").
37. *See, e.g.*, Theodore P. Lianos & Quirino Paris, *American Agriculture and the Prophecy of Increasing Misery*, 54(4) AM. J. AGRIC. ECON. 570-77 (1972) ("Rapid changes, such as introduction of new technology, increasing degree of mechanization, and increasing farm size have characterized the American agricultural sector in the post-war period.").
38. U.S. Dep't. Agric. Econ. Research Serv., *20th Century Transformation*, *supra* note 4, at 9.
39. *Id.* at 9-11; *see also* U.S. Dep't. Agric. Econ. Research Serv., *History of Agricultural Price-Support and Adjustment Programs 1933-84*, *supra* note 15, at 17-28.
40. DENNIS KEENEY, INST. FOR AGRIC. & TRADE POL'Y & LONI KEMP, THE MINNESOTA PROJECT, HOW TO MAKE IT WORK: REQUIRED POLICY TRANSFORMATIONS FOR AGROECOSYSTEM RESTORATION 5 (2004), *available at* http://www.iatp.org/files/421_2_36936.pdf.
41. *Id.*
42. *Id.*
43. U.S. Dep't. Agric. Econ. Research Serv., *History of Agricultural Price-Support and Adjustment Programs 1933-84*, *supra* note 15, at 29.
44. *Id.*
45. A.V. KREBS, THE CORPORATE PAPERS: THE BOOK OF AGRIBUSINESS 404, 428 (1992) (describing "get big or get out" as the slogan of "the Earl Butz school" of agribusiness).

nization.[46] In 1972, he rounded out his vision by urging farmers to "plant from fencerow to fencerow" to maximize yields of commodity crops (namely corn and soy), while simultaneously assisting the agribusiness lobby in eliminating a 40-year-old conservation program that had been a mainstay of the nation's agricultural policy since the original 1933 farm bill.[47] As a result, "[f]armers who had maintained wild or semi-wild borders around and between fields (in accordance with the best practices [recommended by] former administrations), tore out shelterbeds, windbreaks, filter strips, and contours."[48] Moreover, in part because of his involvement in charting a new course in the legislation, Secretary Butz hailed the 1973 farm bill, as described above, as "'an historic turning point in the philosophy of farm programs in the United States . . . [because of] [i]ts emphasis on maintaining or increasing output was in marked contrast to earlier programs to curtail production of wheat, corn, upland cotton, and tobacco."[49] Accordingly, by shifting the focus primarily from supply control to an even heavier reliance on artificial price supports for commodity farmers, the 1973 farm bill achieved Secretary Butz's "historic" and radical "turning point" for farmers and consumers alike. For the first time in nearly 40 years, agricultural progress began to once again be measured more by crop yield maximization and surpluses, and less by a well-rounded set of criteria that included not only crop production but also conservation, nutritional assistance, and rural stability—i.e., the key tenets of the original farm bill.[50]

While Secretary Butz's term at USDA was relatively short, his vision fundamentally altered the nation's farming system. Because many farmers heeded Butz's call in the early 1970s to fully embrace capital-intensive methods of mechanized farming on larger plots of land in an effort to drive efficiency, they had little choice but to stay the course in order to recoup their upfront investments that were made to participate in the modern agricultural system.[51] This paid off in the short term, as "[t]he mid-1970s "were favorable years for American farmers."[52] Net farm incomes reached, and remained at, historically high levels throughout the middle of the decade.[53] Towards the end of the decade, however, the long-term consequences of this gamble would leave "many farmers overcapitalized . . . [because] [h]igh farm prices set off a scramble for farmland that drove land values up," coupled with the fact that "[g]reater dependence on export markets made commodities more vulnerable to sudden price swings due to economic or political events in other parts of the world."[54]

Nonetheless, these warning signals were not enough to steer agriculture policy in a different direction heading into the 1980s. For example, Congress enacted the Food and Agriculture Act of 1977,[55] which "continued the dual system of target prices [of commodities] at higher rates than loan levels to allow crops to move freely in international trade."[56] In effect, the legislation heavily leveraged its encouragement of maximized yields on the assumption that foreign countries would continue to purchase our growing surpluses at similar rates to those seen during the Butz era. But agricultural prices remained stagnant or decreased, leading to reduced farm income in the late 1970s, forcing Congress to pass the Emergency Assistance Act of 1978 for the dual purpose of instituting a moratorium on foreclosures by the Farmers Home Administration and creating a $4 billion emergency loan program for struggling farmers.[57] It was in this tepid economic climate in the farming sector that the fledgling agribusiness model of the 1970s led to, or at least found itself mired in, a new crisis in our nation's breadbasket.

---

46. Leslie A. Duram, Encyclopedia of Organic, Sustainable, and Local Food at xviii (2010).
47. U.S. Dep't. Agric. Econ. Research Serv., *History of Agricultural Price-Support and Adjustment Programs 1933-84*, *supra* note 15, at 29-31; Agricultural and Consumer Protection Act of 1973, Pub. L. No. 93-86, 87 Stat. 221 (1973), *available at* http://www.nationalaglawcenter.org/assets/farmbills/1973.pdf.
48. Imhoff, *supra* note 17, at 38-39.
49. U.S. Dep't. Agric. Econ. Research Serv., *History of Agricultural Price-Support and Adjustment Programs 1933-84*, *supra* note 15, at 29.
50. U.S. Dep't. Agric. Econ. Research Serv., *20th Century Transformation*, *supra* note 4, at 5 (Figure 4 of this report shows that the number of commodity crops produced per farm remained steady for the first half of the twentieth century at approximately four to five crops per farm before declining sharply in the 1970s to less than three crops per farm. In 2002, the number dipped even more steeply as the average neared only one commodity crop produced per farm.).
51. *See generally* U.S. Dep't. Agric., Econ. Research Serv., *History of Agricultural Price-Support and Adjustment Programs 1933-84*, *supra* note 15, at 30-31.
52. *Id.* at 31.
53. *Id.*
54. *Id.*
55. Food and Agricultural Act of 1977, Pub. L. No. 95-113, 91 Stat. 913 (1977), *available at* http://www.nationalaglawcenter.org/assets/farmbills/1977-1.pdf.
56. U.S. Dep't. Agric., Econ. Research Serv., *History of Agricultural Price-Support and Adjustment Programs 1933-84*, *supra* note 15, at 32.
57. *Id.* at 36.

4. The 1980s Farm Crisis and Congress' Response

The economic problems of the late 1970s were exacerbated by political tensions on the global scale that came to a head in 1980, when President Jimmy Carter partially suspended agricultural exports to the Soviet Union, one of the largest importers of U.S. surplus commodities.[58] Things worsened in 1982, when a worldwide recession caused overall agricultural exports to decline for the first time in eight years, rendering the 1981 farm bill, the Agricultural and Food Act of 1981,[59] unable to provide adequate support for many smaller farms.[60] With farm income far below levels needed by many to recoup the investments made years earlier in both land and capital-intensive farm machinery, what unfolded next was a farm crisis unlike anything seen since the Great Depression. Starting in 1982, "[l]oan delinquencies grew and farmland values leveled off after tripling over the course of a decade."[61] Combined farm debt was more than $200 billion by the early 1980s.[62] The federal government's attempts to generally reduce the nation's deficit made the situation for farmers worse: "[t]he Federal Reserve Board attacked inflation by raising interest rates, and Congress cut taxes without cutting government expenditures thus causing massive budget deficits . . . [and] [t]hese policies hurt farmers by strengthening the dollar, which in turn made U.S. exports less competitive, and by increasing farmers' borrowing costs."[63]

By the middle of the decade, the worst possible scenarios that could have resulted from the coalescing crisis were occurring with disturbing frequency, as described by one observer:

> By 1985, the U.S. farm economy was in a full-blown depression. Farmers filed for bankruptcy in record numbers. Farm foreclosures reached levels not seen since the 1930s. Businesses dependent on agriculture suffered. Sixty-eight agricultural banks failed in that year alone. Equipment dealers and chemical suppliers went under. Migration from the countryside increased dramatically. Between 1981 and 1985, the state of Iowa [alone] lost 33,000 people.[64]

Congress took immediate steps in the 1985 farm bill to respond to the alarming trend of significantly reduced farm income,[65] in an effort to arrest not only the financial downturn in the agricultural sector but also the exorbitant rate of midwestern foreclosures, bankruptcies, and even suicides that marked a turning point in our farming history.[66] Seizing on this watershed moment, Congress enacted the Food Security Act of 1985, which continued the policy trend by emphasizing maintenance of high target prices for crops and utilizing foreign markets to export significant volumes of surplus crops.[67]

However, there were critical changes in the legislation, including the introduction of marketing loans as an additional form of farm revenue because they allowed producers to repay marketing loans at rates far lower than the contracted repayment rate, ensuring that the difference between what a farmer actually repaid and what he was contractually obligated to repay served as a significant income subsidy for farmers.[68] Perhaps most important, the 1985 farm bill included several new conservation programs with broad reaches that "were at least partially in response to soil erosion problems created during the 1970s when farmers put many acres of marginal and highly erodible farmland into production to meet perceived world food demand."[69] These programs included the Conservation Reserve Program, Sodbuster, and Swampbuster (some of which are discussed in more detail in Chapter 2).[70] These conservation programs were both

---

58. *Id.*
59. Agricultural and Food Act of 1981, Pub. L. No. 97-98, 95 Stat. 1213 (1981), *available at* http://www.nationalaglawcenter.org/assets/farmbills/1981-1.pdf.
60. U.S. Dep't. Agric. Econ. Research Serv., *History of Agricultural Price-Support and Adjustment Programs 1933-84*, *supra* note 15, at 40.
61. *Id.*
62. Allen H. Olson, *Federal Farm Programs—Past, Present, and Future—Will We Learn From Our Mistakes*, 6 Great Plains Nat. Res. L.J. 1, 15 (2001).
63. *Id.*
64. *Id.* at 16.
65. Food Security Act of 1985, Pub. L. No. 99-198, 99 Stat. 1354 (1985), *available at* http://www.nationalaglawcenter.org/assets/farmbills/1985-1.pdf.
66. *See* Stephan J. Goetz & David L. Debertin, *Rural Population Decline in the 1980s: Impacts of Farm Structure and Federal Farm Programs*, 78(3) Am. J. Agric. Econ. 517-29 (Aug. 1996), *available at* http://www.jstor.org/pss/1243270 (explaining that "[b]etween 1980 and 1990, certain farming-dependent counties lost over 20% of their residents").
67. Olson, *Federal Farm Programs*, *supra* note 62, at 16-17.
68. *Id.* at 17.
69. *Id.*
70. *Id.*

environmentally and financially sensible, considering that they would prevent farming of our nation's most marginal (and often the most ecologically sensitive) lands while simultaneously limiting the production of surplus crops and thereby stabilizing prices at a higher level than would otherwise be the case.[71]

The 1985 farm bill marked a crossroads in our nation's agricultural policy, capping off roughly one-half century of policy that had remained relatively static from its inception in 1933 through the 1960s, changed in a dramatic and arguably precarious direction during the 1970s and early 1980s, and had been jolted by the consequences of that gamble in the 1980s farm crisis. The 1985 strategic reform program joined the best of each of the previous programs and goals in an effort to maintain American agricultural prosperity and efficiency without destroying the country's resources and rural communities in the process. The new compromise policy, which was far more balanced than its predecessor, quickly returned dividends to our nation's farmers. Not only did the 1985 farm bill put an end to the farm crisis, but in the years after its enactment farm income increased significantly, rising steadily each year between 1986 and 1990, during which time "[a]gricultural exports increased from $26 billion in 1986 to $40 billion in 1990."[72]

As with the retaliation against the Soviet Union in 1980, events in the international arena once again heavily influenced domestic farm policy in the late 1980s and early 1990s; this time, it was the eight-year negotiations on the General Agreement on Tariffs and Trade (GATT), which would eventually lead to the creation of the World Trade Organization (WTO) in 1995.[73] Both the Reagan Administration and the Administration of George H.W. Bush viewed expanded foreign trade as a means of reducing the price tag of our nation's farm subsidy framework, as increased foreign trade could serve as a way to give more bargaining power back to individual farmers, therefore reducing their need for government price supports to fill the market shortfall.[74] As one commentator noted, the result of a hard stance on this during negotiations "meant, of course, that the United States would eventually have to reduce or eliminate some of its own farm subsidies . . . [and thus] [o]ur GATT commitments would influence changes to U.S. farm policy made in the 1990 and 1996 farm bills."[75]

5. Recent History: Decoupling and Beyond

On the heels of the watershed 1985 farm bill and in the midst of important trade negotiations, the 1990s brought calls for further agricultural reform that mostly played out in efforts to "decouple," or segregate, a farmer's planting decisions from federal subsidy payments.[76] In the 1990 farm bill,[77] Congress primarily adhered to programs it adopted in the 1985 farm bill, but made tweaks to both conservation programs and income support programs to ensure increased efficiency and enhanced flexibility in both sets of programs.[78] As a commentator noted at the time, "[t]he significance of this move in the direction of flexibility [of domestic farm policy], despite the many conditions surrounding it, is potentially far reaching" because it could not only save significant federal tax dollars, but also, "if continued, will lead over time to more general decoupling."[79] The 1990 farm bill also created additional conservation programs in the Wetlands Reserve Program and the Water Quality Incentives Program, reinforcing Congress' commitment to environmental protection in the farming sector by providing incentives for certain on-farm activities to conserve wetlands and ensure clean water.[80] Some scholars have since pointed out that despite the value of the myriad programs authorized by the 1990 farm bill, the legislation marked a continuation of an escalating price tag for our nation's agricultural policy even though an express goal of Congress in fashioning the bill

---

71. U.S. Senate, THE UNITED STATES SENATE COMMITTEE ON AGRICULTURE, NUTRITION, AND FORESTRY: 1825-1998, at Chapter 8, S. Doc. 105-24, *available at* http://www.access.gpo.gov/congress/senate/sen_agriculture/ch8.html.
72. Olson, *supra* note 62, at 17-18.
73. *Id.* at 17.
74. *Id.*
75. *Id.*
76. Laurie Erdman & C. Ford Runge, *Review: American Agricultural Policy and the 1990 Farm Bill*, at 2 (1991), *available at* http://ageconsearch.umn.edu/bitstream/13593/1/p91-02.pdf.
77. Food, Agriculture, Conservation, and Trade Act of 1990, Pub. L. No. 101-624, 104 Stat. 3359 (1990), *available at* http://www.nationalaglaw-center.org/assets/farmbills/1990-1.pdf.
78. Erdman & Runge, *supra* note 76, at 2-10.
79. *Id.* at 11.
80. *Id.* at 19-20.

was to reduce the government's farm budget.[81] The inability to reduce the agriculture budget is no less important today than it was in 1990, as the agricultural budget under the farm bill generally continues to increase.

The close of the 20th century saw two important events that heavily shaped today's agricultural policy framework (described in Chapter 2 of this book): (1) the completion of the Uruguay Round of GATT negotiations, and (2) the 1996 farm bill's attempt to implement a strategy to decouple domestic farm subsidies on a wide scale. In 1994, the GATT negotiations, involving 123 nations, concluded after a nearly eight-year negotiation process, which, among other things, resulted in a "classification scheme whereby domestic policies [are] characterized by the extent to which they were considered to be trade-distorting."[82] Under this scheme, "[p]olicies that were considered to be minimally trade-distorting, such as conservation programs, domestic food aid, and research and extension expenditures were termed to be 'green box' policies and not subject to limits on overall domestic support."[83] In contrast, policies that were deemed to have far more serious implications for distorting international agricultural prices were labeled as "amber box" policies, and all GATT nations were required to reduce amber box subsidies by approximately 20% compared to each respective country's 1986-1988 average, and thereafter were required to stay below a certain threshold based on the nation's initial figure.[84]

To carry out these new international trade obligations and satisfy these mandates, Congress began to once again revamp the farm bill by shifting financial incentives from amber-box programs (such as nonrecourse marketing loans and countercyclical payments that are tied directly to the number of acres farmed for specified commodities) to green-box programs (such as conservation programs and direct payments, which are not directly tied to acres farmed because of planting flexibility rules).[85] However, a deep divide in Congress thwarted the promulgation of a farm bill in 1995, as different factions of key midwestern legislators battled over whether to essentially stay the course with the seemingly efficient but costly programs adopted in the 1985 and 1990 farm bills or to jettison that approach in favor of an attempt to radically transform the agricultural subsidy system by guaranteeing certain farm program payments regardless of the actual number of acres farmed in a given year (i.e., decoupling planting decisions in a given year from the financial incentives paid per acre).[86] What finally emerged in 1996 was legislation that entirely eliminated many programs that had existed since the 1933 farm bill (e.g., target prices, deficiency payments) and "terminated all supply management programs" in an effort to decouple many of the central programs in our nation's farm policy.[87] Importantly, however, the 1996 farm bill—known as the Federal Agriculture Improvement and Reform Act[88]—did not significantly reduce the overall cost of our farm policy; it simply shifted the manner in which the same farmers would receive incentives from supply-oriented subsidies to direct payments based not on present acres farmed but rather on past program acres, conservation program dollars, and other subsidies that were not supply-oriented.[89]

It was within this framework that U.S. agricultural policy entered the 21st century and which, in large measure, remained unchanged through the first two farm bills of the new millennium (as discussed in Chapter 2). While the decoupling efforts of the 1996 bill have ultimately failed to pay down the price tag of agricultural policies, in part because many of the decoupled subsidies were never truly decoupled from per-acreage farm production,[90] the current farming system has been heavily influenced by the 1996

---

81. *Id.* at 20.
82. Barry K. Goodwin & Ashok K. Mishra, *Are "Decoupled" Farm Program Payments Really Decoupled? An Empirical Evaluation*, 88 Am. J. Agric. Econ. 73, 74 (2006), *available at* http://naldc.nal.usda.gov/download/6851/PDF.
83. *Id.*
84. *Id.*
85. *See* Keeney & Kemp, How to Make It Work, *supra* note 40, at 12.
86. Olson, *supra* note 62, at 20.
87. *Id.* at 20-21.
88. Federal Agriculture Improvement and Reform Act of 1996, Pub. L. No. 104-127, 110 Stat. 888 (1996), *available at* http://frwebgate.access.gpo.gov/cgi-bin/getdoc.cgi?dbname=104_cong_public_laws&docid=f:pub l127.104.pdf.
89. *See* Keeney & Kemp, How to Make It Work, *supra* note 40, at 12 ("[T]he supply management elements of farm policy were banished [by the 1996 farm bill]. And the same 40% of farmers who always got subsidies continued to get them—now in a regular check tied only to their past crops."). The 1996 farm bill also introduced the Environmental Quality Incentives Program, which consolidated several pre-existing conservation programs, in an attempt to make the programs more efficient, and to provide technical and financial assistance to farmers. *See, e.g.,* Olson, *supra* note 62, at 22.
90. Goodwin & Mishra, *supra* note 82 (arguing that the 1996 farm bill did not actually decouple many program payments from crop production because "a fully decoupled payment is one for which the level of payment is fixed and guaranteed and thus is not influenced by ex-post

reforms—many of which carried through to the 2002 and 2008 farm bills[91]—and is best viewed as the uneven outgrowth of not only the historical safety net programs of early farm legislation, but also the more recent approaches to first maximize production at all costs and then subsequent measures to enhance efficiency while moving away from supply management and maintaining a focus on conservation. These fits and starts, in addition to other societal changes, have led to a profitable, but vastly different American agricultural economy than that which existed when President Roosevelt first enacted the farm bill in 1933 as a temporary effort to save the American family farm.

B. How Agricultural Policy Has Shaped the Modern Farm Economy, and How Historical Trends Can Help Us Predict the Policy Future

One of the most dramatic changes encouraged by U.S. agricultural policy over the past half-century has been the consolidation of farm markets in an effort to promote efficiency and maximize profits, both at the individual farm level and at the macrolevel with processing and distribution. In 1935, shortly after the original farm bill was enacted, there were 6.8 million farms in the United States with an average size of 155 acres.[92] By the start of the 21st century, there were only 2.1 million farms with an average size of 441 acres.[93] The total number of farms declined by 70% in less than 70 years, but the amount of land in agricultural production stayed relatively constant, and in some states actually increased.[94] Table 1.1 illustrates this trend in commodity-heavy Iowa.

Table 1.1[95]
Consolidation of Farms in Iowa, 1900–1997

Year	Total Farms	Farms <50 Acres	Farms 50–500 Acres	Farms >1,000 Acres
1900	228,622	35,941	192,341	340
1950	203,159	29,103	173,802	254
1997	90,972	29,642	55,443	5,887

The total number of farms in Iowa only decreased 10% between 1900 and 1950, but then decreased by more than 55% between 1950 and 1997.[96] This means that in the pre-farm bill era and in the first 17 years after the first farm bill when small farm "safety net" protection was a key goal of agricultural policy, overall farm loss was quite minimal. However, in the post-World War II era, when crop maximization was at the zenith of its prominence in American farm policy, the overall number of farms plummeted more than 55% between 1950 and 1997. Even more telling of the on-the-ground impacts of the post-war farm bills is the increase in large farms in Iowa (more than 1,000 acres) by 2,000% between 1950 and 1997 after actually decreasing between 1900 and 1950.[97] In contrast, mid-sized farms (between 50 and 500 acres) declined nearly 70% between 1950 and 1997 after experiencing only a 10% drop between 1900 and 1950.[98]

These trends of farm consolidation in Iowa, which have been mirrored throughout the Midwest over the past century, can be largely traced back to agricultural policies adopted by the federal government to control crop supply and crop prices.[99] Indeed, on a nationwide basis, "[a]bout 8% of farms account for 75%

---

realizations of market conditions (e.g. low prices or area yields)").
91. Farm Security and Rural Investment Act of 2002, Pub. L. No. 107-171, 116 Stat. 134 (2002), *available* at http://frwebgate.access.gpo.gov/cgi-bin/getdoc.cgi?dbname=107_cong_public_laws&docid=f:publ171.107.pdf.
92. U.S. Dep't. Agric., Econ. Res. Serv., Structure and Finances of U.S. Farms: 2005 Family Farm Report, *available at* http://www.ers.usda.gov/publications/eib-economic-information-bulletin/eib66.aspx.
93. *Id.*
94. *Id.*
95. Keeney & Kemp, A New Agricultural Policy, *supra* note 1, at 9 tbl.1.
96. *Id.*
97. *Id.*
98. *Id.*
99. CRS, Farm Commodity Programs, *supra* note 34, at 2 ("When farm programs were first authorized in the 1930s, most of the 6 million farms in the United States were small and diversified. Policymakers reasoned that stabilizing farm incomes using price supports and supply controls would help a large part of the economy (25% of the population lived on farms) and assure abundant food supplies. In recent decades, the face

of farm sales (these 175,000 farms had average sales over $1 million)."[100] Likewise, broader consolidation of farm markets is also the direct result of such policies, as illustrated in Table 1.2, which defines "very highly concentrated" where the four largest processors control more than 50% of the entire market and "concentrated" where the four largest processors control 30-50% of the market.

Table 1.2[101]
U.S. Market Share Controlled by Four Largest Agricultural Processors, by Commodity

Very Highly Concentrated	Market Share (%)	Concentrated	Market Share (%)
Beef Packing	84%	Pork Production	49%
Soybean Crushing	71%	Animal Feed Processing	34%
Pork Packers	64%		
Flour Milling	63%		
Broiler Production	56%		
Turkey Production	51%		

The United States has undeniably entered a new era where the iconic family farm idealized by Thomas Jefferson and Franklin Roosevelt has taken on a much different, if not altogether dissimilar, meaning from the family farm of yesteryear. At the close of the 20th century, nearly 90% of all farm bill subsidies in the form of direct payments, countercyclical payments, crop insurance payments, and other incentives paid to farmers, whether family farms or otherwise, were devoted to only fives crops—corn, cotton, wheat, rice, and soybeans.[102] The nearly two million remaining farmers who opt to produce other crops receive little to no assistance from the federal government and, in many cases, must survive primarily on income earned from off-farm pursuits as a supplement to their on-farm income.[103]

Long debated in the agricultural community is whether farm bill policies have hastened an exodus from rural America or slowed such trends. As agroeconomists Stephan Goetz and David Debertin have noted:

> Farm program payments continue over time largely because of the political influence of their beneficiaries: the owners of resources used to produce the outputs covered by the payments. One argument for continuing these programs is that the payments preserve family farms and maintain the viability of rural economies, as the program payments are spent and re-spent within local communities. Agricultural economists have argued that farm program payments may have reduced, but not reversed, the outflow of residents from rural areas.[104]

Regardless of the answer, the rural economy and the farming sector that bolsters it remain important parts of our national identity but will face some of its largest hurdles in the coming decades. For example, the exodus that gradually continues to diminish rural population numbers has already in some cases reached a point where community functionality is difficult as diverse employment opportunities are eliminated and the local tax base and related public services are depleted—all of which gives way, as described in the following chapters, to a system of farming that deemphasizes natural resource conservation and protection.[105]

---

of farming has changed. Farmers now comprise less than 2% of the population. Most agricultural production is concentrated in fewer, larger, and more specialized operations.").

100. *Id.*
101. Timothy A. Wise, *Identifying the Real Winners From U.S. Agricultural Policies* 3 (Global Dev. and Env't Inst. Working Paper No. 05-07, 2005), *available at* http://www.ase.tufts.edu/gdae/Pubs/wp/05-07RealWinnersUSAg.pdf.
102. *See* Dennis A. Shields, Congressional Research Service, *Federal Crop Insurance: Background and Issues*, at 2 (2010) ("Four crops—corn, cotton, soybeans, and wheat—accounted for three quarters of total enrolled acres [for crop insurance payments]."); *see also* CRS, Farm Commodity Programs, *supra* note 34, at 3 ("Federal support exists for about two dozen farm commodities representing nearly one-third of gross farm sales. Five crops (corn, cotton, wheat, rice, and soybeans) account for about 90% of these payments.").
103. CRS, Farm Commodity Programs, *supra* note 34, at 2 ("Most of the country's 2 million farms are part-time, and many operators rely on off-farm jobs for most of their income.").
104. Goetz & Debertin, *supra* note 66, at 517.
105. Thomas L. Dobbs, *Working Lands Agri-Environmental Policy Options and Issues for the Next United States Farm Bill* 9 (Econ. Staff Paper 2006-3, 2006), *available at* http://repec-sda.sdstate.edu/repec/sda/pdf/sp060003.pdf.

Moreover, historical trends demonstrate that our agricultural policies and consequent farm consolidation have resulted in an ever-aging farmer population; there are currently three times as many farmers over the age of 65 as there are under the age of 35, with an average farmer age of 57.1 years old.[106] Thus, particularly with the United States recently becoming a net importer of food, history dictates that agricultural policies can and should be crafted to eliminate barriers to access (such as loan eligibility and certification fees) for beginning farmers, especially since starting farmers generally create smaller, more diversified farms than established operators, and tend to favor more ecologically sustainable farming methods than their longer established counterparts.[107]

In addition, history has shown, for the most part, that conservation of natural resources must be an integral part of American agricultural policy. Despite the attempts to incorporate sweeping conservation programs in recent farm legislation, agriculture continues to have a powerful impact on the environment (as described in detail in Chapters 3-7). Therefore, with history as a guide, our conservation programs must continue to enhance protection for natural resources while simultaneously allowing farmers enough flexibility to remain profitable.

Agriculture policy in the 20th century provides glimpses of future international farming and economic issues. The GATT negotiations of the late 1980s and early 1990s paved the way for the prevailing WTO agricultural trade system, which will prove critical in finding markets for surplus foods produced by American farmers.[108] Recent history has also shown how intimately our nation's agricultural supply and demand, which is set in large part by farm bill subsidy, crop insurance, and loan programs, strongly influences foreign farm markets, and therefore can greatly impact land use and immigration far from home.[109] Finally, on both an international and domestic scale, historical trends illustrate a notable shift in the types of foods produced when the farm bill was first enacted in 1933 (whole foods and grains with little processing) and in the early 21st century (raw materials to be turned into highly processed foods), leading to the plethora of potential health concerns that must necessarily be addressed in 21st century farm legislation since, historically, farm bills have failed to provide systematic guidance for food production that supports public health.[110]

Conclusion

The 20th century marked the most active era of agricultural policy in U.S. history, ushered in by the emergency response to the 1930s farm crisis by New Deal agricultural legislation. While the initial farm bill was intended as a temporary fix to an urgent situation, its quinquennial passage has become a fixture of political commitment to an important sector of American society, and indeed one that affects the nation in a profound, albeit often unnoticeable, way. Because the underlying goals of the farm bill have changed drastically since 1933, two large questions loom in light of that policy history: (1) what level of federal involvement should exist in 21st century domestic agriculture, including with respect to natural resource protection; and (2) which legislative programs are successful in achieving the country's farming and food goals, as well as its dual goal of conservation, and how can we enhance their efficacy in a cost-efficient manner? The remainder of this book attempts to answer these difficult political and societal questions through the lens of environmental protection, by explaining in detail the components of the current farm bill (Chapter 2), analyzing the ecological impacts of the modern farming system encouraged by our nation's agricultural policy (Chapters 3-7), examining the interplay between agriculture and existing environmental and related laws (Chapters 8-13), and concluding with concrete proposals to reform agricultural policy that serve as models of how to ensure a more productive and yet more sustainable farming and food system (Chapters 14-18).

---

106. U.S. Dep't. Agric., Econ. Research Serv., *Beginning Farmers and Ranchers*, at 8 (2009), *available at* http://ers.usda.gov/publications/eib-economic-information-bulletin/eib53.aspx.
107. *Id.*; *see also* Nat'l Sustainable Agric. Coal., *Farming for the Future: A Sustainable Agriculture Agenda for the 2012 Food & Farm Bill*, at 13 (2012), *available at* http://sustainableagriculture.net/wp-content/uploads/2008/08/2012_3_21NSACFarmBillPlatform.pdf.
108. *See* William S. Eubanks II, *A Rotten System: Subsidizing Environmental Degradation and Poor Public Health With Our Nation's Tax Dollars*, 28 Stan. Envtl. L.J. 213, 234-39 (2009).
109. *Id.*
110. *See generally id.* 213-310 (analyzing in detail the public health impacts of our nation's agricultural policy).

Chapter 2
An Overview of the Modern Farm Bill
Mary Jane Angelo and Joanna Reilly-Brown

In the United States, an array of federal laws governs agricultural and food issues, but the most comprehensive collection of agricultural and food policy legislation is found in what is commonly referred to as the "farm bill." The farm bill covers a wide range of federal food and farming programs and provisions under the purview of the U.S. Department of Agriculture (USDA). The U.S. Congress passes a new omnibus farm bill roughly every five years, and the new legislation typically amends, reauthorizes, or repeals provisions of previous farm bills.[1]

The farm bill includes myriad programs, ranging from school lunches to biofuels. The most relevant programs related to the impacts of agriculture on the environment, however, are ones providing subsidies to farmers, which tend to encourage industrialized farming practices that are environmentally harmful, and those providing financial incentives for certain conservation practices.[2] These programs significantly influence what crops farmers choose to grow, how they grow them, and what environmental impacts result from these decisions.[3] Due to its constantly evolving, omnibus nature, the farm bill both stimulates cooperation between often-conflicting interest groups and creates intense competition among groups with differing priorities for farm bill programs and between producers of different commodities.[4]

Although the basic structure of the farm bill has remained intact over the past 70-plus years, several significant changes have been made, numerous programs have been added, and the breadth of issues covered by the farm bill has expanded to encompass emerging agricultural interests such as conservation, organic production, and bioenergy.[5]

Nevertheless, the largest components of farm bill policy continue to be the price support and income support programs that encourage large-scale monoculture industrialized agricultural practices.[6] Beginning with the Agricultural Adjustment Act of 1933, the United States has had a long history of subsidizing and regulating its agricultural sector. A product of the New Deal era, the 1933 legislation aimed to control crop prices by decreasing supply, a feat achieved by paying farmers to produce less.[7] The series of farm bills that followed in the subsequent seven decades—15 pieces of legislation in all—evolved into the country's comprehensive agricultural policy, tackling a variety of goals, from price support to conservation.[8]

The most recent farm bill, enacted in 2008, contains a labyrinth of complex, piecemeal, and often contradictory agricultural, energy, and conservation subsidy programs with a price tag of almost $284 billion.[9]

Portions of this chapter have been adapted from, with permission, Mary Jane Angelo, *Corn, Carbon, and Conservation: Rethinking U.S. Agricultural Policy in a Changing Global Environment*, 17 Geo. Mason L. Rev. 593 (2010).

1. Renée Johnson, Cong. Research Serv., What Is the "Farm Bill"?, 1 (2008), *available at* http://www.nationalaglawcenter.org/assets/crs/RS22131.pdf [hereinafter CRS, What Is the Farm Bill].
2. John H. Davidson, *The Federal Farm Bill and The Environment*, 18-SUM Nat. Resources & Env't 3, 37 (2003); Jason J. Czarnezki, *Food, Law & The Environment: Informational and Structural Changes for a Sustainable Food System*, 31 Utah Envtl. L. Rev. 263, 266 (2011); William S. Eubanks II, *A Rotten System: Subsidizing Environmental Degradation and Poor Public Health With Our Nation's Tax Dollars*, 28 Stan. Envtl. L.J. 213, 251-73 (2009).
3. Julie Foster, *Subsidizing Fat: How The 2012 Farm Bill Can Address America's Obesity Epidemic*, 160 U. Pa. L. Rev. 235, 255 (2011).
4. CRS, What Is The Farm Bill, *supra* note 1, at 4.
5. *Id.* at 1.
6. Eubanks, *supra* note 2.
7. Agricultural Adjustment Act of 1933, Pub. L. No. 73-10, §8(1), 48 Stat. 31, 34 (1933).
8. *See* CRS, What Is The Farm Bill, *supra* note 1 (discussing the variety of programs encompassed within the farm bill).
9. Renée Johnson et al., Cong. Research Serv., The 2008 Farm Bill: Major Provisions and Legislative Action (2008), 1, 8, *available at* http://www.nationalaglawcenter.org/assets/crs/RL34696.pdf [hereinafter CRS, The 2008 Farm Bill].

Commodity subsidies under the 2008 farm bill include a large number of complex programs including price support programs and income support programs. While Congress has historically defended these farm support programs as necessary to ensure that the United States has access to an affordable, safe food supply,[10] such programs have also been criticized as being ineffective, costly, international-trade distorting, and environmentally destructive due to their encouragement of large-scale industrial agriculture.[11]

This chapter provides an overview of the policies, subsidies, and trends associated with the current farm bill, discusses the provisions and programs that ostensibly impact the natural environment, and identifies perceived gaps or weaknesses within its programs that have the ability to produce harmful environmental consequences. Proposals for environmentally beneficial changes that could be made to future farm bills are also discussed, as well as challenges (budgetary and otherwise) to implementing those changes. The chapter concludes with a discussion of proposed changes to the farm bill currently being debated, which are primarily focused on reducing subsidies in response to the ongoing government budgetary problems.

A. The Food, Conservation, and Energy Act of 2008: Key Provisions and Programs Impacting the Environment

The most recent omnibus farm bill is the Food, Conservation, and Energy Act of 2008.[12] The Act reauthorized and makes small changes to programs carried through from previous farm bills.[13] Some new programs were added, but the basic structure of previous farm bills was maintained.[14] The Act contains 15 titles providing support for commodity price and income supports, conservation, trade, energy, horticulture and organic agriculture, and crop insurance, among other programs.[15] The 2008 bill added five titles that had not been in the 2002 farm bill, which contain provisions to address horticulture, livestock, and organic products issues (including mandatory funding to support organic production and block grants for specialty crops), commodity futures, crop insurance and disaster assistance, and assorted tax and trade provisions.[16]

Many provisions in the 2008 farm bill have the ability to beneficially or adversely influence, both directly and indirectly, the impacts of agriculture on the environment. For example, the various subsidy programs for large-scale commodity production incentivize farming practices that maximize production often at the expense of natural resource preservation, while the conservation subsidy programs are designed to protect natural resources by encouraging farmers to set aside sensitive lands or participate in voluntary working lands programs.

1. Title I: Commodities

Three permanent laws, as amended, give USDA the authority to operate farm commodity programs: the Agricultural Adjustment Act of 1938,[17] the Agricultural Act of 1949,[18] and the Commodity Credit Corporation (CCC) Charter Act of 1948.[19] These laws are typically amended through the omnibus farm bills in response to market or budgetary concerns. Every farm bill has an expiration date, and, if Congress fails to enact a new farm bill upon expiration of an old one, the commodities programs revert to the permanent laws mentioned above.[20] It is therefore highly desirable for Congress to enact a new farm bill when an old one expires because the permanent laws support eligible commodities at rates significantly higher than

---

10. CRS, What Is The Farm Bill, *supra* note 1, at 1.
11. William S. Eubanks II, *The Sustainable Farm Bill: A Proposal for Permanent Environmental Change*, 39 ELR 10493, 10504 (May 2009).
12. Food, Conservation, and Energy Act of 2008, Pub. L. No. 110-246, 122 Stat. 923 (2008).
13. CRS, The 2008 Farm Bill, *supra* note 9, at 5.
14. *Id.* at 7.
15. *Id.* at 6.
16. *See id.* (summarizing the 15 titles contained in the 2008 farm bill).
17. Agricultural Adjustment Act of 1938, Pub. L. No. 75-430, 52 Stat. 31.
18. Agricultural Act of 1949, 7 U.S.C. §1431.
19. Commodity Credit Corporation Charter Act of 1948, 15 U.S.C. §714.
20. Jim Monke, Cong. Research Serv., Farm Commodity Programs in the 2008 Farm Bill, 3 (2008), *available at* http://www.nationalaglawcenter.org/assets/crs/RL34594.pdf [hereinafter CRS, Farm Commodity Programs].

current ones, and many commodities that are currently supported would likely be ineligible under the permanent laws.[21]

The 2008 farm bill mostly continues the farm commodity price and income support program framework of the 2002 bill, although some changes to program eligibility requirements and loan rates for some commodities were made, and the new Average Crop Revenue Election (ACRE) Program, discussed below, was added.

The primary types of subsidies in Title I of the 2008 farm bill are *income supports* and *price supports*.[22] The goal of income supports is to keep farmer income high regardless of prices they get for crops.[23] The goal of price supports is to keep prices that farmers get for crops at a high and stable level.[24]

In general, to be eligible to receive farm support payments, a farmer must both produce an eligible commodity and meet the statute's definition of a producer.[25] The 2008 farm bill defines "producer" as "an owner, operator, landlord, tenant, or sharecropper that shares in the risk of producing a crop on a farm and is entitled to share in the crop available for marketing from the farm, or would have shared had the crop been produced."[26]

Crops eligible for federal support include roughly two dozen commodities that represent approximately one-third of gross farm sales.[27] Five of these crops—corn, cotton, rice, soybeans, and wheat—account for nearly 90% of farm support payments.[28] More than 60% of farm support payments go to 10% of recipients.[29]

To be eligible to receive payments, a farmer must also comply with certain provisions related to conservation and planting flexibility. Modern farming operations typically involve a combination of both rented and owned land, usually under either a cash rent or share rent arrangement.[30] Under a cash rental contract, the landlord is not eligible to receive program payments because the tenant pays a fixed amount of rent and bears all the risks and benefits of the farming operations.[31] Under a share rental contract, however, both the landlord and the tenant are eligible to share in the government subsidy because, while the tenant supplies most of the labor and machinery, the landlord supplies the land and sometimes machinery or management.[32] Both the landlord and the tenant therefore bear the risks and benefits of crop production under a share rental arrangement and, as such, both are eligible to receive payments.[33]

Because the United States is subject to the agricultural trade policies of the World Trade Organization (WTO), the way a particular farm commodity program is classified under WTO policies is important. All WTO member countries are required to submit annual reports of their farm program expenditures to the WTO, which subjects these disbursements to specific spending limits depending on whether the program is considered trade distorting (classified as "amber box" when reporting to the WTO) or "decoupled" (classified as "green box").[34] For programs in the United States that are considered trade distorting under the WTO (i.e., amber box), the total spending limit is $19.1 billion per year.[35] There is no spending limit for programs that are not classified as trade distorting or for programs that are "decoupled" (i.e., green box programs).[36] A commodity program is characterized as "decoupled" when the amount of financial incentive received under the program is not tied to the volume of crop produced by the farmer.[37]

---

21. *Id.* at 4-25.
22. For a detailed summary of the commodity programs in the 2008 farm bill, see *id.* at 3.
23. *Id.* at 1.
24. *Id.*
25. Food, Conservation, and Energy Act of 2008, Pub. L. No. 110-246, 122 Stat. 923, §1001(13)(A) (2008).
26. Food, Conservation, and Energy Act of 2008, Pub. L. No. 110-246, §1001(13)(A), 122 Stat. 1666 (2008).
27. CRS, Farm Commodity Programs, *supra* note 20, at 3.
28. *Id.*
29. *Id.*
30. *Id.* at 4.
31. *Id.*
32. *Id.*
33. *Id.*
34. *Id.* at 5.
35. *Id.*
36. *Id.* at 4.
37. Economic Research Service, U.S. Dep't of Agric., *Farm and Commodity Policy: What Is Meant by Decoupling?*, http://www.ers.usda.gov/topics/farm-economy/farm-commodity-policy/what-is-meant-by-decoupling.aspx (last visited Aug. 13, 2012).

Table 2.1 provides a summary of the commodity crop programs contained in the 2008 farm bill, including information on type of support, eligibility, conditions, payment amounts, and WTO classification for each subsidy program.

Table 2.1
Summary of Farm Bill Commodity Crop Programs[38]

Program	Type of Support	Eligibility	Basis of Subsidy	Conditions	Maximum Amount	WTO Classification
Direct Payment	Income support	Historical production of commodity crops	National payment rate, planting history, and historical yields	Do not have to grow crops to receive, but cannot grow fruit or vegetables.	$40,000 per person per year	Green box
Counter-Cyclical Payment	Income support	Historical production of commodity crops	Difference between target prices and national average farm prices	Do not have to grow crops to receive, but cannot grow fruit or vegetables.	$65,000 per person per year	Amber box
Average Crop Revenue Election (ACRE) Program	Income support	Acres planted in commodity crops	State-based guaranteed and state yields	Farmer must experience crop revenue loss. Cannot receive countercyclical payments, and direct payments and loan amounts are reduced.	$65,000 per farm per year. Can enroll multiple farms	Amber box
Marketing Loan Assistance	Price support	Production of commodity crops	Loan provides guarantee minimum price	When market prices fall below loan amount, farmer forfeits crop and keeps loan amount.	No maximum	Amber box

a. Income Support Programs

Income support programs make up the largest category of commodity subsidies.[39] They include direct payments, countercyclical payments, and the ACRE program.[40]

i. Direct Payments

Direct payments are fixed annual payments based on a national payment rate that vary depending on an individual farm's planting history and historical yields.[41] Direct payments are made to producers in proportion to a farm's "base acres" for a particular commodity, measured as a constant average of the farm's

---

38. The source for the information contained in Table 2.1 is CRS, FARM COMMODITY PROGRAMS, *supra* note 20.
39. *Id.* at 5–11.
40. U.S. Dep't of Agric., Economic Research Service, *2008 Farm Bill Side-by-Side, Title I: Commodity Programs*, http://www.ers.usda.gov/FarmBill/2008/Titles/TitleIcommodities.htm (last visited Apr. 10, 2012) [hereinafter *Title I: Commodity Programs*]. The 2008 farm bill allows the 2002 farm bill's direct payment program to remain intact with a few minor changes. In particular, the 2008 bill provides that payment acres for crop years 2009-2011 are 83.3% and eliminates the three-entity rule (limited number of farms from which a person could receive program payments—full payment directly and up to one-half payment for two additional entities) for payment limits. *Id.*; *see also* Center for Rural Affairs, *Overview of the 2008 farm bill*, http://www.cfra.org/newsletter/2008/05/overview-2008-farm-bill (last visited Mar. 8, 2010) (discussing the three-entity rule).
41. DENNIS A. SHIELDS ET AL., CONG. RESEARCH SERV., FARM SAFETY NET PROGRAMS: ISSUES FOR THE NEXT FARM BILL, 3 (2010), *available at* http://www.cnie.org/NLE/CRSreports/10Oct/R41317.pdf [hereinafter CRS, FARM SAFETY NET PROGRAMS].

"eligible historical production of commodity crops."[42] The covered commodity crops are wheat, feed grains, cotton, rice, and oilseeds.[43] The payment amount is the product of the payment rate (statutory), the payment acres of the covered commodity, and the historical payment yield for the commodity.[44] The payment acres, by statute, were 83.3% of base acres in crop years 2009-2011 and 85% of base acres in crop years 2008 and 2012.

For example, for 2012, the statutory payment rate for corn was $0.28 per bushel, and the payment acres were 85% of the farm's base acreage for the covered commodity. If a farm had 1,000 base acreage for corn and we assume a payment yield of 110 bushels of corn per acre, the farmer would be eligible to receive a direct payment under this calculation:

$$850 \text{ acres (payment acres 85\% of 1000 acres)} \times 110 \text{ (assumed payment yield)} \times \$0.28 \text{ (payment rate per bushel)} = \$26{,}180 \text{ per person per year}$$

The payment rate varies by commodity type and does not fluctuate with market prices.[45] Direct payments are limited to $40,000 per person per crop year, and they must be linked to a person or legal entity, either directly or indirectly.[46] In order to receive direct payments, farmers have flexibility in what commodities they choose to plant, but they are required to adhere to statutory conservation provisions that are the functional equivalent of best-management practices.[47] One of the most interesting aspects of the direct payment program is that the producer does not need to actually grow a covered crop to get a payment for that commodity. For example, if the producer historically grew soy on a farm's base acres, she can now grow corn on the soy base acres and still get the subsidy for soy. It is even possible to get the subsidy for leaving the farmland fallow. However, to receive this subsidy, growers may not grow fruit, vegetables, or wild rice.[48]

Direct payments are considered less trade distorting than other farm support programs because the payments are constant and the program allows for planting flexibility.[49] For this reason, direct payments are considered "decoupled" and the United States classifies them as "green box" when annually reporting agricultural subsidies to the WTO. The WTO classifies subsidies that do not distort trade (i.e., "decoupled") as "green box", and allows green box subsidies without limits.[50] Such subsidies therefore do not count against the United States' WTO subsidy ceilings. The classification of direct payments as "green box" subsidies has been subject to challenge,[51] however, because planting flexibility rules, discussed below, maintain restrictions on the planting of certain fruits and vegetables, direct payments were found to be accurately classified.[52]

Under the direct payment program, farmers have planting flexibility because they may receive direct payments for planting crops that are not covered by the program on program base acres. Farmers are restricted, however, from planting vegetables, fruits, or wild rice on base acres. This restriction was developed as a protection for producers of unsubsidized fruits and vegetables to decrease competition from subsidized producers of crops covered by the direct loan program.[53] For WTO accounting purposes, however, the planting restriction on fruits and vegetables makes the United States' classification of direct payments as "green box" subsidies vulnerable to challenge as being "inconsistent with the rules of a minimally distorting subsidy."[54]

---

42. *Title I: Commodity Programs, supra* note 40.
43. Food, Conservation, and Energy Act of 2008, Pub. L. No. 110-246, 122 Stat. 1665, §1001(4) (2008).
44. Food, Conservation, and Energy Act of 2008, Pub. L. No. 110-246, 122 Stat. 1670, §1103(c) (2008).
45. CRS, Farm Safety Net Programs, *supra* note 41, at 4.
46. Food, Conservation, and Energy Act of 2008, Pub. L. No. 110-246, §1603(b)(1).
47. CRS, Farm Safety Net Programs, *supra* note 41, at 4.
48. *Id.*
49. CRS, Farm Commodity Programs, *supra* note 20, at 6.
50. World Trade Organization, Agriculture Negotiations: Background Fact Sheet, *Domestic Support in Agriculture: The Boxes*, http://www.wto.org/english/tratop_e/agric_e/agboxes_e.htm (last visited Oct. 26, 2012).
51. Brazil and Canada brought WTO cases against the U.S. farm programs in 2007, charging that direct payments are improperly classified as green box subsidies. For a detailed description of the Brazil and Canada cases, see Randy Schnepf, Cong. Research Serv., Brazil's and Canada's WTO Cases Against U.S. Agricultural Direct Payments (2010), *available at* http://www.nationalaglawcenter.org/assets/crs/RL34351.pdf.
52. CRS, Farm Commodity Programs, *supra* note 20, at 6.
53. *Id.* at 24.
54. *Id.*

ii. Countercyclical Payments

Under the countercyclical payment program, automatic payments are made to producers with "eligible historical production of commodity crops" when market prices for covered crops fall below statutory target prices.[55] Countercyclical payments are considered a "safety net" for farmers in years with low crop prices. The covered crops are wheat, feed grains, cotton, rice, legumes, and oilseeds. Countercyclical payments depend on the relationship between government-determined target prices and national average farm prices.[56] When the current effective commodity price for a covered crop falls below the target price, farmers and landowners participating in the program become eligible to receive a payment based on their farm's "historical acreage and yield."[57] The countercyclical payment rate is the amount by which the target price of each commodity exceeds its effective price. The effective price for covered commodities, except rice, is calculated by adding (1) the national average market price received by producers during the marketing year, or the commodity national loan rate, whichever is higher, and (2) the direct payment rate for the covered commodity. The payment acres, by statute, are 85% of base acres. The limit of countercyclical payments is $65,000 per person per year.[58]

For example, the statutory target price for corn is $2.63 per bushel. If in a particular year the effective price for corn is higher than $2.63 per bushel, countercyclical payments would not be available for corn in that year. If in a particular year, the effective price is lower than $2.63 per bushel, countercyclical payments would be available. In 2004, for example, corn farmers were eligible for countercyclical payments because the effective price of corn was lower than the target price. The countercyclical payment rate is the amount by which the target price of each commodity (in this case $2.63) exceeds the effective price (the higher of the national average market price or the national loan rate, which in 2004 was $2.06 per bushel for corn plus the direct payment rate, which was $0.28 per bushel of corn). In this example, the effective price was $2.34. Thus, the payment rate would be $0.29, the target price ($2.63) minus the effective price ($2.34). The countercyclical payment for each crop year is based on 85% of the farm's base acreage. For a 1,000 acre farm in 2004, therefore, the calculation would be:

$$850 \text{ acres (payment acres 85\% of 1000 acres)} \times 110 \text{ (assumed payment yield)} \times \$0.29 \text{ (payment rate)} = \$27,115 \text{ per person per year.}^{59}$$

Countercyclical payments are made in proportion to a farm's base acres and historical yield (i.e., yields attained for a particular crop in previous years), but they do not depend on current production.[60] Thus, while the rate formula depends on market prices, farmers are not actually required to produce any of the covered commodities.[61] Countercyclical payments are therefore partially decoupled because the payment rate is decoupled from acreage and yield, but still depends on market prices. The United States classifies countercyclical payments as "amber box" when reporting its agricultural subsidies to the WTO.[62] Amber box subsidies are considered to distort trade, and are therefore limited in size under the WTO.[63]

iii. Average Crop Revenue Election Program

Under the Average Crop Revenue Election (ACRE) Program, which was introduced in the 2008 farm bill, producers may make an irrevocable election (once enrolled, a farm is permanently enrolled in ACRE until the end of the 2012 crop year) to receive a state-based revenue guarantee equal to 90% of benchmark state

---

55. *Id.* at 9.
56. CRS, Farm Safety Net Programs, *supra* note 41, at 6.
57. *Id.*
58. U.S. Dep't. Agric., Economic Research Service, *Farm and Commodity Policy: Program Provisions: Payment Limitations*, http://www.ers.usda.gov/topics/farm-economy/farm-commodity-policy.aspx (last visited Apr. 10, 2012).
59. This example is based on information contained in the document Direct and Countercyclical Program, which is available at http://www.usda.gov/documents/DIRECT_AND_%20COUNTER_CYCLICAL_PROGRAM.pdf (last visited Aug. 15, 2012).
60. CRS, Farm Commodity Programs, *supra* note 20, at 9.
61. *Id.* at 10.
62. *Id.*
63. World Trade Organization, Agriculture Negotiations: Background Fact Sheet, *Domestic Support in Agriculture: The Boxes*, *supra* note 50.

yield multiplied by the ACRE program guarantee price for the crop year.[64] The ACRE program covers wheat, feed grains, cotton, rice, legumes, and oilseeds, and payment is based on planted acres rather than base acres.[65] A farmer who operates multiple farms may elect to enroll one or more farms in the ACRE program.[66] A producer electing to receive ACRE payments for a particular farm becomes ineligible to receive countercyclical payments, direct payments are reduced by 20%, and marketing assistance loan rates are reduced by 30%.[67] This program provides a revenue guarantee each year based on state market prices and average state yields, and is designed to protect farmers from revenue losses for covered crops, regardless of the cause.[68] Participating farmers receive payments for all covered crops on the enrolled farm, with the amount of payment for each crop calculated separately.[69] Payments are limited to $65,000 per farm in addition to the 20% reduction in direct payments and 30% reduction in marketing loan rates.[70]

Participation in the ACRE program has been relatively low since its introduction in the 2008 farm bill.[71] During the 2009 crop year, only about 8% of the total number of eligible farms, representing less than 13% of the total program base acres, elected to participate in the program.[72] The complexity of the payment determination process as well as difficulties determining whether enrollment would be advantageous for individual farmers are reportedly issues that have limited program participation.[73] Critics of the ACRE program have also argued that the program can result in duplicate payments in situations where low yields or prices result in a farm's receiving both crop insurance indemnities and ACRE program payments.[74]

Although not specifically designed to be conservation programs, the direct payments, countercyclical payments, and the ACRE program require producers to agree to carry out certain conservation-related practices to qualify for the payments. During the crop year in which a producer receives payments he must agree to designate for conservation uses or set aside land if a majority of it is highly erodible[75]; not to produce an agricultural commodity on a wetland that has been drained, dredged, filled, leveled, or converted in any way for production of an agricultural commodity[76]; to plant any crop on base acres except fruits, vegetables (excluding mung beans and pulse crops), or wild rice; to use the farmland for agricultural or conservation uses; and to control noxious weeds and follow sound agricultural practices.[77]

b. Price Support Programs

i. Marketing Assistance Loan Program

The Marketing Assistance Loan (MAL) program is a nonrecourse loan program that allows farmers to pledge their harvested crops as collateral that can be forfeited with no penalty. The program benefits farmers by providing interim financing when the loan is requested and a guaranteed minimum price for qualifying crops.[78] Under the MAL program, producers of various crops—wheat, corn, grain sorghum, barley, oats, upland cotton, extra long staple cotton, long and medium grain rice, soybeans, other oilseeds, peanuts, wool, mohair, honey, dry peas, lentils, and small and large chickpeas—are allowed to receive a nonrecourse loan at a commodity-specific loan rate per unit of production.[79] Producers pledge their production as collateral for the loan. Producers may then store their crops and sell their production when mar-

---

64. *Title I: Commodity Programs, supra* note 40.
65. *Id.*
66. CRS, Farm Safety Net Programs, *supra* note 41, at 7.
67. *Title I: Commodity Programs, supra* note 40.
68. CRS, Farm Safety Net Programs, *supra* note 41, at 7.
69. *Id.*
70. *Title I: Commodity Programs, supra* note 40.
71. CRS, Farm Safety Net Programs, *supra* note 41, at 7.
72. *Id.*
73. *Id.*
74. *Id.*
75. The conservation requirements are found in subtitle B of Title XII of the Food Security Act of 1985, 16 U.S.C. §3811.
76. The wetland protection requirements are found in subtitle C of Title XII of Food Security Act of 1985, 16 U.S.C. §3821.
77. U.S. Dep't. Agric., Farm Service Agency, FSA Handbook: Direct and Countercyclical Program and Average Crop Revenue Election for 2009 and Subsequent Crop Years (2009), *available at* http://www.fsa.usda.gov/Internet/FSA_File/1-dcp_r03_a03.pdf.
78. CRS, Farm Commodity Programs, *supra* note 20, at 12.
79. *Title I: Commodity Programs, supra* note 40.

ket conditions are more favorable.[80] Prior to loan maturity, if the local market price for the crop (known as the "posted price") is higher than the loan rate, the producer may sell the crop, pay off the loan, and keep the profits. If, on the other hand, the posted price is lower than the amount of the loan, the producer may pay off the loan at that price and keep the difference or forfeit the pledged crop to the Commodity Credit Corporation at loan maturity.[81] In contrast to the direct payment and countercyclical payment income support programs, which base payments on historical yields and acres and are thus not dependent on current production, MAL program benefits depend on the entire crop produced, such that no benefits are received in the event of a crop loss.[82] In years when market prices are lower than the loan rate, this program in essence amounts to the federal government purchasing crops at above market value. Thus, growers have the security that they can either take advantage of market prices in good years or simply "sell" their crops to the government at above market value in bad years.[83] Because of the glut of many commodity crops most years, the latter frequently occurs. Thus, the federal government is left with vast quantities of surplus commodities for which it must either find a market or find some method of disposal.

Farmers may choose to receive loan deficiency payments (LDPs) as an alternative to taking out a marketing loan. The LDP option allows farmers to receive the benefits of the loan program by providing them the opportunity to sell covered crops for a cash payment without pledging their commodity as collateral for the loan.[84]

Marketing loans are not decoupled because, unlike direct or countercyclical payments, they provide price guarantees on commodities actually produced. Marketing loans therefore depend upon both market prices and current production, and the United States classifies them as "amber box" when reporting its agricultural subsidies to the WTO.[85]

ii. Dairy and Sugar Price Support Programs

The 2008 farm bill reauthorized existing price support programs for producers of both dairy and sugar, with some changes. The bill extended the previous indirect dairy price support program, which provided indirect price support to dairy farmers through government purchases of surplus dairy products at a specified price of $9.90 per hundredweight (cwt) of farm milk, through the end of 2012.[86] The bill modified the dairy support program, however, to provide direct price support through required government purchases of surplus dairy products, including nonfat dry milk, butter, and cheese, at individual statutorily mandated minimum prices, as opposed to the $9.90 per cwt overall price level used under the indirect price support program.[87] This modification from indirect to direct price support was designed to reduce the program's vulnerability to WTO limitations on indirect price support programs.[88] The sugar price support program provides indirect support through import quotas, which restrict sugar imports, and domestic marketing allotments, which limit the amount of sugar processors can sell domestically.[89] In addition, the 2008 bill mandates a sugar-for-ethanol program, which requires USDA to purchase domestically produced sugar to be sold to bioenergy producers for ethanol production.[90] This provision was designed to address potential U.S. sugar surpluses caused by relaxed restrictions on sugar imports from Mexico under the North American Free Trade Agreement (NAFTA).[91] Other price support provisions in the 2008 bill increased

---

80. CRS, Farm Commodity Programs, *supra* note 20, at 12.
81. CRS, Farm Safety Net Programs, *supra* note 41, at 7.
82. *Id.*
83. Foster, *supra* note 3, at 251.
84. CRS, Farm Commodity Programs, *supra* note 20, at 12.
85. *Id.*
86. Ralph M. Chite & Dennis A. Shields, Cong. Research Serv., Dairy Policy and the 2008 Farm Bill, 7 (2009), *available at* http://www.nationalaglawcenter.org/assets/crs/RL34036.pdf.
87. *Id.* at 9.
88. *Id.*
89. Remy Jurenas, Cong. Research Serv., Sugar Policy and the 2008 Farm Bill, 1 (2008), *available at* http://www.nationalaglawcenter.org/assets/crs/RL34103.pdf.
90. *Id.* at 10.
91. *Id.* at 3-4.

the guaranteed prices for raw sugar and refined beet sugar, and mandated a market share of 85% for the domestic sugar production sector.[92]

One of the most significant differences between the existing income support programs and the MAL nonrecourse loan program is that, while direct payments are limited to $40,000 per person per year and countercyclical and ACRE payments are limited to $65,000 per person per year, nonrecourse loan payment amounts are not capped under the 2008 bill. The 2002 farm bill had placed a $75,000 per person per year limit on these loans, limiting the maximum total amount of commodity payment that could be received to $360,000 per farm per year. Under the 2008 farm bill, the limit for direct and countercyclical payments continues to be $210,000 per farm couple (e.g., a husband and wife running a farm per year), but there is no longer a limit on nonrecourse marketing loans.[93] In other words, a husband and wife running a farm could have each received up to $40,000 in direct payments and $65,000 in countercyclical payments for a total of $210,000 per couple per year under the 2002 Act and they can continue to do so under the 2008 bill. However, under the 2008 Act, the $75,000 per person per year nonrecourse marketing loan cap would no longer apply to the couple. Thus, at least in theory, under the 2008 Act farming couples could receive considerably more than the previous maximum of $360,000 per farm per year.

2. Title II: Conservation

The 2008 farm bill provides a number of incentive programs designed to conserve natural resources.[94] It reauthorized a majority of the conservation programs found in previous bills, modified several others, and created a number of new programs. The conservation title generally expanded requirements for program eligibility and receipt of technical assistance under most programs to include forested and managed lands, specified natural resource areas, and pollinator habitat, among others. Coverage for producers under most programs was also expanded to include producers of specialty crops and farmers transitioning to organic production.

The 2008 farm bill's conservation programs include voluntary land retirement or set asides, voluntary working lands programs, voluntary farmland protection programs, and mandatory conservation requirements linked to accepting other subsidies.[95] Land retirement programs generally offer cost-sharing assistance and annual payments to farmers who set aside land from crop production to achieve conservation goals such as the reversion of agricultural land back into forests, wetlands, and grasslands.[96] Working lands programs seek to improve agricultural land management practices by providing technical and financial assistance to farmers.[97]

a. Land Retirement and Easement Programs

The major title II land retirement and easement programs include the Conservation Reserve Program, the Wetlands Reserve Program, the Grasslands Reserve Program, and the Farmland Protection Program, among others.

i. Conservation Reserve Program

The Conservation Reserve Program (CRP), which appeared for the first time in 1985 in the Food Security Act, is one of the largest conservation programs in the farm bill in terms of acreage enrolled and total annual funding.[98] The CRP is a voluntary land retirement program that offers annual payments and cost-

---

92. *Id.* at 2, 7.
93. CRS, THE 2008 FARM BILL, *supra* note 9, at 13.
94. For a detailed discussion of the conservation provisions in the 2008 farm bill, see TADLOCK COWAN & RENÉE JOHNSON, CONG. RESEARCH SERV., CONSERVATION PROVISIONS OF THE 2008 FARM BILL (2009), *available at* http://www.nationalaglawcenter.org/assets/crs/RL34557.pdf [hereinafter CRS, CONSERVATION PROVISIONS].
95. 16 U.S.C. §§3831-35a (2006 & Supp. 2009); *see also* Erin Morrow, *Agri-Environmentalism: A Farm Bill for 2007*, 38 TEX. TECH. L. REV. 345, 392 (2006) (arguing that U.S. farm policy should strive to protect the environment and that such a policy is feasible).
96. CRS, CONSERVATION PROVISIONS, *supra* note 94, at 3.
97. *Id.* at 3-4.
98. U.S. DEP'T OF AGRIC., ECONOMIC RESEARCH SERVICE, THE CONSERVATION RESERVE PROGRAM: ECONOMIC IMPLICATIONS FOR RURAL AMERICA 1 (2004), *available at* http://www.ers.usda.gov/publications/aer834/aer834.pdf.

sharing assistance to participants that establish long-term, resource-conserving plant cover on environmentally sensitive land. The purpose of the CRP program is to improve and conserve water, soil, and wildlife resources on agricultural lands by providing financial and technical support for the conversion of environmentally sensitive lands to long-term vegetative cover.[99] Pursuant to the CRP, the federal government enters into contracts of 10-15 years with farmers to retire lands that are highly erodible or environmentally sensitive from agricultural uses. Participants must establish long-term resource-conserving plant cover on retired environmentally sensitive land. As of October 2007, 34.6 million acres of former farmland had been enrolled in the CRP.[100] The 2008 farm bill caps CRP enrollment at 32 million acres, approximately 7.2 million acres lower than the previous cap.[101]

The 2008 bill made some changes to the CRP, including allowing consideration of state, regional, and national conservation initiatives; requiring participating producers to manage enrolled lands in accordance with a conservation plan; and allowing installation of wind turbines on enrolled land.[102] The 2008 bill also amended the CRP pilot program for wetland and buffer acres such that each state can enroll up to 100,000 acres of eligible land up to a maximum of one million acres nationally. Lands eligible for this program include wetlands used for crop production in three of the immediately preceding 10 crop years; land upon which wetlands will be constructed to manage runoff from fertilizers; and lands devoted to commercial aquacultural production.[103]

ii.   Wetlands Reserve Program

The Wetlands Reserve Program (WRP), is a voluntary program to help owners of eligible lands restore and protect wetlands and wildlife on their property.[104] The program provides cost sharing and long-term or permanent easements for restoration of wetlands on agricultural land. Landowners have three enrollment options.[105] Under the first option, participants grant a perpetual conservation easement on enrolled acreage. In exchange, they receive 100% of the easement value and up to 100% of restoration costs. Under the second option, participants grant a 30-year easement on enrolled acres and receive 75% of the easement value and up to 75% of restoration costs. The third option does not involve the granting of a conservation easement. Instead, a participant who agrees to restore or enhance wetland functions and values enters into a cost-share agreement with the federal government in which the government provides assistance for the costs of wetland restoration or enhancement. More than 2.3 million acres of wetlands had been enrolled in the WRP as of 2010.[106] The 2008 farm bill increases the WRP enrollment cap from the previous 2.275 million acres to over three million acres and expands eligibility for the program to include lands that provide habitat for specific wildlife species and certain types of tribal and privately owned wetlands, croplands, and grasslands.[107]

iii.   Grasslands Reserve Program and Farmland Protection Program

The Grassland Reserve Program (GRP) provides landowners assistance for restoring grassland, rangeland, pastureland, and shrubland, and for conserving virgin grasslands with easements or long-term rental agreements. Livestock grazing and hay production are still allowed.[108] The 2008 bill modified the terms and conditions of contracts and easements under the GRP to allow fire presuppression management activities on enrolled lands.[109]

---

99. CRS, Conservation Provisions, *supra* note 94, at 4.
100. Nathaniel Kale, A Brief Economic Survey of the USDA Conservation Reserve Program 5 (Apr. 2009) (unpublished professional paper, University of Minnesota), *available at* https://conservancy.umn.edu/bitstream/49111/1/Kale,Nathaniel.pdf.
101. CRS, The 2008 Farm Bill, *supra* note 9, at 17.
102. CRS, Conservation Provisions, *supra* note 94, at 17.
103. *Id.* at 5.
104. 16 U.S.C. §§3837-3837f (2006 & Supp. 2009).
105. 16 U.S.C. §3837(b)(2).
106. U.S. Dep't of Agric., Natural Resources Conservation Service, *Wetlands Reserve Program*, http://www.nrcs.usda.gov/wps/portal/nrcs/main/national/programs/easements/wetlands/ (last visited Oct. 26, 2012).
107. CRS, Conservation Provisions, *supra* note 94, at 6.
108. 16 U.S.C. §3838n-q.
109. CRS, Conservation Provisions, *supra* note 94, at 6.

The Farmland Protection Program (FPP) provides financial assistance for the purchase of easements by state, tribal, or local governments and nonprofit organizations in order to maintain the agricultural production and use of land.[110] The 2008 bill changed the purpose of the FPP from protecting topsoil to protecting land's potential for agricultural use by restricting nonagricultural uses.[111] In other words, easements purchased by state, tribal, and local governments under the pre-2008 FPP contained requirements designed to protect topsoil such as anti-erosion and soil conservation measures, whereas easements purchased under the 2008 FPP contain requirements designed to limit conversion of farmland to nonagricultural uses.

b. Working Lands Conservation Programs

Working lands programs provide incentives for farmers who voluntarily choose to employ specified conservation practices in their farming operations. The major working lands programs under the conservation title include the Environmental Quality Incentives Program, the Conservation Stewardship Program, the Agricultural Management Assistance Program, and the Wildlife Habitat Incentives Program.

i. *Environmental Quality Incentives Program*

The Environmental Quality Incentives Program (EQIP) is a voluntary program for agricultural producers and forestry managers. Contracts under the program can run 1-10 years.[112] The program provides technical and financial assistance and cost sharing for conservation and environmental improvements and practices on eligible lands used for agricultural production. The 2008 farm bill expanded the EQIP program to provide assistance for practices that enhance and conserve soil, ground and surface water, air quality, and energy.[113] The program was also amended in 2008 to cover wetlands, forestlands, grasslands, and other types of land and natural resources that provide support for at-risk wildlife species.[114] The 2008 bill reduced the per-entity EQIP payment limit to $300,000 during any six-year period, with exceptions for cases determined by USDA to be of special environmental significance.[115]

A number of subprograms are included under EQIP, including the Conservation Innovation Grants Program, which was modified by the 2008 bill to allow grants to cover air quality issues associated with agriculture (including emissions of greenhouse gases),[116] and the new Agricultural Water Enhancement Program, which provides payments to producers who carry out agricultural water-enhancement activities.[117] The purpose of the program is to promote ground and surface water conservation and improve water quality on agricultural lands.[118]

ii. *Conservation Stewardship Program*

The Conservation Stewardship Program (CSP), a modification of the former Conservation Security Program, provides payments to producers for adopting and maintaining conservation activities.[119] Under the new CSP program, participants enter into five-year contracts, and payments are predicated on meeting or exceeding a predetermined stewardship threshold, defined as the level of environmental and conservation management required to "improve and conserve the quality and condition of at least one resource concern."[120] Resource concerns are identified on a state, state area, or watershed level and can include such issues as water quality, wildlife habitat, biodiversity, soil quality, soil erosion, water quantity, energy, and

---

110. 16 U.S.C. §3838h-j.
111. CRS, Conservation Provisions, *supra* note 94, at 6.
112. 16 U.S.C. §3839.
113. CRS, Conservation Provisions, *supra* note 94, at 7.
114. *Id.*
115. *Id.*
116. *Id.* at 8.
117. 16 U.S.C. §3839aa-9.
118. CRS, Conservation Provisions, *supra* note 94, at 7.
119. 16 U.S.C. §§3838d, 3838e. The program was repealed in 1996, but the 2008 farm bill reenacted the Program. *See* Food, Conservation, and Energy Act of 2008, Pub. L. No. 110-246, §2301, 122 Stat. 923, 1768 (2008) (codified at 16 U.S.C. §§3838d-3838g).
120. CRS, Conservation Provisions, *supra* note 94, at 8.

air quality.¹²¹ Participants must also meet the stewardship threshold for at least one state-identified priority resource concern for a particular watershed or area of special concern within the state.¹²²

Payments are based on actual costs incurred for installation of the conservation measure, the income foregone by the producer, and the expected environmental gain.¹²³ Participants may also receive supplemental payments for crop rotations that provide specific environmental benefits, such as reducing the need for irrigation and improving soil fertility.¹²⁴ Participation is limited to producers who have addressed at least one resource concern at the time of application and who agree to address at least one more priority resource concern by the end of the contract.¹²⁵ The new CSP program adds requirements for monitoring and evaluation of the stewardship plan to assess environmental effectiveness and expands eligibility requirements to include provision of technical assistance to producers of specialty and organic crops.¹²⁶

iii. Wildlife Habitat Incentive Program

The Wildlife Habitat Incentives Program (WHIP) provides cost sharing and technical assistance to landowners and producers that develop and improve fish and wildlife habitat.¹²⁷ Participants enter into agreements of 5-10 years that provide both technical assistance and up to 75% cost-sharing assistance for the implementation of habitat-improving activities on eligible lands.¹²⁸ The 2008 farm bill limited WHIP eligibility to "the development of wildlife habitat on private agricultural land, nonindustrial private forest land, and tribal lands."¹²⁹ The bill also gave priority to projects that address specific issues raised by state, regional, and national conservation initiatives.¹³⁰

iv. Agricultural Management Assistance Program

The Agricultural Management Assistance (AMA) Program is a voluntary program that provides cost-share and incentive payments to agricultural producers that address issues of water management, water quality, and erosion control by using conservation measures in farming operations.¹³¹ The conservation measures can include construction of water-management structures or irrigation structures; planting of trees for windbreaks or for water quality improvement; production diversification; and conservation practices (soil erosion control, integrated pest management, or transition to organic farming).¹³² AMA is available only to states in which participation in the Federal Crop Insurance Program is historically low.¹³³ Because of this requirement, participation in AMA is limited to farmers in Connecticut, Delaware, Hawaii, Maine, Maryland, Massachusetts, Nevada, New Hampshire, New Jersey, New York, Pennsylvania, Rhode Island, Utah, Vermont, West Virginia, and Wyoming.¹³⁴

Additional conservation programs provide cost sharing and technical assistance to farmers implementing measures geared toward conservation of specific resources or types of land. The Conservation of Private Grazing Lands program provides technical and educational assistance for conservation and enhancement of private grazing lands.¹³⁵ The Grassroots Source Water Protection Program provides funding for projects designed to prevent surface and groundwater pollution from affecting drinking water.¹³⁶ The Emergency

---

121. NATIONAL SUSTAINABLE AGRICULTURE COALITION, FARMER'S GUIDE TO THE CONSERVATION STEWARDSHIP PROGRAM 6 (2011), *available at* http://sustainableagriculture.net/wp-content/uploads/2011/09/NSAC-Farmers-Guide-to-CSP-2011.pdf.
122. Jody M. Endres, *Agriculture at a Crossroads: Energy Biomass Standards and a New Sustainability Paradigm?*, U. ILL. L. REV. 503, 534 (2011).
123. CRS, CONSERVATION PROVISIONS, *supra* note 94, at 8.
124. *Id.* at 9.
125. 16 U.S.C. §3838f(a).
126. CRS, CONSERVATION PROVISIONS, *supra* note 94, at 9.
127. 16 U.S.C. §3839bb.
128. CRS, CONSERVATION PROVISIONS, *supra* note 94, at 8.
129. Food, Conservation, and Energy Act of 2008, Pub. L. No. 110-246, 122 Stat. 1666, §2602(a)(1) (2008).
130. CRS, CONSERVATION PROVISIONS, *supra* note 94, at 9.
131. 7 U.S.C. §1524(b).
132. *Id.* §1524(b)(2).
133. U.S. Dep't of Agric., *Agricultural Management Assistance*, http://www.nrcs.usda.gov/wps/portal/nrcs/main/national/programs/financial/ama (last visited Aug. 13, 2012).
134. *Id.* §1524(b)(1).
135. 16 U.S.C. §3839bb.
136. CRS, CONSERVATION PROVISIONS, *supra* note 94, at 9.

Conservation Program provides assistance to farmers whose farmland is damaged by natural disasters.[137] The Cooperative Conservation Partnership Initiative directs the federal government to work with state and local governments, Indian tribes, producer associations, farmer cooperatives, institutions of higher education, and nongovernmental organizations with a history of addressing conservation issues to provide technical and financial assistance to producers in all the conservation programs except (1) The Conservation Reserve Program; (2) The Wetlands Reserve Program; (3) The Farmland Protection Program; or (4) The Grassland Reserve Program.[138]

While farm bill conservation programs certainly encourage conservation practices, they do not address the overarching environmental concerns associated with industrial commodity production—unsustainability due to high energy (i.e., fossil fuel) inputs, widespread environmental harms caused by chemical outputs (i.e., fertilizer and pesticides), and loss of biodiversity and ecological integrity due to large-scale monoculture production.[139] The largest programs, the CRP and WRP, are land set-aside programs and thus do not address in any way the manner in which farming is carried out. All of the programs are voluntary with strict limits on the types and amounts of lands that can be enrolled and on the types of practices that qualify for the subsidies. Moreover, the amount of money devoted to these conservation programs pales in comparison to the money expended on commodity subsidy programs. For example, the Congressional Budget Office estimates that the costs for the 2008 conference agreement on the farm bill (FY2008-2012) are approximately $42 billion for Title I commodity programs as opposed to approximately $24 billion for all Title II conservation programs.[140]

3. Title IX: Energy

With the intense focus in recent years on both climate change and the desire for U.S. energy independence, scientists and policymakers have searched for alternative energy sources that are domestically produced to lower contributions to climate change compared to fossil fuels. In response to the recognition that agriculture could play a critical role in supplying the nation's renewable energy, Congress added the first farm bill energy title to the Farm Security and Rural Investment Act of 2002.[141] The 2002 bill included $800 million in funding for programs designed to help rural small businesses and agricultural producers invest in projects that involve renewable energy and energy efficiency.[142] The 2002 energy title also included funding for research and development of energy from biomass and subsidies for the production of ethanol and other biofuels.[143]

One of the major alternative energy supplies that has been heavily subsidized by the federal government is corn ethanol.[144] Corn ethanol production has increased from approximately 175 million gallons in the early 1980s to almost 6.5 billion gallons in 2007, making it one of the fastest growing industries in the United States.[145] Often touted a "renewable" or "alternative" energy,[146] the use of ethanol as a major source of fuel is not without controversy.[147] The rapid acceleration in corn ethanol production is at least in part

---

137. 16 U.S.C. §§2201-2205.
138. 16 U.S.C. §3843.
139. Eubanks, *supra* note 2.
140. CRS, What Is the Farm Bill, *supra* note 1, at 2.
141. *See* Farm Security and Rural Investment Act of 2002, Pub. L. No. 107-171, tit. 9, 116 Stat. 134, 475-85 (2002) (codified at 7 U.S.C. §§8101-8108).
142. John N. Moore & Kale Van Bruggen, *Agriculture's Fate Under Climate Change: Economic and Environmental Imperatives for Action*, 86 Chi.-Kent L. Rev. 87, 96 (2011).
143. *Id.*
144. Gary D. Libecap, *Agricultural Programs With Dubious Environmental Benefits: The Political Economy of Ethanol*, in Agricultural Policy and the Environment 89, 89 (Rodger E. Meiners & Bruce Yandle eds., 2003) (quoting David Pimentel).
145. James A. Duffield et al., *Ethanol Policy: Past, Present and Future*, 53 S.D. L. Rev. 425, 425 (2008). *See also* Karl R. Rabago, *A Review of Barriers to Biofuel Market Development in the United States*, 2 Envtl. & Energy L. & Pol'y J. 211, 212 (2008) (describing the remaining barriers to full commercial success for biofuels in the United States). Corn is not the only plant, or even the only vegetable, that can be used to make ethanol. Jose C. Escobar et al., *Biofuels: Environment, Technology and Food Security*, 13 Renewable & Sustainable Energy Reviews 1275, 1278 (2008). However, corn is the major ethanol raw material in the United States. *Id.* at 1280. Other countries, such as Brazil, produce large quantities of ethanol from other plants, such as sugar and palm. *See id.* at 1284.
146. *See, e.g.*, Ethanol Promotion and Information Council, Ethanolfacts.com, http://www.ethanolfacts.com (characterizing ethanol as renewable) (last visited Oct. 6, 2011).
147. *See, e.g.*, Christopher Jensen, *Critics Find Flaws in the Case Made by Ethanol Advocates*, N.Y. Times, May 7, 2009, http://www.nytimes.com/2009/05/10/automobiles/10CRITICS.html?_r=1 (critiquing the movement advocating use of ethanol as a source of fuel).

attributable to the heavy subsidies that have been provided since the 1970s.[148] With the recent focus on finding alternative sources of energy, corn ethanol subsidy programs have proliferated. Both the Energy Policy Act of 2005[149] and the Energy Independence and Security Act of 2007[150] created additional incentives for ethanol development.

The 2008 farm bill expanded and extended the energy provisions of the 2002 farm bill and provided additional funding for renewable energy,[151] asserting "a stronger federal commitment to farm-based energy."[152] These renewable energy provisions range from regulatory measures to subsidies.[153] In some instances, significant additions were made to the previous farm bill; in others, existing provisions are simply renewed.[154] Although some of the programs target more efficient and sustainable forms of renewable energies such as cellulosic biofuels,[155] most of the programs provide financial and other incentives for the development of corn ethanol. Many of the programs apply generally to "biofuels" or "biobased products." Unfortunately, most of these programs do not distinguish between alternative energy sources that provide a net energy benefit and those, such as corn ethanol, that take more fossil fuel to make than they provide.[156]

a. Biofuels-Related Programs

The 2008 farm bill significantly expanded existing biofuels-related programs:[157]

- The Biobased Markets Program[158] established a label—"USDA Certified Biobased Product"—available to producers of biobased products, and a process for certification. The program also established a preference for biobased products in federal procurement and set forth guidelines for intermediate ingredients and feedstocks for such procurement.

- The Biorefinery Assistance Program[159] provides grants on a competitive basis to eligible entities for development, construction, and retrofitting of demonstration-scale biorefineries. It also guarantees loans of up to 30% of the cost of development and construction of a biorefinery.

- The Repowering Assistance Program[160] provides grants to existing biorefineries for installation of new systems that use renewable biomass to reduce or eliminate the use of fossil fuels.

- The Bioenergy Program for Advanced Biofuels[161] provides payments to eligible producers of advanced biofuels to support and expand their production. Payment is based on quantity and duration of production and net nonrenewable energy content of the advanced biofuel.

- The Biodiesel Fuel Education Program[162] provides competitive grants to eligible entities to educate governmental and private entities that operate vehicle fleets and the public about the benefits of biodiesel fuel use.

---

148. *See* Wallace Tyner, *The U.S. Ethanol and Biofuels Boom: Its Origins, Current Status, and Future Prospects*, 58 Bioscience 646 (2008); *see also* Robert W. Hahn, *Ethanol: Law, Economics, and Politics*, 19 Stan. L. & Pol'y Rev. 434, 437-45 (2008) (describing how federal subsidies have driven the development of the ethanol fuel industry in the United States); Libecap, *supra* note 144, at 89.
149. Energy Policy Act of 2005, Pub. L. No. 109-58, 119 Stat. 594 (codified as amended at 42 U.S.C. §15801).
150. Energy Independence and Security Act of 2007, Pub. L. No. 110-140, 121 Stat. 1492.
151. CRS, The 2008 Farm Bill, *supra* note 9, at 31.
152. Moore & Van Bruggen, *supra* note 142, at 97.
153. U.S. Dep't of Agric., *2008 Farm Bill Renewable Energy Provisions*, http://www.usda.gov/documents/FB08_Pub_Mtg_Renew_Energy_Factsheet.pdf.
154. U.S. Dep't of Agric., Economic Research Service, *2008 Farm Bill Side-by-Side, Title IX: Energy*, http://www.ers.usda.gov/farm-bill-resources.aspx (last visited Oct. 26, 2012).
155. Unlike corn ethanol biofuel, which is derived from the starch in corn grain, cellulosic biofuels are derived from fibrous cellulosic plant sources such as leaves. For more information on cellulosic biofuels, see Ecological Society of America, The Sustainability of Cellulosic Biofuels, *available at* http://www.esa.org/pao/policyActivities/Sustainability%20of%20Cellulosic%20Biofuels%20handout%206.11.pdf.
156. Bruce A. McCarl & Fred O. Boadu, *Bioenergy and U.S. Renewable Fuels Standards: Law, Economics, Policy/Climate Change and Implementation Concerns*, 14 Drake J. Agric. L. 43, 59 (2009).
157. For a detailed summary of the federal bio-fuel incentives, see Megan Stubbs, Cong. Research Serv., Renewable Energy Programs in the 2008 Farm Bill (2010), *available at* http://assets.opencrs.com/rpts/RL34130_20101208.pdf [hereinafter CRS, Renewable Energy Programs].
158. 7 U.S.C. §8102 (Supp. II 2008).
159. *Id.* §8103.
160. *Id.* §8104.
161. *Id.* §8105.
162. *Id.* §8106.

- The Rural Energy for America Program[163] provides grants and financial assistance to agricultural producers and rural small businesses for energy audits, energy development assistance, energy efficiency improvements, and renewable energy systems. It also provides competitive grants to eligible entities that help agricultural producers and rural small businesses become more energy efficient and use renewable energy technologies and resources. The Program provides loan guarantees for up to 75% of the cost of purchasing renewable energy systems and making energy efficiency improvements.

- The Biomass Research and Development Initiative[164] requires the secretaries of agriculture and energy to coordinate promotion and development policies and procedures for biofuels and biobased products. It provides competitive grants, contracts, and financial assistance to eligible entities to research, develop, and demonstrate the methods, practices, and technologies used to produce biofuels and biobased products.

b. Cellulosic Biofuels Programs

For the first time, the 2008 farm bill included a number of key provisions intended to accelerate the development of advanced (primarily cellulosic, i.e., derived from wood, grasses, or inedible parts of plants) biofuels.[165] Advanced biofuels include those derived from the cellulosic structural materials of plants; renewable biomass; animal, plant, or food wastes; organic matter; and other fuels produced from cellulosic biomass. The 2008 energy title provides over $1 billion in support and financial incentives to promote the production of cellulosic biofuels, including grants and loan guarantees intended to encourage investment in advanced biofuel technologies and production of cellulosic feedstocks.[166]

i. Biomass Crop Assistance Program

The Biomass Crop Assistance Program (BCAP),[167] which appears for the first time in the 2008 farm bill, is the United States' first program offering subsidies for the production of energy biomass crops.[168] BCAP provides financial assistance to producers of eligible crops in a designated BCAP project area to establish and produce eligible crops, and to collect, harvest, store, and transport eligible material for use in a biomass conversion facility. The term "eligible crop" includes renewable biomass, but excludes crops eligible to receive payments under Title I of the 2008 farm bill and invasive or noxious plants. In order to receive payments for the production of renewable biomass, producers are required to comply with land use restrictions that prohibit cropping on lands with native vegetation or on land that receives payments under the conservation, wetland, or grassland reserve programs.[169] To be eligible for BCAP payments, producers must also comply with forestry and agronomic biomass practices mandated by the statute and regulations, implement and follow a conservation or forest stewardship plan in accordance with BCAP guidelines, and comply with conservation requirements for wetlands and highly erodible lands.[170]

ii. Other Cellulosic Biofuels Programs

In addition to BCAP, the 2008 farm bill contains numerous other programs designed to encourage the production of cellulosic biofuels:

---

163. *Id.* §8107.
164. *Id.* §8108.
165. *See Implications of the U.S. Farm Bill for Cellulosic Ethanol Development* (Journalists Roundtable) 4.2 INDUS. BIOTECHNOLOGY 131 (2008) (discussing the cellulosic biofuel provisions of the 2008 farm bill).
166. CRS, RENEWABLE ENERGY PROGRAMS, *supra* note 157.
167. 7 U.S.C. §8111 (Supp. II 2008).
168. Endres, *supra* note 122, at 513.
169. *Id.* at 514.
170. *Id.*

- The Rural Energy Self-Sufficiency Initiative[171] helps eligible rural communities to increase energy self-sufficiency by providing grants to eligible rural communities to conduct energy assessments, and to formulate and analyze ideas for reducing energy usage.

- The Feedstock Flexibility Program for Bioenergy Producers[172] requires the secretary of agriculture to purchase raw or refined sugar or in-process sugar that would otherwise be forfeited to the Commodity Credit Corporation (a government owned and operated corporation that is authorized to buy, sell, lend, make payments, and engage in other activities to stabilize, support and protect farm income and prices) and sell it to bioenergy producers to use in the production of bioenergy.

- The Forest Biomass for Energy, §12,[173] authorizes a competitive research and development program to encourage use of forest biomass for energy.

- The Community Wood Energy Program, §13,[174] provides grants to state and local governments of up to $50,000 to assess available feedstocks necessary to supply a community wood energy system and the long-term feasibility of supplying and operating a community wood energy system. It provides competitive grants to state and local governments to acquire or upgrade community wood energy systems.

- The Sun Grant Program[175] provides grants to six regional sun grant centers (North-Central Center, Southeastern Center, South-Central Center, Western Center, Northeastern Center, and Western Insular Pacific Subcenter) to enhance national energy security through development, distribution, and implementation of biobased energy technologies; promote diversification and environmental sustainability of agricultural production in the United States using biobased energy and product technologies; promote economic diversification in rural areas of the United States using biobased energy and product technologies; and improve coordination and collaboration by the Departments of Agriculture and Energy and colleges and universities to enhance efficiency of bioenergy and biomass research and development programs.

- The Cellulosic Biofuel Producer Credit[176] provides a tax credit up to $1.01 to any taxpayer for each gallon of qualified cellulosic biofuel production.

- The Modification of Alcohol Credits Program[177] reduces tax credit after annual production or importation of ethanol reaches 7.5 million gallons.

Cellulosic biofuels appear to be preferable to corn-based biofuels in that they do not require intense fossil fuel and water inputs.[178] Because there are still many uncertainties about the potential impacts of cellulosic biofuels, however, they should be pursued with caution and not perceived as a panacea. The concerns associated with planting large areas of plants for cellulosic biofuels depend in large part on what land uses

---

171. 7 U.S.C. §8109.
172. *Id.* §8110.
173. *Id.* §8112.
174. *Id.* §8113.
175. *Id.* §8114.
176. 26 U.S.C. §40(b)(6) (2006).
177. *Id.*
178. Berk Akinci et al., *The Role of Bio-fuels in Satisfying US Transportation Fuel Demands*, 36 ENERGY POL'Y 3485, 3488 (2008). For further reading on the benefits of biomass and cellulosic biofuels, see generally Dennis R. Becker et al., *Assessing the Role of Federal Community Assistance Programs to Develop Biomass Utilization Capacity in the Western United States*, 11 FOREST POL'Y & ECON. 141 (2009); R.H.V. Corley, *How Much Palm Oil do We Need?*, 12 ENVTL. SCI. & POL'Y 134 (2009); L. Leon Geyer et al., *Ethanol, Biomass, Biofuels and Energy: A Profile and Overview*, 12 DRAKE J. AGRIC. L. 61 (2007); Jose Goldemberg & Patricia Guardabassi, *Are Biofuels a Feasible Option?*, 37 ENERGY POL'Y 10 (2009); Robert R. Harmon & Kelly R. Cowan, *A Multiple Perspectives View of the Market Case for Green Energy*, 76 TECH. FORECASTING & SOC. CHANGE 204 (2009); Mark Murphey Henry et al., *A Call to Farms: Diversify the Fuel Supply*, 53 S.D. L. REV. 515 (2008); Timo Kaphengst et al., *At a Tipping Point? How the Debate on Biofuel Standards Sparks Innovative Ideas for the General Future of Standardisation and Certification Schemes*, 17 J. CLEANER PRODUCTION S99 (2009); Lian Pin Koh & Jaboury Ghazoul, *Biofuels, Biodiversity, and People: Understanding the Conflicts and Finding Opportunities*, 141 BIOLOGICAL CONSERVATION 2450 (2008); Li Lu et al., *The Role of Marginal Agricultural Land-Based Mulberry Planting in Biomass Energy Production*, 34 RENEWABLE ENERGY 1789 (2009); David Nicholls et al., *International Bioenergy Synthesis—Lessons Learned and Opportunities for the Western United States*, 257 FOREST ECOLOGY & MGMT. 1647 (2009); Rudolf M. Smaling, *Environmental Barriers to Widespread Implementation of Biofuels*, 2 ENVTL. & ENERGY L. & POL'Y J. 287 (2008); Gail Taylor, *Biofuels and the Biorefinery Concept*, 36 ENERGY POL'Y 4406 (2008); Tobias Wiesenthal et al., *Biofuel Support Policies in Europe: Lessons Learnt for the Long Way Ahead*, 13 RENEWABLE & SUSTAINABLE ENERGY REVIEWS 789 (2009).

they are replacing. If cellulosic plants replace fields currently occupied by industrial commodity crops, the environmental and energy benefit could be significant. However, if the same acreage of industrialized commodity crops continues to be grown and additional natural lands are converted to grow cellulosic biofuel crops, there would likely be additional environmental harms that must be taken into consideration.

4. Title X: Horticulture and Organic Agriculture

Interest in promoting the environmental and human health benefits associated with organic farming has expanded exponentially in the United States in recent decades.[179] Congress passed the Organic Foods Production Act, which authorized the creation of the USDA National Organic Program (NOP), in 1990.[180] The NOP is responsible for establishing standards for the production and processing of organic foods and for administering the "USDA Organic" product certification program for organic foods. The NOP defines organic farming as "a production system that is managed in accordance with the [Organic Foods Production] Act and regulations . . . to respond to site-specific conditions by integrating cultural, biological, and mechanical practices that foster cycling of resources, promote ecological balance, and conserve biodiversity."[181] This definition incorporates the idea that organic agricultural production is "an approach to food production based on biological methods that avoid the use of synthetic crop or livestock production inputs . . . and a broadly defined philosophical approach to farming that puts value on resource efficiency and ecological harmony."[182]

The 2008 farm bill created the first ever Horticulture and Organic Production farm bill title. It contains a number of provisions that support organic agricultural production, most of which reauthorize organic programs contained in the 2002 farm bill and provide mandatory funds for their support. The Certification Cost-Sharing Program provides eligible producers or handlers located in qualified states up to $750 each as reimbursement for organic product certification costs.[183] This program caps the amount of federal assistance for certification costs at 75% of certification costs per producer.[184] To be eligible for reimbursement under this program, the organic production or handling operation must comply with NOP organic production or handling regulations and have completed the USDA certification process between October 1, 2008, and September 30, 2009.[185] Section 2801 of the 2008 conservation title provides $1.5 million in mandatory funds under the Agricultural Management Assistance Program, discussed above, specifically to provide certification cost-sharing assistance to organic farmers located in Connecticut, Delaware, Hawaii, Maryland, Massachusetts, Maine, Nevada, New Hampshire, New Jersey, New York, Pennsylvania, Rhode Island, Utah, Vermont, West Virginia, and Wyoming.[186]

Other provisions related to organic production can be found elsewhere in the 2008 farm bill. As mentioned above, the Organic Conversion Cost-Sharing Program makes grants available under EQIP to assist farmers in converting from conventional to organic agricultural production. Section 12023 of the 2008 farm bill's crop insurance and disaster assistance title (Title XV) mandates a review of federal crop insurance procedures for conventional versus organic crops.[187] This provision was included in response to organic producers' dissatisfaction with coverage of their crops under the crop insurance program.[188] Based

---

179. David Pimentel et al., *Environmental, Energetic, and Economic Comparisons of Organic and Conventional Farming Systems*, 55 BIOSCIENCE 573 (2005). For further reading on the environmental and health benefits of organics, see generally INTERNATIONAL TRADE CENTRE (UNCTAD/WTO) & RESEARCH INSTITUTE OF ORGANIC AGRICULTURE (FiBL), THE CONTRIBUTION OF ORGANIC AGRICULTURE TO CLIMATE CHANGE ADAPTATION AND MITIGATION (2009), available at http://www.ifoam.org/growing_organic/1_arguments_for_oa/environmental_benefits/pdfs/IFOAM-CC-Mitigation-Web.pdf; Alyson E. Mitchell et al., *Ten-Year Comparison of the Influence of Organic and Conventional Crop Management Practices on the Content of Flavonoids in Tomatoes*, 55 J. AGRIC. & FOOD CHEMISTRY 6154 (2007); K. Brandt et al., *Agroecosystem Management and Nutritional Quality of Plant Foods: The Case of Organic Fruits and Vegetables*, 30 CRITICAL REVIEWS IN PLANT SCIENCES 177 (2011); Walter J. Crinnion, *Organic Foods Contain Higher Levels of Certain Nutrients, Lower Levels of Pesticides, and May Provide Health Benefits for the Consumer*, 15 ALTERNATIVE MED. REV. 4 (2010).
180. Organic Foods Production Act of 1990, 7 U.S.C. §§6501 et seq. The NOP is discussed in detail in Chapter 13 of this book.
181. 7 C.F.R. §205.2.
182. RENÉE JOHNSON, CONG. RESEARCH SERV., ORGANIC AGRICULTURE IN THE UNITED STATES: PROGRAM AND POLICY ISSUES, 1 (2008), available at http://www.fas.org/sgp/crs/misc/RL31595.pdf [hereinafter CRS, ORGANIC AGRICULTURE].
183. *Id.* at 9.
184. *Id.*
185. *Id.* at n.25.
186. *Id.* at 9.
187. Food, Conservation, and Energy Act of 2008, Pub. L. No. 110-246, 122 Stat. 923, §12023(10)(B) (2008).
188. CRS, ORGANIC AGRICULTURE, *supra* note 182, at 10.

on the results of the review, the Federal Crop Insurance Corporation will "work to reduce or eliminate premium surcharges on policies for organic producers."[189] The 2008 farm bill also contains provisions funding organic research and data collection and increasing the organic sector's access to programs under the conservation, credit, and trade titles of the bill.[190]

B. Environmental Implications of Farm Bill Programs

The current system of farm bill agricultural subsidies, found mostly in the farm support programs in Title I of the 2008 bill, creates perverse incentives by encouraging farmers to produce commodity crops at a large scale in order to receive financial benefits under the bill's programs.[191] The farm support and subsidy programs begun in the 1930s to address emergencies created by the Great Depression were intended to be temporary.[192] But the agricultural subsidy system has not only persisted but has thrived and expanded. While the commodity subsidy programs in the original 1933 farm bill may have addressed the imminent need to stabilize farm prices and prevent a collapse of the U.S. agricultural system, the decades-long expansion of these programs in ways that encourage overproduction of certain crops has distorted the market by providing incentives that are in many cases antithetical to modern concerns about climate change, energy independence, and environmental degradation. The commodity subsidy programs that most tend to incentivize large-scale industrialized monocultural production are the price support programs, such as the Market Assistance Loan Program, which encourage farmers to grow large quantities of commodity crops regardless of whether a market exists for those crops.[193] Through these programs, farmers are ensured high prices for their crops regardless of market conditions,[194] and farmers maximize profits by maximizing yields.[195] Income support programs, on the other hand, do not directly encourage maximum yields of commodity crops; however, the limitations on growing fruits and vegetables on eligible acreage can indirectly have the same result by encouraging large-scale monocultures of commodity crops,[196] which require high inputs of pesticides, fertilizers, water, and fossil fuels.[197]

The programs offering subsidies for corn ethanol production found in the energy title of the 2008 farm bill only exacerbate the problems associated with subsidizing commodity crop production. Use of biofuels, including corn ethanol, is projected to grow exponentially in the 21st century.[198] Proponents point to economic stimulation and job creation, bolstering of domestic corn prices, energy security and independence, and reduction of harmful pollutants as proof that ethanol's role in the nation's energy portfolio should continue to grow.[199] Critics, however, note that the use of corn for ethanol drives up world food prices and agricultural land use.[200] Opponents also point to the inputs of ethanol—including fossil fuels, fertilizers, and pesticides—as being environmentally costly.[201] Others argue that the current focus on biofuels, especially corn ethanol, could inhibit the development of other alternative technologies that could better address the problems associated with dependence on fossil fuels.[202]

Although the 2008 farm bill conservation programs are obviously steps in the right direction by providing incentives for farmers to conserve various types of lands and resources, they do very little to discourage agriculture's potentially extensive impacts on the environment. In focusing on land set-asides and work-

---

189. *Id.*
190. *Id.*
191. Davidson, *supra* note 2, at 37; Foster, *supra* note 3 at 241.
192. Eubanks, *supra* note 11, at 10494 (discussing President Roosevelt's New Deal enactment of the first farm bill as a temporary fix for the Depression-era decline in U.S. small farm well-being).
193. Melanie J. Wender, *Goodbye Family Farms and Hello Agribusiness: The Story of How Agricultural Policy Is Destroying the Family Farm and the Environment*, 22 Vill. Envtl. L.J. 141, 164-65 (2011).
194. *Id.*
195. Foster, *supra* note 3, at 252.
196. *Id.* at 247.
197. Davidson, *supra* note 2, at 37.
198. Ayhan Demirbas, *Biofuels Sources, Biofuel Policy, Biofuel Economy and Global Biofuel Projections*, 49 Energy Conversion & Mgmt. 2106, 2114 (2008).
199. *See* American Coalition for Ethanol, *Ethanol 101*, http://www.ethanol.org/index.php?id=34&parentid=8 (last visited Oct. 26, 2012).
200. Duffield et al., *supra* note 145, at 425.
201. John Boardman et al., *Socio-Economic Factors in Soil Erosion and Conservation*, 6 Envtl. Sci. & Pol'y 1, 4-5 (2003); Pål Börjesson, *Good or Bad Bioethanol From a Greenhouse Perspective—What Determines This?*, 86 Applied Energy 589, 589-91 (2009).
202. *See* Börjesson, *supra* note 201, at 593; Michael B. Charles et al., *Public Policy and Biofuels: The Way Forward?*, 35 Energy Pol'y 5737, 5737-38 (2007).

ing lands practices instead of providing incentives to address the environmental degradation associated with farming practices, these largely voluntary programs do comparatively little to mitigate or reform the environmental concerns associated with industrial commodity production. While the inclusion of conservation programs within the farm bill is encouraging because it signifies policymakers' recognition that the environmental consequences of our agricultural system should be addressed, such programs are only a first step toward addressing these concerns.

Similarly, the 2008 farm bill provisions providing support for producers converting to organic agriculture are steps in the right direction because of recognized benefits to health and the environment,[203] but these programs could do much more to produce increased environmental benefits.[204] Organic agriculture, characterized by more sustainable farming practices and decreased use of chemical inputs, certainly produces fewer environmental and human health impacts than conventional farming production.[205] The farm bill organic agriculture programs, however, only provide financial incentives for costs related to converting to organic production or becoming certified under the federal certification program. While these programs indirectly encourage environmentally beneficial farming practices by providing incentives for organic conversion and certification, the programs are unlikely to produce industry-wide changes in overall farming practices unless they are amended to receive more funding, to encourage broader participation, and to apply more broadly to organic production than just certification and conversion.

C. Looking Toward the Future: Trends and Issues for Upcoming Farm Bill Legislation

Because Congress passes a new farm bill roughly every five years, policymakers constantly debate which issues should be included in or removed from the next farm bill. Much of this debate centers on many of the issues discussed in this chapter, such as the cost and extent of farm support, conservation, and other programs. The types of issues and programs debated and ultimately given funding in a farm bill are undoubtedly influenced by the overall political climate at the time the new bill is being developed. Issues likely to be central to upcoming farm bill legislative debates include budgetary considerations, trade considerations, and policy considerations.

1. Budgetary Considerations

In times of federal surplus, such as during the development of the 2002 farm bill, it is often relatively easy to maintain existing programs and add new programs to the omnibus farm bill because the federal budget situation allows for increased spending. When the federal budgetary situation is restricted, however, as is the case for the development of the 2013 bill, the number and scope of new programs is likely to be far smaller, and funding for any new programs will likely have to be offset by reductions to existing programs. The restricted federal budgetary climate means that the amount of funding allocated to the next farm bill will be based on the Congressional Budget Office's baseline projection of the bill's costs and on "varying budgetary assumptions about whether programs will continue."[206]

For example, while some programs have baseline funding beyond the end of the 2008 farm bill, 37 that received mandatory funding under the 2008 bill do not have budgetary baseline (i.e., automatic) funding beyond fiscal year 2012, which ended on September 30, 2012.[207] This means that a new farm bill would not have baseline budgetary funding to continue these programs. Accordingly, if policymakers want to continue these programs in the 2012 farm bill, they will have to fund them through $9 billion to $10 billion in offsets and reductions to other programs.[208] The 37 provisions without baseline funding past FY2012 span 12 out of the 15 titles in the 2008 farm bill. The energy title contains the most provisions without baseline

---

203. Pimentel et al., *supra* note 179. For further reading on the environmental and health benefits of organics, see supra note 179.
204. Davidson, *supra* note 2, at 36-38.
205. Pimentel et al., *supra* note 179.
206. Jim Monke, Cong. Research Serv., Previewing the Next Farm Bill: Unfunded and Early-Expiring Provisions 1 (2010) [hereinafter CRS, Next Farm Bill], *available at* http://www.nationalaglawcenter.org/assets/crs/R41433.pdf.
207. *Id.*
208. *Id.* at 2.

funding, followed by conservation, nutrition, and horticulture and organic agriculture.[209] Three farm bill provisions, the WRP, the BCAP, and the agricultural disaster assistance program, account for nearly 75% of the $9-10 billion total.[210]

Given the current ongoing governmental budgetary crisis, there is potential for the upcoming farm bill to include massive cuts to conservation programs, particularly those without baseline funding beyond FY2012. Indeed, such massive cuts to mandatory conservation programs had already been included, at the time of this writing, in the House's FY2012 agricultural appropriations bill.[211] Despite these proposed cuts in funding, both Republican and Democratic members of the House Agriculture Committee's Subcommittee on Conservation, Energy, and Forestry demonstrated broad bipartisan support for farm bill conservation programs during recent hearings.[212] Even if the 2013 farm bill maintains current funding for conservation programs and instead makes massive cuts to direct payment programs, elimination of direct payments could reduce the number of farms subject to conservation compliance.[213] The farm bill budgetary debate will therefore likely include some mix of cuts to both conservation and farm support programs, with Congress making the difficult decisions as to how to achieve a desirable balance between these programs.

2. Trade Considerations

Another policy constraint that promises to impact future farm bills involves the trade policies mandated by the WTO, by which the United States, as a WTO member, has agreed to abide. Because the United States is one of the world's largest agricultural producing and trading nations, U.S. agricultural policy is constantly evaluated against the WTO Agreement on Agriculture (AA) and Subsidies and Countervailing Measures (SCM) rules. Agricultural programs or policies found to be in violation of these rules may be subject to challenge by other WTO members, as occurred in the so-called "Brazil cotton case," in which a WTO panel ruled against the U.S. cotton export credit guarantee program.[214] As a result of the WTO panel ruling, the United States is now expected to bring the program into compliance with WTO rules or be subject to WTO sanctions.[215] Because of the importance of complying with WTO rules, a key issue likely to be raised in future farm bill debates is how new and existing farm programs will affect the United States' WTO commitments.

Under the WTO, member countries' agricultural policies are evaluated under AA spending limits and SCM adverse-effects determinations. The AA classifies farm support programs as either amber box (i.e., trade distorting), green box (minimally trade distorting), blue box (payments under production-limiting programs based on historical data or fixed numbers of livestock), or de minimis (program spending so minimal it is deemed benign).[216] Under the AA, the United States is limited to spending no more than $19.1 billion per year on support programs classified as trade-distorting amber box subsidies.[217] Programs classified as green box, blue box, or de minimis are exempted from spending limits.[218] An amber box program may still be subject to WTO challenge under the SCM rules, however, even if the program stays within its designated WTO spending limit under the AA.[219] One such challenge for U.S. farm commodity programs that is of likely concern under the SCM rules is compliance with "actionable subsidy" rules, which are broadly defined as those subsidies that cause "adverse effects" to other member countries' trade interests and agricultural markets.[220] When measured against the WTO criteria for establishing the existence of adverse effects, it appears that all major U.S.-subsidized program crops could potentially be vulnerable to

---
209. *Id.* at 3.
210. *Id.* at 6-9.
211. National Sustainable Agriculture Coalition, House Agriculture Subcommittee Holds Hearing on Conservation Programs (July 8, 2011), http://sustainableagriculture.net/blog/conservation-subcommittee-hearing/ (last visited Oct. 26, 2012).
212. *Id.*
213. National Sustainable Agriculture Coalition, House Hearing on Commodity and Disaster Programs (July 28, 2011), http://sustainableagriculture.net/blog/house-hearing-disaster-program/ (last visited Oct. 26, 2012).
214. For more information on the Brazil Cotton Case, see Randy Schnepf, Cong. Research Serv., Brazil's WTO Case Against the U.S. Cotton Program, *available at* http://www.nationalaglawcenter.org/assets/crs/RL32571.pdf.
215. *Id.*
216. World Trade Organization, *Agriculture Negotiations: Background Fact Sheet, Domestic Support in Agriculture: The Boxes, supra* note 50.
217. CRS, Next Farm Bill, *supra* note 206, at 21.
218. *Id.* at 23.
219. *Id.* at 21-22.
220. *Id.* at 22.

WTO challenge, although such challenges are rarely filed due to expense and political issues.[221] Because the United States is obligated to abide by these WTO regulations, whether new and existing agricultural programs comply with limits on spending, produce adverse effects, or qualify as exempt from these limitations will likely be key issues in future farm bill debates.[222]

3. Policy Considerations

In addition to budget and trade issues, many policy questions are likely to be raised time and again during the development of future farm bills. Chief among these ongoing policy issues is the debate surrounding farm support programs, including the nature and extent of the farm bill subsidy programs. Supporters contend that subsidies are necessary to provide farmers a financial safety net in response to the volatility of the agricultural market.[223] Critics of agricultural subsidies, on the other hand, question the continued need for the antiquated subsidy system and contend that funding would be better spent advancing conservation goals or promoting practices that increase agricultural productivity while decreasing environmental degradation.[224] Deficit reduction proposals specifically targeting agricultural subsidies have already begun to surface in the 2012/2013 farm bill debate.[225]

Farm bill subsidies for biofuel production are likely to come under increased scrutiny in future years, as demand for biofuel sources, especially corn for the production of ethanol, increases in response to the high degree of farm bill support offered for producers of biofuels. In the year 2009 alone, biofuels subsidies totaled nearly $6 billion.[226] The demand for corn has expanded by nearly 30% in the past decade due to its increased use for corn ethanol production. This increased demand has contributed to rising prices for both corn and other field crops, as the demand for corn has caused an expansion of planted corn into nontraditional crop areas.[227] If the biofuel subsidy programs continue in future farm bills, it is likely that they will continue to be subject to the same criticisms levied at the other farm bill agricultural price and income subsidies, discussed above.

Conclusion

The omnibus farm bill contains the majority of U.S. agricultural and food policy. Farm bill policies influence virtually every aspect of agriculture, from the decision to grow certain crops, the amount of crops grown, the manner (industrial, organic, or otherwise) in which the crops are grown, and the ingredients in processed foods. Many provisions contained within the farm bill impact the environment directly or indirectly. Farm support programs and subsidies contained in the farm bill primarily reward large-scale industrial production of a few commodity crops and thus incentivize agricultural practices that greatly exacerbate issues such as dependence on foreign fossil fuels, climate change, and overall environmental degradation. The growth of biofuel subsidy programs under the most recent farm bill produces similar adverse environmental impacts. While farm bill conservation and organic production programs provide some countervailing environmental benefits, these programs do not go far enough toward encouraging sustainable farming practices to balance the potential harms flowing from practices subsidized through farm support programs. What emerges from this evaluation is a picture of a complex, outdated, and arguably flawed agricultural policy that can result in serious impacts on the conservation of energy, water resources, and other natural resources, and exacerbate the effects of climate change.

---

221. *Id.* For more information on possible WTO challenges to U.S. farm subsidy programs, see RANDY SCHNEPF, CONG. RESEARCH SERV., POTENTIAL CHALLENGES TO U.S. FARM SUBSIDIES IN THE WTO: A BRIEF OVERVIEW (2007), *available at* http://www.nationalaglawcenter.org/assets/crs/RS22522.pdf.
222. CRS, NEXT FARM BILL, *supra* note 206, at 20.
223. Karl Beitel, *U.S. Farm Subsidies and the Farm Economy: Myths, Realities, Alternatives*, FOOD FIRST/INSTITUTE FOR FOOD AND DEVELOPMENT POLICY BACKGROUNDER (June 2005), *available at* http://www.foodfirst.org/files/pdf/backgrounders/subsidies.pdf; Jerry McReynolds, *No Better Investment Than Farm Safety Net*, WICHITA EAGLE, Apr. 24, 2011, *available at* http://www.kansas.com/2011/04/24/1820520/no-better-investment-than-farm.html.
224. Foster, *supra* note 3, at 256-58; Mark Bittman, *Don't End Agricultural Subsidies, Fix Them*, N.Y. TIMES OPINIONATOR, Mar. 1, 2011, http://opinionator.blogs.nytimes.com/2011/03/01/dont-end-agricultural-subsidies-fix-them/.
225. National Sustainable Agriculture Coalition, House Hearing on Commodity and Disaster Programs, *supra* note 213.
226. CRS, FARM SAFETY NET PROGRAMS, *supra* note 41, at 19.
227. *Id.*

Chapter 3
The Environmental Impacts of Industrial Fertilizers and Pesticides
Mary Jane Angelo and Seth Hennes

Although pesticides and fertilizers have been used virtually since the origin of agriculture, it was not until the second half of the 20th century that the use of vast quantities of new synthetic pesticides and fertilizers became the mainstay of modern conventional farming.[1] During World War II, many new chemicals were synthesized for the first time, often as part of programs to develop new chemical weapons. The same properties that made these new chemicals effective poisons against enemy troops also made them effective poisons against pest species. Accordingly, many of these new chemicals began to be used as agricultural pesticides.

During the latter half of the 20th century, the Green Revolution transformed agriculture and produced extremely high crop yields but depended on heavy inputs of fertilizers, pesticides, and water. As with most technological advances, there were unanticipated consequences to the Green Revolution. While high crop yields increased the availability of food to a growing global population, high inputs of pesticidal chemicals and synthetic fertilizers resulted in an array of impacts on the environment and human health. This chapter reviews the history of the use of pesticides and fertilizers used to produce high yields in agriculture and the potential environmental risks posed by their widespread use.

A. History of Pesticide and Fertilizer Use

Over the past 50 years, U.S agriculture has undergone a dramatic transformation, due in large part to the technological advances of the Green Revolution, which promoted production practices that would maximize crop yields. Led by American agronomist Norman Borlaug, the Green Revolution occurred between the 1940s and the late 1970s.[2] It refers to a series of technology, research, and development policy initiatives aimed at feeding hungry people around the world through the development of high-yield crop varieties using seed hybridization and other agricultural techniques.[3] To produce higher per acre farm yields, human labor was supplanted by technology and a reliance on large inputs of fossil fuel and mechanized farm equipment.[4] New government policies encouraged high-yield farming of commodity crops by linking subsidy payments to production levels. High-yield farming was further promoted by an increase in government funding for research and development and the creation of a vast network of the agriculture extension

---

Portions of this chapter have been adapted from, with permission, Mary Jane Angelo, *Corn, Carbon, and Conservation: Rethinking U.S. Agricultural Policy in a Changing Global Environment*, 17 Geo. Mason L. Rev. 593 (2010); Mary Jane Angelo, *The Killing Fields: Reducing the Casualties in the Battle Between U.S. Endangered Species and Pesticide Law*, 32 Harv. Envtl. L. Rev. 96 (2008); Mary Jane Angelo, *Regulating Evolution for Sale: An Evolutionary Biology Model for Regulating the Risks Posed by Genetically Modified Organisms*, 42 Wake Forest L. Rev. 93 (2007); and Mary Jane Angelo, *Embracing Uncertainty, Complexity and Change to Protect Ecological Integrity: An Eco-Pragmatic Reinvention of a First Generation Environmental Law*, 33 Ecology L.Q. 105 (2006).

1. Mary Jane Angelo, *Corn, Carbon, and Conservation: Rethinking U.S. Agricultural Policy in a Changing Global Environment*, 17 Geo. Mason L. Rev. 605-09 (2010).
2. Alan L. Olmstead & Paul W. Rhode, *Adapting North American Wheat Production to Climatic Challenges, 1839-2009*, 108 Proceedings of the National Academy of Sciences 480, 483 (2011).
3. *Id.*
4. Angelo, *Corn, Carbon, and Conservation*, *supra* note 1; William S. Eubanks II, *A Rotten System: Subsidizing Environmental Degradation and Poor Public Health With Our Nation's Tax Dollars*, 28 Stan. Envtl. L.J. 213, 269-70 (2009) [hereinafter Eubanks, *A Rotten System*].

service to educate and train farmers in high-yield commodity farming.[5] The Green Revolution is estimated to have resulted in an increase of more than 150% in farm production over the past 60 years.[6]

The predominant U.S. agricultural production system that grew out of the Green Revolution is what is often referred to as "industrialized agriculture," which is characterized by large-scale monocultures[7] (the cultivation of one crop over a large area), limited crop varieties, heavy use of synthetic chemicals and other inputs, and the separation of animal and plant agriculture.[8] Large-scale monocultures and other industrial agricultural practices require large inputs of chemical pesticides and fertilizers in large part because these practices tend to eliminate or drastically reduce the natural nonchemical pest control and soil nutrient enhancements that were integral parts of agriculture prior to the Green Revolution.[9] Monocultures eliminate the diversity, and thus the natural forces that can keep pest populations in check, that formerly occurred through intercropping, crop sequencing, or crop rotation.[10] Consequently, large monocultures depend on chemical pesticide and fertilizer inputs to control pests and enhance soil fertility.[11] Each of these features, alone and in combination, has the potential to contribute to a variety of environmental, human health, and socioeconomic impacts.

Synthetic chemical pesticides developed during World War II were spectacularly effective at controlling a wide variety of pests, and they quickly began to be used throughout the United States and other regions of the world. Estimates of global chemical pesticide use show that more than 1,600 types of pesticides are currently available.[12] More than five billion pounds of pesticides, with a value of almost $40 billion, are used annually in the world.[13] Pesticide use in the United States accounts for approximately 22% of global pesticide usage, with more than one billion pounds of pesticides with a value of approximately $12 billion used annually.[14] U.S. exports to other countries exceed 450 million pounds of pesticides per year.[15] The use of chemical pesticides has more than doubled since the time large-scale environmental regulation was instituted in the early 1970s. Pesticide use in the United States has almost tripled since Rachel Carson published *Silent Spring* in the early 1960s.[16] Farms use enormous quantities of pesticides every year, and since 1979 agriculture has accounted for approximately 80% of pesticide use in the United States.[17] The Pesticide Action Network has pointed out that starting in 2004, there was a significant growth in worldwide pesticides sales.[18] The current global pesticide market of approximately $40 billion per year[19] is expected to grow by approximately 3% per year, reaching approximately $52 billion per year by 2014.[20]

The rapid worldwide adoption of synthetic chemical pesticides began during World War II with the development of two primary categories of chemical insecticides: the *organochlorine*s and the *organophosphates*. The organochlorines, which include the notorious pesticide DDT,[21] were first considered highly desirable because, while very toxic to a broad range of invertebrates, they are not highly acutely toxic to humans or other mammals.[22] Organochlorine pesticides such as DDT are credited with saving thousands

---

5. Angelo, *Corn, Carbon, and Conservation*, supra note 1, at 602; Eubanks, *A Rotten System*, supra note 4, at 251-52.
6. U.S. Dep't. Agric., Econ. Res. Service, *Agricultural Productivity in the United States*, http://www.ers.usda.gov/data-products/agricultural-productivity-in-the-us.aspx (last updated July 5, 2012) ("The level of U.S. farm output in 2008 was 158 percent above its level in 1948.").
7. For a discussion of the global reliance on monoculture farming, see Helena Norberg-Hodge, *Global Monoculture: The Worldwide Destruction of Diversity*, in THE FATAL HARVEST READER: THE TRAGEDY OF INDUSTRIAL AGRICULTURE 58 (Andrew Kimbrell ed., 2002).
8. Union of Concerned Scientists, *Industrial Agriculture: Features and Policy*, http://www.ucsusa.org/food_and_agriculture/science_and_impacts/impacts_industrial_agriculture/industrial-agriculture-features.html (last revised May 17, 2007); *see also* Kelley R. Tucker, *Wildlife Harvest*, in THE FATAL HARVEST READER, *supra* note 7, at 208, 221 (discussing the impacts of agriculture on wildlife).
9. H.F. VAN EMDEN & M.W. SERVICE, PEST AND VECTOR CONTROL at 41-42 (2004).
10. Id.
11. Id.
12. Clive A. Edwards, *The Impact of Pesticides on the Environment*, in THE PESTICIDE QUESTION: ENVIRONMENT, ECONOMICS AND ETHICS 13 (David Pimentel & Hugh Lehman eds., 1993).
13. Arthur Grube et al., U.S. EPA, *Pesticide Industry Sales and Usage 2006 and 2007*, http://www.epa.gov/opp00001/pestsales/ (last visited Oct. 25, 2012).
14. Id.
15. Edwards, *supra* note 12, at 13.
16. JASON CLAY, WORLD AGRICULTURE AND THE ENVIRONMENT 49 (2004) (citing Kimbrell 2002).
17. J.B. Ruhl, *Farms, Their Environmental Harms, and Environmental Law*, 27 ECOLOGY L.Q. 263, 283 (2000).
18. Pesticide Action Network, *Myths About Pesticides*, http://www.panna.org/science/myths (last visited Oct. 25, 2012).
19. Id.
20. Id.
21. DDT is the abbreviation for synthetic insecticide, 1, 1, 1-trichloro-2, 2-bis (p-chlorophenyl) ethane. ROBERT E. PFADT, FUNDAMENTALS OF APPLIED ENTOMOLOGY 227, 755 (3d ed. 1978).
22. Edwards, *supra* note 12, at 14.

of lives from insect-borne diseases during World War II.[23] Furthermore, these pesticides are extremely persistent in the environment, which makes them highly effective for long-term effective pest control. Their downfall came, however, when the long-term ecological consequences of that very persistence became apparent, specifically that the pesticides accumulate in living tissues and bioconcentrate as they move through the food chain.[24] This resulted in serious impacts to predators at the top of the food chain, including the bald eagle. Consequently, the U.S. Environmental Protection Agency (EPA) cancelled the registration for most uses of DDT in 1972,[25] and most other organochlorine pesticides were either banned or severely restricted during the 1970s and 1980s in the developed countries of the world.[26]

The other major category of pesticides developed during World War II is organophosphate pesticides, initially developed as wartime nerve gases,[27] which were well-suited as biological warfare agents because they are quick-acting neurological poisons in mammals, including humans. Likewise, they act rapidly to kill insects and other pest species. However, due to their high acute toxicity, organophosphates actually pose a greater immediate threat to humans, fish, and wildlife than do many of the organochlorines, such as DDT.[28] Many organophosphates kill rapidly upon contact, whether through ingestion, breathing, or mere skin exposure.[29]

The organophosphates, which are far less persistent in the environment than are organochlorines, became the pesticides of choice in the United States after most organochlorine pesticides were banned or severely restricted. Organophosphates remain the largest category of chemical insecticide in use in the United States today.[30] In addition to posing risks of acute poisoning to farm workers, these pesticides have been implicated in a large number of avian and wildlife poisonings.[31] Given their potentially extreme toxicity, the large quantities released into the environment each year, and the fact that these pesticides are used with the express purpose of killing or disrupting living organisms, it is not surprising that threats to wildlife remain.

Other categories of chemical pesticides include the synthetic pyrethroids and carbamates. Pyrethrum is a naturally occurring pesticide derived from chrysanthemum flowers.[32] Synthetic pyrethroids are synthetically produced versions of pyrethrum.[33] These synthetic pyrethoids have the environmental benefit of low mammalian toxicity and low environmental persistence.[34] Nevertheless, they are highly toxic to a broad range of invertebrates, including many beneficial insects.[35] They are also highly toxic to fish and other aquatic organisms.[36] Carbamates are more persistent than organophosphates in the environment and are generally broad-spectrum, having adverse impacts on many different groups of organisms.[37] Other commonly used pesticides include nematicides, which not only are of high mammalian toxicity and broad-spectrum but also are very transient in soil; herbicides, which generally are not highly toxic to mammals

---

23. Andrew P. Morriss & Roger E. Meiners, *Property Rights, Pesticides, and Public Health: Explaining the Paradox of Modern Pesticide Policy*, 14 Fordham Envtl. L. Rev. 1, 6 (2002).
24. Stevens Indus., Inc., 1 E.A.D. 9, 16 (June 2, 1972) (Consolidated DDT Hearings).
25. Id.
26. Edwards, *supra* note 12, at 14. A number of international agreements exist to restrict the use of persistent organic pollutants, such as the organochlorine pesticides. For a detailed discussion of such agreements, see Mary Jane Angelo et al. (Center for Progressive Reform), *Reclaiming Global Environmental Leadership: Why the United States Should Ratify Ten Pending Environmental Treaties* (2011). *See also* Michael P. Walls, International Chemicals Update 2005, SK 058 A.L.I.-A.B.A. Course of Study Materials 661 (2005). It should be noted that there has been a recent resurgence in efforts to loosen restrictions on DDT due to its potential use in combating malaria and other insect-borne diseases. See, e.g., Morriss & Meiners, *supra* note 23.
27. Edwards, *supra* note 12, at 15.
28. Pesticide Action Network UK, *DDT Factsheet*, http://www.pan-uk.org/pestnews/Actives/ddt.htm (last visited Oct. 25, 2012) (comparing the acute toxicities of DDT and of the organophosphate parathion).
29. Extension Toxicology Network, *Pesticide Information Profiles*, http://extoxnet.orst.edu/pips/parathio.htm (last visited Oct. 25, 2012) (describing long-term effects of parathion on humans and a variety of animals).
30. See Grube et al., *supra* note 13.
31. Id.
32. U.S. Dept. of Labor, Occupational Safety and Health Administration, *Pyrethrum*, http://www.osha.gov/dts/sltc/methods/organic/org070/org070.html (last visited Oct. 25, 2012).
33. Id.
34. Beyond Pesticides, Synthetic Pyrethroids: Chemical Watch Factsheet, 1-2, *available at* http://www.beyondpesticides.org/pesticides/factsheets/Synthetic%20Pyrethroids.pdf.
35. Id.
36. Id.
37. Whitney Cranshaw, *Classes of Pesticides Used in Landscape/Nursery Pest Management*, in Tactics and Tools for IPM, at 42, *available at* http://www.entomology.umn.edu/cues/Web/042ClassesOfPesticides.pdf.

but travel easily in water where they may be toxic to fish and aquatic organisms; and fungicides, which vary greatly in their toxicity.[38]

Although, from an ecological standpoint, narrow-spectrum pesticides are preferable, broad-spectrum synthetic pesticides continue to dominate U.S. pesticide usage.[39] This phenomenon is at least in part attributed to the time and costs associated with bringing a new pesticide to the market.[40] Pesticide manufacturers may be inclined to develop broad-spectrum pesticides with greater market opportunities in order to maximize profits during the commercial life of a pesticide.[41]

In the past 20 years, the biotechnology sector of the pesticide industry has undergone tremendous growth. Naturally existing microbes have been genetically modified to make them toxic to insects and other pests.[42] In addition, agricultural crop plants themselves have been genetically modified to produce substances that have pesticidal effects,[43] which pose novel ecological risks by virtue of their ability to reproduce and spread in the environment (as described in more detail in Chapter 6).

Throughout much of history, the nutrients needed to grow agricultural crops were provided by planting crops in soils rich in nutrients. Where nutrient-rich soils were not available, and as the desire to increase crop yields grew, supplemental nutrients were primarily provided by the use of animal wastes (manure) and compost as fertilizers.[44] In addition, nutrients produced by plants were frequently used as "green manure."[45] Many traditional farming practices, such as crop rotation, cover cropping, especially with nitrogen-fixing plants such as legumes, and intercropping helped to preserve and enhance soil nutrients.

One of the most significant changes that occurred in the evolution to large-scale industrial agriculture was the separation of animals from plants on the farm. Today, corn has replaced grass as the primary cow feed.[46] Consequently, many cattle ranchers have replaced open-range grazing with corn production and feed their animals a mostly corn-based diet in confined feedlots.[47] Aside from ethical issues of raising animals in the conditions of the modern concentrated animal feeding operation, these practices have resulted in significant water pollution.[48] Animal wastes, which once could be readily used as fertilizers for crops grown on the same farm as the animals that created the waste, now have no use, and the vast quantities of concentrated animal waste have become a serious source of water pollution.[49] What once was a win-win situation—animal wastes fertilized the crops that fed the animals in a relatively "closed loop" system with relatively insignificant pollution resulting—has now become a substantial environmental problem.[50] Michael Pollan, the author of the best-selling *The Omnivores' Dilemma*,[51] has opined:

> [T]o take animals off farms and put them on feedlots is to take an elegant solution—animals replenishing the fertility that crops deplete—and neatly divide it into two problems: a fertility problem on the farm and a pollution problem on the feedlot. The former problem is remedied with fossil-fuel fertilizer; the latter is remedied not at all.[52]

---

38. *Id.*
39. WILLIAM H. RODGERS, ENVIRONMENTAL LAW 407-08 (West, 2d ed. 1994). Narrow-spectrum pesticides are those that only affect a small group of organisms, whereas broad-spectrum pesticides are those that affect a wide range of organisms.
40. *Id.*
41. *Id.* at 407-09.
42. *See* Microbial Pesticides; Experimental Use Permits and Notifications, 58 Fed. Reg. 5878 (proposed Jan. 22, 1993).
43. *See* Regulations Under the Federal Insecticide, Fungicide, and Rodenticide Act for Plant-Incorporated Protectants, 66 Fed. Reg. 37772 (July 19, 2001) (codified at 40 C.F.R. pts. 152, 174).
44. James F. Parr & Sharon B. Hornick, *Agricultural Use of Organic Amendments: A Historical Perspective*, 7 AM. J. ALTERNATIVE AGRIC. 181-89 (1992).
45. MICHAEL POLLAN, THE OMNIVORE'S DILEMMA: A NATURAL HISTORY OF FOUR MEALS (2006).
46. *Id.*
47. *Id.* at 66. Currently, corn is the primary feed grain in the United States, accounting for more than 90% of total feed grain produced and used, http://www.ers.usda.gov/Briefing/Corn/background.htm.
48. Bruce Yandle & Sean Blacklocke, *Regulating Concentrated Animal Feeding Operations: Internalization or Cartelization?*, at 45, 48-49, *in* AGRICULTURAL POLICY AND THE ENVIRONMENT (Roger E. Meiners and Bruce Yandle, eds. 2003).
49. *See* Eubanks, *A Rotten System, supra* note 4, at 260.
50. *See id.* at 259-60.
51. Pollan, *supra* note 45.
52. Michael Pollan, *Farmer in Chief*, N.Y. TIMES MAG., Oct. 12, 2008 (paraphrasing Wendell Berry).

B. Environmental Risks of Pesticides

Scientists estimate that as many as 10 million species, or 99% of the earth's wild biodiversity, not including cultivated and weedy species, are in a "precarious condition."[53] Causes and contributors to the decline of so many species include indirect habitat destruction through clearing for agriculture and development, the spread of nonnative invasive species, pollution, overharvesting of species, and disease.[54] Although direct habitat destruction is undoubtedly the leading contributor of species loss (estimated as being implicated in 85% of U.S. species decline), pollution, including pesticide pollution, is also implicated in 24% of U.S. species decline.[55] Pesticide poisoning of fish and wildlife is a significant factor in species decline.[56]

Because pesticides, by definition, are intended to kill or disrupt living organisms and because they are intentionally released into the environment, often in large quantities over large areas, it is not surprising that pesticides pose a wide array of potential risks to individual species as well as to overall ecosystem function. Many pesticides are broad-spectrum, affecting diverse species, including many nontarget organisms[57] Others are more narrowly targeted to pest species. But even the narrowly targeted pesticides may have significant impacts on nontarget species closely related to the intended targets.[58] Some pesticides persist in the environment for weeks, months, and even years, while others break down relatively quickly.[59] Moreover, living organisms vary significantly in their susceptibility to pesticides.[60] The potential ecological risks of pesticide use depend on a number of factors, including toxicity or other hazards of the pesticide, method of application, persistence in the environment, amount used, and susceptibility of nontarget organisms.[61] Very little data are available on the environmental effects of pesticide usage on many species, so the ecological risks of pesticides cannot be easily described or quantified. Nevertheless, some generalizations can be made.

Many pesticides in current use in the United States and other parts of the world are acutely toxic to nontarget mammals, birds, reptiles, amphibians, fish, and invertebrates.[62] Birds and other wildlife may be exposed through direct spraying, ingesting pesticide granules, drinking water contaminated by pesticides, or eating prey organisms contaminated by pesticides.[63] While the banning and severe restriction of certain pesticides such as DDT over the past 30 years has dramatically reduced certain risks to wildlife, many risks remain.[64]

Several recent studies suggest that the bans on DDT and other organochlorines have not ended the pesticide threat. In 2004, the Center for Biological Diversity (CBD) reported that EPA-approved pesticides

---

53. David S. Wilcove, The Condor's Shadow: The Loss and Recovery of Wildlife in America xiv (1999).
54. Id.
55. Id. at 8.
56. In one study of the decline of fish species in the United States, Canada, and Mexico, it was determined that the destruction of physical habitat was implicated in 73% of the decline, the displacement by introduced species was implicated in 68% of the decline, the alterations of habitat by chemical pollutants was implicated in 38% of the declines, hybridization with other species and subspecies was implicated in 38% of the declines, and overharvesting was implicated in 15% of the declines. The numbers add up to more than 100 because more than one factor is implicated in many of the fish population declines. Edward O. Wilson, The Diversity of Life 253-54 (1992). Thus, while pesticide usage in itself may not directly destroy habitat (although clearing for agriculture certainly does), chemical pesticides may be a significant contributor to species decline, and pesticidal GMOs, which pose risks of spread in the environment similar to non-indigenous species release, may also be important contributors. Wilcove, *supra* note 53, at 8.
57. See Rodgers, *supra* note 39 and accompanying text.
58. See Edwards, *supra* note 12, at 17-24.
59. Id. at 17.
60. Id. at 18.
61. Id.
62. For a detailed discussion of the risks pesticides cause to wildlife species, see Comments on the Proposed Joint Counterpart Endangered Species Act Section 7 Consultation Regulations. Letter from Defenders of Wildlife and 29 other commenters to Gary Frazer, Assistant Director for Endangered Species, U.S. Fish & Wildlife Serv., and Phil Williams, Chief of Endangered Species Div., Nat'l Oceanic and Atmospheric Admin. Fisheries (Apr. 16, 2004) (on file with author).
63. The New York State Department of Environmental Conservation found that a number of different avian species, such as screech owls, red-tailed hawks, American kestrels, and other raptors, died as a result of eating small rodents that had consumed rat poison. Laura A. Haight, *Local Control of Pesticides in New York: Perspectives and Policy Recommendations*, 9 Alb. L. Envtl. Outlook 37, 51 (2004).
64. For example, when roughly 10,000 dead birds were tested for the presence of West Nile virus in 2000, the New York State Department of Environmental Conservation determined that pesticides and other chemicals were actually responsible for more bird kills than the virus. Haight, *supra* note 63, at 51. As further evidence of the effects on bird populations, studies have shown substantially higher nesting rates of birds, as well as significantly higher bird abundance and avian species richness, on organic farms compared to conventional farms that use synthetic pesticides. Nancy A. Beecher et al., *Agroecology of Birds in Organic and Nonorganic Farmland*, 16 Conservation Biology 1620 (2002).

currently were putting more than 375 threatened and endangered species at risk.[65] The report summarized the existing data on pesticide-related harm to aquatic life, birds, and other wildlife, including protected species.[66] It also described the problems associated with pesticide-contaminated waterways, soils, and biota, as well as pesticide spray drift.[67] The report also included a detailed description of the endocrine-disrupting effects associated with many pesticides.[68]

CBD is not alone in its concerns over pesticide impact on wildlife. The American Bird Conservancy (ABC) estimates that more than 15 million birds per year will die as a result of exposure to pesticides.[69] Although significant, the Conservancy points out that this number is down from an estimate of 67 million birds killed by pesticides per year in the United States in 1992. It attributes the decline to the cancellation in recent years of the registrations of over a dozen mostly organophosphate pesticides such as fenthion, chlorfenapyr, and ethyl parathion, which posed significant risks to birds and other wildlife.[70] Although fish, bird, and other wildlife poisonings from exposure to pesticides are fairly frequent and widespread,[71] the U.S. government rarely publishes data on wildlife deaths from pesticide exposure. It is therefore necessary to rely on the reports of advocacy organizations for these statistics. One database tracking pesticide-related bird mortality lists over 400,000 reported deaths caused by 4,000 pesticide poisoning incidents.[72] Due to known underreporting of bird deaths, it is likely that actual mortality from pesticide poisonings is substantially greater.[73] The organophosphate and carbamate pesticides appear to be the greatest cause of these deaths.[74]

Other less visible species are also at considerable risk from pesticide exposure. For the past decade, considerable concern and debate existed in the scientific community over the worldwide decline of amphibians. Significant data now support a conclusion that certain pesticides, such as the herbicide atrazine, may be contributing to the world-wide decline in amphibian populations. For example, in 2002, the organization Californians for Alternatives to Toxics filed suit seeking an order requiring the state's Department of Pesticide Regulation to reevaluate the state registration of pesticide products containing the active ingredients malathion, chlorpyrifos, diazinon, methidathion, endosulfan, chorothalonil, and trifluralin. The lawsuit contended that these pesticides may be responsible for significant population declines of several species of amphibians in the Sierra Nevada mountains.

Until recently, many studies on the effects of pesticides on amphibians have been puzzling because pesticide levels in nature tend to be much lower than levels found to be lethal in the laboratory setting.[75] A recent study sheds new light on this dilemma.[76] Scientists have determined that the combination of the pesticide carbaryl and stress from the presence of predators was more lethal in certain amphibian species than the pesticide by itself. In other words, a synergistic effect appears to be at work between pesticides and predators, making the combination of the two more lethal than the sum of the parts and resulting in even low concentrations of pesticides in nature being highly lethal to amphibians. Of course, amphibians in nature must cope with other stressors, such as the presence of predators, in addition to the stress of pes-

---

65. Brian Litmans & Jeff Miller, Center for Biological Diversity, *Silent Spring* Revisited: Pesticide Use and Endangered Species (2004), *available at* http://www.biologicaldiversity.org/swcbd/Programs/science/pesticides/ [hereinafter CBD Report].
66. *Id.* at 6-9, 16-44.
67. *Id.* at 1-5.
68. *Id.* at 10-15.
69. American Bird Conservancy, *Pesticides and Birds*, http://www.abcbirds.org/abcprograms/policy/toxins/pesticides.html (last visited Oct. 25, 2012).
70. *Id.*
71. *See* American Bird Conservancy, *AIMS Database*, http://www.abcbirds.org/abcprograms/policy/toxins/aims/aims/doc.cfm (last visited Oct. 25, 2012). AIMS is a cooperative program between the American Bird Conservancy and EPA. The AIMS database tracks incidents of pesticide exposure affecting wild birds.
72. *Id.*
73. *See* David Pimentel et al., *Assessment of Environmental and Economic Impacts of Pesticide Use*, *in* The Pesticide Question: Environment, Economics, and Ethics 66 (David Pimentel & Hugh Lehman eds., 1993). Bird deaths are underreported for a number of reasons. First, sick or dying birds typically fly away from the area where they were poisoned and often seek shelter in a hidden location. Second, bird carcasses are quickly carried away by predators and scavengers. Finally, humans often fail to report deaths, either because they are not aware that there is reason to do so, or they want to avoid potential legal liability for contributing to the bird death.
74. *See* Edwards, *supra* note 12, at 27; Pimentel, *supra* note 73, at 66.
75. *Id. See* Rick A. Relyea, *Predators Make Pesticides More Lethal*, Conservation in Prac., Spring 2004, at 5 (excerpting Rick A. Relyea, *Predator Cues and Pesticides: A Double Dose of Danger for Amphibians*, 13 Ecological Applications 1515 (2003)).
76. Carlos Davidson et al., *Spacial Tests of the Pesticide Drift, Habitat Destruction, UV-B, and Climate-Change Hypotheses for California Amphibian Declines*, 16 Conservation Biology 1588 (2002).

ticides. Accordingly, this study demonstrates that amphibians in nature may be significantly more sensitive to pesticides than they are in the sterile, isolated confines of the research laboratory.[77]

Although the most obvious adverse effects of pesticide use are those on humans and large animals such as mammals and birds, the most significant adverse effects of pesticides are likely to be those on invertebrates, which are closely related to target pest species.[78] Casualties from this "friendly fire" are widespread in the invertebrate world.[79] For example, the insect Order Lepidoptera contains not only many species of pest moths but also many nonpest butterfly species. These butterfly species may be beneficial pollinators and may be aesthetically pleasing, colorful, and interesting. Also, Lepidoptera contains a number of butterfly species listed as threatened or endangered under the federal Endangered Species Act.[80] Pesticides used to kill pest moth species generally do not discriminate within the Lepidoptera order and will also kill nonpest, beneficial butterflies including endangered species.[81] Mosquito control pesticides have been indicted as one of the threats to the continued survival of the endangered Miami Blue butterfly over the past few decades.[82]

One example of this friendly fire of pesticides in the environment is their toxicity to bees. Honeybees and wild bees are vital parts of agriculture systems, providing pollinator services for about one-third of U.S. and world crops.[83] Bees can be killed by insecticides meant for other insects, but also can transmit these and other chemicals into their products, including honey, pollen, wax, and royal jelly.[84] Bee mortality can result from direct contact with pesticides while gathering nectar or pollen and can strike individuals away from the colony. Alternatively, and more dangerously, bees carrying pesticides in nectar or pollen may transport these chemicals back to the colony, causing mass mortality and collapse of the hive.[85] When viewed with the ecosystem services bees provide in mind, this effect of pesticides is particularly pernicious. The loss of pollination services coupled with the direct loss of honeybees is a significant problem caused by pesticides.[86]

Perhaps more important than direct acute effects on nontarget organisms are the chronic effects of pesticides on growth, physiology, reproduction, and behavior.[87] Much less is known about these chronic effects than is known about acute effects.[88] Even a pesticide not toxic enough to kill an organism can have significant sublethal effects on the organism by affecting its life span, growth, physiology, behavior, and reproduction.[89] Moreover, pesticides have been documented to have significant indirect effects on nontarget organisms by reducing the populations of animals or plants that serve as food or cover for other species.[90]

One of the most insidious risks posed by pesticides is the tendency of certain synthetic pesticides to mimic hormones, such as estrogen, in humans and wildlife. Only recently has science begun to understand these complex effects.[91] Estrogen-mimicking substances include a number of pesticides as well as a wide

---

77. *Id.*
78. Malcolm L. Hunter Jr., Fundamentals of Conservation Biology 156-57 (1996).
79. *See* May Berenbaum, *Friendly Fire*, Wings: Essays on Invertebrate Conservation, Spring 2004, at 8.
80. 50 C.F.R. §17.11.
81. *See* Hunter, *supra* note 78, at 157-58.
82. Jaret C. Daniels & Thomas C. Emmel, *Florida's Precious Miami Blues*, Wings: *Essays on Invertebrate Conservation*, Spring 2004, at 3. This issue of the Xerces Society publication was devoted exclusively to butterfly conservation; four out the five articles listed pesticides as a significant contributor to butterfly population declines. *See id.* Moreover, recent studies demonstrate a reduction in the abundance of nontarget butterflies on conventional farms as compared to butterflies on organic farms. D.J. Hole et al., *Does Organic Farming Benefit Biodiversity?*, 122 Biological Conservation 113 (2004). 50 C.F.R. §17.11.
83. Pimentel, *supra* note 73, at 97.
84. Zaneta Barganska & Jacek Namiesnik, *Pesticide Analysis of Bee and Bee Product Samples,* 40 Critical Reviews in Analytical Chemistry 159 (2010).
85. *Id.* at 160.
86. Pimentel, *supra* note 73, at 98.
87. Edwards, *supra* note 12, at 24.
88. *Id.*
89. *Id.* For example, extreme low doses of some pesticides have been determined to disrupt honeybees' homing flight behavior, thereby adversely affecting pollination. Helen M. Thompson, *Behavioural Effects of Pesticides in Bees—Their Potential for Use in Risk Assessment,* 12 Ecotoxicology 317 (2003).
90. Edwards, *supra* note 12, at 28-29.
91. Although the term "environmental estrogen" was coined in the 1970s, not until the past 15 years were any scientific studies conducted to support the hypothesis that environmental exposure to certain synthetic chemicals could cause estrogenic effects. For a detailed discussion of the risks of endocrine disrupting chemicals and the legal shortcomings in addressing such risks, see generally Noah Sachs, *Blocked Pathways: Potential Legal Responses to Endocrine Disrupting Chemicals,* 24 Colum. J. Envtl. L. 289 (1999). *See also* Theo Colburn et al., Our Stolen Future: Are We Threatening Our Fertility, Intelligence and Survival? A Scientific Detective Story (1997); Matthew P. Longnecker et al., *The Human Health Effects of DDT (Dichlorodiphenyltrichloroethane) and PCBs (Polychlorinated Biphenyls) and an Overview of Organochlorines in Public Health,* 18 Rev. Pub. Health 211 (1997); Louis J. Guillette et al., *Developmental Abnormalities of the Gonad and*

variety of other products in common use, such as toiletries, spermicides, and plastics.[92] Exposure to these compounds, particularly when the exposure occurs to a fetus or young children, has been correlated with a substantial number of effects in humans, including decreased sperm counts, breast and testicular cancer, endometriosis, deformed or stinted reproductive organs, neurological defects, and low birth weights.[93] These substances have also been implicated in numerous wildlife harms, such as deformed alligators and reproductive difficulty in birds, fish, and mammals.[94]

These estrogenic effects can be extremely complex, unpredictable, and difficult to understand.[95] For example, DDT exposure has caused female gulls to begin sharing nests with other females rather than males, and young male gulls to have grossly feminized reproductive tracts.[96] Moreover, a large number of studies on various species of fish exposed to estrogenic compounds have shown effects such as increased time to maturity, smaller gonads, and reduced fertility.[97] Similarly, declines in the reproductive rates of mammals, such as minks, have been linked to ingesting fish contaminated with estrogenic substances.[98] The U.S. Fish and Wildlife Service has reported that between 1985 and 1990, 67% of male Florida panthers were born with one or more undescended testicle compared with only 14% a decade earlier.[99] Although the phenomenon is not fully understood, scientists suspect a link with exposure to estrogenic substances in the environment. Perhaps the most widely cited examples of endocrine dysfunction in wildlife are the feminization of alligators and occurrence of masculinized female fish in Florida. Although the estrogenic effects of certain pesticides are only beginning to be understood, Rachel Carson was prescient when she predicted such effects in her 1962 book *Silent Spring*:

> A substance that is not a carcinogen in the ordinary sense may disturb the normal functioning of some part of the body in such a way that malignancy results. Important examples are the cancers, especially of the reproductive system, that appear to be linked with disturbances of the balance of sex hormones . . . [t]he chlorinated hydrocarbons are precisely the kind of agent that can bring about this kind of indirect carcinogenesis.[100]

Another concern with pesticides is the uncertainty regarding their effects on ecologically significant microorganisms. Very little is known about the complex ecology of microorganisms.[101] Although few studies have been conducted to evaluate whether most types of pesticides pose significant risks to microorganisms, soil fumigants—designed to destroy soil microorganisms and applied at very high doses—may pose substantial risks to beneficial microorganisms.[102] For example, the killing of soil microbes and invertebrates resulting from pesticide use may actually cause crops to become more susceptible to disease and thereby reduce crop growth. In addition, populations of nitrogen-fixing organisms may be reduced, thereby requiring higher levels of fertilizer application.[103] Critical ecological services provided by microorganisms, including decomposition, may also be affected by certain pesticides.[104]

The ecological risk from pesticide exposure is exacerbated by the tendency of certain pesticides to undergo a phenomenon known as bioaccumulation, which became widely recognized during the 1960s as a result of Rachel Carson's book. Carson explained how DDT and other organochlorine pesticides can persist in the environment for years if not decades and can also accumulate in the tissue of animals and humans.[105] These pesticides accumulate in animals on the bottom of the food chain and then pass from prey to predator, eventually resulting in very high concentrations in top predators, a phenomenon known as biomagnification. Pesticides that persist, accumulate, and biomagnify are especially insidious in that

---

Abnormal Sex Hormone Concentration in Juvenile Alligators From Contaminated and Control Lakes in Florida, 8 ENVTL. HEALTH PERSP. 680 (1994); David Crewset et al., *Endocrine Disruptors: Present Issues, Future Directions*, 75 Q. REV. BIOLOGY 243 (2000).
92. Sachs, *supra* note 91, at 302-07.
93. *Id.* at 293-98.
94. *Id.*
95. *Id.* at 300.
96. *See* Susan M. Salvatore, *Estrogens in the Environment*, 69 FLA. B.J. 39, 42 n.35 (1995).
97. *Id.* at 39 n.36.
98. *Id.* at 42 n.37.
99. *Id.* at 42 n.38.
100. RACHEL CARSON, SILENT SPRING 235 (40th anniv. ed., First Mariner Books 2002).
101. Edwards, *supra* note 12, at 18.
102. *Id.*
103. *Id.* at 31.
104. Pimentel, *supra* note 73, at 68-69.
105. CARSON, *supra* note 100, at 21-23.

they can adversely affect organisms far removed in both time and space from the original release of the pesticide into the environment.[106]

Although agricultural systems are not natural systems per se, they are generally in close proximity to natural ecosystems and often contain within their borders sizable natural and seminatural ecosystems.[107] Thus, pesticide usage in agricultural systems may negatively affect ecosystems within the farm boundaries as well as contaminating nearby ecosystems by pesticide runoff in water, drift through the air, or movement of contaminated organisms.

As described above, loss of invertebrate nontarget species may be the greatest risk from pesticide use. Equally concerning are the ecological and economic disruptions that frequently occur as a result of nontarget predators and parasites being killed by pesticides. Many pest populations are naturally kept in check by organisms that feed on pest species. If these predators or parasites are eliminated or greatly reduced in number, the populations of pest species will explode, or new pest species will be created.[108]

Furthermore, the U.S. Department of Agriculture (USDA) has warned of an "impending pollinator crisis," due in part to pesticide use.[109] Pollinators at risk include commercial bees and a number of wild pollinators, including wild bees and various bird and bat pollinators.[110] A number of other studies reveal substantial risks and a lack of full understanding regarding the extent of pesticide risks to wildlife.[111]

In sum, despite the cancellation of DDT and its relatives in the 1970s and 1980s, pesticides continue to pose significant risks to birds and other wildlife. The pesticides that replaced the banned organochlorines, while not bioaccumulating, are more acutely toxic. Consequently, relatively large numbers of animals, including threatened and endangered species, continue to be adversely affected by the use of EPA-approved pesticides.

In the past 20 years, a completely new set of risk concerns have emerged regarding the use of pesticidal genetically modified organisms (GMOs). Although many of the risk considerations for biotechnology pesticides are similar to those of traditional chemical pesticides, these new pesticides pose a number of novel risks.[112] One of the most significant novel risks for pesticidal GMOs is the potential for the spread of the living organism or the organism's genetic material.[113] For example, plants can reproduce sexually and/or asexually and the genetic material introduced into the plant enabling the plant to produce pesticidal substances could spread through agricultural or natural ecosystems.[114] If a plant that produces a pesticide can spread in the environment or can spread its genetic material to other plants, a greater potential would exist for exposure to nontarget organisms than would be for a pesticide produced in a plant that can only grow in a limited geographic area or does not have the ability to cross-fertilize with other plants.[115] This is a particular concern for pesticides produced in plants that have wild relatives in the United States.[116] If these wild relatives acquire the ability to produce the pesticide though cross-fertilization, many additional

---

106. Hunter, *supra* note 78, at 156. Other risks posed by pesticides have only recently begun to be studied. For example, in recent years the extent of atmospheric transport of pesticides has come to light. Edwards, *supra* note 12, at 32-33. Moreover, the pesticide methyl bromide has been determined to be a significant contributor to the thinning of the stratospheric ozone layer. *See* Protection of Stratospheric Ozone: Process for Exempting Critical Uses From the Phaseout of Methyl Bromide, 69 Fed. Reg. 76982 (Dec. 23, 2004) (to be codified at 40 C.F.R. pt. 82).
107. Hunter, *supra* note 78, at 276.
108. *Id.* at 158. An example of new pest creation resulting from pesticide use is the bollworm, which is now a major economic pest of cotton. *Id.* Although the bollworm existed previously, it was not a pest until pesticides used to control the boll weevil, another pest of cotton, killed the natural enemies of the bollworm, allowing its population to explode. *Id.*
109. CBD Report, *supra* note 65.
110. *See* Pimentel, *supra* note 73, at 58-60.
111. *See, e.g.*, Lawrence J. Blus & Charles J. Henny, *Field Studies on Pesticides and Birds: Unexpected and Unique Relations*, 7 Ecological Applications 1125 (1997) (finding shortcomings with existing field testing of pesticides on birds and unexpected toxic effects and routes of exposure of certain organophosphate pesticides); Andrew Ogram & Yun Cheng, *Final Report: Biological Breakdown of Pesticides in Lake Apopka North Shore Restoration Area Soil in a Mesocosm Experiment*, St. Johns River Water Management District Special Publication SJ2007-SP1 (2007) (demonstrating the complexity of pesticide breakdown in soils and under a variety of conditions); *see also* Ruhl, *supra* note 17, at 263, 274-93, 337-38 (describing the environmental hazards of the farming industry, the consequences of pesticide use, and the lack of strong environmental regulation of agriculture).
112. For a detailed discussion of the potential risks and benefits of pesticidal GMOS, see Mary Jane Angelo, *Regulating Evolution for Sale: An Evolutionary Biology Model for Regulating the Risks Posed by Genetically Modified Organisms*, 42 Wake Forest L. Rev. 93 (2007). Further discussion of the risks of GMO crops can be found in Chapter 6 of this book. *See also* Mary Jane Angelo, *Genetically Engineered Plant Pesticides: Recent Developments in the EPA's Regulation of Biotechnology*, 7 U. Fla. J.L. & Pub. Pol'y 257 (1996).
113. David J. Earp, Comment, *The Regulation of Genetically Engineered Plants: Is Peter Rabbit Safe in Mr. McGregor's Transgenic Vegetable Patch?*, 24 Envtl. L. 1633, 1666-69 (1994).
114. *Id.*
115. *Id.*
116. *Id.* at 287.

nontarget organisms could be exposed to the pesticide.[117] One of the most cited concerns regarding pesticidal GMOs is the potential for development of "superweeds" through the outcrossing of pesticidal GMOs to wild relatives.[118] Outcrossing occurs when one variety or species of a plant pollinates another species or variety to form a hybrid of the two. Development of such a superweed has the potential to substantially disrupt agricultural and natural ecosystems.

Perhaps, the most serious concern with pesticidal GMOs stems from the uncertainty of their risks. Although the risk of GMO release creating a new superweed or disrupting the balance of natural ecosystems may be small, the consequences could be disastrous and irreversible.[119] The precise nature and magnitude of the risk are difficult to predict because of the almost infinite variety of potential GMOs, the ability of GMOs to reproduce and spread, the complexity inherent in natural ecosystems, and the dearth of long-term data on the effects of GMOs.[120]

C. Environmental Risks of Fertilizers

In addition to its reliance on chemical pesticides, industrial agriculture also depends on the use of synthetic fertilizers to produce high crop yields. Three major components of any fertilizer are nitrogen (N), phosphorous (P), and potassium (K).[121] Fertilizers used to maximize yields in industrial agriculture typically contain nutrients such as phosphorus and ammonium nitrate.[122] While the chemistry behind the creation, application, and absorption is quite complex, it has been long understood that plant growth is "limited by the nutrient that is present in the environment in the least quantity relative to plant demands for growth,"[123] a concept known as Liebig's Law of the Minimum. The two nutrients that have been found to be the most limiting for terrestrial plants are N and P.[124] As with pesticides, a marked increase in fertilizer production has occurred in the last 50 years. Increased human population and subsequent demand for greater quantities of agricultural crops have driven these production increases.[125] Abandonment of traditional crop rotation methods for monocultured crop plantings have driven the need for fertilizers even higher, as the natural cycles for N and P have been disrupted.[126]

The movement of pollution from agricultural lands through runoff, or return flows from irrigation, is known as nonpoint source discharge. In the United States, nonpoint sources of N and P are the main inputs of these nutrients to most surface waters.[127] Nonpoint source pollution comes from many sources that are diffuse and difficult to identify; for example, the runoff from a shopping mall parking lot following a heavy rain contains many pollutants that have leaked from cars but are not necessarily attributable to specific responsible parties. Under the Clean Water Act, nonpoint source discharges are any discharges that do not meet the definition of point source discharge. A point source is defined in the Clean Water Act as any "discernible, confined and discrete conveyance . . . from which pollutants are or may be discharged";

---

117. The potential for a GMO or its genetic material to spread from one plant to another raises additional risk issues beyond those of exposure to humans and nontarget organisms. One potential risk of biotechnology products parallels the risk of the introduction of any nonnative species into a new environment. Earp, *supra*, note 113, at 1666-69. Even very small genetic manipulations can significantly change an organism's ability to survive and flourish in a particular ecosystem. *Id.* Examples abound regarding the disastrous but unpredicted effects of introducing nonnative species into the environment, displacing native species. *See* Judy J. Kim, *Out of the Lab and Into the Field: Harmonization of Deliberate Release Regulations for Genetically Modified Organisms*, 16 Fordham Intl. L.J. 1160 (1993). Introducing GMOs into the environment could have similar impacts. *See* David J. Earp, Comment, *The Regulation of Genetically Engineered Plants: Is Peter Rabbit Safe in Mr. McGregor's Transgenic Vegetable Patch?*, 24 Envtl. L. 1633, 1653 (1994). One of the most significant risks is the risk of a genetically engineered plant becoming a weed or pest itself or outcrossing to related species to create new weeds or pests. *Id.* at 1654-55. Once released into the environment, the spread of a GMO may be extremely difficult, if not impossible, to control. *Id.*
118. For example, the ability to produce a pesticide that makes a plant resistant to insect or viral pests can be spread to a wild relative and subsequently passed on to the relative's subsequent generations. Consequently, the wild relative, by virtue of its newly acquired ability to resist insects or viruses, has the potential to become a hardy weed, or superweed.
119. *See* John Charles Kunich, *Mother Frankenstein, Doctor Nature, and the Environmental Law of Genetic Engineering*, 74 S. Cal. L. Rev. 807, 818 (2001).
120. Celeste Marie Steen, *FIFRA's Preemption of Common Law Tort Actions Involving Genetically Engineered Pesticides*, 38 Ariz. L. Rev. 763 (1996).
121. U.S. EPA, *Nutrient Management and Fertilizer*, http://www.epa.gov/oecaagct/tfer.html (last visited Oct. 25, 2012).
122. Eubanks, *A Rotten System*, *supra* note 4, at 284.
123. V.H. Smith et al., *Eutrophication: Impacts of Excess Nutrient Inputs on Freshwater, Marine, and Terrestrial Ecosystems*, 100 Envtl. Pollution 179, 180 (1999).
124. *Id.*
125. *Id.* at 179.
126. *Id.*
127. S.R. Carpenter et al., *Nonpoint Pollution of Surface Waters With Phosphorus and Nitrogen*, 8 Ecological Application 559, 561 (1998).

an example would be an industrial pipe spilling chemical waste into a river or other water body.[128] Agricultural stormwater discharges and return flows from irrigation are specifically excluded from this definition.[129] The 2000 National Water Quality Inventory identified nonpoint source pollution as the primary source of water-quality impacts on surveyed rivers and lakes, the second leading source of degradation to wetlands, and a large source of contamination to groundwater and estuaries.[130] In 2009, the State-EPA Nutrients Innovations Task Group published *An Urgent Call to Action*, which evaluated a number of recent studies by the EPA Scientific Advisory Board, the National Research Council, and others, and concluded that nutrients now pose significant water quality and public health concerns across the United States.[131] The report described how significant increases in nutrient pollution in U.S. water bodies over the past 50 years have significantly impacted "drinking water supplies, aquatic life and recreational water quality."[132] Nutrient pollution is the cause of water quality impairment for approximately 20% of river and stream miles (approximately 80,000 miles), 22% of lake acres, and 8% of bay and estuary miles.[133] About one-third of the nation's streams have been found to have high concentrations of phosphorous or nitrogen. The report states that "[n]utrient pollution from row crop agricultural operations, a byproduct of excess manure and chemical fertilizer application, is the source of many local and downstream adverse nutrient-related impacts."[134]

Both N and P fertilizers are applied in various forms to agricultural crops. They can either be organic, composed of enriched plant and animal matter, or inorganic, meaning that they are made from synthetic chemicals or minerals.[135] Often, fertilizers are applied in excess of the absorption capacity and nutrient needs of the treated crops.[136] Additionally, if the application is timed poorly and the weather patterns are not identified properly, a rainfall quickly following fertilizer application can wash away the nutrients before the crops can absorb them.[137] This excess can accumulate in soil, move into surface waters, migrate into groundwater, or enter the atmosphere via chemical processes.[138]

As with pesticides, the use and overuse of fertilizers has, in some cases, led to unintended environmental consequences. Scientific studies demonstrate that agricultural intensification via increased chemical fertilizer and other inputs is directly linked to increased environmental damage.[139] Large quantities of these compounds are carried in rain runoff into water bodies, where they stimulate plant growth that results in overgrowth of algae.[140] When algae become overabundant, it depletes oxygen and reduces sunlight penetration, resulting in a condition referred to as eutrophication.[141] The process of eutrophication manifests itself initially as an increase in the amount of aquatic plant life and algae.[142] While seemingly benign at this stage, eutrophication can lead to serious and irreversible long-term damage to aquatic resources. Eutrophication has several major effects on lakes, reservoirs, rivers, and coastal oceans:

1. Increased biomass of phytoplankton;

2. Shifts in phytoplankton to bloom-forming species that may be toxic or inedible;

---

128. Clean Water Act, 33 U.S.C. §1362. As defined in the Clean Water Act, "[t]he term 'point source' means any discernible, confined and discrete conveyance, including but not limited to any pipe, ditch, channel, tunnel, conduit, well, discrete fissure, container, rolling stock, concentrated animal feeding operation, or vessel or other floating craft, from which pollutants are or may be discharged. This term does not include agricultural stormwater discharges and return flows from irrigated agriculture."
129. *Id.*
130. U.S. EPA, *National Water Quality Inventory: 2000 Report*, Aug. 2000, *available at* http://water.epa.gov/lawsregs/guidance/cwa/305b/2000report_index.cfm.
131. An Urgent Call to Action: Report of the State-EPA Nutrient Innovations Task Group (Aug. 2009) [hereinafter State-EPA Report].
132. *Id.* at 1-2.
133. *Id.* at 5.
134. *Id.* at 17.
135. Roger C. Funk, International Society of Arboriculture New England Chapter, Comparing Organic and Inorganic Fertilizers, *available at* http://www.newenglandisa.org/FunkHandoutsOrganicInorganicFertilizers.pdf.
136. V.H. Smith et al., *supra* note 123.
137. U.S. EPA, Doc. No. 841-F-05-00, Protecting Water Quality From Agricultural Runoff (2005), *available at* http://water.epa.gov/polwaste/nps/upload/2005_4_29_nps_Ag_Runoff_Fact_Sheet.pdf.
138. V.H. Smith et al., *supra* note 123.
139. Jan Lewandrowski et al., *The Interface Between Agricultural Assistance and the Environment: Chemical Fertilizer Consumption and Area Expansion*, 73 Land Econ. 407 (1997).
140. Eubanks, *A Rotten System, supra* note 4, at 255-56.
141. *Id.*
142. *Id.*

3. Increases in blooms of gelatinous zooplankton (in marine environments);
4. Increased biomass of benthic and epiphytic algae;
5. Changes in macrophyte species composition and biomass;
6. Death of coral reefs and loss of coral reef communities;
7. Decreases in water transparency;
8. Taste, odor, and water treatment problems;
9. Oxygen depletion;
10. Increased incidence of fish kills;
11. Loss of desirable fish species;
12. Decreases in perceived esthetic value of the water body.[143]

These effects range from mere inconvenience or nuisance to being potentially lethal for humans, animals, and plants. When nutrient-laden water finds its way to estuarine areas, it can create "dead zones" in areas that previously had high fish and aquatic organism productivity.[144]

One of the most infamous instances of harmful eutrophication occurs yearly in the northern portion of the Gulf of Mexico, where both the Mississippi and Atchafalaya Rivers spill into the gulf. Approximately 66% of the nitrogen entering the gulf can be traced back to agricultural activities in the Mississippi River basin.[145] As nutrients, primarily N and P but also silicon and other sediments, get washed into the tributaries of the Mississippi River through agricultural and urban runoff, they ultimately flow downriver until they reach the Gulf of Mexico.[146] These nutrients feed blooms, or mass growth, of algae, bacteria, and cyanobacteria.[147] The blooms die off as their nutrient capacity is reached and also as zooplankton consume them, and the dead organic material and zooplankton waste migrate to the seabed.[148] Decomposition of dead plant and animal material on the seabed utilizes oxygen "that cannot be renewed from surface waters because of strong stratification of fresh and salt water."[149] This depletion of oxygen is known as hypoxia. Hypoxia in the northern Gulf of Mexico has created an area that averages 16,500 square kilometers (10,250 square miles) along the coast referred to colloquially as the "dead zone," referring to the inability to capture fish, shrimp, and crabs in bottom-dragging trawls when the oxygen concentration falls below a critical level in water near the seabed.[150] While there is some debate as to whether N or P is the most culpable fertilizer contributing to the dead zone in the Gulf of Mexico, it is widely held that reducing both total N and P entering the Gulf is necessary to prevent the negative environmental impacts associated with eutrophication.[151]

While the Gulf of Mexico dead zone is the largest of the hypoxic zones in the United States, there are many other areas affected by eutrophication. For example, the Chesapeake Bay, previously one of the most productive estuaries in the country, has been severely degraded by high nutrient loading.[152] Serving as the watershed for parts of six states (Delaware, Maryland, New York, Pennsylvania, Virginia, and West Virginia) and all of the District of Columbia, the Bay encompasses 4,480 square miles.[153] The Chesapeake

---

143. Carpenter, *supra* note 127, at 561.
144. Eubanks, *A Rotten System*, *supra* note 4, at 256.
145. *Id.*
146. J.A. Downing et al., *Gulf of Mexico Hypoxia: Land-Sea Interactions*, Council for Agricultural Science and Technology Task Force Report No. 134 (1999).
147. Virginia H. Dale et al., *Hypoxia and the Northern Gulf of Mexico—A Brief Overview*, in Hypoxia in the Northern Gulf of Mexico 1 (Bruce N. Anderson et al. eds., Springer 2010).
148. *Id.*; J.A. Downing et al., *supra* note 146, at 5.
149. J.A. Downing et al., *supra* note 146, at 5.
150. Nancy Rabalais et al., *Gulf of Mexico Hypoxia, a.k.a. "The Dead Zone,"* 33 Ann. Rev. Ecology & Systematics 235, 236 (2002).
151. Walter K. Dodds, *Nutrients and the "Dead Zone": The Link Between Nutrient Ratios and Dissolved Oxygen in the Northern Gulf of Mexico*, 4 Frontiers Ecology & Env't 211 (2006).
152. Rita Cestti et al., Agriculture Non-Point Source Pollution Control: Good Management Practices—The Chesapeake Bay Experience 1 (World Bank 2003).
153. Chesapeake Bay Program: A Watershed Partnership, *Bay Barometer: A Health and Restoration Assessment of the Chesapeake Bay and Watershed in 2008*, Chesapeake Bay Program (Mar. 2009), *available at* http://www.chesapeakebay.net/content/publications/cbp_34915.pdf.

serves as a habitat for approximately 3,600 species of plants, fish, and other animals, and produces approximately 500 million pounds of seafood per year.[154] Population growth in the region led to a decrease in the water quality of the Bay, brought about primarily by the use of agricultural chemicals and the growth in livestock numbers.[155] Agricultural runoff is the single largest source of pollution in the Bay.[156] Pesticides, fertilizers, and sediments from agricultural runoff led to decreased dissolved oxygen levels, increased turbidity, and eventually to substantial harm to sea grasses, fish, and shellfish.[157] Twenty-five percent of the land within the Chesapeake Bay watershed is devoted to agriculture, with approximately 87,000 farms covering 8.5 million acres.[158] Agriculture contributes 42% of nitrogen and 46% of phosphorus to the pollution loads entering the Bay.[159]

Despite a plan by the Chesapeake Bay Program to reduce agriculture pollution, the water quality of the Chesapeake Bay remains poor. Farms utilize conservation practices like "nutrient management plans, cover crops, vegetation buffers, conservation tillage, and animal manure and poultry litter controls" in an effort to reduce nutrient loading into the Bay.[160] Although the poor state of the Bay was first recognized in the 1970s and 1980s, and efforts were undertaken to reverse the effects of eutrophication,[161] there has been relatively little success, as noted in a 2008 report by the Chesapeake Bay Program:

> Despite small successes in certain parts of the ecosystem and specific geographic areas, the overall health of the Chesapeake Bay did not improve in 2008. The Bay continues to have poor water quality, degraded habitats and low populations of many species of fish and shellfish. Based on these three areas, the overall health averaged 38 percent, with 100 percent representing a fully restored ecosystem.[162]

Another aquatic ecosystem that has been subjected to serious environmental harms caused by agricultural fertilizers is the Florida Everglades. The slow moving waters of the Everglades densely covered in tall grasses has been called the "River of Grass" and has experienced eutrophication from fertilizers, particularly phosphorus from large-scale agricultural operations. Historically, the Everglades was an oligotrophic, or nutrient deprived, and oxygen-rich aquatic ecosystem that received nutrient inputs through rainfall.[163] Any phosphorus inputs from Lake Okeechobee would be filtered out by sawgrass prior to reaching what is now the water conservation areas and Everglades National Park.[164]

In 1948, the Everglades Agricultural Area (EAA) was created by the Central and Southern Florida Project for Flood Control and Other Purposes.[165] To establish this area, 470,000 acres were set aside for agricultural purposes, and commercial agriculture (primarily sugar cane and winter vegetables) was able to flourish in the EAA.[166] In the mucky soils typical of the drained Everglades, aerobic bacteria oxidize organic plant remains after harvesting.[167] The remaining inorganic materials "contain high concentrations of nitrogen and phosphorus, and these minerals, in turn, are pumped south with the excess water that is routinely removed from the EAA."[168] This pumping, along with additional phosphorus fertilizer runoff, creates eutrophication. One negative effect of excess phosphorus is the displacement of formerly abundant sawgrass with cattails, a plant that crowds out other native plants and is disfavored for nesting by wading birds and other wildlife.[169] A study of alligator holes in proximity to anthropogenic phosphorus inputs (canals and pumps) demonstrated more cattail domination and decreased biodiversity.[170]

---

154. *Id.*
155. Cestti, *supra* note 152.
156. *Id.*
157. *Id.*
158. Chesapeake Bay Program, *supra* note 153, at 10.
159. *Id.* at 24.
160. *Id.*
161. Cestti, *supra* note 152.
162. Chesapeake Bay Program, *supra* note 153, at 6.
163. Steven M. Davis, *Phosphorus Inputs and Vegetation Sensitivity in the Everglades, in* Everglades: The Ecosystem and Its Restoration 357, 359 (Steven M. Davis & John C. Ogden eds., St. Lucie Press 1994).
164. *Id.*
165. Thomas E. Lodge ed., The Everglades Handbook: Understanding the Ecosystem 222-23 (2d ed. 2005).
166. David McCally, The Everglades: An Environmental History 170-71 (Raymond Arsenault & Gary R. Mormino eds., University Press of Florida 1999).
167. *Id.* at 172.
168. *Id.*
169. Davis, *supra* note 163, at 357.
170. Michelle L. Palmer & Frank J. Mazzotti, *Structure of Everglades Alligator Holes*, 24 Wetlands 115, 120 (2004).

Eutrophication also affects periphyton, one of the primary biomass producers in the Everglades.[171] Periphyton acts as a primary source of food for small consumers, like fish and invertebrates.[172] Disruption of periphyton communities leads to a break down in the natural food web that served as the basis for the oligotrophic Everglades and demonstrates the slow progress made in Everglades restoration.

D. Other Considerations Prompted by Pesticide and Fertilizer Use

Pesticides and fertilizers not only pose risks to the environment, but they both also pose additional risks to human health and well-being. There are indirect considerations that must be taken into account based on the production and use of pesticides and fertilizers, as well as the constant need to consider the potential lethality of their use or misuse. These considerations have prompted some scientists and policymakers to increase their efforts to find viable alternatives to the use of pesticides and fertilizers.

1. Carbon Footprint/Energy Usage

Pesticides and fertilizers can negatively impact the environment well before any postapplication contamination. The intensive use of fossil fuels in the production of fertilizer helps make agriculture one of the largest sources of greenhouse gases (GHG) in the world.[173] In combination, pesticides and fertilizers account for 45-55% of the total energy consumption in worldwide agriculture.[174] Carbon footprint accounting attempts to quantify the amount of GHGs emitted from all sources involved in the production, transport, and use of those fertilizers and pesticides, in a process known as a life-cycle assessment. According to the National Academy of Sciences, agricultural production is estimated to account for approximately 10-12% of all human-made GHG emissions.[175] In the United States, the application of synthetic fertilizers contributes approximately 304 million pounds of GHG emissions each year.[176] The predominant contributors to GHG emissions in agriculture, other than emissions related to land use changes and livestock production, are related to fertilizer application.[177] Other significant contributions of GHG come from the production of fertilizers and pesticides,[178] which are estimated to add approximately 480 million tons of GHG emissions to the atmosphere annually.[179] Others suggest that agricultural GHG emissions may be as much as 33% of global GHG emissions.[180]

In addition to GHG emissions, fertilizer production and use also are large consumers of energy.[181] Natural gas accounts for 93% of the total energy requirement, the remainder being electricity and steam energy.[182] In addition to production, an accurate carbon footprint must include energy consumption for packaging, transporting, and applying these fertilizers.[183] The total energy consumption for these stages of the life cycle of these fertilizers is immense.

Pesticides are made from products derived from fossil fuels, such as petroleum or natural gas.[184] The amount of energy required to produce the active ingredients for herbicides, insecticides, and fungicides is

---

171. Evelyn E. Gaiser et al., *Periphyton Responses to Eutrophication in the Florida Everglades: Cross-System Patterns of Structural and Compositional Change*, 51 LIMNOLOGY & OCEANOGRAPHY 617, 624 (2006).
172. Pamela Brown & Alan L. Wright, Institute of Food and Agricultural Science, *The Role of Periphyton in the Everglades*, 2009, available at http://edis.ifas.ufl.edu/pdffiles/SS/SS52200.pdf.
173. Mahadev G. Bhat et al., *Energy in Synthetic Fertilizers and Pesticides: Revisited*, Department of Agricultural Economics and Rural Sociology, University of Tennessee, Jan. 1994, at 2, available at http://www.osti.gov/bridge/servlets/purl/10120269-p6yhLc/webviewable/10120269.pdf.
174. *Id.*
175. Jennifer Burney et al., *Greenhouse Gas Mitigation by Agricultural Intensification*, PROC. NAT'L ACAD. SCI. (June 29, 2010).
176. Meredith Niles, *Sustainable Soils: Reducing, Mitigating, and Adapting to Climate Change With Organic Agriculture*, 9 SUSTAINABLE DEV. L. & POL'Y 19-20 (2008).
177. Burney et al., *supra* note 175.
178. *Id.*
179. Niles, *supra* note 176, at 20.
180. *Id.* at 19 (citing JESSICA BELLARBY ET AL., GREENPEACE INTERNATIONAL, COOL FARMING: CLIMATE IMPACTS OF AGRICULTURE AND MITIGATION POTENTIAL 5 (2008), available at http://www.greenpeace.org/international/Global/international/planet-2/report/2008/1/cool-farming-full-report.pdf).
181. Bhat et al., *supra* note 173, at 26.
182. *Id.* at 27. For phosphorus, the weighted average energy requirements were 4.52 GJ/mt (Gigajoule/metric ton) and potash was 4.80 GJ/mt. Electricity and steam are the main sources of energy in phosphorus production.
183. *Id.* at 29. For nitrogen, phosphorus, and potash, 8.60, 9.80, and 7.30 GJ/mt, respectively, were used in packaging, transportation, and application.
184. *Id.* at 30.

also immense.[185] These active ingredients must be formulated into usable products prior to use, and the most common forms are emulsifiable oil, wettable powder, and granules. Production of each of these formulations requires different amounts of energy.[186]

2. Groundwater Contamination

Groundwater contamination by pesticides and fertilizers is a major problem because one-half of the human population obtains its water from wells and because residues from certain pesticides can remain in groundwater for a long time.[187] Factors contributing to potential pesticide groundwater leaching include the aquifer recharge rate, soil type, depth of aquifer, nitrate contamination, and soil pH.[188] Manure spreading and other fertilizing efforts have the potential to leach nitrates to groundwater, an impact that can be exacerbated by poor irrigation practices. One recent study of pesticide contamination of groundwater concludes that "soil and water conservation BMPs [best management practices] will not control pesticide leaching," but integrated pest management, crop rotation and improved application practices could help.[189] Some of the most commonly found pesticides in groundwater are arsenic, aldicarb, alachlor, and atrazine.[190] Additionally, soil fumigants like ethylene dibromide, dichloropropene, and dibromo-chloropropane (which are all toxic chemicals) are often detected in groundwater.[191]

Groundwater contamination can also lead to a condition known as methemoglobinemia, or blue baby syndrome.[192] Infants who are fed formula prepared with nitrate-contaminated drinking water can develop a bluish tint to their skin, become irritable and lethargic, and die if not treated promptly.[193] Nitrate-contaminated water, when ingested, will promote the production of nitrites that produce methomoglobia, a compound incapable of binding molecular oxygen and transporting it through the body.[194] Currently, infants at risk are those who are fed formula mixed with nitrate-contaminated well water.[195] Each year in the United States, there are anywhere between 500 and 1,200 violations of Safe Drinking Water Act maximum contaminant levels of nitrates in public drinking water supplies, and there appears to be an increase in the number of violations over the past decade.[196] Moreover, many Americans obtain their drinking water from private wells. Nitrate was detected in approximately 72% of the 2,100 private wells sampled by the U.S. Geological Survey between 1991 and 2004.[197] Methemoglobinemia can also affect livestock, depending on the age of the animals and concentrations of nitrate in contaminated water.

3. Sustainable Pest Management and Fertilization

Over the past few decades, scientists, policymakers, farmer advocacy organizations, environmentalists, and others have called for sustainable agriculture—new approaches to replace the modern agriculture that relies heavily on industrial pesticide and fertilizer inputs. Sustainable agriculture focuses on three main objectives: environmental health, economic profitability, and social and economic equity.[198] The Union of Concerned Scientists explains that

---

185. *Id.* at 32. The energy required for herbicides, insecticides, and fungicides is 214.93 GJ/mt, 245.06 GJ/mt, and 356.39 GJ/mt, respectively.
186. *Id.* The energy requirement for each is 20, 30, and 10 GJ/mt, respectively.
187. David Pimentel, *Environmental and Economic Costs of the Application of Pesticides Primarily in the United States*, in Integrated Pest Management: Innovation-Development Process 89, 100 (R. Peshin & A.K. Dhawan eds., Springer Science 2009).
188. W.F. Ritter, *Pesticide Contamination of Ground Water in the United States—A Review*, 25 J. Envtl. Sci. & Health 1, 25 (1990).
189. *Id.* at 25-26.
190. *Id.*
191. *Id.* at 4.
192. Itzel Galaviz-Villa et al., *Agricultural Contamination of Subterranean Water With Nitrates and Nitrites: An Environmental and Public Health Problem*, 2 J. Agric. Sci. 17, 18 (2010).
193. Lynda Knobeloch et al., *Blue Babies and Nitrate-Contaminated Well Water*, 108 Envtl. Health Persp. 675, 675 (2000).
194. Frank R. Greer & Michael Shannon, *Infant Methemoglobinemia: The Role of Dietary Nitrate in Food and Water*, 116 Pediatrics 784, 784 (2005).
195. *Id.* at 785.
196. State-EPA Report, *supra* note 131, at 3.
197. *Id.*
198. U.C. Davis College of Agricultural and Environmental Sciences, Sustainable Agriculture Research and Education Program, *What Is Sustainable Agriculture?*, http://www.sarep.ucdavis.edu/concept.htm.

sustainable agriculture views a farm as a kind of *ecosystem*—an "agroecosystem"—made up of elements like soil, plants, insects, and animals. These elements can be enriched and adjusted to solve problems and maximize yields. This integrated approach is both practical and scientific: it relies on modern knowledge about the interactions within natural systems, as well as cutting-edge technologies, to achieve its results. It is a powerful approach that can produce high yields and profits for farmers while protecting human health, animal health and the environment.[199]

The underlying principle of sustainability is the desire to meet current needs of society while still preserving sufficient resources for future generations to meet their needs.[200]

The Union of Concerned Scientists has identified five techniques of sustainable agriculture: crop rotation, cover crops, soil enrichment, natural pest predators, and biointensive integrated pest management.[201] Crop rotation is the practice of growing different crops in the same field over a period of time,[202] which discourages the buildup of pests that can occur when one crop is continuously grown in the same location. Sustainable farming uses cover cropping (planting specific crops on fields between plantings of the primary crop so that fields do not remain bare), which reduces soil erosion and weed growth, and enhances soil nutrients.[203] Rich healthy soil is critical to sustainable farming. Rather than allowing soils to be depleted and then relying on synthetic fertilizers, sustainable agriculture seeks to maintain and enhance soil richness by plowing under cover crops and using natural fertilizers such as composted animal and waste.[204] Soil richness is also maintained by abstaining from heavy use of pesticides, which kill beneficial soil-inhabiting organisms.[205] In lieu of synthetic pesticides, sustainable farming practices are geared toward maintaining a healthy ecosystem, which allows natural predators and parasites of crop pests to thrive and keep pest populations in check.[206] If natural pest control is not adequate, integrated pest management practices, which employ a range of biological and cultural control practices, as well as narrowly targeted synthetic chemical pest control, are used.[207] To achieve an agricultural system that universally uses these alternative methods, it is necessary to develop regulatory and incentive-based tools that require or promote these practices.

Conclusion

The use and overuse of pesticides and fertilizers can have serious and sometimes irreversible dangers. While pesticides protect against insects and other vectors that may cause significant crop damage, spread contagious disease, or otherwise disrupt human comfort, much of their danger revolves around the inherent toxicity of these chemicals and poor management practices in their use and application. Fertilizers help ensure that crops will be healthy and plentiful, but they too are subject to poor management and application processes. If efforts to restore the ecological harms by these substances are to succeed, the continued use of pesticides and fertilizers needs to be better managed and new measures to prevent runoff and leaching enacted. As important as better management of chemical substances, however, is the adoption of agricultural practices that utilize a range of biological and cultural pest management and fertilizer approaches, thereby minimizing reliance on synthetic chemical inputs. The challenge we face is how to revise current policies to ensuring affordable and healthful food while moving toward an agricultural system that is environmentally, economically, and socially sustainable.

---

199. Union of Concerned Scientists, *Sustainable Agriculture—A New Vision*, http://www.ucsusa.org/food_and_agriculture/solutions/big_picture_solutions/sustainable-agriculture-a.html (last visited Oct. 25, 2012). For further descriptions of sustainable agriculture, see John H. Davidson, *Agriculture*, in STUMBLING TOWARD SUSTAINABILITY 347, 360-62 (John C. Dernbach ed., 2002).
200. Davidson, *supra* note 199, at 360; *see also* Union of Concerned Scientists, *Sustainable Agriculture Techniques*, http://www.ucsusa.org/food_and_agriculture/science_and_impacts/science/sustainable-agriculture.html (last visited Oct. 25, 2012) [hereinafter *Sustainable Techniques*].
201. *Sustainable Techniques*, *supra* note 200.
202. *Id.*; *see also* John Boardman et al., *Socio-Economic Factors in Soil Erosion and Conservation*, 6 ENVTL. SCI. & POL'Y 1-3 (2003) (discussing runoff and soil erosion associated with industrial agriculture).
203. *Sustainable Techniques*, *supra* note 200.
204. *Id.*
205. *Id.*
206. *Id.*
207. *Id.*

Chapter 4
Agricultural Irrigation
John H. Davidson

Irrigation is a key element in any analysis of domestic and global agriculture. It has been a central part of agriculture for millennia, and is known to have been practiced by indigenous peoples in the American Southwest as early as 100 B.C.[1] Ruins of early irrigation canals are still visible in the vicinity of the Gila and Salt Rivers in what is now Arizona; the canals served to irrigate lands in the Pima and Papagos pueblos.[2] Similarly, in what is now New Mexico, the Indians in the Rio Grande Valley irrigated their farms long before the Spaniards arrived at the end of the 15th century.[3]

Despite these and numerous other efforts throughout the Southwest, large-scale irrigation was not successful until the Mormons entered the Salt Lake Valley in Utah in 1847.[4] By 1850, 926 Utah farms involving a total of 16,133 acres were under irrigation.[5] By 1860 there were 77,219 irrigated acres of farmland growing crops that were vital to the survival of the community.[6] The Mormon achievement is the first large agricultural economy in the West dependent entirely upon artificial irrigation.[7] The central engineering method was the efficient and sustainable diversion and conveyance of canyon streams to distant valley farmlands.[8] The Mormon success provided the model from which the first generations of irrigation in the United States grew.[9]

Irrigation is today a feature of agriculture in all 50 American states.[10] Over 55 million acres are irrigated, and nearly 50 million of those acres are on farms of 200 acres or more. Just under 30 million of those acres are on farms with more than 1,000 acres.[11] In 1997, the West held 78% of the country's irrigated land, but recent trends forecast faster growth in the East.[12] From 1987 to 1997, irrigated land increased by 14% in the West, and 38% in the East.[13] Substantial increases in irrigation must be anticipated as land prices escalate, the total amount of land available for agriculture diminishes, demand for all crops increases, costs of irrigation infrastructure are lowered, and financial capital concentrates.[14]

When done well, irrigation has only minor environmental impacts.[15] It can conserve water, allow full utilization of prime soils, increase productivity of existing tilled acres, enhance the production of essential human foods, and provide resiliency during periods of drought.[16] Irrigation has the potential to help agriculture shift toward sustainability—using resources to serve both present and future generations. That this

---

1. DESIGN AND OPERATION OF FARM IRRIGATION SYSTEMS 8 (G.J. Hoffman et al. eds., 2d ed., 2007) [hereinafter DESIGN].
2. NATIONAL RESEARCH COUNCIL, COMM. ON THE FUTURE OF IRRIGATION, A NEW ERA FOR IRRIGATION 9 (Nat'l Academy Press, 1996) and J.L. SAX ET AL., LEGAL CONTROL OF WATER RESOURCES 326-27 (4th ed. 2006).
3. *See generally* S. CRAWFORD, MAYORDOMO: CHRONICLE OF AN ACEQUIA IN NORTHERN NEW MEXICO (1988).
4. SAX ET AL., *supra* note 2, at 327.
5. United States Seventh Census, 1850, 1006 (1853).
6. United States Eighth Census, 1860, Agriculture, 180 (1864).
7. *Id.*
8. *Id.*
9. *Id.*
10. CHARLES A. JOB, GROUNDWATER ECONOMICS 203 (2010).
11. U.S. Dep't. of Agric., National Agricultural Statistics Service, *2008 Farm and Ranch Irrigation Survey*, *in* THE CENSUS OF AGRICULTURE (2008).
12. JOB, GROUNDWATER ECONOMICS, *supra* note 10, at 203.
13. *Id.*
14. LAWRENCE J. MACDONNELL, FROM RECLAMATION TO SUSTAINABILITY: WATER, AGRICULTURE, AND THE ENVIRONMENT IN THE AMERICAN WEST 1-9 (1999).
15. *Id.*
16. *Id.*

chapter is discussing environmental problems of irrigation, however, simply reflects what is well known—irrigation is not always, or even regularly, done well. Or, to state things in a more positive way: we are still learning how to irrigate our crops in an environmentally beneficial manner.

As often occurs with natural resources management, the potential for harm to the environment became apparent as the result of a specific disaster, and for irrigation that case was provided by the Kesterton Wildlife Refuge in California's San Joaquin Valley. There, from 1971 to 1978, man-made ponds were supplied entirely by fresh water. As the result of operational changes, the supply by 1982 was entirely irrigation drainage water.[17] In 1982, it was observed that the irrigation drainage was increasing selenium concentrations in the refuge's ponds, resulting in reproductive failure and death in some species of the aquatic life, including waterfowl.[18] The pollution at Kesterton was the result of both natural and human factors.[19] Selenium is found naturally in soils like many salts. Intensified irrigation with the installation of subsurface drains without consideration of the effects of the content and disposal of drainage water led to serious water quality impacts.[20] As summarized in a subsequent report of the National Research Council: "The underlying issue is clear: irrigation, like many other uses of water, degrades water quality for water users. The contaminant of concern and the severity of impacts may vary, but the phenomenon of irrigation-induced water quality contamination can no longer be ignored."[21]

As a caveat, it should be said that any thoughtful consideration of surface water irrigation as practiced in the United States requires recognition that it is not simply the application of water to plants; it is the system that provides the basis for an economy, a way of life, and which gave rise to the settlement of the American West.[22] Irrigation has its own distinct social, economic, political, and agronomic history, and is often the result of generations of extraordinary human ingenuity and perseverance. Many of the political and social institutions now familiar in farm country—irrigation districts in particular—are the result of efforts to develop irrigation agriculture. Much of this history and achievement is lost to view in our urbanized society, but it is unwise to discuss irrigation without taking this special history into account. Many of the most visible controversies in irrigation country may have been, in part, a result of failure to recognize this special history and culture.[23]

A. Irrigation in Practice

To understand how irrigation impacts the environment today, it is necessary, first, to appreciate how irrigation practices have evolved over the past century. It is important also to understand the purposes of field drainage and the engineering techniques used.

1. Background

Early irrigation works in the United States were rudimentary. Small diversion structures were built up at the side of flowing streams, allowing water to spill into ditches through which water flowed by gravity to fields at lower elevations.[24] Incrementally, such irrigation works evolved into interrelated projects requiring elaborate ditch systems and construction of enhanced storage at the point of diversion.[25] The early surface system had complete access to the streams—there were few conflicting claimants for the water—and, as a result, tended to be wasteful, losing vast amounts to seepage and evaporation, problems that persist to this day.[26] From this early experience, it became apparent that successful surface water irrigation requires large capital expenditures as well as institutions by which to organize irrigating landowners.

---

17. National Research Council, Comm. on Irrigation-Induced Water Quality Problems, *Irrigation-Induced Water Quality Problems: What Can Be Learned From the San Joaquin Valley Experience* 1 (1989).
18. *Id.*
19. *Id.*
20. National Research Council, *supra* note 17.
21. *Id.* at 2.
22. National Research Council, Comm. on the Future of Irrigation, *A New Era for Irrigation* 2 (1996).
23. *See* A. Dan Tarlock & Holly D. Doremus, *Fish, Farms, and the Clash of Cultures in the Klamath Basin*, 30 Ecol. L.Q. 279 (2003).
24. MacDonnell, *supra* note 14, at 1-9.
25. *Id.*
26. *Id.*

Three significant landmarks now appear in the history of irrigation development in the United States. The first of these is judicial sanction of California's Wright Act in 1896.[27] Physically and economically, it is usually unfeasible for an independent farm enterprise to construct large-scale surface irrigation without the cooperation of surrounding landowners. In recognition of this, all state legislatures have enacted enabling laws for special districts that provide generous police and financing powers to supplement the land development enterprise. Today, rural America is governed by thousands of local government units that function apart from the standard county, municipal, and school district entities.[28] In regions where surface water irrigation prevails, these districts are quite often the most significant units of local government.[29]

The consistent theme in the slow development of western irrigation is the lack of capital. Some early mutual companies met with success, especially in Utah, but as a general matter few privately owned water companies were able to amass capital sufficient to construct irrigation on a meaningful scale.[30] The Wright Act changed that and became the model for rural special districts across the land.[31] The Act empowered irrigation districts to levy property assessments, issue bonds, and be controlled by local landowners.[32] The key, however, was that the Act gave districts the power to include unwilling landowners within the district boundaries and to exercise the power of eminent domain.[33] Landowners resented being forced into districts against their will, and it took a strong opinion from the U.S. Supreme Court to settle the matter.[34] Governmental in form, and exercising governmental powers, special districts serve predominantly private purposes. This fact was recognized by Justice Potter Stewart in 1981, when he wrote that "though the state legislature has allowed water districts to become nominal public entities in order to obtain inexpensive financing, the districts remain essential business enterprises, created and chiefly benefiting a specific group of landowners."[35]

The second landmark is passage in 1902 of the federal reclamation law, usually referred to as the Newlands Act, and also known as the Reclamation Act.[36] Beginning in the 1890s, western farmers—now more stoutly organized in their special districts—encouraged the federal government to become involved in the delivery of water to farmers. After years of failed experiments with half-measures, Congress enacted the Newlands Act, under which the federal government itself would construct irrigation projects and deliver water to western farmers.[37] The extent of the impact of this program is summarized accurately in a recent volume on water law:

> Under the guidance of the Bureau of Reclamation, . . . the federal reclamation program has radically transformed the West over the last century. The Bureau has built over 600 dams, 16,000 miles of canals and aqueducts, 280 miles of tunnels, 37,000 miles of laterals, 50 hydroelectric generators, and 140 pumping stations. Through local water districts, the Bureau currently supplies water to roughly 10 million acres of cropland—about half of all the land irrigated by surface water in the West—as well as to 30 million domestic users. The federal reclamation program, however, has generated considerable criticism. Reclamation water is heavily subsidized by federal taxpayers and, in many parts of the West, goes to large farmers who often have enjoyed other federal farm subsidies. Few Bureau projects have ever paid for themselves, even if you ignore the interest that the federal government absorbs on every project. Numerous Bureau projects, moreover, have caused sizeable environmental damage.[38]

The third major development was the application of well-drilling technology to groundwater pumping. Prior to the 1940s, the technology was crude and incapable of developing flows of water sufficient to

---

27. *Public Water Supply Organizations*, in 2 WATER AND WATER RIGHTS §27 (R. Beck & A. Kelley eds., 2011 edition).
28. *See, e.g.*, S.N. Bretsen & P.J. Hill, *Irrigation Institutions in the American West*, 25 U.C.L.A. J. ENVT'L L. & POL'Y 283 (2006) and Harriet M. Hageman, *Irrigation Districts as Public Corporations*, 32 WYO. LAW. 16 (2009).
29. *Id.*
30. John H. Davidson, *South Dakota's Special Water Districts—An Introduction*, 36 S.D. L. REV. 499, 539-49 (1991).
31. *Id.*
32. *Id.*
33. *Id.*
34. Fallbrook Irrigation Dist. v. Bradley, 164 U.S. 112 (1896). The Court held that inclusion and taxation of nonirrigable lands with an adequate water supply and within incorporated cities did not violate the Due Process Clause.
35. Ball v. James, 451 U.S. 355 (1981).
36. MACDONNELL, *supra* note 14, at 143-44.
37. 32 Stat. 388.
38. SAX ET AL., *supra* note 2, at 687-88. A.K. Kelley, *Federal Reclamation Law*, in 4 WATER AND WATER RIGHTS §41.02 (Robert Beck ed., Repl. Vol. 2011).

support widespread use in farm fields.[39] In the 1940s, well drillers began using drill bits similar to those employed in oil fields, resulting in wider holes.[40] During the same period, irrigators acquired high-speed turbine pumps, which could push the water up the well in large quantities.[41] With these tools in hand, development of groundwater sources for irrigation expanded rapidly, first in the West and moving steadily eastward through the plains to the East Coast.[42] The ability to pump water at the site of use, combined with the development of efficient center-pivot sprinkler systems, allowed for irrigation without the need for complex surface water systems of ditches, laterals, and canals. In 2003, there were 401,193 irrigation wells that were in use in the United States.[43]

2. Field Drainage: Purposes

It has been said that every irrigation project eventually becomes a drainage project. Ditches for the collection of waste water and its redistribution to other users, and for the drainage of what would otherwise become seeped and boggy lands, are commonly constructed as part of modern projects, or as additions to older projects, by drainage districts or by the irrigation district that distribute the water.[44]

As there is a recognized hydrologic cycle, there is a parallel irrigation cycle. Irrigation water passes through the soil-crop system, and drainage carries off salts, nutrients, pesticides, and trace elements.[45] Much of the irrigation water that enters the plants is transpired back into the atmosphere.[46] A smaller portion is incorporated into plant tissues, and the remainder leaves the soil-crop system by leaching, runoff, seepage, or subsurface drainage.[47] Because a significant portion of all soils under irrigation are subject to "waterlogging" of soils, artificial drainage is necessary.[48] Irrigation is short-lived in the absence of adequate drainage.[49] California alone has millions of miles of agricultural drains.[50]

Because successful land drainage is an essential component of irrigation, it is described here in some detail. It should be noted that the drainage techniques described here apply equally to nonirrigated field agriculture as well.

Agricultural drainage has been defined as the art and science of removing water from land to enhance agricultural operations.[51] One engineering treatise states: "The main objective of agricultural land drainage is to remove excess water in order to improve the profitability of farming the land."[52] Another definition states: "The objective of drainage in agriculture is to create between the soil surface and the water table a partially saturated zone of optimum quality and extent for exploitation by plants and for the management of the soil and crops by the farmer."[53]

There are several incentives for constructing drainage structures on and beneath agricultural land. First, some soils, either due to their structure or their topography, are waterlogged during a portion of the growing season. In this condition, plant roots do not receive adequate oxygen, soil is compacted, and crop growth is hindered; drainage can correct this problem.[54]

Second, constructed drainage may lengthen the crop growing season on a particular farm.[55] When fields are slow to lose the moisture that builds up after the spring thaw or heavy rains, the farmer must delay

---

39. Job, Groundwater Economics, *supra* note 10, at 204.
40. *Id.*
41. *Id.* at 204-05.
42. *Id.*
43. *Id.* at 204.
44. Frank J. Trelease & George A. Gould, Water Law Cases and Materials 232-33 (4th ed. 1986).
45. Nat'l Research Council, *supra* note 17, at 3.
46. *Id.*
47. Nat'l Research Council, Committee on Long-Range Soil and Water Conservation, *Soil and Water Quality: An Agenda for Agriculture* 57 (1993).
48. Nat'l Research Council, *supra* note 17, at 3-4.
49. *Id.* at 364. Western Water Policy Review Commission, *Water in the West: Challenge for the Next Century* 2-31-32 (June 1998).
50. L.N. Smith & L.J. Harlow, *Regulation of Nonpoint Source Agricultural Discharge in California*, 26 A.B.A., Nat'l Res. & Env't 28 (2011).
51. L.K. Smedema & D.W. Rycroft, Land Drainage: Planning and Design of Agricultural Drainage Systems 39 (1984).
52. Drainage for Agriculture 311 (J. Van Schilfgaarde ed., 1974). *See also* G.O. Schwab et al., Soil and Water Conservation Engineering 1 (3d ed. 1981).
53. Drainage for Agriculture, *supra* note 52, at 311.
54. Field Drainage: Principles and Practice 21 (D. Castle et al. eds., 1985).
55. Drainage for Agriculture, *supra* note 52, at 7.

planting, weed control, harvesting, and other field work.[56] If the land is seeded to pasture, there are delays in turning livestock in; drainage can correct this problem.[57]

Third, drainage allows farmers to bring into production land that nature has otherwise claimed as swamp, wetland, slough, or marsh, in the case of humid lands (and arid valley floors in the case of irrigation).[58] Despite the resulting loss to water conservation and wildlife habitat, the opportunity to "make land" is an inviting prospect for the landowner, particularly when agricultural land values are high.[59]

Fourth, drainage allows farmers to improve the productivity of land already in production. For example, land that is naturally wet, and has supported only grass, may, after drainage, be brought into row-crop production.[60]

Fifth, agricultural drainage pipe systems are an essential engineering feature of most organized irrigation projects. Land under irrigation is exposed to the risk of waterlogging with resultant leaching of chemical salts into the plant root zone.[61] By placing drainage pipes beneath the root zone, the risk of salinity is reduced. Salinity control may also be an objective of drainage on farms not served by irrigation.[62]

The farmer who drains land is influencing the hydrologic cycle by accelerating the flow of excess water from the land before it damages the soil structure and affects the crop.[63] This practice is well accepted and is closely associated with good land husbandry.[64] The economic incentive for drainage by the individual landowner is compelling, for it presents an opportunity, at relatively low cost, to increase the production of a capital asset that is already owned.[65] Despite the substantial acreage of land already drained in this country, the combination of intense economic pressure on farmers to improve production in order to achieve profitability, as well as the ever-improving efficiency and diminishing cost of drainage technology, creates an economic environment in which a rapid expansion of drainage is under way.[66] Technical engineering sources hold that excess water continues to be a "major problem" on 25% of U.S. cropland.[67]

Moreover, as more of America's productive farmland accumulates in the hands of larger operating entities, demands for profit accelerate the expansion of agricultural land drainage.[68] The potential for problems is worldwide, as so many countries are feeling pressure to bring new land into production in order to achieve domestic food requirements and compete for export markets.[69]

3. Field Drainage: Techniques

Taken in the aggregate, the engineering features of drainage are far from simple, and represent extensive and precise interference with natural hydrologic systems. Their purpose is to move water steadily from where it is not wanted—farm fields—to watercourses that will carry it away. A good portion of this water would, under natural conditions, remain in the soil and never reach a watercourse. This is particularly true in the Great Plains where, although annual rainfall is not great, much of the topography is such that soils retain the greatest part of precipitation rather than lose it to streams through runoff.[70] Hence, agricultural field drainage not only speeds the flow of water to rivers and streams, it also augments flows in those rivers and streams well beyond natural levels. Typical drainage is not unlike a municipal sewer collector system. A large number of small pipes carry flows to larger conduits where the waters are gathered for disposal. The waters are collected on the surface and the subsurface of the land. Surface water passes rather quickly over

---

56. *Id.*
57. *Id.*
58. Field Drainage: Principles and Practice, *supra* note 54, at 20; Schwab et al., *supra* note 52, at 1.
59. *Id.*
60. Field Drainage: Principles and Practice, *supra* note 54, at 20.
61. Drainage for Agriculture, *supra* note 52.
62. *Id.*
63. Field Drainage: Principles and Practice, *supra* note 54, at 69.
64. *See, e.g.,* Gross v. Connecticut Mutual Life Insurance Company, 361 N.W.2d 259 (S.D. 1985).
65. *Id.*
66. Jon R. Luoma, *Twilight in Pothole Country*, Audubon, Sept. 1985, at 68, 75.
67. Schwab et al., *supra* note 52, at 5.
68. Drainage for Agriculture, *supra* note 52, at xv.
69. *Id.*
70. *Id.*

the soil without infiltrating it. As it does so, it picks up suspended and soluble material. Subsurface water moves slowly through the soil, and in so doing, leaches chemicals from it.[71]

Typical agricultural field drainage is accomplished by a combination of field shaping and leveling, as well as surface and subsurface pipes and drains. Each system will reflect the topography, climate, soil type, cropping pattern, and economics of a particular farm.[72] Surface ditches and pipe drains, in combination with open channels, are the most frequent methods used.[73] Because drainage rarely honors surveyor's lines, it is customary for neighboring landowners to cooperate in developing drainage, and large special municipal drainage districts are commonplace.[74]

Subsurface drains are placed in the ground directly below the root zone. They may be spaced as closely as every four feet but typically are found in spacings of more than 20 to 30 feet.[75] The pipe drains may be concrete, burned clay tile, corrugated plastic tubing, or other perforated conduit.[76] Corrugated steel is used where there is a heavy soil load or unstable soils, and to provide a stable outlet into open ditches.[77]

Installation is by a variety of trenching methods. Recent advances in the technology of installation have made field drainage more available than ever before.[78] Rolled plastic pipe can now be buried on the move by trenching "pipe trains," which simultaneously dig the trench, install the pipe at the correct angle, and cover it.[79] Lateral drains are intended to receive water directly from the soil and pass it on to main lines, which gather the flows. Main lines can be either surface or subsurface conduits.[80]

The patterns for the layout of subsurface drains are usually described in four categories: (1) the herringbone, used in areas that have a concave surface or a narrow draw with the land sloping to it from either direction; (2) the gridiron, which is similar to the herringbone except the laterals enter only from one side; (3) the cutoff or interceptor, which is normally placed over the upper edge of a wet area; and (4) the random design, used in smaller isolated areas.[81] The outlets for subsurface pipe drains are usually operated on gravity principles although pump drains are frequently used.[82]

The benefits of subsurface drains are that they do not interfere with farming, make more efficient those soils that do not drain naturally, aerate the soil, remove salts and other toxics from the root zone, and reduce surface runoff.[83]

Surface drains (open ditches) are an essential complement to subsurface drainage.[84] Open channels provide outlets for tile and surface drains and also carry off surface waters.[85] They are generally earth-lined and drain much larger areas.[86] The open drain is excellent for rapid removal of large quantities of water, and also enjoys great cost advantages over covered drains.[87]

A common use of the open channel is to connect wetlands to drains so that the wetland areas can be farmed. So used, the channel will rarely be more than one or two yards deep, and may be so shaped that it can be farmed or, at least, allow for the easy passage of farm vehicles. From an engineering standpoint, a properly designed open ditch should provide a velocity of flow that does not allow scouring or sedimentation; this requires sufficient capacity, stable side slopes, and correct hydraulic grade.[88] Channels, of course, may require substantial structures of concrete or other materials, the purpose of which is channel stabilization.[89]

---

71. DRAINAGE FOR AGRICULTURE, *supra* note 52, at 490.
72. *Id.* at 93.
73. SCHWAB ET AL., *supra* note 52, at 8.
74. *Supra* note 28.
75. SCHWAB ET AL., *supra* note 52, at 8.
76. *Id.*
77. *Id.* at 318-19.
78. *Id.* at 348-50.
79. *Id.*
80. *Id.* at 319.
81. *Id.* at 321-22. *See also* SMEDEMA & RYCROFT, LAND DRAINAGE, *supra* note 51, at 50-95.
82. SCHWAB ET AL., *supra* note 52, at 322-23.
83. *Id.* at 314.
84. DRAINAGE FOR AGRICULTURE, *supra* note 52, at 101.
85. SCHWAB ET AL., *supra* note 52, at 290.
86. *Id.*
87. DRAINAGE FOR AGRICULTURE, *supra* note 52, at 99.
88. *Id.*
89. *Id.*

Earth embankments and farm ponds are sometimes used as part of surface land drainage schemes. They normally serve to hold back water so that it may be used for other purposes such as stock watering or irrigation.[90] Additionally, they serve the advantageous role of keeping waters from main channels where it can contribute to flooding.[91]

Random field drains are best suited for draining scattered depressions or potholes. The location and direction of these drains will be dictated largely by the topography. Side slopes are generally as flat as possible so that tillage operations can be performed through the channels. Erosion in the channel is generally not a problem because the grades are flat. Spoil from the channel should be spread or moved into the depression to reduce the depth of the drain.

Where fields are flat with slopes less than 2%, a system of parallel drains is often used. Where dead furrows are left by plowing lands at the same location for several years, the drainage system is known as "bedding."

* * *

Drainage has, of course, been practiced for more than 2,000 years. Discussions of it can be found in Roman history and that of most other significant civilizations.[92] In England, a royal charter was granted in 1252 to a board of commissioners that sat as a court to maintain drainage, carry out drainage improvements, and resolve land drainage disputes.[93] But it is only since the second half of the 19th century—with the development of drainage engineering theory and manufactured drain tile, combined with extravagant public subsidies—that it has become commonplace. Moreover, with the importance of irrigation in the American West and the growth of irrigation in the eastern United States as well, drainage is of increasing significance everywhere. Observers of American farming are keenly aware that drainage engineering programs are an integral part of most successful farms, and that successful farming regions will inevitably boast of many organized irrigation and drainage districts. The 1969 agricultural census indicated that organized municipal drainage projects provided drainage over 90 million acres of land.[94] Landowners invested in drainage systems on over 29 million acres of farmland from 1975 to 1977.[95] For those same years, owners of 24 million acres of farmland reported investing about $2.25 billion for drainage.[96] About 15% of all owners making drainage investments in the 1975-1977 period participated in some type of drainage district.[97] It is estimated that 54 million acres of land presently in agricultural production in the United States could be made more productive by the application of drainage practices. About two-thirds of the land yet to be drained is in the South.[98]

A continuing objective of drainage systems is to drain wetlands in order to bring new land into production. A recent report points out that even before 1920 nearly one-half-million acres of North Dakota wetlands had been drained, and they continue to be drained at the rate of 20,000 acres per year.[99] Minnesota, one of the states most blessed with surface water, has lost 90% of its natural wetlands to agricultural drainage.[100]

B. The Potential Environmental Effects of Agricultural Irrigation

Agricultural irrigation can have negative impacts on the environment, principally caused by runoff from farmlands and overpumping of groundwater. Runoff water carries sediments and chemicals into surrounding streams. Excessive pumping of groundwater can deplete the water table, lessen surface resources, and damage the ecosystems dependent on them.

---

90. DRAINAGE FOR AGRICULTURE, *supra* note 52, at 101.
91. *Id.*
92. *See generally* H.H. Wooten & L.A. Jones, *The History of Our Drainage Enterprises*, in U.S.D.A. WATER: THE YEARBOOK OF AGRICULTURE 478 (1955).
93. FIELD DRAINAGE: PRINCIPLES AND PRACTICE, *supra* note 54, at 222.
94. Douglas Lewis & Thomas McDonald, *Improving U.S. Farmland*, 482 U.S.D.A. Agricultural Information Bulletin (Nov. 1984).
95. *Id.*
96. *Id.*
97. *Id.*
98. *Id.*
99. *Id.*
100. *Id.*

1. Runoff From Farm Fields

Land cultivation is a major polluting activity that, among other things, increases the amount of soil erosion.[101] Water from farm fields is also the primary carrier of pollutants from farmland. Drainage water or irrigation return flows carry sediment and chemicals, changing the quality of the drainage water.[102]

All natural waters and soils contain chemical salts, which drainage water will collect and concentrate. Although the collection of salts in drainage water is most often associated with irrigation return flows where salts have the best chance to concentrate in evaporating desert water, this phenomenon is also associated with drainage of humid lands.[103] Thus, irrigation drainage in particular exacts an environmental price by degrading water quality along the disposal routes, and in closed basins it "can render the terminus biologically uninhabitable."[104] Where there are seasonal low stream flows, as in the arid West, pollution problems result because depleted stream flows are comprised mostly of irrigation runoff.[105] And it is reported that irrigation return flows are the most common source of pollution in national wildlife refuges.[106]

Sediment is the major nonpoint (unregulated) pollutant of American waters.[107] One effect of most drainage systems is to accelerate the flow of water during spring thaw or immediately following rainfall. Waters that would naturally be retained in fields or flow quite slowly toward watercourses are gathered rapidly and cast into watercourses. As flows accumulate in open channels, the soil is scoured, and sediment loads increased.[108] Soil particles in water not only indicate loss of soils by erosion, but they carry attached to them most chemicals found in the soils. Sediment directly damages waters, but these waters also carry nutrients and pesticides from fields. In the long term, it is the presence of fertilizers, agricultural chemicals, and trace materials attached to the sediment that make agricultural drainage a major source of water pollution.[109] Ironically, that which is a benefit in the field can cause problems when transmitted to surface and groundwater.

2. Groundwater Pumping

Groundwater is a principal source not only for irrigation but also for municipal and industrial use. The overall supply is vast—the largest source of freshwater—but its availability varies from place to place, depending upon the depth of the supply, geologic conditions, and the volume of water needed at a site.[110] Its availability at the place of use and its widespread availability mean that it is a resource that is vital to the health and the economy of the country. Despite that, and considering its vital importance, the impacts of groundwater are not widely understood.

The principal effect of irrigation from groundwater sources is connected to the potential for overuse. For obvious reasons, irrigation from groundwater is most likely to occur in arid regions where the opportunity for recharge from natural precipitation is small and where high demand is present.[111] As a result, withdrawals may not be balanced by recharge, and the groundwater levels can be lowered until they reach unsustainable depths.[112] In the United States, around 10 million acres of land are irrigated from groundwater sources in which pumping exceeds recharge and water tables are declining.[113] Often associated with

---

101. John C. Keene, *Managing Agricultural Pollution*, 11 Ecology L.Q. 135, 137 (1983). *See also* Robert W. Adler, *Water Quality and Agricultural: Assessing Alternative Futures*, 25 Environs Envtl. L. & Pol'y J. 77, 78 (2002); James Stephen Carpenter, *Farm Chemicals, Soil Erosion, and Sustainable Agriculture*, 13 Stan. Envtl. L.J. 190, 209-10 (1994); John H. Davidson, *Little Waters: The Relationship Between Water Pollution and Agricultural Drainage*, 17 ELR 10074 (Mar. 1987); J.B. Ruhl, *Farms, Their Environmental Harms, and Environmental Law*, 27 Ecology L.Q. 263, 274-92 (2000); David Zaring, *Agriculture, Nonpoint Source Pollution, and Regulatory Control: The Clean Water Act's Bleak Present and Future*, 20 Harv. Envtl. L. Rev. 515, 516-21 (1996). The landmark article remains N. William Hines, *Agriculture: The Unseen Foe in the War on Pollution*, 55 Cornell L. Rev. 740 (1970).
102. Nat'l Research Council, *supra* note 17, at 3-4; Nat'l Research Council, *supra* note 22, at 10.
103. *Id.*
104. Nat'l Research Council, *supra* note 22, at 73.
105. *Water in the West*, *supra* note 49, at 2-30.
106. Nat'l Research Council, *supra* note 22, at 73.
107. *Id.*
108. John H. Davidson, *Factory Fields: Agricultural Practices, Polluted Water and Hypoxic Oceans*, 9 Great Plains Nat. Resources J. 1 (2004).
109. *Id.*
110. Job, Groundwater Economics, *supra* note 10, at 38.
111. *Id.*
112. *Id.*
113. Design, *supra* note 1, at 83.

regional declines in groundwater levels is subsidence as a result of a decline in the elevation of the land surface, a phenomenon that is particularly noticeable in California's Central Valley, parts of Arizona, central Florida, and the Texas Gulf Coast.

Groundwater is also typically connected to surface formations of springs, wetlands, and flowing streams and rivers. This connection is now viewed as one more essential stage in the hydrologic cycle that affects both the quantity and quality of supply in both sources.[114] Overuse by pumping of the underlying groundwater formations can, as a result, lower the levels and even eliminate these surface resources and the ecosystems associated with them.[115]

Water table decline along coastal areas may allow saltwater intrusion into groundwater aquifers.[116] When water table levels are reduced sufficiently, saltwater can move inland making the groundwater unfit for use.[117] This is a particularly difficult problem for coastal cities that overlie freshwater sources but which must limit use in order to avoid saltwater intrusion.

The most prominent environmental effect of groundwater pumping is the overuse of existing supply. However the principles of sustainable development may be applied in other contexts, with groundwater, the concern is with balancing current use with reliable annual recharge. To the extent that usage exceeds recharge, the groundwater source is being "mined." If the imbalance continues, the current usage cannot be sustained.

Regional and local examples of groundwater overuse are numerous. Perhaps, the most notable is the Ogallala Aquifer, which underlies the central plains states of South Dakota, Nebraska, Wyoming, Colorado, Kansas, Oklahoma, New Mexico, and Texas.[118] More than 90% of the water pumped from the Ogallala irrigates about one-fifth of all U.S. cropland, accounting for 30% of all groundwater used for irrigation in the United States.[119] The aquifer, however, is being depleted in many areas. Six percent of the aquifer has dropped to an unusable level that can no longer be pumped.[120] If irrigation continues to draw water from the aquifer at the same rate, about 6% of the aquifer will be used up in 25 years.[121]

The first effect of aquifer mining is the increased cost of pumping as wells must be annually deepened.[122] A second effect is that farmland is removed from production. One estimate is that 20% of irrigated land over the Ogallala has returned to dry land use due to groundwater mining.[123] Finally, of course, continued mining must mean that the rich supply of farm production that has been generated in the High Plains will eventually be lost. If predictions about a drying climate over the Ogallala hold, recharge will be reduced further.[124] The potential water scarcity suggested by the Ogallala example is a reminder that where irrigation from groundwater exceeds annual recharge, the supply of food now produced by irrigation must eventually be reduced. As a result, by overusing the aquifer today, we are creating an unsustainable level of food production.[125]

3. Responding to Environmental Impacts: Law and Policy

The system of riparianism in the East and prior appropriation in the West were developed as property doctrines, but during the second half of the last century statutory permit systems were laid over the traditional property rules.[126] As competition for water increased nationwide, administrative permit processes became the fora for considering the extent to which elements such as "reasonable use," "beneficial use," and "public

---

114. Job, Groundwater Economics, *supra* note 10, at 51.
115. *Id.*
116. Design, *supra* note 1, at 84.
117. *Id.*
118. Job, Groundwater Economics, *supra* note 10, at 12.
119. *Id.*
120. *Id.*
121. *Id.*
122. *Id.*
123. *Id.*
124. *Id.*
125. *Id.*
126. Sax et al., *supra* note 2, at 12-15.

interest" would expand to accommodate increased municipal demand as well as wildlife and environmental concerns.[127]

As irrigation and population growth in the east increased demand, the eastern states inevitably began to draw on elements of prior appropriation, such as temporal priority and concerns over wasteful use. The western states, in turn, are drawing on aspects of riparianism, exemplified by concerns over protecting areas of origin from transbasin diversion schemes.

Over the last few decades, new themes have been introduced into water law and policy and are achieving various degrees of ascendency among the states.[128]

One theme is water as a public trust with water held by the state as custodian, charged with managing water in a manner that maximizes overall benefit to society.[129] A second and competing theme is water as an economic commodity, subject to full-cost pricing and trading in an open, private marketplace.[130] The third, and also competing theme, is that of water as a basic human right to which all people are entitled to access. Although these three themes appear to be incompatible, it may be that each has a role, depending upon circumstances.[131] As demand increases each of these themes will find voice as specific water contests are resolved.

C. Other Factors Affecting the Impact of Irrigation

1. Shift to Large, Integrated Farm Operations

The evolution of agriculture toward larger units of production is particularly evident in irrigated agriculture, where well-financed, integrated, and diversified farm operations are becoming normal.[132] A variety of factors favor this trend, including a decline in the overall farm population, a federal farm policy that favors large producers, and the ability of large producers to bear market risk.[133] Larger units are also better positioned to acquire and benefit from technology because irrigation is a capital-intensive enterprise most likely to benefit large farming entities with vast acreage available for farming.

2. Conversion of Rangeland to Irrigation

In recent years, a convergence of developments have led to a pattern of rapid conversion of native prairie grasslands to croplands and, in many cases, irrigated croplands. These lands are concentrated in the Dakotas, eastern Montana, Nebraska, and Kansas. Until recently, the best agricultural use of these rich native grasslands was assumed to be grazing; the topsoils are thin and the quality of the soils marginal, often sandy. The native grasses support a healthy grazing agriculture in association with a diverse ecosystem of wetlands, small rivers, and prairie. The advent of federal subsidies for ethanol production, the desire of landowners for access to federal crop benefits, new seeds suitable for marginal lands, and a steady reduction in the costs of technology have led to a rapid conversion to crops. The result is a spread of irrigation to lands that may not be well-suited to it, with the possibility of new regions of polluted runoff.[134]

3. Conversion of Wetlands to Irrigated Farmland

Irrigation can lead to the elimination of wetlands, playa lakes, prairie potholes, lakes, and flowing streams, all of which are vital to the survival of natural systems and wildlife of all kinds. Surface water and groundwater sources are typically connected—draw down one and you draw down the other. Surface wetlands and potholes are typically only the point where fully charged groundwater formations spill onto the sur-

---

127. *Id.*
128. *Id.*
129. *Id.* at 590-605.
130. *Id.* at 19.
131. *See* Barton H. Thompson Jr., *Water as a Public Community*, 95 Marq. L. Rev. 17 (2011).
132. Davidson, *supra* note 108, at 28-30.
133. Nat'l Research Council, *supra* note 22, at 173.
134. U.S. Government Accountability Office, *Farm Program Payments Are an Important Factor in Landowner's Decisions to Convert Grassland to Cropland* (Sept. 2007); *see also* Scott Stephens, *Plowing the Prairie*, Ducks Unlimited (July-Aug. 2005).

face. Thus, annual irrigation diversion from surface streams or pumping of underlying groundwater carries with it the likelihood that naturally occurring ecosystems will be depleted.

Irrigation is simply a more intense form of field agriculture. To farmers, any wetland, pothole, or minor wet slope is an impediment of efficient tillage.[135] Similarly, as the value of land and crops increases, the economic incentive to eliminate such "wet spots" increases as well. Irrigation is just one part of this dilemma. With increased demand for every use, surface water ecosystems are always the first to be placed at risk.[136]

As these surface water resources are depleted, the impact on aquatic species and migratory wildlife is direct and obvious.[137] Recurrent drought and flood are cycles to which aquatic species are adopted, but permanent elimination of the surface waters on which they depend means gradual elimination of the species themselves.

4. Climate Change and Variability

The example of the Ogallala Aquifer provides a useful discussion point for climate change and irrigation. In the case of the Ogallala, we can conclude that our farm production in the High Plains is artificially high because of overuse of the water supply and that production must necessarily decline in the future. Warming in the climate has the clear potential to impact water supplies, leading again to a reduction in farm production.[138]

The working example is the American West, which is not only the base of the nation's abundant irrigation production, but also the region that is being affected most by changed climate. As reported recently, "When compared to the 20th century average, the West has experienced an increase in average temperature during [2004-2007] that is 70 percent greater than the world as a whole."[139] The resulting decrease in snowpack, less snowfall, earlier snow melt, and reduced summer flow will not only disturb ecosystems and diminish reserves for municipal and industrial use, but will reduce agricultural production in all categories.[140] Thus, climate issues point to a conclusion that irrigation in the West may at some point be unsustainable.

D. Irrigation's Challenges and Opportunities

1. Reducing Polluted Return Flows

Treatment of polluted return flows is considered unlikely due to high cost as well as the need to develop disposal mechanisms for the high salt content waste byproduct.[141] Source control is sometimes accomplished by retiring land in which the salt load has become excessive and changing the cropping and use on the land.[142] Another approach is to alter the method of application from surface flow to spray or drip technologies. Recycling and diluting the water and the use of subsurface trickle technology can reduce the volume of drainage to under 10% of the amount of the water applied.[143] Each of these approaches is burdened, however, by high capital costs.

Polluted return flows may, perhaps, best be treated as land use problems, as is the case with so many of what we refer to as nonpoint sources. Although Congress has elected not to regulate nonpoint sources, it has correctly recognized that effective corrective measures will incorporate best-management practices (BMPs),[144] land use controls, and watershed management. BMPs recognize that national, or even regional, technology-based effluent standards cannot work a cure. Since nonpoint sources are the result of activities

---

135. *Id.*
136. *Id.*
137. *Id.*
138. It is clear that a warmer climate would accelerate the hydrologic cycle, increasing the rates of precipitation and evapotranspiration. *A New Era for Irrigation, supra* 22, at 74-75. Whether increased precipitation will occur at times and places necessary to support irrigation agriculture is the troubling question.
139. Rocky Mountain Climate Organization, *The West's Changed Climate* (Mar. 2008).
140. *Id.*
141. *Water in the West, supra* note 49, at 2-32.
142. *Id.*
143. *Id.*
144. 40 C.F.R. §§122.2, 130.2.

as various as human activity itself, controls must take the form of land management plans that consider the unique circumstances of any given plot of land.

Watershed management, like land use controls, has an inevitable role to play. Nonpoint sources are generated by human activity on the land but are often carried to watercourses by return flows from irrigation, precipitation, and diffused water. Nonpoint sources will be controlled not by any one landowner, but by a majority of landowners in a watershed who cooperate to implement a common plan.[145]

It is suggested that because so much of irrigation agriculture is organized in sophisticated irrigation districts—particularly in the West where the problem of polluted runoff is most evident—that the irrigation districts should be empowered to develop pollution-control plans suitable to their particular location and geology.[146]

2. Water Conservation

Whether the goal is to reduce return flows, increase water available for nonagricultural use, or increase overall food production, the real opportunity is for water conservation and more efficient use. Because agriculture in the West, the Great Plains, and Midwest developed prior to most other water uses of any scale, irrigators have had first call; they have been at the front of the line. From 1902 forward, new irrigation projects sponsored by the Bureau of Reclamation were not only numerous and vast in scale, they were also designed to serve irrigated agriculture first, with other uses servient and secondary.[147] The first effect of this was a tendency toward overuse, leading to waste and excessive runoff. In addition, these projects also integrated an enormous financial subsidy into irrigated agriculture, which eliminated any necessities for conservation and efficiency in water use.

In recent decades, the favored place at the front of the water use line that irrigation enjoyed for so long has been challenged as the West's great new cities and industry seek water, recreation grows as a major economic use, and the public calls for ecosystem protections. A new era is underway in which a balance among water use, water pricing, and ecosystems is sought.[148]

The heavy subsidies for irrigation waters from Bureau of Reclamation projects insulated users from water markets, which in the absence of Bureau programs would have increased the value of the water.[149] The gradual emergence of water markets offers an attractive mechanism for achieving efficiency in water use by applying high-value water to high-value crops while providing nonagricultural users the opportunity to acquire supplies at the higher cost.[150] Markets should also serve to increase the incentives for conservation. By raising the value of water, markets might also decrease unnecessary agricultural runoff and resulting sources of pollution.

Conclusion

Irrigation will play a pivotal role as agriculture responds to the many challenges outlined in this chapter. Done well, it will be a larger part of the solution. If it continues in the old ways, it will present a serious drag on innovation.

Irrigation has historically enjoyed first call on water resources, leading to overuse. Today, that water is sought by urban, industrial, recreational, and environmental users, while the overall supply in key regions is facing potential reduction as the result of global warming and climate change. While it strives to increase production of essential foods, irrigation agriculture is challenged to do so with less water and while reducing the pollution that is part of field runoff. In early stages, this may be achieved by moving from lower-value, water-consumptive crops to high-value, water-conserving crops. Subsequent stages will require stronger measures, ranging from water-saving technologies to abandonment of irrigated agriculture in water-scarce regions such as the High Plains.

---

145. *Water in the West, supra* note 49, at 6-21.
146. *See* John H. Davidson, *Using Special Water Districts to Control Nonpoint Sources of Water Pollution*, 65 CHI-KENT L. REV. 503 (1990).
147. MACDONNELL, *supra* note 14, at 1-9.
148. *Id.*
149. *Id.*
150. *Id.*

In the larger sense, the issue is that of increased reliance on an ever more finite natural resource. In this, it is hardly distinguishable from other examples such as ocean fisheries, prime farmland, and forests. Is it possible to absorb increased competition for a limited resource while establishing a sustainable level of food production and providing for survival of some portion of water-dependent ecosystems? A positive answer will involve institutional, technological, and social change to a degree that is not yet in sight.

In the larger sense, the issue is that of increased reliance on an ever more finite natural resource. In this it is hardly distinguishable from other examples such as ocean fisheries, prime farmland, and forests. Is it possible to absorb increased competition for a limited resource while establishing a sustainable level of food production and providing for survival of some portion of water-dependent ecosystems? A positive answer will involve institutional, technological, and social change to a degree that is not yet in sight.

Chapter 5
The Industrializtion of Animal Agriculture: Connecting a Model With Its Impacts on the Environment

Hannah M.M. Connor

Globally, the production and consumption of livestock, including swine and cattle, and poultry products is growing.[1] Global meat production more than doubled from over 136 million tons in 1980 to over 285 million tons in 2007,[2] while U.S. meat production followed similar growth trends.[3] Increases in the supply of livestock and poultry products can be linked directly to structural changes in production systems that allow more animals to be raised in confinement over a shorter period of time and with comparably lower cost and uncertainty.[4] The expansion of animal agriculture is, in part, a response to global population growth and changes in economic conditions, as well as changes in social attitudes with respect to the consumption of animal products. It is also attributable to the increased accessibility and the decreased cost of these products.[5]

As costs to producers and consumers have declined, however, secondary environmental and public health costs have arisen and, in many cases, undermined many of the benefits of the confinement production model. Specifically, studies show that livestock and poultry confinement operations can directly impact the environment through water pollution and water scarcity, climate change, air pollution, land degradation, loss of biodiversity, and environmental justice implications, and can indirectly impact the environment through the increased demand for monoculture feed crop production.[6]

In connecting changes in the production of animals with their impacts on the environment, this chapter begins with an account of the transformation and consolidation of the livestock and poultry industries, both globally and domestically. The chapter then describes the commonly utilized operational designs, including waste management and disposal systems, and concludes with an examination of the relationship between confined animal production and the environment.[7]

---

1. *See* H. Steinfeld et al., Food and Agriculture Organization of the United Nations (FAO), Livestock's Long Shadow: Environmental Issues and Options 7-12 (2006), *available at* http://www.fao.org/docrep/010/a0701e/a0701e00.htm [hereinafter Livestock's Long Shadow].
2. *See* FAO, The State of Food and Agriculture: Livestock in the Balance 15, tbl. 3 (2009), *available at* http://www.fao.org/docrep/012/i0680e/i0680e00.htm [hereinafter Livestock in the Balance].
3. R.U. Halden & K.J. Schwab, Pew Comm'n on Indus. Farm Animal Prod., Environmental Impact of Industrial Farm Animal Production 14 (2006), *available at* http://www.ncifap.org/bin/s/y/212-4_EnvImpact_tc_Final.pdf [hereinafter Pew Environmental Impacts] (referencing U.S. Dep't. of Agric. (USDA), Rise in Meat Production (2007), *available at* http://usda.mannlib.cornell.edu/usda/nass/MeatAnimPr//2000s/2006/MeatAnimPr-04-27-2006.pdf.).
4. *See* Pew Environmental Impacts, *supra* note 3, at 5 (For example, one confined poultry operation can now raise 25,000 to 50,000 chickens "to market weight within a few weeks by automated feeding apparatuses dispensing a growth-optimizing diet usually supplemented with antimicrobials.").
5. *See* Livestock's Long Shadow, *supra* note 1, at 7-12; Pew Environmental Impacts, *supra* note 3, at 5.
6. *See, e.g., id.*; Food & Water Watch, Turning Farms Into Factories: How the Concentration of Animal Agriculture Threatens Human Health, the Environment, and Rural Communities 5 (2007), *available at* http://www.pigbusiness.co.uk/pdfs/Food-and-Water-Watch-Farms-to-Factories.pdf(referencing F. Henry, *Mega-Farms Stoke Worries Over Waste Spills*, Plain Dealer, Oct. 9, 2005 (Cleveland, Ohio); Iowa Dep't of Natural Resources, Environmental Services Division, Manure (2003); Iowa Dep't of Natural Resources, Environmental Services Division, Manure (2004); Iowa Dep't of Natural Resources, Environmental Services Division, Manure (2005); Iowa Dep't of Natural Resources, Environmental Services Division, Manure (2006)).
7. While a few concerns related to livestock and poultry slaughtering facilities will additionally be addressed, those facilities are not the focus of this chapter.

A. Growth and Consolidation of the Livestock and Poultry Industry

Traditional animal agriculture relies on a diversified farming approach in which animals are raised primarily in a forage and pasture-based model, and the size of the livestock herd is highly dependent on the operation's land and grazing resources.[8] Sometimes referred to as a "family farm," "[m]ost [traditional] farms were independently owned and managed by families," and livestock were considered "'mortgage lifters,' adding value to crops grown on the farms."[9] In many cases, a farm's income was derived through the production of both livestock and cash crops.[10]

As international demand for meat, dairy, and poultry products has increased,[11] however, the production sectors have "responded [to this demand] . . . mainly through intensification rather than expansion."[12] In other words, as consumption has increased, the corresponding production sectors have grown vertically, consolidating production into larger, more intensive operations, instead of relying on the expansion and propagation of smaller, traditional farms. This process of intensification and consolidation is often referred to as the "verticalization" or "industrialization" of animal agriculture.[13]

U.S. corporations, operating both domestically and abroad, have been integral in establishing and buttressing the use of consolidation practices in livestock and poultry production.[14] Indeed, the introduction in the 1930s of "highly mechanized" swine slaughterhouses to the American landscape paired with the growth of "vertically integrated" meat packing companies are often credited as the root of industrial animal production.[15] Although hog meatpackers initiated the industrialization trend, it was the adaptation and refinement of that model to poultry production practices that fully integrated the confinement model into modern industrial animal production.[16]

Specifically, poultry production companies recognized that while growth and slaughter are each distinct practices, a similar mechanization and uniformity could be applied to and merged within the overall business approach. For this type of industrialization to be successful, though, a number of substantive changes had to be made to the traditional farming model, including changes to the structure, control, and basic approach of farming systems.[17] In effect, to maximize profits, the industry altered the traditional farming approach to animal agriculture by controlling the supply and specializing the means of production.

On average, "specializing the means of production" refers to the act of increasing the number of animals being raised at one operation, confining the animals to one or a number of highly controlled "barns" or feedlots, and converting the animals to a feed-based diet.[18] The resultant system is "characterized by a switch from the traditional animal agriculture's reliance on animal biology and behavior and skilled husbandry to the new animal agriculture's reliance on mechanized techniques of restricting and controlling animals' behavior and biological processes."[19] Ostensibly, these new operations are intended to reduce risk by lowering primary transaction costs, reducing the amount of space necessary for each animal, accruing

---

8. *See* LIVESTOCK IN THE BALANCE, *supra* note 2, at 26-29, 33; MARLENE HALVERSON, INSTITUTE FOR AGRICULTURE AND TRADE POLICY, THE PRICE WE PAY FOR CORPORATE HOGS 15 (2000), *available at* http://www.iatp.org/documents/the-price-we-pay-for-corporate-hogs [hereinafter THE PRICE WE PAY] (citing D. Houghton, The Vanishing American Hogman, SUCCESSFUL FARMING, H1-H8 (1984)).
9. *Id.*
10. *Id.*
11. For example, in 2005 the world's population was estimated at 6.5 billion and was increasing by an estimated 76 million persons annually. Based on this continued rate of expansion, the U.N. projects the worldwide population to expand to 9.1 billion by 2050 and 9.5 billion by 2070. *See* LIVESTOCK'S LONG SHADOW, *supra* note 1, at 7 (citing U.N. Dep't of Economic and Social Affairs, World Population Prospects, The 2004 Revision, New York, NY (2005), *available at* http://www.un.org/esa/population/publications/sixbillion/sixbilpart1.pdf). *See also* LIVESTOCK'S LONG SHADOW, *supra* note 1, at 20.
12. *Id.* at 6.
13. PEW ENVIRONMENTAL IMPACTS, *supra* note 3, at 2.
14. *See* LIVESTOCK'S LONG SHADOW, *supra* note 1, at 19 ("Large multinational firms are becoming dominant in the meat and dairy trade, both in the developed world and in many developing experiencing fast livestock sector growth."); *id.* at 279 (finding that industrialization of animal agriculture was observed "in the EU and North America from as early as the 1960s, and in emerging economies since the 1980s and 1990s.").
15. PEW COMMISSION ON INDUSTRIAL FARM ANIMAL PRODUCTION, PUTTING MEAT ON THE TABLE: INDUSTRIAL FARM ANIMAL PRODUCTION IN AMERICA 5, 6 (2008), *available at* http://www.ncifap.org/bin/e/j/PCIFAPFin.pdf [hereinafter PUTTING MEAT ON THE TABLE].
16. M. Broadway & D. Stull, *The Wages of Food Factories*, 18 FOOD AND FOODWAYS 43-65, 47 (2010) [hereinafter *The Wages of Food Factories*].
17. *See, e.g.*, PUTTING MEAT ON THE TABLE, *supra* note 15.
18. *Id.*
19. THE PRICE WE PAY, *supra* note 8, at 16 (referencing I. Ekesbo, *Animal Health Implications as a Result of Future Livestock and Husbandry Developments*, 20 APPLIED ANIMAL BEHAV. SCI. 95-104 (1988)).

the benefits of market ownership, and providing a means for the industrialized operator to generate a consistent and uniform product.[20]

Establishment of the industrialized production model transformed animal agriculture. According to the U.S. Department of Agriculture (USDA), "the number of large farms that raise animals has increased 234 percent, from about 3,600 in 1982 to almost 12,000 in 2002."[21] Indeed, estimates show that, while as few as 5% of U.S. livestock operations rely on confinement practices, those operations produce upwards of 80% of the animals used for domestic animal-related food products.[22] Similar shifts are occurring internationally.[23]

However, for the farming community, these shifts come at a price.[24] First, "[t]he shift to intensive production systems is accompanied increasing size of operation, driven by economies of scale. Despite an overall growth of the sector, this is only achieved at the cost of pushing numerous small- and middle-scale producers and other agents out of business."[25] For example, in the swine sector, the number of swine producers in this country fell from an estimated 2.1 million in 1950[26] to just over one million by 1965 [27] to 98,460 by 1999[28] — a reduction of more than 95%. Over the same half-century, average sales per operation skyrocketed from 31 hogs per farm to 1,100 per farm, with a mere 105 farms possessing an estimated 40% of the U.S. hog inventory by the year 2000.[29]

Second, as dependence on the industrial model has increased, so has reliance on specialized growing facilities and operators.[30] Specifically, upon recognizing the hierarchical benefit of this model, a number of large livestock and poultry companies (also known as "integrators"), such as American corporations Smithfield Foods, Tyson Foods, and Perdue Farms, began to step back from the role of raising the animals for slaughter or use to the controlling role of contracting out the production of the animals to smaller industrialized livestock and poultry growers ("producers").[31] Under the resultant- and often very strict- contract, "[t]he grower does not own the animals and frequently does not grow the crops to feed them. The integrator (company) controls all phases of production, including what and when the animals are fed."[32] Today, the industrial poultry sector and most of the industrial swine sector are dominated by the integrator-producer contractual model, with the model continuing to spread across different animal sectors.[33]

---

20. *See* LIVESTOCK'S LONG SHADOW, *supra* note 1, at 20 ("Vertical integration allows not only for gains from economies of scale. It also secures benefits from market ownership and from control over product quality and safety, by controlling the technical inputs and processes at all levels."); *The Wages of Food Factories, supra* note 16, at 43.
21. U.S. GOV'T ACCOUNTABILITY OFFICE (GAO), GAO-08-944, CONCENTRATED ANIMAL FEEDING OPERATIONS: EPA NEEDS MORE INFORMATION AND A CLEARLY DEFINED STRATEGY TO PROTECT AIR AND WATER QUALITY FROM POLLUTANTS OF CONCERN 4-5 (2008), *available at* http://www.gao.gov/new.items/d08944.pdf [hereinafter GAO-08-944].
22. D. GURIAN SHERMAN, UNION OF CONCERNED SCIENTISTS, CAFOS UNCOVERED: THE UNTOLD COSTS OF CONFINED ANIMAL FEEDING OPERATIONS 2 (2008), *available at* http://www.ucsusa.org/assets/documents/food_and_agriculture/cafos-uncovered.pdf.
23. LIVESTOCK'S LONG SHADOW, *supra* note 1, at 278 ("[A]n estimated 80 percent of total livestock sector growth comes from industrial production systems.").
24. *Id.* at 19 ("The combined effect of economic gains from lowering transaction costs by vertical integration, and more favourable tax regimes for larger enterprises, tends to disadvantage independent and small-scale producers severely.").
25. *Id.* at 279.
26. THE PRICE WE PAY, *supra* note 8, at 15 (referencing D. Houghton, *The Vanishing American Hogman*, SUCCESSFUL FARMING H1-H8 (1984)).
27. *The Wages of Food Factories, supra* note 16, at 45 (referencing K.M. Thu, *The Centralization of Food Systems and Political Power*, 31(1) CULTURE & AGRIC. 13-18, 15 (2009)).
28. THE PRICE WE PAY, *supra* note 8, at 15 (referencing U.S. Dep't. of Agric., National Agric. Stat's Serv., Hogs and Pigs Report, Mt An 4, 12-99. (1999)).
29. THE PRICE WE PAY, *supra* note 8, at 15.
30. By 2002, only 1% of livestock confinement operations in the United States was diversified. R. HOPPER, IZAAK WALTON, LEAGUE OF AMERICA, GOING TO MARKET: THE COST OF INDUSTRIALIZED AGRICULTURE 5 (2002), *available at* http://www.iwla.org/index.php?ht=action/GetDocumentAction/i/939 (referencing Kellogg et al., U.S. Dep't of Agric., Natural Resources Conservation Service, Economic Research Service, *Manure Nutrients Relative to the Capacity of Cropland and Pastureland to Assimilate Nutrients: Spatial and Temporal Trends for the United States* (2000)).
31. U.S. EPA, National Pollution Discharge Elimination System Permit Regulations and Effluent Limitation Guidelines and Standards for Concentrated Animal Feeding Operations; Proposed Rule, 66 Fed. Reg. 2960, 3024 (Jan. 12, 2001) [hereinafter 2001 Proposed Rule] (explaining that under the vertical integration model, a large corporation, such as a slaughtering facility, a meat packing plant, or an integrated food manufacturing facility, will retain ownership of the animals and/or will "exercise[ ] substantial operational control over the type of production practices used at the CAFO," but instead of raising their own animals, the corporation will subcontract, often using very stringent contract terms, with smaller farmers to grow the animals until harvest or slaughter).
32. PUTTING MEAT ON THE TABLE, *supra* note 15, at 5, 6.
33. *Id.* ("Today, the swine and poultry industries are the most vertically integrated, with a small number of companies overseeing most of the chicken meat and egg production in the United States."); *The Wages of Food Factories, supra* note 16, at 45 (citing M.A. Grey, "Those Bastards Can Go to Hell!" Small-farmer resistance to vertical integration and concentration in the pork industry, 59 HUM. ORG. 169-76, 169 (2000) (By 2000, "seven of every ten [hog] was being grown under contract.")).

Finally, much of the growth in this sector has been developed in geographically limited areas, especially in regions of North Carolina, Georgia, Iowa, and the Chesapeake Bay. As a result, "[i]n contrast to traditional small farms, industrial production systems involve large concentrations of animals in confined spaces and the generation of enormous quantities of solid, liquid, and gaseous wastes in small geographic areas."[34] In 2006, for example, North Carolina's swine population, mostly grown in a small region of eastern North Carolina, outnumbered its human residents,[35] and for every person living in Iowa in 2006, there were five-and-a-half hogs.[36] Limiting confinement practices to discrete geographic areas not only limits the majority of jobs related to animal agriculture to those regions, but it also elevates environmental justice concerns and intensifies pollution potential by multiplying localized pollution sources.

B. Livestock Production Models

Environmental damage from livestock production results from the pairing of consolidation practices with a business model that devalues environmental welfare by employing limited waste treatment and management practices, employing geographic clustering, countenancing practices that cause the emission of aerial pollutants into the ambient environment, and supporting the propagation of monoculture feed crop production.[37] In many circumstances, it is the design, operation, and maintenance of a concentrated animal production facility that will dictate the extent to which the facility will impair the surrounding environment. Therefore, to understand the environmental impacts of these facilities, it is first necessary to understand and appreciate standard operational design.

Confinement operations, including waste management systems, exist in a variety of sizes and designs.[38] On average, however, most domestically operated[39] facilities are comprised of two interrelated and environmentally significant elements: the production area and the land application area.[40] The production area includes the animal confinement area (usually in the form of confinement buildings or feedlots) and the areas immediately surrounding the confinement structure, including the manure storage area, the raw materials storage area, and the waste containment area.[41] The land application area is the area where generated animal wastes may be applied.[42]

The following sections describe a representative range of confinement models, focusing on the three animal types that are often raised using confinement practices:[43] beef and dairy cattle, swine, and poultry.[44]

1. Production Area

To produce maximum yield, animal confinement operations generally raise animals in one or a number of confinement buildings or feedlots,[45] replace animal grazing and foraging with deliberate feeding prac-

---

34. M. Tajik et al., *Impact of Odor From Industrial Hog Operations on Daily Living Activities*, 18(2) New Solutions 193-205, 194 (2008).
35. *See* N.C. Dep't of Agric. and Consumer Services, Agric. Statistics Division, *North Carolina Hogs Down, U.S. Inventory Up Slightly* (2006); U.S. Census Bureau, *State and County Quick Facts* (2006), http://quickfacts.census.gov/qfd/states/37000.html.
36. Food & Water Watch, Turning Farms Into Factories, *supra* note 5, at 5 (referencing Iowa Dep't of Agric. and Land Stewardship, *Facts About Iowa Agriculture*, http://www.iowaagriculture.gov/quickfacts.asp).
37. *See, e.g.*, Livestock's Long Shadow, *supra* note 1; Putting Meat on the Table, *supra* note 15.
38. *See* U.S. GAO, GAO/RCED-99-205, Animal Agriculture: Waste Management Practices 6 (1999), *available at* http://www.gao.gov/archive/1999/rc99205.pdf [hereinafter GAO/RCED-99-205].
39. *Id*. at 2 (noting that most "animal waste management practices used in other major livestock and poultry production countries are similar to those used by U.S. farmers").
40. *See* 40 C.F.R. §§122.23(b)(3), (8) & (e).
41. *See* 40 C.F.R. §122.23(b)(8) ("The animal confinement area includes but is not limited to open lots, feedlots, confinement houses, stall barns, free stall barns, milkrooms, milking centers, cowyards, barnyards, medication pens, walkers, animal walkways, and stables. The manure storage area includes but is not limited to lagoons, runoff ponds, storage sheds, stockpiles, under house or pit storages, liquid impoundments, static piles, and composting piles. The raw materials storage area includes but is not limited to feed silos, silage bunkers, and bedding materials. The waste containment area includes but is not limited to settling basins, and areas within berms and diversions which separate uncontaminated storm water. Also included in the definition of production area is any egg washing or egg processing facility, and any area used in the storage, handling, treatment, or disposal of mortalities.").
42. *See* 40 C.F.R. §122.23(b)(3).
43. Operations maintaining other species, such as geese, emus, ostriches, llamas, mink, bison, and alligators, may be additionally recognized by EPA as confinement operations. U.S. EPA, EPA 833-F-12-001, NPDES Permit Writers' Manual for Concentrated Animal Feeding Operations 2-6 (2012), *available at* http://cfpub.epa.gov/npdes/afo/info.cfm#guide_docs [hereinafter NPDES Permit Writers' Manual].
44. *See* 40 C.F.R. §§122.23(b)(1), (2), (4)-(6).
45. Some operations only confine their poultry and livestock for a portion of the animal's time at the operations. For those operations, the determination of whether the operation should be considered a "confinement" operation can be variable, depending on personal objective. However,

tices, and fully systematize operational waste management systems.[46] Many of these essential elements occur within the production area. As a result, the production area possesses many of the significant structural components that guide the degree and type of pollutant that may be released from a confinement operation.

a. Confinement Structure

Adapted from traditional barn design, the confinement structure is a roofed building or feedlot in which animals are housed or confined.[47] To facilitate constricted space allowances, confinement structures are often designed to replace tradition stalls with crates, cages, pens, or contracted stalls.[48] These structures use minimal or no bedding materials, such as straw, sawdust, woodchips, or sand,[49] and have flooring that is often made of concrete and mesh or slotted metal so that the confinement structure can be cleaned and animal and process wastes can be removed without moving the confined animals.[50] To keep air from becoming stagnant within the structure, to dry animal wastes,[51] and to remove generated gasses and other aerial contaminants (including carbon dioxide, ammonia, hydrogen sulfide, and dusts), confinement buildings often have large ventilation fans installed in their walls or ceilings.[52]

b. Waste Management System

USDA conservatively estimates that confinement operations "generate about 500 million tons of manure annually (as excreted)."[53] A large dairy cow, for example, can produce an astonishing 120 pounds of wet manure per day,[54] while the average hog can produce 3,000 pounds of solid manure and over 5,000 gallons of liquid manure annually, and the average poultry broiler house can produce as much as 200 tons of chicken waste, including discarded biological waste, feathers, and bedding, per year.[55] By comparison, the U.S. Environmental Protection Agency (EPA) estimates that the U.S. population of roughly 285 million people produces only about 150 million tons (wet weight) of sanitary waste per year, or about one-third the size of the wastes produced by confinement operations.[56]

Consequently, a confinement operation must have an animal waste management system capable of collecting, storing, ideally treating, and eventually disposing of these produced wastes, including biological wastes, urine and feces, and associated process wastes (also known as "process wastewater"[57]), such as wash

---

as a good reference point, the regulations to the Clean Water Act contain one of the most widely used and recognized definitions. Under that definition, a farm becomes and animal feeding operation if the poultry or livestock "have been, are or will be" confined "for a total of *45 days or more in any 12-month period.*" 40 C.F.R. §122.23(b)(1) (emphasis added).

46. An exception would be a poultry house where the waste is immediately transferred to a third party upon collection. However, even in this situation, since poultry waste is often only collected every few month or sometimes years, the operation must manage the waste as it accumulates in the confinement buildings and lots. Collins et al., Natural Resource, Agriculture, and Engineering Service (NRAES) Cooperative Extension, NRAES-132, POULTRY WASTE MANAGEMENT HANDBOOK 1, 3, 5-6. Ithaca, NY (1999) [hereinafter POULTRY WASTE MANAGEMENT HANDBOOK].
47. PUTTING MEAT ON THE TABLE, *supra* note 15, at 33.
48. *See id.* at 5-6, 33.
49. L.S. BULL ET AL., PEW COMMISSION ON INDUSTRIAL FARM ANIMAL PRODUCTION, RECENT CHANGES IN FOOD ANIMAL PRODUCTION AND IMPACTS ON ANIMAL WASTE MANAGEMENT 10 (2006), *available at* http://www.ncifap.org/bin/u/v/PCIFAP_FW_FINAL1.pdf [hereinafter PEW RECENT CHANGES].
50. Impermeable surfaces, such as concrete flooring or slotted metal flooring over a concrete pit, "facilitate waste handling and the recycling of wastewater." U.S. EPA, EPA-833-B-04-001, *NPDES Permit Writers' Guidance Manual and Example NPDES Permit for Concentrated Animal Feeding Operations* 3-4, 3-6 (2003), *available at* http://cfpub.epa.gov/npdes/afo/info.cfm [hereinafter NPDES PERMIT WRITERS' GUIDANCE].
51. POULTRY WASTE MANAGEMENT HANDBOOK, *supra* note 46, at 4 ("Circulation fans to move air over the manure mass can assist in moisture evaporation, increasing the dryness of the manure.").
52. *See, e.g.,* 2001 Proposed Rule, *supra* note 31, at 2993 ("Ventilation flows through the house from the roof down over the birds and into the pit over the manure before it is forced out through the sides of the house. The ventilation drys the manure as it piles up into cones.").
53. U.S. EPA, National Pollution Discharge Elimination System Permit Regulations and Effluent Limitation Guidelines and Standards for Concentrated Animal Feeding Operations; Final Rule, 68 Fed. Reg. 7176, 7180 (Feb. 12, 2003) [hereinafter 2003 Rule].
54. U.S. EPA, Region 9, *EPA Region 9: Animal Waste* (2002), http://www.epa.gov/region9/animalwaste/problem.html.
55. *The Wages of Food Factories, supra* note 16, at 45 (citing D.D. STULL & M.J. BROADWAY, SLAUGHTERHOUSE BLUES: THE MEAT AND POULTRY INDUSTRY IN NORTH AMERICA 31, 58, 134 (Wadsworth 2004)).
56. 2003 Rule, *supra* note 53, at 7180. *See also* Burkholder et al., *Impacts of Waste From Concentrated Animal Feeding Operations on Water Quality*, 115(2) ENVTL. HEALTH PERSP. 308 (2007), *available at* http://www.ncbi.nlm.nih.gov/pmc/articles/PMC1817674/.
57. "*Process wastewater* means water directly or indirectly used in the operation of the CAFO for any or all of the following: spillage or overflow from animal or poultry watering systems; washing, cleansing, or flushing pens, barns, manure pits, or other CAFO facilities; direct contact swimming, washing, or spray cooling of animals; or dust control. Process wastewater also includes any water which comes into direct contact with any raw materials, products or byproducts including manure, litter, feed, milk, eggs, or bedding." 40 C.F.R. §412.2(d).

water, bedding, and hair or feathers.[58] However, accepted confined animal waste management methods are very different from human waste treatment methods. For, while human waste is required to go through an intensive treatment and monitoring process (the Clean Water Act even commits an entire subchapter to human waste and wastewater management),[59] concentrated animal wastes typically go through a much reduced waste treatment regimen, the efficacy of which depends on the established operational waste management system.[60]

The selection of a waste management system for a confinement operation is often dependent on waste type, waste quantity, cost, climate, and operational water use.[61] After cost, the principal determinative factor is usually waste type, which typically falls into one or more of three categories: solid waste, semi-solid waste, and liquid waste.[62] Within these categories, most confinement operations, especially of the same animal type, rely on a few representative systems. While a variety of "alternative" waste treatment technologies exists that alleviate one or a number of environmental concerns, few of these technologies are regularly in use, and they are not the accepted industry standard.[63]

i. *Liquid and Slurry Waste Management Systems*

In liquid and slurry manure management systems, the animal excretes its "wet" waste product, in the form of urine and feces, onto the floor of the confinement structure.[64] The waste is then washed, scraped, or flushed,[65] usually with water, "through slats in the building floor into a series of trenches and pipes" and into a storage impoundment.[66] The impoundment is typically a contained storage structure or an open pit (sometimes also referred to as a "pond" or a "lagoon") that is situated underneath of or directly adjacent to the confinement building.[67] Once impounded, liquid waste may be treated,[68] depending on the waste management system, and will remain in the storage system until it is utilized or disposed of, often through land application.[69] While regional differences, such as climate and local conventions, can significantly affect manure handling methods, most swine operations utilize a liquid waste management system.[70]

---

58. *See* GAO/RCED-99-205, *supra* note 38, at 7.
59. The reason for such intensive treatment of human waste is safety of human health and environmental preservation. *See, e.g.*, S.M. Khopkar, *Environmental Pollution Monitoring and Control*. New Delhi: New Age International (2004). *See also* 42 U.S.C. §§1281-1301, 1317, 1323, 1342, 1345; 40 C.F.R. §§133, 257, 403, 501, 503.
60. *See generally* Pew Recent Changes, *supra* note 49.
61. Poultry Waste Management Handbook, *supra* note 46, at 1 ("The character of the waste involved, the level of moisture in the waste, the soil type, and other site conditions are important factors in determining the suitability of any particular waste management system."); GAO/RCED-99-205, *supra* note 38, at 6.
62. *See* GAO/RCED-99-205, *supra* note 38, at 7 ("The choice of collection method and storage structure depends, in part, on the volume and moisture content of the waste being handled."); 2001 Proposed Rule, *supra* note 31, at 2988-93; S.W. Melvin et al., Iowa State University Extension, Pork Industry Handbook Publication PIH-67, *Swine Waste Management Alternatives* (1989).
63. *See, e.g.*, RTI International, Project Number 08252.000.2003, *Benefits of Adopting Environmentally Superior Swine Waste Management Technologies in North Carolina: An Environmental and Economic Assessment* (November 2003); J.H. Robbins, Waterkeeper Alliance, *Understanding Alternative Technologies for Animal Waste Treatment: A Citizen's Guide to Manure Treatment Technologies* (Feb. 2005); Pew Recent Changes, *supra* note 49.
64. M.A. Mallin, *Impacts of Industrial Animal Production on Rivers and Estuaries: Animal-Waste Lagoons and Sprayfields Near Aquatic Environments May Significantly Degrade Water Quality and Endanger Human Health*, 88 Am. Scientist 2, 4 (2000).
65. "A flushing system uses fresh or recycled water to move manure from the point of deposition or collection to another location." NPDES Permit Writers' Guidance, *supra* note 50, at 3-8.
66. Mallin, *Impacts of Industrial Animal Production*, *supra* note 64, at 2, 4.
67. NPDES Permit Writers' Manual, *supra* note 43, at 2-8 ("An AFO is considered to have a liquid-manure handling system if it uses pits, lagoons, flush systems (usually combined with lagoons), or holding ponds, or has systems such as continuous overflow watering, where the water comes into contact with the manure and" poultry waste).
68. For example, as applied to liquid waste impoundments, "[t]here are three broad categories of earthen manure storage: earthen manure storage, anaerobic lagoons, and aerobic lagoons. In contrast to the lagoon systems, earthen manure storages are not designed to encourage microbial decomposition and treatment of the manure.... In the absence of significant microbial treatment degradation, manure nutrients are largely conserved." T.L. Richard et al., Iowa State University, *Management and Maintenance of Earthen Manure Structures: Implications and Opportunities for Water Quality Protection*, 18(6) Applied Engineering Agric. 727-34 (2002).
69. 2001 Proposed Rule, *supra* note 31, at 3010, 3031-32.
70. *See* Pew Recent Changes, *supra* note 49, at 15. *See also* 2001 Proposed Rule, *supra* note 31, at 2991 ("Most confinement hog operations use one of three waste handling systems: flush under slats, pit recharge, or deep underhouse pits. Flush housing uses .. water ... to remove manure from sloped floor gutters or shallow pits.... Flushing occurs several times a day. Pit recharge systems are shallow pits under slatted floors with 6 to 8 inches of pre-charge waters. The liquid manure is pumped or gravity fed to a lagoon approximately once a week. Deep pit systems start with several inches of water, and the manure is stored under the house until it is pumped out for field application.").

ii. Solid or "Dry" Waste Management Systems

On the other side of the moisture scale is the solid, or "dry," waste management system. Predominantly employed by confined poultry feeding operations,[71] these systems are designed so that animals "are housed on dirt or concrete floors that have been covered with a bedding material such as wood shavings," and are to excrete their waste products directly onto that surface.[72] Alternately, poultry operations that house birds in cages, such as many egg layer operations, may have the waste excreted by the bird fall onto a conveyor belt that will move that waste to another part of (or outside of) a confinement building.[73] Upon excretion, the droppings, which for birds are drier in composition due to an evolutionary adaptation designed to aid in flight, mixes with the bedding to form a waste product often referred to in poultry production as "litter."[74] Litter will accumulate in the confinement structure until being removed to a temporary pile outside of the building, to a nearby permanent roofed structure, to a litter-spreading machine, or to another location outside of the confinement property.[75]

The litter is typically removed from the confinement structure on a rolling basis, the frequency of which is often determined by the operation's nutrient management plan, the moisture content of the waste, "the quantity of the manure or manure litter in the house and the amount of remaining storage space available."[76] On average, poultry confinement operations will conduct a "whole house" litter removal, sometimes known as a "scrape," at least once a year.[77] After removal of the dry waste from the confinement structure (or from the drying system), the waste will then remain stored until it is removed and utilized, often through land application.[78] However, it should be noted that EPA only allows operations to retain the "dry litter" system designation if waste that has been removed from the confinement building is covered or used immediately and is not exposed to excessive precipitation.[79]

While non-poultry operations may also employ a dry waste system, those operations often must utilize a drying system to remove the urine and other moisture from the fecal matter.[80] Because drying is resource-intensive, application of this waste management system to wetter wastes is usually limited.

iii. Semisolid Waste Management Systems

Finally, waste can be managed in a semisolid state. Often, semisolid waste management is used by animal operations in concert with either a solid waste management system or a liquid waste management system.[81] Cattle operations, for example, often rely on waste management systems that utilize both a semisolid and either a liquid or a solid waste component.[82] This is because, as ruminants, cattle produce waste that is highly variable in nature. For example, standard cattle waste has a "solid content of about 12% and tends to act as a slurry; however, it can be handled as a semisolid or a solid if bedding is added."[83]

---

71. *See id.* at 3011. *See also* POULTRY WASTE MANAGEMENT HANDBOOK, *supra* note 46, at 17 (discussing that poultry operations can use dry or liquid waste management systems. In poultry operations, "[w]aste management systems are dependent upon the type of layer, pullet, broiler/turkey, or breeder house design involved.").
72. 2001 Proposed Rule, *supra* note 31, at 3011. *See also*, POULTRY WASTE MANAGEMENT HANDBOOK, *supra* note 46, at 17 (discussing that with egg-laying operations, laying hens are usually "kept in cages" that are often contained in "high rise houses where the birds are kept on the second floor and the manure drops to the first floor. . . . Manure can usually be stored in high rise houses for up to a year before requiring removal.").
73. *Id.*
74. Poultry "litter" includes excrement, bedding, water, soil, grit, feathers, nonutilized dietary minerals, errant feed, and "any other waste material that is part of the waste stream from the production houses." POULTRY WASTE MANAGEMENT HANDBOOK, *supra* note 46, at 3.
75. *Id.* at 20 ("The type of system . . . is dependent on the quantity of manure to be handled, manure moisture content, the frequency of timing of manure movement, the capital investment required, and outside environmental and social factors.").
76. *Id.*
77. 2001 Proposed Rule, *supra* note 31, at 3011.
78. *See, e.g.,* GAO/RCED-99-205, *supra* note 38.
79. *See* NPDES PERMIT WRITERS' MANUAL, *supra* note 43, at 8-2 (discussing that often an operation can store litter uncovered outside for up to 15 days before the waste management system converts to a liquid waste management system, and that "operations that stack or pile manure in areas exposed to precipitation are considered to have liquid-manure handling systems. That includes operations that remove litter from the confinement area and stockpile or store it uncovered in remote locations for even a day.").
80. 2001 Proposed Rule, *supra* note 31, at 2988-93.
81. *Id.*.
82. *See id.* at 2988-90 (showing that operationally beef feedlots may use either or both a solid or liquid waste management system, but veal operations and large dairy operations are more likely to employ a liquid waste management system).
83. *Id.* at 2989.

To be semisolid, waste must have a solid content ranging from 10 to 16%.[84] Solid and semisolid waste management systems are often utilized to minimize water use and to reduce the quantity of manure that must be handled and stored.[85]

2. Land Application

Regardless of whether waste is produced as a solid, semisolid, or liquid,[86] once it is collected and potentially treated, some quantities of it may evaporate into the surrounding aerial environmental, some of it may be removed from the property (often for third-party use[87]), or some or all of it may be applied to nearby crops or fields in a process called "land application."[88] Land application is the act of either topically applying or injecting waste and wastewater onto or into land, such as crop, fields, and pastures, usually for the purpose of waste disposal, soil conditioning, or fertilization.[89] As stated by EPA, at a confinement operation the land application area is "any land to which . . . [the operation's] manure and wastewater is applied . . . that is under the control of the . . . [operation's] owner or operator, whether through ownership or a lease contract."[90] Illustratively, USDA has estimated that approximately 83% of concentrated beef cattle operations practice land application.[91]

An operation that land applies its waste product may transport its waste and wastewater to the land application area through a number of different mechanisms. For example, "[s]olids and semisolids are typically transported using mechanical conveyance equipment, pushing the waste down alleys, and transporting the waste in solid manure spreaders,"[92] while "[l]iquids and slurries are [usually] transferred through open channels, pipes, or in a portable liquid tank."[93] Once arriving at the designated land application site, waste and wastewater is usually applied to crops or land using a spraying device (similar to a large sprinkler or other irrigation mechanism); through spreading equipment, usually for more solid waste types;[94] or through below surface injection, which can be used to "mitigate odor problems and volatilization of ammonia."[95] As noted by EPA, waste application or injection rates are expected to be dependent on the nutrient needs of the receiving fields and crops, including site-specific conditions that could affect acceptable waste absorption and use.[96]

C. Environmental Impacts of Industrial Animal Production

Industrial animal production can be appreciated for its application of basic capitalistic business principles to animal gestation and growth; however, it can also be questioned for the ethical quandary that it creates and for its hefty externalized costs of production, principally the hidden costs that it levies on nearby communities and the environment. Even when analyzed solely within an economic framework, the confinement model has a significant limitation in that it largely discounts the fundamental byproducts of animal production, including solid, liquid, and gaseous waste. As summarized by USDA:

---

84. *Id.*
85. *Id.* ("In a liquid or slurry system, the manure is typically mixed with flushing system water from lagoons; the milking center effluent is usually mixed in with the animal manure in the lagoon or in the manure transfer system to ease pumping.")
86. *See, e.g., id.* at 3011 ("land application remains the primary management method for significant quantities of poultry litter (including manure generated from facilities using 'dry' systems)").
87. *See id.* at 3032.
88. *Id.* at 3010, 3031. *See also* J.A. Stingone & S. Wing, *Poultry Litter Incineration as a Source of Energy: Reviewing the Potential for Impacts on Environmental Health and Justice*, 21(1) New Solutions 27-47, 28 (2011) (finding that alternatives to land application do exist, including composting, gasification, and direct combustion/incineration of wastes).
89. *See* 2001 Proposed Rule, *supra* note 31, at 3010, 3031; GAO/RCED-99-205, *supra* note 38, at 8; Pew Recent Changes, *supra* note 49, at 12-13.
90. 2001 Proposed Rule, *supra* note 31, at 3010.
91. Pew Recent Changes, *supra* note 49, at 13 (referencing U.S. Dep't of Agric., Animal and Plant Health Inspection Service, Veterinary Services, Centers for Epidemiology and Animal Health, National Animal Health Monitoring System, N327.0500, *Part I: Baseline Reference of Feedlot Management Practices 1999* (2000)).
92. 2001 Proposed Rule, *supra* note 31, at 2988.
93. *Id.* at 2991.
94. *See* GAO/RCED-99-205, *supra* note 38, at 8-9 ("Irrigation equipment can be used to pump liquid waste from storage structures onto fields; dry waste is usually applied with a tractor-drawn manure spreader.").
95. 2001 Proposed Rule, *supra* note 31, at 2988.
96. *See id.* at 3032-33.

While the transformation benefits society via lower food prices, it is not without costs. Large confined herds concentrate large quantities of manure, which must be removed from housing facilities, stored, and then moved to spread on crops and pasture land. Animal manure contains nutrients like nitrogen, phosphorus, and potassium, and can therefore replace commercial fertilizers. But if not properly managed, manure can pose environmental risks. Excess nutrients do not contribute to further crop growth, but instead may damage air and water resources. Manure also contains bacterial pathogens that can pose direct threats to animal and human health.[97]

When animal wastes and byproducts are not adequately managed and disposed, especially in the large quantities produced at animal confinement operations, they can impact the natural environment on several levels, including through the air, water, and land. In 1995, for example, "25 million gallons of hog waste—more than twice the volume of the Exxon Valdez oil spill—overflowed from a waste storage lagoon in Onslow County, North Carolina," and drained into the New River, devastating the river and its ecosystem and "causing [among other things] a 'massive fish kill.'"[98] In Iowa, between 1992 and 2002, approximately 329 manure spills occurred, "killing some 2.6 million fish"; between 2003 and 2006, hog operations in the state suffered at least 122 waste spills.[99]

At a minimum, these examples show that spills can both go into waterbodies and onto surrounding lands. Animal confinement operations impact the environment in a variety of profound ways—through water pollution, water scarcity, air pollution, contributions to climate change, and land degradation. This section discusses water pollution and water scarcity, air pollution and global warming, land degradation, loss of biodiversity, and environmental justice.[100]

1. Water Pollution and Water Scarcity

Worldwide, environmental and economic health relies on fresh water.[101] At the same time, clean water is becoming increasingly scarce, and impacts to water quality—including impairment of human use and enjoyment, species impacts, drinking water shortages, and the aqueous spread of disease—are found in all societies.[102] International organizations such as the Food and Agriculture Organization of the United Nations project that water-related disputes, including violent conflict and war, are likely to increase greatly in the near future.[103]

a. Water Scarcity

Confinement operations require abundant freshwater resources to service production systems and provide water for confined animals.[104] In the production systems, for example, studies show that a dairy operation using a flush-based waste management system can use more than 150 gallons of water per cow per day, and

---

97. J.M. MacDonald & W.D. McBride, U.S. Dep't of Agric., Economic Research Service, Bulletin 43, *The Transformation of U.S. Livestock Agriculture: Scale, Efficiency, and Risks* 3 (2009), *available at* http://www.ers.usda.gov.
98. R. Innes, *The Economics of Livestock Waste and Its Regulation*, 82 Am. J. Agric. Econ. 97-117, 97 (2000) (referencing T. Vulknia et al., *Swine Odor Nuisance, Voluntary Negotiation, Litigation, and Regulation: North Carolina's Experience*, 1 Choices 26-29 (1996); *Hog Heaven and Hell*, U.S. News & World Rep., Jan. 22, 1996).
99. Food & Water Watch, Turning Farms Into Factories: How the Concentration of Animal Agriculture Threatens Human Health, the Environment, and Rural Communities 5 (2007), *available at* http://documents.foodandwaterwatch.org/FarmsToFactories.pdf (referencing F. Henry, *Mega-Farms Stoke Worries Over Waste Spills*, Plain Dealer, Oct. 9, 2005; Iowa Dep't of Natural Resources, Environmental Services Division, Manure (2003); Iowa Dep't of Natural Resources, Environmental Services Division, Manure (2004); Iowa Dep't of Natural Resources, Environmental Services Division, Manure (2005); Iowa Dep't of Natural Resources, Environmental Services Division, Manure (2006)).
100. *See, e.g., id.*; Livestock's Long Shadow, *supra* note 1.
101. *See* Livestock's Long Shadow, *supra* note 1, at 125 ("Freshwater resources are the pillar sustaining development and maintaining food security, livelihoods, industrial growth, and environmental sustainability throughout the world.") (reference omitted).
102. *Id.* at 125-26 ("Only 2.5 percent of all water resources are fresh water. . . . [70] percent [of which] are locked up in glaciers, and permanent snow [polar caps for example] and the atmosphere. . . . [Because of this, m]ore than 2.3 billion people in 21 countries live in water stressed basins . . .[, and m]ore than one billion people do not have access to clean water.") (reference omitted); 2003 Rule, *supra* note 53, at 7181.
103. *Id.* at 6 ("Environmental degradation is often associated with war and other forms of conflict. Throughout history, peoples and nations have fought over natural resources such as land and water. By increasing the scarcity of these resources, environmental degradation increases the likelihood of violent conflict.") (reference omitted).
104. *Id.* at 126 ("The agricultural sector is the largest user of freshwater resources. In 2000, agriculture accounted for 70 percent of water use and 93 percent of water depletion worldwide.") (reference omitted); *id.* at 128 ("Livestock's use of water and contribution to water depletion trends are high and growing."); *id.* ("Water-use for drinking and servicing animals is the most obvious demand for water resources related to livestock production."); *id.* at 129.

that "a 5,000-swine [confinement operation] may use an estimated 340 million gallons of flushwater each year."[105] Depending on the operation, some of this water can be reclaimed and reused, but most process water is lost through contamination, evaporation, or abandonment in the waste storage structure.[106]

Further, in calculating animal-related water consumption and representative use, a "wide range of interrelated factors," such as "the animal species; the physiological condition of the animal; the level of dry matter intake; the physical form of the diet; . . . and the ambient temperature of the production system," should be considered.[107] Studies indicate that "eighty-seven percent of freshwater withdrawn in the United States from surface and groundwater resources is used in agriculture."[108] Of that, a substantial portion is used in concentrated animal agriculture where "water consumption per animal can exceed that of traditional animal raising practices by up to a factor of five."[109] As the world's demand for meat, poultry, and dairy products intensifies, production models will shift to a greater reliance on shared water resources.[110]

b. Water Pollutants and Water Quality

Science links animal confinement operations to the impairment, sometimes extreme, of water resources. EPA estimates that pollutants from animal operations have impaired over 24,616 river and stream miles in the United States,[111] and that "[o]ver one-quarter of U.S. surface water contamination from agricultural sources has been attributed to livestock, with agricultural sources overall found to be a source of contamination in almost three-quarters of rivers and streams and about one-half of lakes and estuaries that have been identified by the [EPA] as environmentally impaired."[112] In many cases, it is believed that water quality will continue to decline unless discharges of pollutants are reduced and ultimately eliminated.[113]

In large part, these impairments are due to the considerable amounts of pollutant-heavy waste and wastewater generated by these operations. For example, studies estimate that one dairy farm with 2,500 cows will produce as much waste as a city with approximately 411,000 residents,[114] and that "a large farm with 800,000 hogs could produce over 1.6 million tons of manure per year, which is one-and-a-half times more than the annual sanitary waste produced by the city of Philadelphia, Pennsylvania—about one million tons—with a population of almost 1.5 million."[115]

Whether solid or liquid,[116] the waste produced at these operations can contain a number of hazardous pollutants, including antibiotics and hormones, biodegradable organics, heavy metals, nutrients, pathogens, pesticides, salts, sediments, and suspended solids.[117] Associated water-quality impairment principally results when concentrated animal waste and wastewater containing these pollutants enter ground and surface waters,[118] usually as the result of improper operation, maintenance, and design of animal waste management systems.[119]

---

105. PEW RECENT CHANGES, *supra* note 49, at 9 (referencing Agricultural Animal Waste Management Field (AWMF), U.S. Dep't of Agric., Natural Resource Conservation Service, *AWMF Handbook*, in J.N. Krider, ed., NATIONAL ENGINEERING HANDBOOK PART 651. Washington, D.C. (1999)).
106. *Id.*
107. LIVESTOCK'S LONG SHADOW, *supra* note 1, at 128-29 (reference omitted).
108. PEW ENVIRONMENTAL IMPACTS, *supra* note 3, at 8 (referencing Pimentel et al., *Water Resources: Agriculture, the Environment, and Society*, 47(2) BIOSCIENCE 97 (1997)).
109. *Id.* at 22 (referencing Chapagain and Hoekstra, UNESCO-IHE, *Virtual Water Flows Between Nations in Relation to Trade in Livestock and Livestock Products*, Value of Water Research Report Series No. 13 (2003)).
110. *Id.*
111. *See* U.S. EPA, National Water Quality Inventory (2000).
112. R. Innes, *The Economics of Livestock Waste and Its Regulation*, 82 AM. J. AGRIC. ECON. 97-117, 97 (2000) (referencing U.S. EPA, EPA 841-R-94-001, *National Water Quality Inventory: 1992 Report to Congress* (1994)).
113. *See* 2003 Rule, *supra* note 53, at 7179-80.
114. U.S. EPA, NATIONAL RISK MANAGEMENT LABORATORY, RISK MANAGEMENT EVALUATION FOR CONCENTRATED ANIMAL FEEDING OPERATIONS 7 (2004), *available at* http://cfpub.epa.gov/si/si_public_record_report.cfm?dirEntryId=85107 [hereinafter U.S. EPA RISK MANAGEMENT].
115. GAO-08-944, *supra* note 21, at 5.
116. *See, e.g.,* U.S. EPA RISK MANAGEMENT, *supra* note 114, at 34.
117. *See id.* at 24; 2003 Rule, *supra* note 53, at 7235.
118. PEW ENVIRONMENTAL IMPACTS, *supra* note 3, at 8 (CAFOs "can impact the water environment by depleting limited freshwater resources and by contaminating surrounding surface and groundwater.").
119. *Id.* (referencing Burkholder et al., *Impacts of Waste From Concentrated Animal Feeding Operations on Water Quality*, 115(2) ENVTL. HEALTH PERSP. 308 (2007) (Contamination occurs "either directly, via intentional discharge of insufficiently treated liquid, or indirectly, via infiltration of contaminants into the groundwater through unlined [or leaking] waste lagoons, as runoff from locations where solid waste is stored or has been disposed of, and from the deposition of airborne contaminants onto surface waters.").

To appreciate the significance of the water-quality problems that can result from an animal confinement operation, it is important to understand the hazardous pollutants associated with animal waste and wastewater. Of these, the primary pollutants of concern, due to both their prevalence and their impact on water quality, are nutrients, including nitrogen and phosphorus; pathogens; sediments; heavy metals; and antibiotics, hormones, and pesticides.[120]

i. Nutrients

Nutrients, particularly nitrogen and phosphorus, exist in very high levels in animal manure.[121] Nutrients are essential elements of animal and plant life,[122] but they become environmentally damaging when too much is generated and released into the environment.[123] The need to balance the nutrient needs of plants and animals against the excess and harmful application of nutrients for fertilization purposes is known as the "nutrient balance."[124]

Nutrient pollution that results from exceeding the nutrient balance at confined animal operations is often linked to two factors: (1) the surplus generation of manure nutrients due to larger populations; and (2) inadequate land area to reasonably utilize applied manure nutrients.[125] Data from USDA indicates that "the amount of nutrients . . . produced by confined animal operations rose about 20 percent from 1982 to 1997. During that same period, cropland and pastureland controlled by these farms declined from an average of 3.6 acres in 1982 to 2.2 acres per 1,000 pounds live weight of animals in 1997."[126] Data further indicates that "[l]arger-sized operations with 1,000 or more animals exceeding 1,000 pounds accounted for the largest share of excess nutrients in 1997. Roughly 60 percent of the nitrogen and 70 percent of the phosphorus generated by these operations must be transported off-site."[127]

If discharged into waters, phosphorus can cause eutrophication,[128] which in turn can lead to "fish kills, reduce biodiversity, [produce] objectionable tastes and odors, increase drinking water treatment costs, and [promote the] growth of toxic organisms" (see Chapter 3).[129] When discharged into waters, nitrogen can similarly cause eutrophication and numerous human health issues, including methemoglobinemia (also known as "blue baby syndrome"). Nitrogen can also be "toxic to [natural] aquatic life and . . . exert[ ] a

---

120. See 2003 Rule, *supra* note 53, at 7181.
121. See *id.* at 7237 ("[A] 1997 study by Smith et al. characterizing special and temporal patterns in water quality identified animal waste as a significant source of in-stream nutrient concentrations in many watershed outlets, relative to other sources, particularly in the central and eastern United States. The findings of the report suggest that livestock waste contributes more than commercial fertilizer use to local total phosphorus yield."). See also 2003 Rule, *supra* note 53, at 7180 ("[b]y sector, USDA estimates that operations that confine poultry account for the majority of on-farm excess nitrogen and phosphorus. Poultry operations account for nearly one-half of the total recoverable nitrogen, but on-farm use is able to absorb less than 10 percent of that amount. . . . This is attributable to not only the limited land area for manure application but also the generally higher nutrient content of poultry manure compared to manure of most other farm animals. . . . Dairies and hog operations are the other dominant livestock types shown to contribute to excess on-farm nutrients, particularly phosphorus.").
122. Plants can, however, only absorb only a certain amount of nutrients, as determined by the plant type and species. For example, "[m]anure nitrogen occurs in several forms, including ammonia and nitrate. Ammonia and nitrate have fertilizer value for crop growth, but these forms of nitrogen can also produce adverse environmental impacts when they are transported in excess quantities to the environment." 2003 Rule, *supra* note 53, at 7235.
123. See S. Wing et al., *The Potential Impacts of Flooding on Confined Animal Feeding Operations in Eastern North Carolina*, 110(4) Envtl. Health Persp. 387-91, 387 (2002).
124. See GAO-08-944, *supra* note 21, at 20 ("A USDA report identified this concern as early as 2000 when it found that between 1982 and 1997 as livestock production became more spatially concentrated that when manure was applied to cropland, crops were not fully using the manure and this could result in ground and surface water pollution from excess nutrients. According to the report, the number of counties where farms produced more manure nutrients, primarily nitrogen and phosphorus, than could be applied to the land without accumulating nutrients in the soil increased. . . . As a result, the potential for runoff and leaching of these nutrients from the soil was high, and water quality could be impaired, according to USDA. Agricultural experts and government officials who we spoke to during our review echoed the findings of USDA.") (citing R.L. Kellogg et al., *Manure Nutrients Relative to the Capacity of Cropland and Pastureland to Assimilate Nutrients: Spatial and Temporal Trends for the United States* (2000)).
125. 2003 Rule, *supra* note 53, at 7180 ("Among the principle reasons for the farm-level excess of nutrients generated is inadequate land for utilizing manure.").
126. *Id.* at 7180.
127. *Id.*
128. "Eutrophication is the enhancement of the natural process of biological production in rivers, lakes and reservoirs, caused by increases in levels of nutrients, usually phosphorus and nitrogen compounds. Eutrophication can result in visible cyanobacterial or algal blooms, surface scums, floating plant mats and benthic macrophyte aggregations. The decay of this organic matter may lead to the depletion of dissolved oxygen in the water, which in turn can cause secondary problems such as fish mortality from lack of oxygen and liberation of toxic substances or phosphates that were previously bound to oxidised sediments." World Health Organization (WHO), Toxic Cyanobacteria in Water: A Guide to Their Public Health Consequences, Monitoring and Management 1.1 (I. Chorus & J. Bartram, eds., 1999), *available at* http://www.who.int/water_sanitation_health/resourcesquality/toxcyanobacteria.pdf.
129. 2003 Rule, *supra* note 53, at 7235.

direct BOD [biological oxygen demand] on the receiving waters, thereby reducing dissolved oxygen levels and the ability of the waterbody to support aquatic life."[130] In both cases, harmful aquatic plants, such as algae, may flourish in nutrient-rich waters, clog water resources, and lead to fishery and ecosystem impairment.[131] Due to their high prevalence in animal waste, nutrients are commonly considered to be the biggest animal-related threat to water quality.[132]

ii. Pathogens and Fecal Contamination

When animals defecate, they shed microorganisms, including disease-causing bacteria, viruses, and protozoa.[133] Many of these microorganisms are pathogenic and can present a significant risk to the health of humans, livestock, wildlife, and the environment.[134] Pathogenic microorganisms can be transmitted through direct contact, such as from one animal to another, or as the result of contact with affected waste.[135]

A single gram of manure may contain billions of fecal bacteria, and has the potential of containing one or more of the over 150 pathogens that are associated with risks to human health, "including six human pathogens that account for more than 90% of food and waterborne diseases in humans."[136] Those six organisms, *Campylobacter spp., Salmonella spp., Listeria monocytogenes, Escherichia coli (E. coli), Cryptosporidium parvum,* and *Giardia lamblia,* are rapidly transmissible and can cause abdominal discomfort, vomiting, other acute gastrointestinal distress, or even death.[137] In particular, pregnant women, the very young, the very old, and individuals with compromised immune systems are most at risk from these pathogens.[138]

EPA research indicates that bacteria-related pollution, including pathogenic pollution, is the "leading stressor in impaired rivers and streams and the fourth leading stressor in impaired estuaries."[139] In 2001, EPA found that at least "36% of rivers were unfit for swimming and/or fishing as the result of pathogenic contamination largely attributed to [confinement] operations," and that "the source waters from which drinking water is obtained for up to 43% of the United States comes from waters that are impaired by pathogenic contamination from" these operations.[140] With about 15% of the United States population obtaining its drinking water from individual wells, and a much greater percentage of the population utilizing waterbodies for fishing and recreation, the discharge of fecal bacteria, including pathogenic microorganisms, from confinement operations can severely impact human health and environmental welfare.[141]

---

130. *Id. See also* S. Wing et al., *The Potential Impacts of Flooding on Confined Animal Feeding Operations in Eastern North Carolina,* 110(4) ENVTL. HEALTH PERSP. 387-91, 387 (2002) ("Excessive nitrogen and phosphorus can lead to eutrophication of rivers and estuaries, where they can promote harmful algae blooms.").
131. *Id.*
132. *See* 2001 Proposed Rule, *supra* note 31, at 3000 (Jan. 12, 2001) ("[N]utrients from agriculture are one of the leading sources of water contamination in the United States.").
133. *See* B.H. Rosen, U.S. Dep't of Agric., Natural Resources Conservation Service, Watershed Science Institute, *Waterborne Pathogens in Agricultural Watersheds* (2000) ("A pathogen is any agent that causes disease in animals or plants. Pathogens may be a bacterium, protozoan, virus, or worm. *Waterborne zoonotic disease* is a term used to describe that is transmitted among animals and humans by water. . . . Most waterborne pathogens are in human and animal feces."); C.P. Gerba & J.E. Smith Jr., *Source of Pathogenic Microorganisms and Their Fate During Land Application of Wastes,* 34(1) J. ENVTL. QUALITY 42-48 (2005).
134. *See, e.g.,* U.S. EPA RISK MANAGEMENT, *supra* note 114, at 24.
135. *See* 2003 Rule, *supra* note 53, at 7236.
136. *Id.* at 7236. *See also* M.D. Sobsey et al., National Center for Manure and Animal Waste Management, *Pathogens in Animal Wastes and the Impacts of Waste Management Practices on Their Survival, Transport and Fate* (2004), *available at* http://www.cals.ncsu.edu/waste_mgt/natlcenter/whitepapersummaries/pathogens.pdf.
137. Death usually only occurs in sensitive subpopulations such as children and the elderly. U.S. EPA, National Primary Drinking Water Regulations: Revisions to the Total Coliform Rule; Proposed Rule, 75 Fed. Reg. 40925, 40928 (July 14, 2010). *See also* 2003 Rule, *supra* note 53, at 7236.
138. *See, e.g.,* Institute of Food Technologists, *Expert Report on Emerging Microbiological Food Safety Issues: Implications for Control in the 21st Century* (2002), *available at* http://www.ift.org/Knowledge-Center/Read-IFT-Publications/Science-Reports/Expert-Reports/Emerging-Microbiological-Food-Safety-Issues/-/media/Knowledge%20Center/Science%20Reports/Expert%20Reports/Emerging%20Microbiological/Emerging%20Micro.pdf.
139. 2003 Rule, *supra* note 53, at 7235 (referencing U.S. EPA, EPA-841-R-02-001, *National Water Quality Inventory, 2000 Report* (2002), *available at* http://water.epa.gov/lawsregs/guidance/cwa/305b/2000report_index.cfm.
140. U.S. EPA RISK MANAGEMENT, *supra* note 114, at 28-9 (referencing U.S. EPA, *Draft Proceedings of the Workshop on Emerging Infectious Disease Agents and Issues Associated With Animal Manures, Biosolids and Other Similar By-Products,* Vernon-Manor Hotel, Cincinnati, Ohio, June 4-6, 2001).
141. *See* U.S. EPA RISK MANAGEMENT, *supra* note 114, at 24 ("Recreational use of . . . streams may . . . bring people into direct exposure to large numbers of potentially pathogenic microorganisms. Several disease outbreaks have been associated with manure contamination of water or food that has been contacted by manure.").

iii. Antibiotic, Hormone, and Pesticide Residues

Eighty percent or more of confinement operations use a broad spectrum of antibiotics and hormones to control the spread of disease, maintain production yields, promote faster growth, or improve the efficiency of feed conversion.[142] In 2009, an estimated 13.3 million kilograms, or just under 29 million pounds, of antibiotics, including penicillins and tetracyclines, were sold or distributed domestically for use in food-producing animals.[143] Comparable estimates for antibiotic usage in human populations are scarce, but conservative estimates from 2001 place the domestic human use at about 4.5 million pounds.[144] Of the amount used in confinement operations, it is estimated that "about 10% [of antibiotics are] used to treat active infections while the remaining nearly 90% [are] used for growth promotion and prophylactic care," also known as nontherapeutic (or subtherapeutic) use.[145]

Once ingested, hormones and antibiotics may be released into the waste management system through urine and excrement (sometimes in a chemically unchanged form).[146] Due in part to their chemical composition and the metabolic cycle, there are a number of concerns related to the discharge of antibiotics and hormone residues from confinement operations into the environment.[147]

Besides the potentially negative side-effects of unintentional human or animal ingestion of waste residues, of critical concern is the possibility that excessive use of antibiotics in food animal production can breed antibiotic resistance.[148] An EPA report explains:

> Antibiotic resistance develops in microbial populations due to the selective pressure exerted on the population by the antibiotic. If the level of antibiotic used is inadequate to completely eliminate the microorganisms from the animals some members of the population will survive. These organisms will continue to increase their resistance to the antibiotic until the antibiotics are no longer effective in controlling populations or diseases. The enzymatic capacity for resistance to antibiotics may be transferred in the environment by different mechanisms. Plasmids may be transferred directly from microorganism to microorganism, by bacteriophages, or upon cell lysis, leading to the uptake of free plasmids by other organisms.... Antibiotics may also be spread throughout the environment via manure and urine.... Since the antibiotics used for animals are often the same for humans, different antibiotics may have to be used to fight the resistant microbes.[149]

Once developed, antibiotic resistance can be passed on to humans in a number of ways, including through direct contact with animals infected with antibiotic-resistant bacteria, secondary contact (with air, water, dust, or soils contaminated with antibiotic-resistant bacteria), and consumption of animal products contaminated with antibiotic-resistant bacteria.[150]

---

142. *See* 2003 Rule, *supra* note 53, at 7236.
143. *See* U.S. Food & Drug Administration, *2009 Summary Report on Antimicrobials Sold or Distributed for Use in Food-Producing Animals* (2010), available at http://www.fda.gov/downloads/ForIndustry/UserFees/AnimalDrugUserFeeActADUFA/UCM231851.pdf.
144. *See* M. Mellon et al., Union of Concerned Scientist, Hogging It: Estimates of Antimicrobial Abuse in Livestock 17-18, 60 (2001), available at http://www.ucsusa.org/assets/documents/food_and_agriculture/hog_front.pdf.
145. U.S. EPA Risk Management, *supra* note 114, at 36. *See also* M. Mellon et al., Union of Concerned Scientist, *Hogging It: Estimates of Antimicrobial Abuse in Livestock* 22 (2001), available at http://www.ucsusa.org/assets/documents/food_and_agriculture/hog_front.pdf (discussing that nontherapeutic use is often defined as the application of antibiotics for uses such as "growth promotion and disease prevention in which animals are treated in the absence of illness."); U.S. Food & Drug Administration, *Draft Guidance: The Judicious Use of Medically Important Antimicrobial Drugs in Food-Producing Animals* 4 (2010), available at http://www.fda.gov/downloads/AnimalVeterinary/GuidanceComplianceEnforcement/GuidanceforIndustry/UCM216936.pdf..
146. *See* 2003 Rule, *supra* note 53, at 7236; Kolpin et al., *Pharmaceuticals, Hormones, and Other Organic Wastewater Contaminants in U.S. Streams*, 36 Envtl. Sci. & Tech. 1202-11 (2002); Daughton, U.S. EPA, Office of Research and Development, *PPCPs as Environmental Pollutants* (2004) (discussing that physiologically animals such as swine are very similar to humans in the way they use and excrete hormones).
147. *Id.*
148. *See* A.R. Burriel, *Resistance to Coagulase-Negative Staphylococci Isolated From Sheep to Various Antimicrobial Agents*, 63 Res. Veterinary Sci. 189-90 (1997); J.M. MacDonald & W.D. McBride, U.S. Dep't of Agric., Economic Research Service, Bulletin 43, *The Transformation of U.S. Livestock Agriculture: Scale, Efficiency, and Risks* 3 (2009), available at http://www.ers.usda.gov; Nijsten et al., *Antibiotic Resistance Among Escherichia Coli Isolated From Fecal Samples of Swine Farmers and Swine*, 37 J. Antimicrobial Chemotherapy 1131-40 (1996).
149. U.S. EPA Risk Management, *supra* note 114, at 37.
150. *See* A. Batt et al., *Occurrence of Sulfonamide Antimicrobials in Private Water Wells in Washington County, Idaho, USA*, 64(11) Chemosphere 1963-71 (2006) (finding that even "low levels of antibiotics [in water resources] can favor the proliferation of antibiotic resistant bacteria."). *See also* C.M. Lathers, *Clinical Pharmacology of Antimicrobial Use in Humans and Animals*, 42 J. Clinical Pharmacology 587-600 (2002) (determining that, in addition to antibiotic resistance to antibiotics used in humans, "bacteria can develop cross-resistance between antibiotics used in veterinary medicine with those of similar structures that are used exclusively in human medicine"); Sapkota et al., *Antibiotic-Resistance Enterococci and Fecal Indicators in Surface Water and Groundwater Impacted by Concentrated Swine Feeding Operation*, 115 Envtl. Health Persp. 1040-45, 1043 (2007); W. Witte, *Medical Consequences of Antibiotic Use in Agriculture*, 279 Sci. 996-97 (1998).

Of additional concern is that hormone dosing at confinement operations can increase the prevalence in waterbodies of endocrine-disrupting chemicals, which can interfere with the synthesis, secretion, transport, binding, action, or elimination of natural hormones in the body.[151] Confinement operations routinely use estrogens, androgens, progesterones, and thyroid hormones, chemicals that are often not fully metabolized during animal digestion.[152] If ingested, either intentionally or unintentionally through impacted waters, these chemicals can disturb natural hormone homeostasis in humans, fish, and wildlife by affecting mood, behavior, and reproduction.[153] For example, confined beef cattle can be fed the anabolic steroid trebolone acetate to promote the growth of muscle.[154] During degradation, this steroid creates a product that is stable in water and highly capable of binding to androgen receptors in fish.[155] Exposure to this compound has been shown to cause the masculinization of female fish, and results in reduced fertility even at low concentrations.[156] The results in fish indicate what can be anticipated in human beings. The United States Geological Society (USGS) considers endocrine disruption that "may occur as a result of exposure to very low levels of hormonally active chemicals" to be an "adverse health effect of concern."[157]

Pesticide residues are also pollutants of concern at confinement operations. At operations, "[p]esticides are applied to livestock [and poultry] to suppress houseflies and other pests."[158] Those pesticides are subsequently transported, with other wastes and wastewater, to waste storage areas. While there has been little research on the quantities of pesticides that are released through discharges, including runoff from confinement operations, EPA believes that they should be "expected to appear in animal waste," and can impact water quality as a result.[159] Contact with pesticides, such as through tainted waters, can be extremely harmful to humans, wildlife, and fish populations.[160]

iv. Heavy Metals

Confined animals are sometimes fed metals, such as arsenic, zinc, or copper, to promote growth and improve feed efficiency.[161] Much like antibiotics and hormones, animals do not fully digest and metabolize heavy metals.[162] As a result, undigested heavy metals end up in concentrated waste products, and if discharged to water resources, they can contribute to "phytotoxicity, groundwater contamination, and deposition in river sediment that may eventually release to pollute the water,"[163] and can cause health problems in exposed persons.[164] In particular, "[e]xposure to arsenic, a known human carcinogen, is a major health concern. In addition to being associated with multiple types of cancer, arsenic exposure, even at low ambient levels, has been found to be associated with cardiovascular disease, diabetes, endocrine disruption, and decreased immunity."[165] In general, zinc and copper are not considered to be as dangerous to humans and

---

151. U.S. EPA RISK MANAGEMENT, *supra* note 114, at 38.
152. *Id.; see also* 2003 Rule, *supra* note 53, at 7236.
153. U.S. EPA RISK MANAGEMENT, *supra* note 114, at 38.
154. *See* E.J. Durhan et al., *Identification of Metabolites of Trebolone Acetate in Androgenic Runoff From a Beef Feedlot*, 114 Suppl. 1 ENVTL. HEALTH PERSP. 65-68 (2006).
155. *Id. See also* Hotchkiss et al., *Fifteen Years After "Wingspread"—Environmental Endocrine Disruptors and Human and Wildlife Health: Where Are We Today and Where We Need to Go.* TOXICOLOGICAL SCI. ADVANCE ACCESS (Feb. 16, 2008).
156. *Id.*
157. R. Hirsch, Associate Director for Water, USGS, Statement Before the Committee on Environment and Public Works Subcommittee on Transportation Safety, Infrastructure Security and Water Quality, Apr. 15, 2008.
158. 2003 Rule, *supra* note 53, at 7236.
159. *Id.*
160. *See, e.g.*, FAO, FAO Plant Production and Protection Paper 197, *FAO Manual on the Submission and Evaluation of Pesticide Residue Data for the Estimation of Maximum Residue Levels in Food and Feed* (2009), *available at* http://www.fao.org/docrep/012/i1216e/i1216e.pdf; U.S. EPA, *Pesticides: Health and Safety, available at* http://www.epa.gov/pesticides/health/human.htm (stating that pesticides "such as the organophosphates and carbamates, affect the nervous system. Others may irritate the skin or eyes. Some pesticides may be carcinogens."); U.S. EPA, *Pesticides: Health and Safety, available at* http://www.epa.gov/pesticides/health/reducing.htm (stating that since pesticides "[b]y their nature [are] substances that in many cases are designed to kill pests, [they] can pose risks to humans and to the environment").
161. *See* PEW ENVIRONMENTAL IMPACTS, *supra* note 3, at 11; U.S. EPA RISK MANAGEMENT, *supra* note 114, at 43.
162. *See* C.D. Church et al., *Occurrence of Arsenic and Phosphorus in Ditch Flow From Litter-Amended Soils and Barn Areas*, 39 J. ENVTL. QUALITY 2080-88, 2080 (2010); U.S. EPA RISK MANAGEMENT, *supra* note 114, at 24.
163. *See* U.S. EPA RISK MANAGEMENT, *supra* note 114, at 43.
164. *See id. See also* 2003 Rule, *supra* note 53, at 7236.
165. J.A. Stingone & S. Wing, *Poultry Litter Incineration as a Source of Energy: Reviewing the Potential for Impacts on Environmental Health and Justice*, 1(1) NEW SOLUTIONS 27-47, 33 (2011) (referencing International Agency for Research on Cancer, Monographs on the Evaluation of Carcinogenic Risks to Humans: Some Drinking Water Disinfectants and Contaminants, Including Arsenic (2004), *available at* http://monographs.iarc.fr/ENG/Monographs/vol84/mono84-6.pdf); M. Vahter, *Health Effects of Early Life Exposure to Arsenic*, 102 BASIC & CLINI-

wildlife as arsenic, but they can cause health problems if they are toxically discharged, either directly or through soil accumulation and erosion, into waters.[166]

v. Sediments, Suspended Solids, and Biodegradable Organics

Sediments are also a pollutant of concern, but by the nature of erosion, sediments are often better discussed in relation to methods of discharge. It should be noted, however, that sediments ultimately are a concern for water quality not only because they can clog water resources, but because they are able to transport many of the pollutants of concern from soils at the confinement operation to water resources. When soils containing animal wastes and wastewater erode into waterbodies, biological oxygen demand can increase, which decreases dissolved oxygen levels.[167] Increased sediment loads can also decrease the habitability of waterbodies for local species by changing the composition and turbidity of waters.[168]

Further, suspended solids are tiny particles of organic matter that result when decaying plant matter, manure, and the other solid elements mixed in with manure, such as spilled feed, bedding materials, hair, and feathers, are discharged into water.[169] The presence of suspended solids causes water to be turbid, "physically hindering the functioning of aquatic plants and animals, . . . providing a protective environment for pathogens . . . [and] limiting the growth of desirable aquatic plants that serve as a critical habitat for fish, shellfish, and other aquatic organisms."[170] Suspended solids cause impairment in a number of U.S. waterways.[171]

Similarly, livestock and poultry waste and wastewater contain biodegradable organic compounds and matter.[172] Once introduced into water resources, these organic compounds are usually consumed by aerobic bacteria and other microorganisms for energy, growth, and reproduction.[173] This digestion process is oxygen-intensive, and can limit the amount of oxygen available for fish and other aquatic creatures.[174] Depressed oxygen levels can cause massive fish kills and decrease biodiversity.[175]

c. Pathways for the Entry of Pollutants Into Water Resources

Animal waste can enter and damage surface and groundwater resources in a number of ways.[176] Generally, pollutants from animal waste find their way into water resources through "surface runoff and erosion, direct discharges to surface water, spills and other dry-weather discharges [including intentional releases and accidental discharges from lagoons], leaching into soil and groundwater, and volatilization of compounds (e.g., ammonia) and subsequent redeposition to the landscape" from either the production area or the land application area.[177] Therefore, pollutants from confinement operations typically enter water

---

CAL PHARMACOLOGY & TOXICOLOGY 204-11 (2008); C.D. Kozul et al., *Low-Dose Arsenic Compromises the Immune Response to Influenza A Infection in Vivo*, 117 ENVTL. HEALTH PERSP. 1441-47 (2009); J.C. Davey et al., *Arsenic as an Endocrine Disruptor: Arsenic Disrupts Retinoic Acid Receptor- and Thyroid Hormone Receptor-Mediated Gene Regulation and Thyroid Hormone-Mediated Amphibian Tail Metamorphosis*, 116 ENVTL. HEALTH PERSP. 165-72 (2008).

166. See 2003 Rule, *supra* note 53, at 7236.
167. U.S. EPA RISK MANAGEMENT, *supra* note 114, at 24.
168. *Id.*
169. See 2003 Rule, *supra* note 53, at 7235.
170. *Id.*
171. U.S. EPA, *Potential Environmental Impacts of Animal Feeding Operations*, http://www.epa.gov/agriculture/ag101/impacts.html.
172. METCALF & EDDY, WASTEWATER ENGINEERING TREATMENT AND REUSE (McGraw-Hill, 4th ed. 2003).
173. See 2003 Rule, *supra* note 53, at 7235.
174. *Id.*
175. *Id.*
176. See GAO-08-944, *supra* note 21, at 5 ("If improperly managed, manure and wastewater from animal feeding operations can adversely impact water quality through surface runoff and erosion, direct discharges to surface water, spills and other dry-weather discharges, and leaching into soils and groundwater."). See also E. Campagnalo et al., *Chemical Assessment of Surface and Groundwater in the Environment Proximal to Large-Scale Swine and Poultry Feeding Operations; A Pilot Investigation*, 299(1-3) THE SCIENCE OF THE TOTAL ENVIRONMENT 89-95 (2002); T. Ciravolo et al., *Pollutant Movement to Shallow Ground Water Tables From Anaerobic Swine Waste Lagoons*, 8 J. ENVTL. QUALITY 126-30 (1979); G.C. Heathman et al., *Land Application of Poultry Litter and Water Quality in Oklahoma U.S.A.*, 40 FERTILIZER RES. 165-73 (1995); M.A. Mallin, *Impacts of Animal Production on Rivers and Estuaries*, 88 AM. SCIENTIST 26-37 (2000); J.A. Stingone & S. Wing, *Poultry Litter Incineration as a Source of Energy: Reviewing the Potential for Impacts on Environmental Health and Justice*, 21(1) NEW SOLUTIONS 27-47, 33 (2011); Stone et al., *Impact of Swine Waste Application on Ground and Stream Water Quality in an Eastern Coastal Plain Watershed*, 41(6) AM. SOC'Y AGRIC. ENGINEERS 1665-70 (1998).
177. 2003 Rule, *supra* note 53, at 7236-37 ("discharges of manure pollutants can originate from animal confinement areas, manure handling and containment systems, manure stockpiles, and cropland where manure is spread. . . . Other reported causes of discharge to surface waters are

resources through three sources: (1) operation and maintenance problems, often associated with the animal waste management system; (2) discharges of pollutants from the land application area; and (3) deposition. Such discharges can be continuous or intermittent.[178]

One of the primary sources of water pollution from confinement operations is the operation's waste management system.[179] Regardless of waste type, discharges from waste management systems can result, either intentionally or unintentionally, from manure spills, leaks, overflows, or other accidental discharges.[180] Once discharges occur, pollutants usually reach surface waters through direct overland flow and runoff[181] or by traveling through ditches or drains into surface waters,[182] and they reach groundwater primarily through leaching.[183]

Intentional discharges can be attributed to the deliberate transfer of wastes from an operation's waste management system to a nearby water resource. Some waste discharges, while perhaps not intentional, are such that it is reasonably foreseeable that they will result in the discharge of pollutants to water bodies. For example, when an operation utilizing a dry manure (litter) management system leaves manure outside and uncovered, one could conclude that rain falling on the manure is reasonably likely to transport pollutants from the waste pile onto neighboring soils, causing leaching and potential groundwater pollution, and possibly into nearby ditches and surface water resources.[184]

On the other hand, water pollution resulting from unintentional discharges from waste management systems can occur, for example, as the result of "overflows from containment systems following rainfall, catastrophic spills from failure of manure containment systems, and washouts from floodwaters when lagoons are sited on floodplains or from equipment malfunction, such as pump or irrigation gun failure, and breakage of pipes or retaining walls."[185] Unintentional discharges can directly impact groundwater resources through systemic damage or rupture, often in the form of leaking.[186] Indeed, a study of 34 functioning swine waste lagoons in North Carolina found that almost 80% of those tested were experiencing moderate to very strong lagoon seepage, indicating a high probably of contamination to nearby groundwater resources.[187]

The impacts to water quality from waste system discharges can be substantial, even in comparison to other discharge pathways, because the pollutants in the waste are often the most concentrated at this stage of waste management.[188] The proper operation and maintenance of waste treatment systems will help prevent systemic failures and discharges from the manure treatment and storage areas.[189] Specifically, adressing several issues can lessen the probability of discharges: discontinuing the "careless transfer of manure to application equipment; improper manure agitation practices; inadequate controls to prevent

---

overflows from containment systems following rainfall, catastrophic spills from failure of manure containment systems, and washouts from floodwaters when lagoons are sited on floodplains or from equipment malfunction, such as pump or irrigation gun failure, and breakage of pipes or retaining walls.").

178. *See id.* at 7216 ("CAFO runoff can be highly intermittent and is usually characterized by very high flows occurring over relatively short time intervals.").
179. *See, e.g.* 2001 Proposed Rule, *supra* note 31, at 2979.
180. *See* 2003 Rule, *supra* note 53, at 7236-37; Burkholder et al., *Impacts to a Coastal River and Estuary From Rupture of a Large Swine Waste Holding Lagoon*, 26 J. ENVTL. QUALITY 1451-66 (1997); R.L. Huffman & P.W. Westerman, *Estimated Seepage Losses From Established Swine Waste Lagoons in the Lower Coastal Plain of North Carolina*, 38 TRANSACTIONS AM. SOC'Y AGRIC. ENGINEERS 449-53 (1994); Mallin et al., *Effects of Animal Waste Spills on Receiving Waters, in* SOLUTIONS: PROCEED. TECH. CONFERENCE ON WATER QUALITY, North Carolina State University, Mar. 19-21, 1996.
181. Runoff is characterized as wastes and wastewater that is applied or otherwise reaches manmade surfaces or soils without getting absorbed and ultimately traveling, or "running," into surface waters. *See* 2001 Proposed Rule, *supra* note 31, at 2979.
182. *See id.*
183. *Id.* at 2980.
184. *See id.* at 3011 ("If [dry] manure is stored in open stockpiles over long periods of time, usually great than a few weeks, runoff from the stockpiles may contribute pollutants to surface water and/or ground water."); 2003 Rule, *supra* note 53, at 7192.
185. 2003 Rule, *supra* note 53, at 7237.
186. *See, e.g., id.* ("It is well established that in many agricultural areas shallow ground water can become contaminated with manure pollutants."); R.L. Huffman & P.W. Westerman, *Estimated Seepage Losses From Established Swine Waste Lagoons in the Lower Coastal Plain of North Carolina*, 38 TRANSACTIONS AM. SOC'Y AGRIC. ENGINEERS 449-53 (1994); D.B. Parker et al., *Seepage From Earthen Animal Waste Ponds and Lagoons—An Overview of Research Results and State Regulations*, 42(2) TRANSACTIONS OF THE ASAE 485-93 (1999).
187. *See* R.L. Huffman, *Seepage Evaluation of Older Lagoons in North Carolina*, 47(5) TRANSACTIONS OF THE ASAE 1507-1512 (2004).
188. J. Burkholder et al., *Impacts of Waste From Concentrated Animal Feeding Operations on Water Quality*, 115 ENVTL. HEALTH PERSP. 308-12 (2006) (referencing S. Wing et al., *The Potential Impacts of Flooding on Confined Animal Feeding Operations in Eastern North Carolina*, 110 ENVTL. HEALTH PERSP. 387-91 (2002)).
189. 2003 Rule, *supra* note 53, at 7215. *See also* 2001 Proposed Rule, *supra* note 31, at 2979 ("In 1997, an independent review of Indiana Department of Environmental Management records indicated that the most common causes of waste releases in that state were intentional discharge and lack of operator knowledge, rather than spills due to sever rainfall conditions.").

burrowing animals and plants from eroding the storage berms and sidewalls; lack of routine inspection of land application and dewatering equipment during lagoon drawdown; and infrequent visual confirmations of adequate [waste storage space]."[190]

Animal wastes are supposed to be land applied at fertilization rates that can be assimilated by receiving soils and crops, often referred to as "agronomic rates."[191] However, the "[a]pplication of manure, litter, and other process wastewaters in excess of the crop's nutrient requirements increases the pollution runoff from fields because the crop does not need these nutrients."[192] Disproportionate land application can impact surface waters and can lead to the leaching of pollutants into groundwater resources.[193] "[R]unoff of animal waste is more likely when rainfall occurs soon after application (particularly if the manure was not injected or incorporated)," but discharge can additionally occur during dry periods if waste or wastewater is overapplied or misapplied.[194]

Land application practices also can contribute indirectly to the discharge of pollutants into water resources through erosion.[195] Erosion occurs when wind and water displace soils. At confinement operations, eroded soils can impair water quality by transporting pollutants, often nutrients, bound to the soil during land application to nearby water bodies.[196] A 1999 report by the Agricultural Research Service found that "phosphorus bound to eroded sediment particles makes up 60 to 90% of phosphorus transported in surface runoff from cultivated land."[197] In addition, eroded soils can impair water quality by clogging water resources.[198]

Extensive reliance on land application practices makes the land application field an especially susceptible route for the discharge of pollutants. Estimates find that 95-99% of waste generated at swine and cattle operations and over 90% of poultry waste is disposed of through land application practices.[199] "In the southern and southeastern regions of the U.S., where lagoons are commonly used for pig manure treatment and storage, effluent is irrigated to nearby cropland using sprinkler systems, such as big gun and center-pivot units,"[200] which can result in wind drift, overapplication, or spraying directly into ditches and waterways. Consistently diminishing land application field size and clustering of confinement operations in small geographic regions increases the concentration of pollutants and further stresses water resources.[201]

---

190. 2003 Rule, *supra* note 53, at 7215.
191. *See id.* at 7210.
192. *Id.*
193. *See, e.g., id.* at 7196 ("EPA noted . . . that the runoff from land application of manure at CAFOs is a major route of pollutant discharge from CAFOs; [and] that in some regions of the country, the amount of nutrients present in land applied manure has the potential to exceed the nutrient needs of the crops."); R.O. Evans, *Subsurface Drainage Water Quality From Land Application of Swine Lagoon Effluent*, 27 Transactions Am. Soc'y Agric. Engineers 473-80 (1984); U.S. Geological Survey, Report 2004-5283, *Geochemistry and Characteristics of Nitrogen Transport at a Confined Animal Feeding Operation in a Coastal Plain Agricultural Watershed, and Implications for Nutrient Loading in the Neuse River Basin, North Carolina, 1999-2002* (2004) (determining at the USGS Lizzie Research Station located in Greene County that increased concentrations of nitrate and other chemical constituents in the groundwater beneath sprayfields. The nitrate concentrations ranged from 10 to 35 mg/L with one concentration as high as 56 mg/L. During the four years of spray applications, groundwater nitrate levels increased by a factor of 3.5.); Hill et al., *Impact of Animal Application on Runoff Water Quality in Field Experiment Plots*, 2(2) Int'l J' Envtl. Res. & Pub. Health 314-21 (2005); J.A. Stingone & S. Wing, *Poultry Litter Incineration as a Source of Energy: Reviewing the Potential for Impacts on Environmental Health and Justice*, 21(1) New Solutions 27-47, 28 (2011); T.J. Sauer et al., *Runoff Water Quality From Poultry Litter-Treated Pasture and Forest Sites*, 29 J. Envtl. Quality 515-21 (2000) ("While poultry litter can be a valuable fertilizer due to its high nutrient content, poor waste management practices, including over-application, can result in nitrogen and/or phosphorus runoff and subsequent eutrophication of nearby waterways.").
194. 2003 Rule, *supra* note 53, at 7236-37.
195. *See* 2001 Proposed Rule, *supra* note 31, at 2979.
196. *See id.*; U.S. EPA Risk Management, *supra* note 114, at 24.
197. 2001 Proposed Rule, *supra* note 31, at 2979.
198. *See* U.S. EPA Risk Management, *supra* note 114, at 24.
199. E. Silbergeld et al., Pew Commission on Farm Animal Production, Antimicrobial Resistance and Human Health 31 (2008), *available at* http://www.pewtrusts.org/uploadedFiles/wwwpewtrustsorg/Reports/Industrial_Agriculture/PCIFAP_AntbioRprtv.pdf [hereinafter PEW Antimicrobial Resistance] (citing U.S. Dep't of Agric., USDA National Program Annual Report (2005), *available at* http://www.ars.usda.gov/research/programs/programs.htm?np_code=206&docid=1333; USDA/APHIS, National Health Monitoring System, *Part I: Feedlot Management Practices* (1995); T.E. Walton, *Swine 2000-Part III: Reference of Swine Health & Environmental Management in the United States* (2002); P. Moore et al., *Poultry Manure Management: Environmentally Sound Options*, 50(3) J. Soil & Water Conservation 321-27 (1995)).
200. J. Arogo et al., *A Review of Ammonia Emissions From Confined Swine Feeding Operations*, 46(3) Transactions of the ASAE 805-17, 810 (2003).
201. GAO-08-944, *supra* note 21, at 5 ("According to agricultural experts and government officials . . . clustering of operations raises concerns that the amount of manure produced could result in overapplication of manure to croplands in these areas and the release of excessive levels of some pollutants that could potentially damage water quality"); *id.* at 22 ("According to North Carolina agricultural experts, excessive manure production has contributed to the contamination of some of the surface and well water in these counties and the surrounding areas. According to these experts, this contamination may have occurred because the hog farms are attempting to disposed of excess manure but have little available cropland that can effectively use it.").

Finally, pollutants can impact water resources through deposition, which occurs when airborne pollutants, such as nitrogen, are returned (deposited) back to the earth.[202] As it relates to water pollution, deposition can be either wet (the atmospheric combination of an aerial pollutant and a precipitant, such as rain, sleet, and snow) or dry (which occurs when particles and gases are deposited onto soil and water resources in the absence of a precipitant).[203] In the case of confinement operations, deposition often results from operational emissions of ammonia gas into the atmosphere.[204] The wet or dry deposition of ammonia nitrogen can impact water quality by increasing nutrient concentrations in receiving waterways, thereby causing the eutrophication of surface waters, oxygen depletion, algal growth, and aquatic species impacts.[205]

2. Air Pollution and Climate Change

Confinement operations produce and emit six primary categories of air pollutants: ammonia; hydrogen sulfide; volatile organic compounds; particulate matter; endotoxins; and greenhouse gases, including carbon dioxide, methane, and nitrous oxide.[206] These air pollutants can impair the ambient environment and harm human populations through "inflammatory, immunologic, irritant, neurochemical, and psychophysiologic mechanisms."[207] This section discusses the air pollutants associated with confinement operations and describes the common sources for those aerial emissions.

a. Air Pollutants of Concern: Types and Impacts

i. Ammonia

Domestic animals are the largest global source of atmospheric ammonia emissions, constituting at least 40% of natural and anthropogenic sources.[208] Domestically, this adds up to almost 2.5 million tons of

---

202. U.S. Geological Survey, Fact Sheet FS-112-00, *Atmospheric Deposition Program of the U.S. Geological Survey* (December 2000).
203. *Id.*
204. *See* 2001 Proposed Rule, *supra* note 31, at 2980 ("Ammonia is very volatile, and can have significant impacts on water quality through atmospheric deposition."); J.T. Walker et al., *Atmospheric Transportation and Wet Deposition of Ammonium in North Carolina*, 34 Atmospheric Env't 3407-18, 3408 (2000).
205. *See* Aneja et al., *Agricultural Ammonia Emissions and Ammonium Concentrations Associated With Aerosols and Precipitation in the Southeast United States*, 108(D4) J. Geophysical Res. 4152-63 (2003); Aneja et al., *Characterization of Atmospheric Ammonia Emissions From Swine Waste Storage and Treatment Lagoons*, 105(D9) J. Geophysical Res. 11535-45 (2000); W.A. Asman et al., *Ammonia: Emissions, Atmospheric Transport and Deposition*, 139 New Phytologist 27-48 (1998); B.K.S. Bajwa et al., *Modeling Studies of Ammonia Dispersion and Dry Deposition at Some Hog Farms in North Carolina*, 58 J. Air & Waste Mgmt. Ass'n 1198-1207, 1198 (2008) ("On a global basis, the amount of nitrogen that enters the biosphere has nearly doubled when compared with the preindustrial times, and a significant component of this increase has been in the form of [ammonia]-nitrogen. In continents with intensive agriculture, atmospheric inputs of reduced nitrogen as [ammonia] and [ammonium] by dry and wet deposition may represent a substantial contribution to the acidification of seminatural ecosystems."); W. Cure et al., North Carolina Department of Environment and Natural Resources, Division of Air Quality, *Status Report on Emissions and Deposition of Atmospheric Nitrogen Compounds From Animal Production in North* 7 (1999), available at http://daq.state.nc.us/monitor/projects/ ("EPA's Second Great Waters Report to Congress reports that more than 40% of the nitrogen(N) entering the Albemarle-Pamlico Sounds is estimated to come from the atmosphere. Comparisons between current and historical levels of N inputs are difficult but if atmospheric inputs are on the rise, then atmospheric depositions could be a major factor contributing to the over-enrichment, or eutrophication, of coastal waters and estuaries."); J.T. Walker et al., *Atmospheric Transportation and Wet Deposition of Ammonium in North Carolina*, 34 Atmospheric Env't 3407-18, 3408 (2000).
206. *See, e.g.,* V.P. Aneja et al., *Effects of Agriculture Upon the Air Quality and Climate: Research, Policy, and Regulation*, 43 Envtl. Sci. & Tech. 4234-40 (2009); K.J. Donham & W.J. Popendorf, *Ambient Levels of Selected Gases Inside Swine Confinement Buildings*, 46(11) Am. Indus. Hygiene Ass'n J. 658-61 (1985); S. Goetz et al., *Measurement, Analysis and Modeling of Fine Particular Matter in Eastern North Carolina*, 58 J. Air & Waste Mgmt. Ass'n 1208-14 (2008); Iowa State University and the University of Iowa. *Iowa Concentrated Animal Feeding Operations Air Quality Study* (Iowa City, IA 2002), available at http://www.ehsrc.uiowa.edu/cafo_air_quality_study.html; National Academy of Sciences, *The Scientific Basis for Estimating Air Emissions From Animal Feeding Operations* (Washington, D.C.: National Academy Press 2002); S.J. Reynolds et al., *Air Quality Assessments in the Vicinity of Swine Production Facilities*, 4 J. Agromed. 37-45 (1997); S.S. Schiffman, *Livestock Odors: Implications for Human Health and Well-Being*, 76 J. Animal Sci. 1343-55 (1998); P.S. Thorne et al., *Concentrations of Bioaerosols, Odors and Hydrogen Sulfide Inside and Downwind From Two Types of Swine Operations*, 6 J. Occupational Hygiene 211-20 (2009); J.T. Walker et al., *Inorganic PM$_{2.5}$ at a U.S. Agriculture Site*, 139 Envtl. Pollutants 258-71 (2006); S. Wing et al., *Air Pollution and Odor in Communities Near Industrial Swine Operations*, 116 Envtl. Health Persp. 1362-68 (2008).
207. S. Wing & S. Wolf, *Intensive Livestock Operations, Health, and Quality of Life Among Eastern North Carolina Residents*, 108(3) Envtl. Health Persp. 233-38, 233 (2000). *See also* S.S. Schiffman, *Livestock Odors: Implications for Human Health and Well-Being*, 76 J. Animal Sci. 1343-55 (1998).
208. *See* V. Aneja et al., *Characterization of Atmospheric Ammonia Emissions From Swine Waste Storage and Treatment Lagoons*, 105(D9) J. Geophysical Res. 11535-45, 11535 (2000); V. Aneja et al., *Emerging National Research Needs for Agricultural Air Quality*, 87(3) EOS 25-36 (2006); A.F. Bouwman et al., *A Global High-Resolution Emission Inventory for Ammonia*, Global Biogeochemical Cycles 11:561-87 (1997); W.H.

ammonia emissions from animal operations alone.[209] On a regional scale, "[n]itrogen emissions estimates for the state of North Carolina (NC) show domestic animals to be the largest statewide contributor of [ammonia], with swine operations present as the primary domestic animal source."[210] While ammonia is emitted by all animal operations to some degree, estimates from 2002 found that the largest producers of domestic ammonia emissions are dairy cattle operations, followed closely by swine operations and broiler poultry operations.[211]

Ammonia, a water soluble colorless gas, is produced during the decomposition of organic nitrogen compounds in livestock and poultry manure.[212] "Nitrogen occurs as both unabsorbed nutrients in animal feces and as either urea (mammals) or uric acid (poultry) in urine."[213] Undigested feed protein, wasted feed, and decomposing dead animals are additional sources of ammonia emissions.[214] Ammonia can be directly aerosolized upon excretion or it can remain in deposited waste to be volatized at a later time.[215]

The decomposition of animal manure occurs continuously, with urine most readily hydrolyzing (a step in waste degradation that results in the release of ammonia) upon excretion.[216] While ammonia emissions occur throughout the year, the degree and quantity of emissions depend on season and geography, with the highest prevalence of emissions occurring during periods of high temperature or high wind.[217] Ammonia emissions also are dependent on animal type, feed type and amount, waste pH, and operational waste management practices.[218]

Ammonia emissions can negatively impact air quality by reducing visibility and by forming inhalable and odorous aerosol particles, which can affect both human and animal health.[219] As noted above, ammonia emissions also contribute to the eutrophication of surface waters, acidification of soils, damage to vegetation and plant biodiversity, and increased contamination of groundwater through deposition.[220] Deposition can consist of nitrogen alone, or it can be more complex, such as when volatized ammonia "react[s] with acids (nitric, sulfuric, and hydrochloric), forming aerosols," or binds with sulfur dioxide.[221] In either case, unlike many air pollutants "[a]mmonia emissions have a relatively short life in the atmosphere before they can enter terrestrial and aquatic ecosystems as a component of wet or dry

---

Schlesinger & A.E. Hartley, *A Global Budget for Atmospheric NH3*, 15 Biochemistry 191-211 (1992); J.T. Walker et al., *Atmospheric Transportation and Wet Deposition of Ammonium in North Carolina*, 34 Atmospheric Env't 3407-18 (2000).

209. *See* U.S. EPA, *National Emissions Inventory—Ammonia Emissions From Animal Husbandry Operations*, Draft Report E-4 (Jan. 20, 2004) [hereinafter *National Emissions Inventory*].
210. *See* J.T. Walker et al., *Atmospheric Transportation and Wet Deposition of Ammonium in North Carolina*, 34 Atmospheric Env't 3407-18, 3408 (2000) (referencing V.P. Aneja, *Summary of Discussion and Research Recommendations*, in Proceedings of the Workshop on Atmospheric Nitrogen Compounds: Emissions, Transport, Transformation, Deposition and Assessment, Raleigh, NC (1997); R.E. Wooten, *Nitrogen Oxides and Ammonia Emission Inventories for North Carolina*, in Proceedings of the Workshop on Atmospheric Nitrogen Compounds: Emissions, Transport, Transformation, Deposition and Assessment, Raleigh, NC (1997).
211. *See National Emissions Inventory*, *supra* note 209, at 4-5.
212. *See id.* at E-1.
213. *Id.*
214. *See* Iowa State University, Practices to Reduce Ammonia Emissions From Livestock Operations, PM 1971a (July 2004); S. Schiffman et al., *Health Effects of Aerial Emissions From Animal Production Waste Management Systems*, in Proceedings of International Symposium: Addressing Animal Production and Environmental Issues, Raleigh, NC (2001) [hereinafter Schiffman Health Effects].
215. *See* Eastern Research Group, U.S. EPA, *Non-Water Quality Impact Estimates for Animal Feeding Operations*. Chantilly, VA (2002).
216. *See National Emissions Inventory*, *supra* note 209, at E-1 ("The volatilization of ammonia from any manure management operation can be highly variable depending on total ammonia concentration, temperature, pH, and storage time.").
217. *See id.*; Y. Liang et al., *Ammonia Emissions From Layer Houses in Iowa*, in International Symposium on Gaseous and Odour Emissions From Animal Production Facilities (Horsens, Jutland, Denmark 2003).
218. *See* P. Yang et al., *Nitrogen Losses From Laying Hen Manure in Commercial High-Rise Layer Facilities*, 43(6) Transactions of the ASAE 1771-80, 1771 (2000) (referencing G. Logsdon, *Ammonia Troubles at Egg Factories*, 30(2) J. BioCycle 62-63 (1989)).
219. *See* V. Aneja et al., *Characterization of Atmospheric Ammonia Emissions From Swine Waste Storage and Treatment Lagoons*, 105 (D9) J. Geophysical Res. 11535-45 (2000); J. Arogo et al., *A Review of Ammonia Emissions From Confined Swine Feeding Operations*, 46(3) Transactions of the ASAE 805-17, 805 (2003); P. Yang et al., *Nitrogen Losses From Laying Hen Manure in Commercial High-Rise Layer Facilities*, 43(6) Transactions of the ASAE 1771-80 (2000) ("High NH_3 concentrations in animal houses may cause decreased production rates and chronic health problems in both animals and human workers.").
220. *See National Emissions Inventory*, *supra* note 209, at 1-3; V. Aneja et al., *Characterizing Ammonia Emissions From Swine Farms in Eastern North Carolina: Part 2—Potential Environmentally Superior Technologies for Waste Treatment*, 58 J. Air & Waste Mgmt. Ass'n 1145-57, 1146 (2008); V. Aneja et al., *Characterization of Atmospheric Ammonia Emissions From Swine Waste Storage and Treatment Lagoons*, 105(D9) J. Geophysical Res. 11535-45 (2000).
221. S. McGinn et al., *Atmospheric Pollutants and Trace Gases: Atmospheric Ammonia, Volatile Fatty Acids, and Other Odorants Near Beef Feedlots*, 32 J. Envtl. Quality 1173-82, 1173 (2003) (references omitted).

deposition."[222] Researchers estimate that 50-85% of ammonia nitrogen tends to be deposited within 60 miles of its emission source.[223]

ii. Hydrogen Sulfide

Hydrogen sulfide is a colorless, odorous gas produced by the decomposition, in this case anaerobic, of sulfur-containing organic matter in animal waste.[224] "The production of hydrogen sulfide is dependent on the outside air temperature, the size of the housing and waste management areas, the air retention time in the housing areas, and the daily sulfur intake of the animals."[225] At concentrated animal production operations, liquid waste impoundments are often the sources of the highest concentrations of hydrogen sulfide.[226] Hydrogen sulfide emissions regularly enter the atmosphere as gaseous sulfur, and in atmospherically converting to sulfuric acid, can return back to the earth in the form of environmentally harmful acid rain.[227] Its presence is readily identifiable by a strong "rotten eggs" odor, and it can contribute to human and animal sensory discomfort, headaches, eye irritation, nausea, and respiratory and cardiovascular irritation.[228]

iii. Endotoxins, Particulate Matter, and Volatile Organic Compounds

Endotoxins, particulate matter (PM), and volatile organic compounds (VOCs) are interrelated aerial pollutants associated with confined animal production. Endotoxins are the lipid component of the outer-membrane of Gram-negative bacteria that are often found in organic dust and prevalent in animal manure.[229] Endotoxins have strong inflammatory properties and can cause a wide range of respiratory health effects in exposed humans, livestock, and wildlife populations.[230]

PM, sometimes referred to colloquially as "dust," can form both directly and indirectly in the atmosphere.[231] Unlike ammonia and hydrogen sulfide, PM is not a distinct chemical entity.[232] Rather, PM "[p]articles are highly complex in size, physical properties, and composition," and are usually categorized according to their physical composition (with PM_{10} representing the large, "coarse" particles and $PM_{2.5}$ representing the much smaller, "fine" particles).[233] Agriculture is a major source of PM, which can be formed from, among other things, animal dander, feed, dried feces, endotoxins, microbial matter, and aerial inter-

---

222. *National Emissions Inventory, supra* note 209, at 1-3.
223. *See* R.J. Barthelmie & S.C. Pryor, *Implications of Ammonia for Fine Aerosol Formation and Visibility Impairment—A Case Study From the Lower Frasier Valley, British Columbia*, 32 ATMOSPHERIC ENV'T 345-52 (1998); A. Fangmeier et al., *Effects of Atmospheric Ammonia on Vegetation—A Review*, 86 ENVTL. POLLUTION 43-82 (1994); D. Fowler et al., *The Mass Budget of Atmospheric Ammonia in Woodland Within 1km of Livestock Buildings*, 102(S1) ENVTL. POLLUTION 343-48 (1998); C. Pitcairn et al., *The Relationship Between Nitrogen Deposition, Species Composition and Foliar Nitrogen Concentrations in Woodland Flora in the Vicinity of Livestock Farms*, 102(S1) ENVTL. POLLUTION 41-48 (1998).
224. *See* C. Copeland, RL32948, *Air Quality Issues and Animal Agriculture: A Primer*, 4 Congressional Research Service 2009 [hereinafter RL32948]; National Research Council Ad Hoc Committee on Air Emissions From Animal Feeding Operations, *Air Emissions From Animal Feeding Operations: Current Knowledge, Future Needs* (National Academies Press 2003).
225. U.S. EPA RISK MANAGEMENT, *supra* note 114, at 65.
226. *Id.*
227. *See* European Environment Agency, *Transboundary Air Pollution*, in ENVIRONMENT IN THE EUROPEAN UNION AT THE TURN OF THE CENTURY 133-54 (2000).
228. *See* RL32948, *supra* note 224, at 4; S. Schiffman et al., *Symptomatic Effects of Exposure to Diluted Air Sampled From a Swine Confinement Atmosphere on Healthy Human Subjects*, 113(5) ENVTL. HEALTH PERSP. 567-76 (2005); Schiffman Health Effects, *supra* note 214.
229. K.J. Donham, *Hazardous Agents in Agricultural Dusts and Methods of Evaluation*, 10 AM. J. INDUS. MED. 205-20 (1986).
230. *See, e.g.*, V.P. Aneja et al., *Effects of Agriculture Upon the Air Quality and Climate: Research, Policy, and Regulation*, 43 ENVTL. SCI. & TECH. 4234-40 (2009); NAT'L ACAD. OF SCIENCES, THE SCIENTIFIC BASIS FOR ESTIMATING AIR EMISSIONS FROM ANIMAL FEEDING OPERATIONS (National Academy Press 2002); P. Attwood et al., *A Study of the Relationship Between Airborne Contaminants and Environmental Factors in Dutch Swine Confinement Buildings*, 48(8) AM. IND. HYGIENE ASS'N J. 745-51 (1987); K.J. Donham et al., *Respiratory Symptoms and Lung Function Among Workers in Swine Confinement Buildings: A Cross-Sectional Epidemiological Study*, 39(2) ARCH ENVTL. HEALTH 96-101 (1984); J. Douwes & D. Heederik, *Epidemiological Investigation of Endotoxins*, 3 (supp. 1) INT'L J. OCCUPATIONAL ENVTL. HEALTH S26-3 (1997); S.J. Reynolds et al., *Air Quality Assessments in the Vicinity of Swine Production Facilities*, 4 J. AGROMED. 37-45 (1997); S.S. Schiffman, *Livestock Odors: Implications for Human Health and Well-Being*, 76 J. ANIMAL SCI. 1343-55 (1998); L. Schinasi et al., *Air Pollution, Lung Function, and Physical Symptoms in Communities Near Concentrated Swine Feeding Operations*, 22(2) EPIDEMIOLOGY 1-8 (2011); P.S. Thorne et al., *Concentrations of Bioaerosols, Odors and Hydrogen Sulfide Inside and Downwind From Two Types of Swine Operations*, 6 J. OCCUPATIONAL HYGIENE 211-20 (2009); S. Wing et al., *Air Pollution and Odor in Communities Near Industrial Swine Operations*, 116 ENVTL. HEALTH PERSP. 1362-68 (2008).
231. *See, e.g.*, J.T. Walker et al., *Inorganic $PM_{2.5}$ at a U.S. Agriculture Site*, 139 ENVTL. POLLUTANTS 258-71 (2006); S. Goetz et al., *Measurement, Analysis and Modeling of Fine Particulate Matter in Eastern North Carolina*, 58 J. AIR WASTE MGMT. ASS'N 1208-14 (2008); M.Y. Menetrez et al., *An Evaluation of Indoor and Outdoor Biological Particulate Matter*, 43 ATMOSPHERIC ENV'T 5476-83 (2009).
232. *See* RL32948, *supra* note 224, at 3.
233. *Id.*

actions of precursor emissions, such as ammonia and hydrogen sulfide.[234] Because of its high prevalence and direct impacts on human and wildlife health, some researchers consider fine PM to be the number one environmental health threat in the United States.[235]

VOCs are known to exist as over 300 identified compounds, and "are formed when the hydrolytic and acetogenic bacteria ferment the organic matter in the waste."[236] Often linked to PM, "VOCs... vaporize easily at room temperature and include a large number of constituents, such as volatile fatty acids, sulfides, amines, alcohols, hydrocarbons, and halocarbons."[237] Confinement operations commonly produce VOCs in the form of acids, phenolic compounds, and aldehydes.[238] VOCs are irritants and can produce some of the unpleasant odors regularly associated with animal agriculture.[239] VOCs can also be a precursor to the formation of fine PM and ozone.[240]

PM, VOCs, and endotoxins can cause very real impairment to human health and the environment. These pollutants can degrade air quality by decreasing visibility and causing haze.[241] In addition, in creating ozone, VOCs can "damage forests, crops, and manmade materials."[242] They also can impact human and wildlife health by transporting pollutants, such as hazardous bacteria, to lungs through inhalation, by increasing morbidity, and by increasing rates of illness, such as respiratory tract infection and cardiovascular diseases.[243]

PM "dust" can also contain potentially harmful microorganisms, sometimes referred to as "bioaerosols," such as antibiotic resistant bacteria, viruses, fungi, and actinomycetes.[244] Recent studies indicate that "[b]acterial concentrations with multiple antibiotic resistance or multidrug resistances were routinely recovered inside and up to 150 [meters] downwind at higher percentages then upwind of the facility."[245] Therefore, similar to concerns associated with the discharge of antibiotic resistant bacteria to water resources, the emission of these bacteria to air resources can propagate antibiotic resistance and lead to human and wildlife health impairments.[246]

iv. Greenhouse Gases, Including Methane, Nitrous Oxide, and Carbon Dioxide

Industrial animal operations emit a significant quantity of greenhouse gases, including methane, nitrous oxide, and carbon dioxide, into the ambient environment. Indeed, at about 18%, industrial animal operations are one of the largest emitters of greenhouse gases globally.[247] Domestically, while the total contribution of livestock and poultry operations to greenhouse gas emissions is not well known, EPA estimates that at least 7.4% of all US greenhouse gas emissions are from agricultural activities (though many estimates place that percentage to be much higher).[248]

---

234. *See, e.g.,* B. Baek & V. Aneja, *Measurement and Analysis of the Relationship Between Ammonia, Acid Gases, and Fine Particles in Eastern North Carolina,* 54 J. Air & Waste Mgmt. Ass'n 623-33 (2004); R.J. Barthelmie & S.C. Pryor, *Implications of Ammonia for Fine Aerosol Formation and Visibility Impairment—A Case Study From the Lower Frasier Valley, British Columbia,* 32 Atmospheric Env't 345-52 (1998); S. Goetz et al., *Measurement, Analysis, and Modeling of Fine Particulate Matter in Eastern North Carolina,* 58 J. Air & Waste Mgmt. Ass'n 1208-14 (2008); J. Hartung, *Dust in Livestock Buildings as a Carrier on Odours,* in Odour Prevention and Control of Organic Sludge and Livestock Farming (V.C. Nielsen et al., eds., Elsevier Science Publications 1986); L. Schinasi et al., *Air Pollution, Lung Function, and Physical Symptoms in Communities Near Concentrated Swine Feeding Operations,* 22(2) Epidemiology 1-8 (2011); J.-H. Shih et al., Resources for the Future, *Air Emissions of Ammonia and Methane From Livestock Operations: Valuation and Policy Options* (2006).
235. *See* J.-H. Shih, *Air Emissions of Ammonia and Methane From Livestock Operations, supra* note 234.
236. U.S. EPA Risk Management, *supra* note 114, at 65. *See also* Schiffman Health Effects, *supra* note 214; S. Schiffman et al., *Quantification of Odors and Odorants From Swine Operations in North Carolina,* 108 Agric. Forest Meteorology 213-40 (2001).
237. RL32948, *supra* note 224, at 4.
238. *See* Schiffman Health Effects, *supra* note 214.
239. *See id. See also* S. Shiffman et al., *Quantification of Odors and Odorants From Swine Operations in North Carolina,* 108 Agric. & Forest Meteorology 213-40 (2001).
240. *See* RL32948, *supra* note 224, at 4.
241. *Id.*
242. *Id.*
243. *Id.*
244. *See* Schiffman Health Effects, *supra* note 214.
245. S.G. Gibbs et al., *Isolation of Antibiotic-Resistant Bacteria From the Air Plume Downwind of a Swine Confined or Concentrated Animal Feeding Operation,* 114(7) Envtl. Health Persp. 1032-37, 1036 (2006).
246. *See, e.g.,* A. Chapin et al., *Airborne Multidrug-Resistant Bacteria Isolated From a Concentrated Swine Feeding Operation,* 113(2) Envtl. Health Persp. 137-42 (2005); S.G. Gibbs et al., *Isolation of Antibiotic-Resistant Bacteria From the Air Plume Downwind of a Swine Confined or Concentrated Animal Feeding Operation,* 114(7) Envtl. Health Persp. 1032-37 (2006).
247. Pew Environmental Impacts, *supra* note 3, at 22 (referencing Livestock's Long Shadow, *supra* note 1, at 390).
248. *See* U.S. EPA, *Inventory of the U.S. Greenhouse Gas Emissions and SInks: 1990-2005* 393 (2007).

Animal confinement operations produce greenhouse gases at three primary stages: (1) gases that are given off by the animal during the digestion process; (2) gases that result from the degradation of wastes in waste management systems, including uncovered waste lagoons and anaerobic digesters; and (3) gases that are emitted during the land application process.[249] With respect to methane production, "[a]n estimated one-half of global methane comes from manmade sources, of which agriculture is the largest source."[250] Indeed, livestock and poultry operations contribute approximately 25% of domestic methane production. In the United States, that adds up to around 1.9 million tons of methane emitted annually from livestock and poultry manure.[251] In part, this is because the "[u]ltimate by-products of anaerobic digestion [a type of waste management system often used, for example, at swine operations] are methane (CH_4) and carbon dioxide (CO_2), with CH_4 making up between 60 and 70% of the biogas."[252] In addition to waste-produced methane, methane gas is also produced through other decomposition and as the result of enteric fermentation, i.e., direct emissions from the animal.[253] Methane can remain in the atmosphere for an estimated 9 to 15 years, and it is an extremely effective heat trapping gas.[254]

Animal operations produce at least 6% of domestic nitrous oxide emissions "via the microbial processes of nitrification and denitrification."[255] The quantity of released nitrous oxide "is a function of the nitrogen content of the manure, the length of time the manure is stored, and the specific type of manure management system used."[256] Like VOCs, nitrous oxide is a precursor to the formation of ozone.[257] In addition, due to its extreme warming effect and its prolonged presence in the atmosphere, nitrous oxide is considered one of the most potent agricultural greenhouse gases.[258]

Finally, confinement operations directly produce the greenhouse gas carbon dioxide during animal breathing and through the "microbial degradation of animal manure under aerobic and anaerobic conditions,"[259] and indirectly as the result of the deforestation of lands for feed crop production, the use of fossil fuels in the production process, and other associated land use changes and degradations, as mentioned more fully in the following sections.[260] Carbon dioxide is most easily linked to climate change because of its heat-trapping qualities, due to its long atmospheric lifetime, and because it is emitted globally in the highest concentrations.[261] While recent estimates attribute approximately 9% of global anthropogenic emissions of carbon dioxide to livestock-related activities, both direct and indirect, the actual contributions of carbon dioxide by the livestock sector as a whole are currently under review.[262]

It is well-accepted that anthropogenic climate change is occurring, and that it is impacting the natural environment at all levels, including land, air and water.[263] Beyond the impacts of greenhouse gases on air quality,

> [i]n general, the faster the changes, the greater the risk of damage exceeding our ability to cope with the consequences. Mean sea level is expected to rise by 9-88cm by 2100, causing flooding of low-lying areas and other damage. Climatic zones could shift poleward and uphill, disrupting forests, deserts, rangelands and other unmanaged ecosystems. As a result, many ecosystems will decline or become fragmented and individual species could become extinct.... Water sources will be affected as precipitation and evaporation patterns change around the world. Physical infrastructure will be damaged.... Economic activities, human settlements, and human health will experience many direct and indirect effects.... [In addition, t]he livestock sector [and the

---

249. Pew Environmental Impacts, *supra* note 3, at 22.
250. RL32948, *supra* note 224, at 4.
251. *See* Livestock's Long Shadow, *supra* note 1, at 98.
252. T.M. DeSutter & J.M. Ham, *Lagoon-Biogas Emissions and Carbon Balance Estimates of a Swine Production Facility*, 34 J. Envtl. Quality 198-206, 198 (2005) (referencing D.P. Chynoweth & P. Pullammanappallil, *Anaerobic Digestion of Municipal Solid Wastes*, in A.C. Palmisano & M.A. Barlaz (eds.), Microbiology of Solid Waste (CRC Press 1996); B.E. Rittman & P.L. McCarty, Environmental Biotechnology: Principles and Applications (McGraw-Hill 2001).
253. *See* RL32948, *supra* note 224, at 4; U.S. EPA, *Sources and Emissions: Where Does Methane Come From?*, http://www.epa.gov/methane/sources.html.
254. *See* Livestock's Long Shadow, *supra* note 1, at 82.
255. RL32948, *supra* note 224, at 4.
256. U.S. EPA Risk Management, *supra* note 114, at 64.
257. *See* RL32948, *supra* note 224, at 4.
258. *See* U.S. EPA Risk Management, *supra* note 114, at 64; Livestock's Long Shadow, *supra* note 1, at 82.
259. U.S. EPA Risk Management, *supra* note 114, at 65. *See also* Putting Meat on the Table, *supra* note 15, at 27.
260. Livestock's Long Shadow, *supra* note 1, at 83, 86.
261. *See* Livestock's Long Shadow, *supra* note 1, at 82, 112.
262. *Id.* This estimate includes deforestation, feed crop production, and animal production.
263. *See id.* at 80.

animals themselves] will also be affected. Livestock products would become costlier if agricultural disruption leads to higher grain prices.[264]

At a minimum, industrialized animal operations can reduce greenhouse gas production through dietary modification and systemic and operational changes to the confinement production system, including to the waste management system, and can mitigate the effects of greenhouse gas production by employing carbon sequestration measures and improving land use practices.[265]

b. Primary Emissions Locations

Air pollutants tend to be released from all areas in the confinement operation, including the confinement structure, the waste storage system, the feed silage, the animal composting structures, and the land application area.[266] Ammonia, for example, will be released from any location on the confinement operation where manure, whether wet or dry, is present and unsealed.[267] In general, however, confinement operations emit pollutants from three primary sources: the confinement structure, including both buildings and lots; the waste storage structure; and the land application area.[268]

The confinement structure often produces a harmful mixture of all of the aerial pollutants of concern.[269] With some exception, this combination is dependent on animal concentration, animal age and type, diet, bedding, and waste management practices.[270] At swine operations, for example, "higher feed intake leads to more [nitrogen] excreted in the urine, which enhances [ammonia] emission."[271] Once emitted into the air, airborne pollutants enter the natural environment passively, though doors or other openings, or actively, through ventilation fans.[272]

Ventilation fans are designed to induce airflow and to transport aerial pollutants, such as dust, debris, and the byproducts of animal digestion, from the structure to the outside environment.[273] In many cases, ventilation fans actually exacerbate the emissions of aerial pollutants from confinement buildings by drying wastes and stimulating waste decomposition.[274] In addition, specifically with respect to VOCs and dust, ventilation fans can be a source for releasing antibiotic resistant bacteria into the environment.[275]

While ventilation systems are a necessary component of the confinement building, associated emissions can be limited by filtration, including biofiltration, and dry scrubbing.[276] Confinement structures also can reduce aerial emissions with solid bedding, diet manipulation, reduction in animal numbers, immediate separation of urine and feces, and application of chemical additives or other dampeners to dry waste products.[277] Once released from the confinement structure, an operation can further limit the release of airborne pollutants from the property by establishing impermeable barriers, including windbreak walls or air dams, and permeable barriers, such as trees and shrubs.[278]

---

264. *Id.* at 80-81 (reference omitted).
265. *See id.* at 112-23.
266. *See* V. Aneja et al., *Characterization of Atmospheric Ammonia Emissions From Swine Waste Storage and Treatment Lagoons*, 105(D9) J. Geophysical Res. 11535-45 (2000); RL32947, *supra* note 228, at 2; L. Schinasi et al., *Air Pollution, Lung Function, and Physical Symptoms in Communities Near Concentrated Swine Feeding Operations*, 22(2) Epidemiology 1-8, 5 (2011); Pew Recent Changes, *supra* note 49, at 20.
267. *See National Emissions Inventory*, *supra* note 209, at E-1.
268. *See* 2003 Rule, *supra* note 53, *at* 7235-37; 40 C.F.R. §122.42(e)(1).
269. *See, e.g.,* K.J. Donham et al., *Environmental and Health Studies in Swine Confinement Buildings*, 10 Am. J. Ind. Med. 289-93 (1986); K.J. Donham et al., *Community Health and Socioeconomic Issues Surrounding Concentrated Animal Feeding Operations*, 115(2) Envtl. Health Persp. 317-20, 318 (2007) (references omitted).
270. *See* J. Arogo et al., *A Review of Ammonia Emissions From Confined Swine Feeding Operations*, 46(3) Transactions of the ASAE 805-17 (2003).
271. *Id.* at 808.
272. Pew Environmental Impacts, *supra* note 3, at 13 ("Airborne . . . emissions arise from both ventilation and passive release.").
273. *See* P. Groot Koerkamp et al., *Concentrations of Emissions of Ammonia in Livestock Buildings in Northern Europe*, 70(1) J. Agric. Engineering Res. 79-95 (1998); Pew Antimicrobial Resistance, *supra* note 199, at 29; H. Takai et al., *Concentrations and Emissions of Airborne Dust in Livestock Buildings in Northern Europe*, 70(1) J. Agric. Engineering Res. 59-77 (1998).
274. Poultry Waste Management Handbook, *supra* note 46, at 4 ("Circulation fans to move air over manure mass can assist in moisture evaporation, increasing the dryness of manure.").
275. Pew Antimicrobial Resistance, *supra* note 199, at 29.
276. *See* RL32948, *supra* note 224, at 5-6; Iowa State University, *Practices to Reduce Ammonia Emissions From Livestock Operations*, PM 1971a (July 2004) ("Biofilters have been developed primarily to reduce emissions from the deep-pit manure ventilation exhausts, and, to a lesser extent, from the building exhaust.").
277. *Id.*
278. *Id.*

Airborne pollutants are emitted from the animal waste management system and the land application area.[279] In both instances, the degree of emissions will be controlled by, among other things, ambient temperature, pH, waste management system, and disposal method.[280] Open waste storage structures, such as open pits, lagoons, and uncovered litter piles, often produce the highest rate of systemic emissions.[281] Emissions rates from waste treatment systems can be reduced by covering or otherwise containing stored wastes.[282]

The land application area is also prone to aerial emissions. "Estimates of whole-farm ammonia emissions suggests that as much as 35 percent of the total ammonia emissions may occur during land application of manure."[283] The land application of waste and wastewater emits aerial pollutants through spraying practices, which readily aerosolize pollutants and hasten aerial emissions through drift, and the decomposition and volatilization of waste products on land surfaces.[284] While it could lead to additional groundwater concerns, the injection of waste and wastewater or the rapid incorporation of sprayed and applied waste into applied soils can reduce emissions from land application.[285]

The impacts of airborne pollutants are felt not only at the operational point of emission, but also in nearby communities and environments. Impacts can manifest directly through inhalation and air quality degradation or indirectly through deposition.[286] Specifically, air pollutants can produce haze; impact atmospheric and ozone resources; influence climate change; impact land and water resources; cause otolaryngologic irritation; compromise respiratory ability; and produce toxicological effects and negative cognitive and emotion responses.[287]

Humans and wildlife are exposed to these pollutants in a variety of ways. First, confinement workers and confined animals can come into contact with concentrated aerial emissions within the production facility.[288] Studies show that "at least 25% of confinement workers suffer from respiratory diseases including bronchitis, mucus membrane irritation, asthma-like syndrome, and acute respiratory distress syndrome.... An additional acute respiratory condition, organic dust toxic syndrome, related to high con-

---

279. *See, e.g.,* S.J. Hoff et al., *Emissions of Ammonia, Hydrogen Sulfide, and Odor Before, During, and After Slurry Removal From a Deep-Pit Swine Finisher,* 56 J. Air & Waste Mgmt. Ass'n 581-90 (2006); R.R. Sherlock et al., *Ammonia, Methane, and Nitrous Oxide Emission From a Pig Slurry Applied to a Pasture in New Zealand,* 31 J. Envtl. Quality 1491-1501 (2002).
280. *See, e.g.,* D. Bussink & O. Oenema, *Ammonia Volatilization From Dairy Farming Systems in Temperate Areas: A Review,* 51 Nutrient Cycling Agroecosystems 19-33 (1998); J. Olesen & S. Sommer, *Modeling Effects of Wind Speed and Surface Cover on Ammonia Volatilization From Stored Pig Slurry,* 27 Atmospheric Env't 2567-74 (1993).
281. *See, e.g.,* G. Hornig et al., *Slurry Covers to Reduce Ammonia Emission and Odor Nuisance,* 73(2) J. Agric. Engineering Res. 151-57 (1999); P. Yang et al., *Nitrogen Losses From Laying Hen Manure in Commercial High-Rise Layer Facilities,* 43(6) Transactions of the ASAE 1771-80, 1771 (2000).
282. *Id.*
283. *See* Iowa State University, Practices to Reduce Ammonia Emissions From Livestock Operations, PM 1971a, at 5 (July 2004); U.S. EPA Risk Management, *supra* note 114, at 66.
284. *See* L.M. Safley et al., *Loss of Nitrogen During Sprinkler Irrigation of Swine Lagoon Liquid,* 40 Biosource Tech. 7-15 (1992); R. Sharpe & L. Harper. *Ammonia and Nitrous Oxide Emissions From Sprinkler Irrigation Applications of Swine Effluent,* 26 J. Envtl. Quality 1703-06 (1997).
285. RL32948, *supra* note 224, at 6.
286. *See, e.g.,* V.P. Aneja et al., *Agricultural Ammonia Emissions and Ammonium Concentrations Associated With Aerosols and Precipitation in the Southeast United States,* 108(D4) J. Geophysical Res. 4152-62 (2003); V.P. Aneja et al., *Characterization of Atmospheric Ammonia Emissions From Swine Waste Storage and Treatment Lagoons,* 105(D9) J. Geophysical Res. 11535-45 (2000); K.S. Bajwa et al., *Modeling Studies of Ammonia Dispersion and Dry Deposition at Some Hog Farms in North Carolina,* 58 J. Air & Waste Mgmt. Ass'n 1198-1207 (2008); J.K. Costanza et al., *Potential Geographic Distribution of Atmospheric Nitrogen Deposition From Intensive Livestock Production in North Carolina, USA,* 398 Sci. Total Env't 76-86 (2008); A.B. Gilliland et al., *Seasonal NH_3 Emission Estimates for the Eastern United States Based on Ammonium Wet Concentrations and an Inverse Modeling Method,* 108(D15) J. Geophysical Res. 4477-87 (2003); S.V. Krupa, *Effects of Atmospheric Ammonia (NH_3) on Terrestrial Vegetation: Review,* 124 Envtl. Pollutants 179-221 (2003).
287. *See, e.g.,* S.S. Schiffman. *Livestock Odors: Implications for Human Health and Well-Being,* 76 J. Animal Sci. 1343-55 (1998).
288. *See, e.g.,* D. Cole et al., *Concentrated Swine Feeding Operations and Public Health: A Review of Occupational and Community Health Effects,* 108 Envtl. Health Persp. 685-99 (2000); Y. Cormier et al., *Effects of Repeat Swine Building Exposures on Normal Naïve Subjects,* 10 Eur. Respiratory J. 1516-22 (1997); K.J. Donham, *The Concentration of Swine Production: Effects of Swine Health, Productivity, Human, and the Environment,* 16(3) Vet. Clinics. N. Am. Food Animal Prac. 559-97 (2000); K.J. Donham et al., *Community Health and Socioeconomic Issues Surrounding Concentrated Animal Feeding Operations,* 115(2) Envtl. Health Persp. 317-20 (2007); K.J. Donham et al., *Synergistic Effects of Dust and Ammonia on the Occupational Health Effects of Poultry Workers,* 8(2) J. Agromed. 57-76 (2002); K.J. Donham et al., *Acute Toxic Exposure to Gases From Liquid Manure,* 24 J. Occupational Med. 142-145 (1982); K.J. Donham et al., *Potential Health Hazards to Agricultural Workers in Swine Confinement Buildings,* 19(6) J. Occupational Med. 383-87 (1977); S.R. Kirkhorn & M.B. Schenker, *Current Health Effects of Agricultural Work: Respiratory Disease, Cancer, Reproductive Effects, Musculoskeletal Injuries, and Pesticide-Related Illnesses,* 8 J. Agric. Safe Health 199-214 (2002); Preller et al., *Lung Function and Chronic Respiratory Symptoms of Pig Farmers: Focus on Exposure to Endotoxins and Ammonia and Use of Disinfectants,* 52(10) Occupational Envtl. Med. 654-60 (1995); M. Schenker et al., *Respiratory Health Hazards in Agriculture,* 158(pt. 2) Am. J. Respiratory Critical Care Med. Suppl. S1-S76 (1998).

centrations of bioaerosols in livestock buildings occurs episodically in more than 30% of swine workers."[289] Reasonably, therefore, the longer a person remains a confinement worker, the higher the probability of respiratory disease.

Second, wildlife and nearby communities can suffer health consequences from aerial emissions released by confinement operations.[290] Studies have "documented excessive respiratory symptoms in neighbors of large scale . . . [operations] relative to comparison populations in low-density livestock-producing areas. The pattern of these symptoms was similar to those experienced by . . . [confinement] workers."[291] In communities bordering confinement operations, for example, "environmental air quality assessments have shown concentrations of hydrogen sulfide and ammonia that exceed U.S. Environmental Protection Agency (U.S. EPA) and Agency for Toxic Substances and Disease Registry recommendations."[292] Health impacts are especially prevalent among vulnerable populations, such as the very young and the elderly.[293]

3. Land Degradation

Confinement operations degrade land resources in a number of ways, but principally through deposition, use, erosion, and climate change. First, land application practices can contaminate soil resources—"[a] . . . pervasive problem caused by the unsuitable, year-round deposition of excess nutrients, chemicals, and pathogens on land in the vicinity of industrial feeding operations."[294] Deposition, in this case, includes both the direct application of wastes and wastewater to soil resources and the aerial deposition of pollutants onto soil resources.

Second, the physical act of producing livestock and poultry utilizes and stresses a large quantity of land resources. Obviously, the operations themselves use and impact land resources through direct production and land application practices. However, in addition to homestead land use, the Food and Agriculture Organization of the United Nations found that, "[d]irectly and indirectly, though grazing and through feedcrop production, the livestock sector occupies about 30 percent of the ice-free terrestrial surface on the planet."[295]

Because industrial production is heavily reliant on feed crop use, all of the concerns associated with monoculture feed crop production should be considered when quantifying the environmental impacts of concentrated animal production.[296] As an illustration, "[c]orn and soybeans, which now are replacing tra-

---

289. K.J. Donham et al., *Community Health and Socioeconomic Issues Surrounding Concentrated Animal Feeding Operations*, 115(2) ENVTL. HEALTH PERSP. 317-20, 318 (2007) (referencing Dosman et al., *Occupational Asthma in Newly Employed Workers in Intensive Swine Confinement Facilities*, 24(4) EUR. RESPIRATORY J. 698-702)).
290. *See, e.g.,* R. Avery et al., *Perceived Odor From Industrial Hog Operations and Suppression of Mucosal Immune Function in Nearby Residents*, 59 ARCH. ENVTL. HEALTH 101-08 (2004); S. Bullers, *Environmental Stressors, Perceived Control, and Health: The Case of Residents Near Large-Scale Hog Farms in Eastern North Carolina*, 33 HUM. ECOLOGY 1-16 (2005); D. Cole et al., *Concentrated Swine Feeding Operations and Public Health: A Review of Occupational and Community Health Effects*, 108 ENVTL. HEALTH PERSP. 685-99 (2000); K.J. Donham et al., *Community Health and Socioeconomic Issues Surrounding Concentrated Animal Feeding Operations*, 115(2) ENVTL. HEALTH PERSP. 317-20 (2007); K. Kilburn, *Exposure to Reduced Sulfur Gases Impairs Neurobehavioral Function*, 90(10) SO. MED. J. 997-1006 (1997); K. Radon et al., *Environmental Exposure to Confined Animal Feeding Operations and Respiratory Health of Neighboring Residents*, 18 EPIDEMIOLOGY 300-08 (2007); S.S. Schiffman et al., *The Effect of Environmental Odors Emanating From Commercial Swine Operations on the Mood of Nearby Residents*, 37 BRAIN. RES. BULL. 369-75 (1995); M. Tajik et al., *Impact of Odor From Industrial Hog Operations on Daily Living Activities*, 18(2) NEW SOLUTIONS 193-205 (2008); L. Schinasi et al., *Air Pollution, Lung Function, and Physical Symptoms in Communities Near Concentrated Swine Feeding Operations*, 22(2) EPIDEMIOLOGY 1-8 (2011); K. Thu, *Public Health Concerns for Neighbors of Large Scale Swine Production*, 8(2) J. AGRIC. SAFE HEALTH 175-84 (2002); K. Thu et al., *A Control Study of the Physical and Mental Health of Residents Living Near a Large-Scale Swine Operation*, 3 J. AGRIC. SAFE HEALTH 13-26 (1997); K. THU & R. DURRENBERGER, EDS. PIGS, PROFITS, AND RURAL COMMUNITIES, State University of NY Press: Albany, NY (1998); S. Wing & S. Wolf, *Intensive Livestock Operations, Health, and Quality of Life Among Eastern North Carolina Residents*, 108 ENVTL. HEALTH PERSP. 233-38 (2000).
291. K.J. Donham et al., *Community Health and Socioeconomic Issues Surrounding Concentrated Animal Feeding Operations*, 115(2) ENVTL. HEALTH PERSP. 317-20, 318 (2007) (referencing K. Thu et al., *A Control Study of the Physical and Mental Health of Residents Living Near a Large-Scale Swine Operation*, 3 J. AGRIC. SAFE HEALTH 13-26)).
292. *Id.* (referencing S.J. Reynolds et al., *Air Quality Assessments in the Vicinity of Swine Production Facilities*, 4 J. AGROMED. 37-45 (1997)).
293. *See, e.g.,* J.A. Merchant et al., *Asthma and Farm Exposures in a Cohort of Rural Iowa Children*, 113 ENVTL. HEALTH PERSP. 350-56 (2005); M.C. Mirabelli et al., *Asthma Symptoms Among Adolescents Who Attend Public Schools That Are Located Near Confined Swine Feeding Operations*, 118(1) PEDIATRICS e66-e75 (2006); M.C. Mirabello et al., *Race, Poverty, and Potential Exposure of Middle School Students to Air Emissions From Confined Swine Feeding Operations*, 114 ENVTL. HEALTH PERSP. 591-96 (2006); S.T. Sigurdarson et al., *School Proximity to Concentrated Animal Feeding Operations and Prevalence of Asthma in Students*, 129 CHEST. 1486-91 (2006).
294. PEW ENVIRONMENTAL IMPACTS, *supra* note 3, at 1.
295. LIVESTOCK'S LONG SHADOW, *supra* note 1, at 4.
296. PEW ENVIRONMENTAL IMPACTS, *supra* note 3, at 14 ("Today, 66% of the grain produced in the US is fed to livestock.") (referencing WORLD RESOURCE INSTITUTE, WORLD RESOURCES 2000-2001: PEOPLE AND ECOSYSTEMS—THE FRAYING WEB OF LIFE (2000)); S. Park & S.L. Egbert,

ditionally used grass as cattle feed, largely are produced in crop monocultures maintained on agricultural land that in many instances is irrigated using groundwater from aquifers whose natural recharge rates are outpaced by this . . . usage."[297] In all, it is not just the quantity of land used but also the degradation of land resources that can be associated with confinement operations.

4. Loss of Biodiversity

Pollution of water, air, and land by animal confinement operations can produce a corresponding impact on both aquatic and terrestrial biodiversity. Confinement operations primarily impact aquatic biodiversity by changes in and reductions to native populations, often as the result of degraded habitat, reduced fertility, species mutation, and mortality; changes in natural food resources; and expansion of nonnative species, often at the expense of native populations.[298]

In favoring monoculture crop production and by utilizing and polluting land resources, industrial animal agriculture can also impact terrestrial biodiversity by restricting genetic diversity; limiting or eliminating habitat, including forest, grassland, and wetland habitat;[299] "increas[ing] vulnerability to large-scale damage by pests;"[300] and introducing invasive species, including the livestock themselves.[301] In particular, as the genetic biodiversity of food crops has become increasingly restricted, diversity in the human diet is also becoming increasingly restricted, such that "the majority of the world's population is now fed by less than 20 staple plant species," which is quite a departure from the over "5,000 different species of plants [that have traditionally] been used as food by humans."[302] Similarly, humans now rely on only 14 domesticated mammalian and bird species to provide 90% of meat for human consumption.[303]

Finally, airborne emissions impact both terrestrial and aquatic biodiversity by harming wildlife health and population numbers, changing species migration patterns, altering vegetative growth rates, and causing species extinction through climate change.[304]

5. Environmental Injustice

Both nationally and internationally, the burden of living with and alongside animal confinement operations is borne disproportionately by underprivileged communities, which, in the United States, are primarily communities of color.[305] In particular, the model of contract meat production "has physically separated key decision makers and many employees from the locality of animal farming operations, a development that has resulted in a loss of accountability and land stewardship as well as a degradation of the quality

---

Assessment of Soil Erodibility Indices for Conservation Reserve Program Lands in Southwest Kansas Using Satellite Imagery and GIS Techniques, 36(6) ENVTL. MGMT. 886-98 (2005).
297. PEW ENVIRONMENTAL IMPACTS, *supra* note 3, at 8.
298. *See id.* at 30 ("Typically, biodiversity loss is caused by a combination of various processes of environmental degradation."); LIVESTOCK'S LONG SHADOW, *supra* note 1, at 273, 196 ("Invasive species can affect native species directly by eating them competing with them, and introducing pathogens or parasites that sicken or kill them or, indirectly, by destroying or degrading their habitat. Invasive alien species have altered evolutionary trajectories and disrupted many community and ecosystem processes."); at 209 ("Over the past four decades, pollution has emerged as one of the most important drivers of ecosystem change in terrestrial, freshwater and coastal ecosystems.").
299. *See id.* at 273, 187 ("Habitat destruction, fragmentation and degradation are considered the major category of threat to global biodiversity.").
300. PEW ENVIRONMENTAL IMPACTS, *supra* note 3, at 30.
301. *See* LIVESTOCK'S LONG SHADOW, *supra* note 1, at 197.
302. LIVESTOCK'S LONG SHADOW, *supra* note 1, at 187 (references omitted).
303. *Id.*
304. *See id.* at 195-96.
305. The environmental justice implications of industrialized animal production are loaded and murky. While this section attempts to topically address the concern, there are deeper implications to this system of injustice that this section does not discuss. For further detail on environmental justice concerns, please refer to the articles cited herein. *See, e.g.,* R.D. Bullard & B.H. Wright, *Environmental Justice for All: Community Perspectives on Health and Research Needs*, 9 TOXICOL. IND. HEALTH 821-41 (1993); A.E. Ladd & B. Edwards, *Corporate Swine and Capitalistic Pigs: A Decade of Environmental Injustice and Protest in North Carolina*, 29 SOC. JUST. 26-46 (2002); M.C. Mirabelli et al., *Asthma Symptoms Among Adolescents Who Attend Public Schools That Are Located Near Confined Swine Feeding Operations*, 118(1) PEDIATRICS e66-e75 (2006); M.C. Mirabello et al., *Race, Poverty, and Potential Exposure of Middle School Students to Air Emissions From Confined Swine Feeding Operations*, 114 ENVTL. HEALTH PERSP. 591-96 (2006); NATIONAL RESEARCH COUNCIL, TOWARD ENVIRONMENTAL JUSTICE: RESEARCH, EDUCATION, AND HEALTH POLICY NEEDS (National Academy Press 1999); M. Tajik et al., *Impact of Odor From Industrial Hog Operations on Daily Living Activities*, 18(2) NEW SOLUTIONS 193-205 (2008); S. Wilson et al., *Environmental Injustice and the Mississippi Hog Industry*, 110 (supp. 2) ENVTL. HEALTH PERSP. 195-201 (2002); S. Wing et al., *Environmental Injustice in North Carolina's Hog Industry*, 108 ENVTL. HEALTH PERSP. 233-38 (2000).

of life in [surrounding] rural communities."[306] Thus, epidemiologists and community welfare specialists argue that the "[d]isproportionate location of CAFOs [concentrated animal feeding operations] in areas populated by people of color or people with low incomes is a form of environmental injustice."[307] Because of these disproportionate impacts, studies show that "[l]ow income communities and populations that experience institutional discrimination based on race have higher susceptibilities to CAFO impacts due to poor housing, low income, poor health status, and lack of access to medical care."[308] Due to pollutants, impacted communities experience human health impairments as well as losses in the "beneficial use" and "quiet enjoyment" of their property.[309]

Conclusion

Through modifications such as physical concentration and increased population size, the industrialization of animal agriculture has encouraged an evolution in animal production models, with most projections indicating significant future reliance on the industrial model to meet global meat, dairy, and poultry demands. The evolution of animal agricultural practices, however, has also resulted in significant increases in the pollution potential of these industrialized operations. To secure a sustainable future for animal agriculture and the environment, that potential should be acknowledged and systemically addressed.

---

306. Pew Environmental Impacts, *supra* note 3, at 11 (referencing K.J. Donham et al., *Community Health and Socioeconomic Issues Surrounding Concentrated Animal Feeding Operations*, 115(2) Envtl. Health Persp. 317-20 (2007); L. Horrigan et al., *How Sustainable Agriculture Can Address the Environmental and Human Health Harms of Industrial Agriculture*, 110(5) Envtl. Health Persp. 455-56 (2002).

307. K.J. Donham et al., *Community Health and Socioeconomic Issues Surrounding Concentrated Animal Feeding Operations*, 115(2) Envtl. Health Persp. 317-20, 318 (2007) (referencing S. Wing et al., *Environmental Injustice in North Carolina's Hog Industry*, 108 Envtl. Health Persp. 225-31(2000)).

308. *Id.*

309. M. Tajik et al., *Impact of Odor From Industrial Hog Operations on Daily Living Activities*, 18(2) New Solutions 193-205, 201 (2008).

Chapter 6
Genetically Modified Organisms and the Environment
Jason J. Czarnezki and Emily Montgomery

As recently as a few decades ago, American farmers and consumers would not have predicted that the nation's fields would soon be dotted—and then covered—with crops that have been genetically modified. Even further from their imagination would be the idea that fish or livestock could be created in a lab and then find its way to the dinner plate. Then, as now, farmers and consumers would be unable to predict with certainty what effect these new technologies might have or whether any effects would be permanent. The age of biotechnology has arrived, drastically changing the landscape of American agriculture. American supermarkets have changed too—most Americans regularly consume genetically modified (GM) food products.[1] Indeed, it is estimated that at least 70% of food on grocery store shelves contains GM products.[2]

Since their introduction in the mid-1990s, there has been an unprecedented rapid rate of adoption of GM organisms by U.S. and global producers.[3] For major crops, the vast majority of acreage is planted with GM seeds rather than non-GM seeds, and GM crops have been planted on more than 88 million acres of U.S. farmland.[4] This astonishing adoption of GM technology has occurred despite uncertainty about the potential social, economic, and environmental impacts of GM organisms. In short, GM products have become firmly established in American agriculture, but many of their implications for the environment, public health, and the economy remain unknown.

This chapter explores the ecological and human health impacts of genetic engineering in the United States, with a focus on GM organisms designed for use in or as food products. It provides readers with introductory information about GM products, the possible benefits and environmental costs linked to their creation and use, and, briefly, how GM products are regulated in the United States (a topic also discussed in Chapter 16). The first part of the chapter focuses on GM plants; the second part on GM animals. Both parts follow identical structures: considering the benefits and costs associated with GM plant or animal use, with an emphasis on environmental implications of bioengineering; providing a brief outline of the applicable regulatory regime; and concluding with case studies that discuss a GM product that has been recently approved or is under consideration for approval. The chapter concludes with recommendations for future research efforts and regulatory improvements that would help to address risks of GM foods.

A. Genetically Modified Plants

GM plants are now a common feature of U.S. agriculture. Despite uncertainty regarding impacts, GM crops and other plants have been widely adopted. Herbicide-resistant corn and soybeans have been most

---

1. Gregory N. Mandel, *Gaps, Inexperience, Inconsistencies, and Overlaps: Crisis in the Regulation of Genetically Modified Plants and Animals*, 45 Wm. & Mary L. Rev. 2167, 2176-77 (2004).
2. *Id.* at 2177.
3. A.M. Shelton et al., *Economic, Ecological, Food Safety, and Social Consequences of the Deployment of Bt Transgenic Plants*, 47 Ann. Rev. Entomol. 845, 847 (2002).
4. Mary Jane Angelo, *Regulating Evolution for Sale: An Evolutionary Biology Model for Regulating Unnatural Selection of Genetically Modified Organisms*, 42 Wake Forest L. Rev. 93, 95 (2007). For more data on the widespread adoption of GM crops, see U.S. Dep't. Agric., Economic Research Service, *Adoption of Genetically Engineered Crops in the U.S.*, http://www.ers.usda.gov/Data/BiotechCrops/.

commonly adopted and now make up a large percentage of overall acreage for their respective crops; for instance, herbicide tolerant soybeans constituted 93% of total acreage of soybeans planted in the United States in 2012 (see Figure 6.1).

Figure 6.1
Adoption of Genetically Engineered Crops in the United States[5]

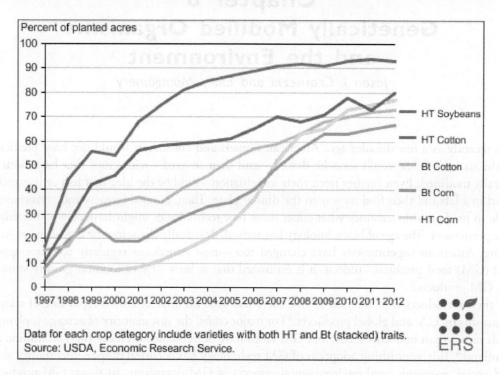

Note: "HT" signifies that a crop contains GM traits for herbicide tolerance. "Bt" signifies that a crop contains GM traits for insect resistance.

The widespread use of GM plants has been an astonishingly rapid development that has left many unsure about the safety of their adoption. Proponents of GM plants point to several possible benefits associated with their use, while others have significant concerns about potential environmental and social impacts.

1. What Is a GM Plant?

A GM plant is one that has been genetically engineered to develop desirable traits sourced from other organisms.[6] Although some form of genetic modification has been prevalent in agriculture for thousands of years through selective breeding, genetic engineering occurs at the cellular or molecular level rather than at the organism level.[7] There are varying definitions of what exactly a GM plant or organism is—and the definition can matter in the regulatory context—but it is generally agreed that a GM plant is one that has been modified by the application of recombinant DNA technology (rDNA).[8] Essentially, rDNA technology involves taking genetic material responsible for a desired trait from one organism and inserting it into

---

5. U.S. Dep't. Agric., Economic Research Service, *Recent Trends in GE Adoption*, July 5, 2012, http://www.ers.usda.gov/data-products/adoption-of-genetically-engineered-crops-in-the-us/recent-trends-in-ge-adoption.aspx (1997-2012 data).
6. Miguel A. Altieri, *The Ecological Impacts of Transgenic Crops on Ecosystem Health*, 6 Ecosys. Health 13 (2000) ("Genetic engineering is an application of biotechnology involving the manipulation of dna and the transfer of gene components between species in order to achieve stable intergenerational expression of new traits.").
7. National Research Council, Biological Confinement of Genetically Engineered Animals 14 (2004).
8. *Id.* at 15.

another organism.⁹ The inserted genetic material will then direct production of specific proteins and result in expression of the desired trait by the target organism.[10] Unlike traditional selective breeding, which is limited to reproductively compatible species, genetic engineering allows scientists to insert genes into an organism from an entirely unrelated species (for instance, inserting genes from a bacterium or insect into an animal), which is known as "transgenic" modification.[11]

The traits that could be introduced to plants via genetic engineering are seemingly endless, and the possibilities test the limits of one's imagination. However, the two most commonly introduced traits in GM plants are herbicide tolerance and resistance to insects.[12] Herbicide tolerance is a trait designed to enable farmers to eliminate preemergent herbicide use (herbicides that prevent seeds from germinating) and instead limit use to postemergent application of a single, broad-spectrum herbicide (herbicides that kill weeds after they have germinated).[13] Insect resistant plants are inserted with genes that promote production of toxins (usually the naturally-occurring Bt toxin) that protect the inserted plant from insect pests.[14] Given the prevalence of insect- and herbicide-resistant plants, they provide a useful starting place for a discussion of the potential benefits and risks associated with commercialization of GM plants.

2. Benefits of GM Plants

The introduction and commercialization of GM crops and other plants offer several potential benefits that range from agricultural efficiency and reduced costs to environmental benefits. Focusing on agricultural and economic advantages, GM crops may produce higher and more reliable yields at a cheaper cost than non-GM counterparts.[15] Costs are decreased by introduction of desirable traits, such as herbicide tolerance or insect resistance, which enables farmers to apply less chemicals to crops, which in turn saves money.[16] The existence of these traits may also promote high yields by reducing the risk that a crop will be decimated by pests.[17] Thus, the combination of reduced input costs and increased yields can result in economic benefits to adopting farmers[18] and decreased costs for consumers.[19] Another possible benefit is their potential to help reduce domestic and international hunger.[20]

Adoption of GM crops may offer some economic benefits at the farm level in developing countries by decreasing costs and increasing yields.[21] And GM crops have the potential to be more nutritious (i.e., by engineering crops to contain higher levels of vitamins and minerals), have a longer shelf life, be better

---

9. Mandel, *supra* note 1, at 2175.
10. *Id.*
11. *Id.*
12. Altieri, *supra* note 6, at 13.
13. *Id.* at 15.
14. *Id.* at 17.
15. Mandel, *supra* note 1, at 2180. Generally, given the lack of internalization of some potential costs, this promise has appeared to play out, though perhaps not at the levels promised by industry. *See, e.g.*, Matin Qaim, *The Economics of Genetically Modified Crops*, 1 Ann. Rev. Resource Econ. 665, 669-74 (2009) (discussing economic benefits associated with GM crop adoption and aggregate global welfare gains). However, it is important to note that such benefits can vary drastically depending on the crop, introduced trait, geographic location, and socioeconomic status of the adopter.
16. Mandel, *supra* note 1, at 2181-82. There is some evidence that a decrease in chemical applications has occurred. *See* Matim Qiam & David Zilberman, *Yield Effects of Genetically Modified Crops in Developing Countries*, 299 Sci. 900 (2003) (finding decreased levels of insecticide applications on Bt-cotton plots in India, resulting in saved economic costs to farmers, and noting that economic and environmental benefits of pesticide savings associated with Bt crops has been documented); Graham Brookes & Peter Barfoot, *GM Crops: The Global Economic and Environmental Impact—The First Nine Years 1996-2004*, 8 AgBioForum 187 (2005) (finding reduced chemical applications); Stuart J. Smyth et al., *Environmental Impacts From Herbicide Tolerant Canola Production*, 104 Agric. Sys. 403, 404-06 (2011) (noting that scholars estimate that GM crops have reduced the intensity and use of chemicals, and finding decreased chemical applications in Canadian Ht canola). However, some scholars estimate that chemical applications have actually increased due to adoption of GM plants. *See* Smyth et al., at 404 (discussing studies estimating an increase in chemical applications associated with GM crops).
17. Mandel, *supra* note 1, at 2181-82. Though this advantage is tempered by the risks of monoculture.
18. *See* National Research Council, The Impact of Genetically Engineered Crops on Farm Sustainability in the United States 174 (2010) (finding that adoption of GM crops resulted in agronomic and economic benefits for farmers that outweighed the higher cost of GM seed, but noting variation among different GM types and the need for increased research on the economic impacts of GM crops).
19. Mandel, *supra* note 1, at 2182.
20. *Id.* This claim is controversial, especially given the distribution problems associated with world hunger (as opposed to supply-side problems) and international resistance to GM crops. Indeed, this criticism is elucidated by the recent decision of several African countries to refuse U.S. food aid based on concerns that the aid contained GM maize. Noah Zerbe, *Feeding the Famine? American Food Aid and the GMO Debate in Southern Africa*, 29 Food Pol'y 593 (2004).
21. *See* Terri Raney, *Economic Impact of Transgenic Crops in Developing Countries*, 17 Curr. Op. Biotech. 174 (2004) (finding that GM technology can benefit farmers in developing countries but noting that benefits depend on factors such as national research and regulatory capacity and input supply systems).

looking, and contain less spray-pesticide residue than non-GM counterparts.[22] It is unclear, however, whether these potential benefits will be realized because the vast majority of GM crops currently in production are designed primarily for either herbicide tolerance or pest resistance, and these other applications are less explored.

GM plants also could offer several environmental advantages, the main one being the potential to reduce application of pesticides and herbicides.[23] The use of GM insect-resistant crops, such as those that produce the Bt toxin, can reduce the need to apply chemical insecticides.[24] Use of GM herbicide-tolerant plants that withstand broad-spectrum herbicides can reduce the need for farmers to apply preemergent herbicides. (However, this does not mean that overall herbicide use will necessarily decrease, as use of broad-spectrum herbicides could increase or take the place of preemergent herbicides). Because preemergent herbicides require incorporation into the soil, reduction of their use can lead to less tillage and enable adoption of conservation or no-tillage practices.[25] Reduced tillage offers several benefits, including reduction of the amount of water runoff from farms to surface water, reduction of erosion, fewer greenhouse gas emissions, and increased carbon sequestration.[26] Additionally, use of GM crops could reduce threats to biodiversity from habitat loss if increased crop yields and improved efficiency on existing agricultural land leads to decreased pressure to plant on undeveloped areas.[27] Finally, future plants could be engineered to be disease-resistant or tolerant of environmental stressors, which could reduce the use of fungicides and chemical fertilizers or reduce the need for irrigation.[28]

3. Risks and Concerns

Although GM plants have benefits, many potential risks must also be considered—information gaps, human health and economic impacts, and environmental concerns. One overarching issue associated with GM plants is scientific uncertainty regarding the risks associated with introduction of GM plants into the environment and human and animal diets.[29]

Genetic engineering is still a new technology, and scientific and regulatory experiences are limited. When GM crops were first planted widely beginning in the 1990s, the potential environmental benefits and risks of the crops went largely unstudied.[30] The state of information has not improved significantly in recent years. Current research primarily relates to the few GM plants that have been widely adopted (such as Bt and herbicide-resistant types of well-understood crops such as corn).[31] It is generally agreed that long-term, longitudinal studies are needed to better understand the effects of GM crops, as well as research pertaining to different GM types and crops that are less understood.[32] However, these long-term studies have not been pursued in much depth because the studies take a long time to conduct and the relative lack of baseline knowledge.[33] Simply put, scientific research on GM crops has failed to keep pace with market and regulatory realities. No clear guidance exists on what information should be included in GM risk assessment, and disagreement continues about what risks should be included or emphasized in such an assessment.[34] Risk assessments are inherently limited because most environmental risks are best assessed

---

22. Mandel, *supra* note 1, at 2183-84.
23. *Id.* at 2184; Jill Hobbs & William Kerr, *Will Consumers Lose or Gain From the Environmental Impacts of Transgenic Crops?*, in ENVIRONMENTAL COSTS AND BENEFITS OF TRANSGENIC CROPS 251 (J.H.H. Wesseler, ed. 2005).
24. *Id.*; NATIONAL RESEARCH COUNCIL, *supra* note 18, at 111 (finding reduced pesticide use or toxicity levels on fields where GM corn, soybean, and cotton are grown). Bt, or *Bacillus thuringiensis*, is an endospore-forming bacillus which is commonly introduced to GM plants to display insecticidal properties. Shelton et al., *supra* note 3, at 848.
25. NATIONAL RESEARCH COUNCIL, *supra* note 18, at 111; Mandel, *supra* note 1, at 2185; Hobbs & Kerr, *supra* note 23, at 251.
26. NATIONAL RESEARCH COUNCIL, *supra* note 18, at 111; Mandel, *supra* note 1, at 2185. *See also* Brookes & Barfoot, *supra* note 16 (finding a reduction in greenhouse gas emissions associated with GM crop adoption due to less fuel use because of less frequent chemical applications and the use of low- or no-till systems); Smyth et al., *supra* note 16, at 407 (finding that GM crop use increases carbon sequestration rates due to use of low- or no-till systems).
27. Mandel, *supra* note 1, at 2186.
28. Hobbs & Kerr, *supra* note 23, at 251.
29. Angelo, *supra* note 4, at 110.
30. NATIONAL RESEARCH COUNCIL, *supra* note 18.
31. *Id.*; Angelo, *supra* note 4.
32. *Id.* at 110-11; NATIONAL RESEARCH COUNCIL, *supra* note 18.
33. Phillip Dale et al., *Potential for the Environmental Impact of Transgenic Crops*, 20 NATURE BIOTECH. 567 (2002).
34. *See, e.g.*, Anthony J. Conner et al., *The Release of Genetically Modified Crops Into the Environment: Overview of Ecological Risk Assessment*, 33 PLANT J. 19, 21 (discussing different considerations for risk assessments); Alan Raybould & Michael Wilkinson, *Assessing the Environmental*

after harm has occurred—a fact that lends itself to favor the precautionary principle with regard to GM crop introductions.[35]

In terms of nonenvironmental risks, most concerns focus on human health and economic impacts. The human health risks are mainly toxicity and allergenicity.[36] Different proteins cause allergic reactions in people, so there is a concern that inserting novel genes that express proteins into a plant could trigger allergic reactions.[37] This could occur either by use of genetic material from a source that is unknown to the human diet or by use of genetic material from a known allergen—for instance, some kind of nut—to produce a crop (like corn) that consumers would have no reason to suspect would contain a known allergen.[38] Elimination of this risk would likely have to involve labeling to advise consumers about the source of genetic material contained in a GM organism.[39] Another risk is that consumption of GM crops could lead to consumption of new toxins or increased levels of naturally occurring toxins.[40] Some worry that GM crops could contain fewer nutrients than non-GM counterparts.[41]

There are also economic concerns. Cross-pollination of GM plants with non-GM crops or, more importantly, organic crops can lead to contamination of the non-GM crop.[42] Contamination may lead to the inability of organic producers to sell their crops as organic or for non-GM growers to market their products as GM-free—resulting in lost sales or decreased revenue.[43]

4. Environmental Risks

Several environmental risks associated with the introduction and maintenance of GM plants are linked with the problem of scientific uncertainty and lack of information, and left unaddressed by regulators. In general, the most prominent environmental concerns relate to (1) weeds and the ability for GM crops to become weeds or for wild weeds to become "superweeds"; (2) insect resistance to Bt crops and the creation of "superbugs"; (3) reduced biodiversity; and (4) effects on nontarget organisms.

First, an herbicide-resistant GM crop (HRC) could itself become a weed if it contains a transgene that confers or enhances weediness.[44] For instance, an HRC could be a "volunteer weed" where it pops up in a field that has been replanted with a rotational crop and the volunteer is resistant to herbicide applications.[45] Another possibility is that GM crops could outcross (also known as cross-pollination, the process by which pollen is carried from one plant to a different plant and a transfer of genes occurs) to related species to create a "superweed" or a wild plant that is resistant to herbicides.[46] Some scientists think that the creation of a superweed is unlikely due to the lack of any competitive advantage that herbicide tolerance would present

---

Risks of Gene Flow From GM Crops to Wild Relatives, in GENE FLOW FROM GM PLANTS 169 (Guy Poppy & Michael Wilkinson, eds., 2005) (discussing what should be included in a risk assessment); David Andow & Angelika Hilbeck, *Science-Based Risk Assessment for Nontarget Effects of Transgenic Crops*, 54 BIOSCIENCE 637 (2004) (same).

35. See Andow & Hilbeck, *supra* note 34, at 637-38 (discussing limits of environmental risk assessments and noting that the use of the precautionary principle is "ancillary" to the U.S. regulatory system).
36. Angelo, *supra* note 4, at 103-04.
37. Mandel, *supra* note 1, at 2190.
38. *Id.*; Angelo, *supra* note 4, at 105.
39. *Id.* For instance, a label might inform buyers that a particular package of tofu contains GM soy that contains genetic material from a peanut, a known allergen.
40. Mandel, *supra* note 1, at 2191.
41. *Id.*
42. Angelo, *supra* note 4, at 108. *See also, e.g.*, Andrew Harris & David Beasley, *Bayer to Pay $750 Million to End Lawsuits Over Genetically Modified Rice*, BLOOMBERG, July 1, 2011, at http://www.bloomberg.com/news/2011-07-01/bayer-to-pay-750-million-to-end-lawsuits-over-genetically-modified-rice.html (discussing litigation over an experimental strain of GM rice which "cross-bred with and 'contaminated' over 30 percent of U.S. ricelands," causing futures prices of U.S. rice to fall).
43. *Id.*
44. Altieri, *supra* note 6, at 16.
45. *Id.* This risk has been somewhat dismissed based on scientific research finding that HRC crops are no more likely to be invasive in agricultural or natural settings than conventional counterparts—however, it is worth acknowledging that such studies have focused on well-established and studied types of HRCs such as corn varieties and that a case-by-case assessment should be used to determine pervasiveness and invasiveness of individual GM crops. Shelton et al., *supra* note 3, at 857; Phillip Dale et al., *supra* note 33, at 569.
46. Altieri, *supra* note 6, at 16. A superweed might occur where a GM crop hybridizes with a sexually compatible wild species and produces hybrids that carry the GM parent's introduced trait. In some cases, the introduced trait might carry a competitive advantage, such as drought tolerance. In the case of a superweed, the hybrid would have the HRC trait of resistance to herbicides, like the GM parent, which would make it difficult to control the spread of the hybrid. *Id.* Recent observations by farmers have noted the development of "superweeds" that are resistant to Monsanto's Roundup Ready herbicide in agricultural areas. Jack Kaskey, *Attack of the Superweed*, BLOOMBERG BUSINESSWEEK, Sept. 8, 2011, at http://www.businessweek.com/magazine/attack-of-the-superweed-09082011.html.

outside of agricultural areas,[47] lack of wild relatives that grow in proximity to GM crops,[48] and differences in chromosome number, phenology, and habitat between GM crops and wild relatives.[49] Still, most studies on this issue point to major established GM crops for which there either are no domestic wild relatives or which have been bred to the point where it is unlikely that they have the capacity to compete with wild species in the environment.[50] Other less-studied minor crops such as rapeseed and barley do have wild relatives and could potentially outcross to related species.[51] Further, traits possessed by non-HRC plants (or additional traits possessed by HRC plants), such as insect resistance or drought tolerance, may confer a competitive advantage if transferred to wild relatives or weed species, which could result in particularly hardy or problematic weeds.[52] Thus, weed-related risks posed by GM crops should not be dismissed due to significant research gaps.

Second, insect-resistant GM plants may cause target insects to develop a resistance to the toxin produced in the GM plant and become a "superbug."[53] Insect-resistant plants continually produce toxins, so pest species are continually exposed to these toxins, which can lead to rapid development of resistance by the insect to the introduced toxin.[54] Some pest species have already developed resistance to Bt.[55] If a large number of common pest species were to develop resistance to pesticides contained in GM plants, such as Bt, all producers of crops targeted by resistant insects would have to turn to alternative, potentially more harmful, insecticides. This is of particular concern for certified organic growers who currently rely on non-engineered forms of microbial *Bt* as important pest control tools. Some solutions identified for delaying resistance development are to produce multitoxin, as opposed to single-toxin, GM plants,[56] and to create refuges by creating patchworks of GM crops alternated with non-GM crops that is kept free of pesticides—thereby attracting insects to the pesticide-free areas and reducing interaction with the GM crops.[57] Based on existing research, it appears that development of resistance to insect-resistant GM plants is a major concern because it could lead to increased pesticide use or the use of more aggressive pesticides, which could be harmful to the environment.

Third, the cultivation of GM crops and other plants could lead to decreased genetic diversity. GM crops could reduce agrobiodiversity by promoting crop uniformity. Though genetic engineering has the potential to create an infinite variety of plants, the trend in research and application has been to promote monoculture of a few prominent GM crops.[58] The ecological drawbacks of monoculture include vulnerability to pests and pathogens and thus increased risk of crop failure.[59] Additionally, contamination of wild and non-GM relatives could change the makeup of plant communities and reduce genetic diversity.[60] Further, introduction of GM herbicide-resistant crops may reduce weed species diversity and ecosystem complexity on GM fields and neighboring areas.[61]

---

47. Altieri, *supra* note 6, at 7.
48. Angelo, *supra* note 4, at 108.
49. Shelton et al., *supra* note 3, at 857.
50. Angelo, *supra* note 4, at 108.
51. *Id.*; Altieri, *supra* note 6, at 17. *See also* NATIONAL RESEARCH COUNCIL, *supra* note 18, at 107-10 (discussing gene flow between GM crops and related weed species and noting that while outcrossing is not an issue for major crops, some crops such as alfalfa and wheat may present a risk of cross-pollination with non-GM plants).
52. Dale et al., *supra* note 33, at 569.
53. Ricarda A. Steinbrecher, *From Green to Gene Revolution: The Environmental Risks of Genetically Engineered Crops*, 26 ECOLOGIST 273, 276 (1996).
54. *Id.*; Angelo, *supra* note 4, at 110.
55. *Id.*; NATIONAL RESEARCH COUNCIL, *supra* note 18, at 111. *See also* Scott Kilman, *Monsanto Corn Under Attack by Superbug*, WALL ST. J., Aug. 29, 2011, at http://online.wsj.com/article/SB10001424053111904009304576532742267732046.html (discussing a study tracking a rootworm's recent development of resistance to Monsanto's Bt SmartStax corn in Iowa and noting that Bt-resistance could lead to increased chemical pesticide application or the need for larger refugia zones).
56. *Id.* at 112.
57. Alteiri, *supra* note 6, at 18 (discussing refugia and noting that the effectiveness of such a strategy would be unlikely given the costs associated with maintaining refuges and the overall prevalence of GM crop monocultures).
58. *Id.* at 15.
59. *Id.*; Mandel, *supra* note 1, at 2197. *See also* Charles Seibert, *Food Ark*, NAT'L GEOGRAPHIC, July 2011 (discussing the recent extinction of many crop varieties and why reliance on only a few varieties, as is typical in modern monoculture, is problematic in light of threats posed by disease or climate change).
60. Altieri, *supra* note 6, at 17.
61. Dale et al., *supra* note 33, at 571.

Fourth, GM plants could have a deleterious influence on nontarget living things such as insects and animals.[62] Given the difficulty of simulating field conditions in a controlled setting, the effect of GM plants on nontarget organisms is hard to predict. Despite this difficulty, many scholars agree that GM plants have the potential to adversely affect nontarget species due to toxicity or secondary effects such as elimination of a food source for predator insects or animals that consume target species.[63] Another concern is that GM plants may adversely affect soil biota. It appears that horizontal transfer of DNA from GM plants to bacteria and soil organisms is not a major issue because plant DNA degrades rapidly in soil systems.[64] However, GM plants and plant litter can influence the composition of microbial communities, which could affect soil health and ecosystem functioning.[65] Studies have not indicated any urgent threats presented to soil communities by GM plants, but the need for further research on the matter, including multiorganism and long-term studies, has been highlighted.[66] In sum, it is unknown whether GM plants pose a threat to nontarget organisms through toxicity or secondary effects, and such threats remain a continuing concern and an area for future research.

5. Regulation of GM Plants

The potential risks associated with GM plants illustrate the need for effective regulation to minimize the possible adverse effects of GM plants and maximize the benefits. However, the U.S. regulatory system for GM plants and other organisms is fraught with "regulatory gaps, overlaps, and inconsistencies" that hinder its overall effectiveness.[67]

Any discussion of GM regulation in the United States begins with the Coordinated Framework for Regulation of Biotechnology. The Framework was promulgated by the White House Office of Science and Technology Policy in 1986 to address the budding biotechnology industry.[68] The Framework was designed to institute a "comprehensive federal regulatory policy" for GM research and products and specified that GM products would be regulated under then-existing laws and regulations instead of developing new laws to address the new technology.[69] The basis for this policy was the government's conclusion that GM products are not fundamentally different from non-GM products or inherently risky, and thus the final *product* of biotechnology should be regulated rather than the process of creating GM products.[70] The Framework also established that three federal agencies have primary responsibility for overseeing GM products: the Food and Drug Administration (FDA),[71] the U.S. Environmental Protection Agency (EPA), and the U.S. Department of Agriculture (USDA).[72] As a result of the policy set out in the Framework, no single law directly addresses GM plants or GM products in general. Instead, as many as 12 statutes, a myriad of regulations, and five different agencies and services play a role in governing GM products.[73]

USDA is the major oversight agency for GM plants and has primary authority over all GM plants except those that are pest-protected.[74] USDA oversees the interstate movement, import, field testing, and release of

---

62. The best-known example of this concern stems from a study reporting that Bt corn pollen could be a hazard to monarch butterfly larvae that consumed milkweed leaves coated with the pollen. Dale, *supra* note 33, at 568; Shelton et al., *supra* note 3, at 860. Ultimately, the concerns posed to monarchs by Bt corn pollen was determined to be negligible because densities of pollen encountered by larvae in the field are too low to present a risk. Dale, *supra* note 33, at 568.
63. *Id.*; Altieri, *supra* note 6, at 18. In contrast, some suggest that effects of Bt crops on nontarget invertebrates are "favorable or neutral." NATIONAL RESEARCH COUNCIL, *supra* note 4, at 111.
64. Shelton et al., *supra* note 3, at 858; Kari Dunfield & James Germida, *Impact of Genetically Modified Crops on Soil- and Plant-Associated Microbial Communities*, 33 J. ENVTL. QUALITY 806, 809 (2004).
65. Dunfield & Germida, *supra* note 64, at 813. Some changes that have been observed include displacement of indigenous populations, suppression of fungal populations, reduced protozoa populations, altered soil enzymatic activity, and increased carbon turnover. Altieri, *supra* note 6, at 19.
66. *See, e.g.*, Shelton et al., *supra* note 3, at 859 (discussing the need for further research on effects of GM plants on soil health).
67. Angelo, *supra* note 4, at 95.
68. Margaret Rosso Grossman, *Genetically Modified Crops and Food in the United States: The Federal Regulatory Framework, State Measures, and Liability in Tort*, in THE REGULATION OF GENETICALLY MODIFIED ORGANISMS: COMPARATIVE APPROACHES 299, 300 (Luc Bodiguel & Michael Cardwell, eds., 2010).
69. *Id.*; Mandel, *supra* note 1, at 2216.
70. Mandel, *supra* note 1, at 2216.
71. FDA authority is generally limited to GM animals.
72. Mandel, *supra* note 1, at 2216.
73. *Id.* at 2228.
74. *Id.* at 2224. A "pest protected" GM plant is one that contains a pesticide, defined under FIFRA as any substance intended for "preventing, destroying, repelling, or mitigating any pest." *Id.* at 2221.

GM plants and generally ensures that they are safe to grow.[75] This oversight is conducted through USDA's Animal and Plant Health Inspection Service (APHIS) under the authority of the Plant Protection Act (PPA).[76] The PPA mandates that APHIS prevent the release and spread of "plant pests," which are defined broadly as organisms that can injure or cause disease or damage (directly or indirectly) in or to any plants or plant parts.[77] Under the PPA, anyone who seeks to "introduce" (by release into the environment, import, or transport) a GM plant must first get approval from APHIS.[78] In general, a GM plant will be considered a "regulated article" under the PPA, and thus subject to USDA authority and unable to be sold in commerce until it is evaluated through field trials and determined to be "unregulated" based on a finding that it poses no plant pest risk.[79] APHIS is required to comply with the National Environmental Policy Act (NEPA) and must assess environmental risks prior to granting a petition for deregulation or permit for field testing.[80] An applicant must petition for deregulation which, if granted, allows the product to move freely in commerce with no further USDA oversight. Deregulation effectively ends USDA authority over a deregulated plant and forecloses the possibility of continued monitoring or intervention where there are unforeseen impacts.[81] As an agency whose overall purpose is to promote domestic agriculture, USDA's approach with respect to GM plants trends toward deregulation.[82]

EPA also plays a role in overseeing GM plants, but its reach is limited to pest-protected plants. In general, EPA has authority to regulate pesticide use, which it has extended to GM plants that have genetic traits introduced to resist insects (for example, the *Bt* pesticide commonly used in GM plants).[83] This authority is pursuant to the Federal Insecticide, Fungicide, and Rodenticide Act (FIFRA), which requires pesticides to be registered with EPA prior to their use or sale.[84] In order to register, an applicant must demonstrate a lack of unreasonable adverse risk to man or the environment.[85] EPA also has authority over pesticide residues in food, including those from pest-protected GM plants, under the Food, Drug, and Cosmetic Act (FFDCA).[86]

FDA is responsible for food safety under the FFDCA,[87] which grants FDA the authority and imposes a statutory duty to regulate food labeling, food additives, and adulterated foods. Regarding adulterated foods, the FFDCA mandates that FDA protect the food system from foods that contain any "poisonous or deleterious substance which may render it injurious to health."[88] Food additives (substances used in food or components of food or that might affect the characteristics of food) require premarket approval and labeling.[89]

While most food additives must undergo extensive premarket safety testing,[90] this is not the case if the food additive is generally recognized as safe. Under the FFDCA, a substance that is intentionally added to food is a food additive and is subject to premarket review and approval by FDA, unless the substance is generally recognized by qualified experts as having been shown to be safe under the conditions of its intended use, or unless the use of the substance is otherwise excluded from the definition of a food additive.[91]

Due to this complex regulatory overlay, rules governing GM plants have been developed in a "piecemeal, haphazard manner," which has led to regulatory overlaps, gaps, and other problems.[92] Understanding these regulatory limits helps one to fully appreciate the scope of issues surrounding GM plants and the effect

---

75. Grossman, *supra* note 68, at 301.
76. *Id.*
77. Angelo, *supra* note 4, at 134.
78. Mandel, *supra* note 1, at 2225.
79. Grossman, *supra* note 68, at 302.
80. *Id.* at 303.
81. Mandel, *supra* note 1, at 2234.
82. Angelo, *supra* note 4, at 142.
83. Mandel, *supra* note 1, at 2221.
84. *Id.* Technically, EPA does not regulate any GM plants themselves, but rather only the inserted pesticidal genetic material. *Id.* at 2223.
85. *Id.*
86. Grossman, *supra* note 68, at 309.
87. 21 U.S.C. §§301 et seq. (2011).
88. 21 U.S.C. §342(a)(1).
89. 21 U.S.C. §348.
90. FDA, Consultation Procedures Under FDA's 1992 Statement of Policy—Foods Derived From New Plant Varieties: Guidance on Consultation Procedures (Revised Oct. 1997) (June 1996), *available at* http://www.fda.gov/food/guidancecomplianceregulatoryinformation/guidancedocuments/biotechnology/ucm096126.htm.
91. *See* FDA, *Generally Recognized as Safe*, http://www.fda.gov/Food/FoodIngredientsPackaging/GenerallyRecognizedasSafeGRAS/default.htm.
92. Mandel, *supra* note 1, at 2230. For a more detailed criticism of the current regulatory regime, see *id.* at 2230-57 and Angelo, *supra* note 4.

that deficiencies of the regulatory system may have in relation to environmental and other risks posed by GM plants.

6. Case Study: Alfalfa

Alfalfa is a perennial plant widely grown to make hay to feed dairy cows, beef cattle, and other animals.[93] Humans generally do not consume alfalfa directly, but may ingest it indirectly by consuming meat or other products from an animal that consumed alfalfa. Alfalfa is grown on over 21 million acres and is the fourth-largest crop by acreage and third most valuable crop in the United States.[94] Monsanto, a large agricultural biotechnology company, developed an alfalfa seed that has been modified to tolerate herbicide. The product—known as Roundup Ready alfalfa—was developed with Forage Genetics to resist Monsanto's own Roundup glyphosate herbicide.[95] Roundup Ready alfalfa allows farmers to spray Roundup herbicide in alfalfa fields without harming the crop itself. Proponents claim that using Roundup and Roundup Ready alfalfa saves time and labor costs for weeding and gives farmers a higher yield of alfalfa per acre, thereby keeping food prices low.[96]

Roundup Ready alfalfa was a source of controversy and a long legal battle that eventually ended in deregulation of the crop. In 2003, Monsanto and Forage Genetics petitioned APHIS for nonregulated status of Roundup Ready alfalfa.[97] APHIS prepared an environmental assessment pursuant to NEPA, and deregulated the alfalfa in 2005 based on a finding of no significant impact.[98] This decision was challenged, and the federal district court in San Francisco vacated the deregulation decision in 2007 and issued a permanent injunction against any deregulation of the alfalfa or planting of the seed based on the conclusion that environmental risks were not properly documented and an environmental impact statement (EIS) was necessary.[99]

The district court's decision was challenged by Monsanto and Forage at the U.S. Court of Appeals for the Ninth Circuit, which affirmed the grant of an injunction until an EIS was completed.[100] In 2010, the U.S. Supreme Court reversed and remanded the case, holding that the order enjoining *any* deregulation did not satisfy the test for granting a permanent injunction and that the respondents could not show irreparable injury would occur if APHIS proceeded with partial deregulation. The Court said, essentially, that the injunctive relief granted was too broad.[101] The case was closely watched because of its implications for the ability to use NEPA when seeking injunctions against industry, particularly in the context of GM crops.[102] Though the injunction was lifted, APHIS still had to conduct an EIS to comply with NEPA before Roundup Ready could be deregulated.[103]

Following this decision, an EIS was completed and Roundup Ready alfalfa was fully deregulated. APHIS issued an EIS in fall of 2010 with two preferred options: deregulation or deregulation with geographical restrictions.[104] In early 2011, it deregulated Roundup Ready based on a finding that there is no plant pest risk posed by the alfalfa.[105] APHIS found that Roundup Ready poses no greater risk than non-GM varieties—in fact, that APHIS found no direct or indirect plant pest risks associated with the GM alfalfa—and that it exhibits no plant pathogenic properties.[106] APHIS also concluded that the GM alfalfa

---

93. Western Organization of Resource Councils (WORC), The Problem With GM Alfalfa 1, Aug. 2005, *available at* http://www.centerforfoodsafety.org/pubs/Alfalfa_WORC_Factsheet.pdf. Monsanto's alfalfa is the first commercial GM perennial crop. *Id.*
94. *Id.*; Andrew Pollack, *U.S. Approves Genetically Modified Alfalfa*, N.Y. Times, Jan. 27, 2011, at http://www.nytimes.com/2011/01/28/business/28alfalfa.html.
95. *Id.*
96. Michael J. Crumb, *Experts: Contamination From GM Alfalfa Certain*, Bloomberg Businessweek, Feb. 7, 2011, at http://www.businessweek.com/ap/financialnews/D9L7QHO00.htm.
97. APHIS, Q&A—Final Environmental Impact Statement Roundup Ready Alfalfa, Dec. 2010, http://www.aphis.usda.gov/publications/biotechnology/content/printable_version/faq_alfalfa_eis.pdf (accessed Apr. 14, 2011) [hereinafter APHIS, Q&A—Final EIS].
98. *Id.*
99. *Id.*
100. Geerston Seed Farms v. Johanns, 570 F.3d 1130 (9th Cir. 2009).
101. Monsanto v. Geertson Seed Farms, 130 S. Ct. 2743, 2759 (2010).
102. Jennifer Koons, *Supreme Court Lifts Ban on Planting GM Alfalfa*, N.Y. Times, June 21, 2010, at http://www.nytimes.com/gwire/2010/06/21/21greenwire-supreme-court-lifts-ban-on-planting-gm-alfalfa-57894.html.
103. Jeffrey L Fox, *GM Alfalfa—Who Wins?*, 28 Nature Biotech. 770 (2010).
104. APHIS, Q&A—Final EIS, *supra* note 98.
105. APHIS, Q&A—Roundup Ready Alfalfa Deregulation, Jan. 2011, http://www.usda.gov/documents/rr_alfalfa.pdf (accessed Sept. 28, 2011).
106. *Id.*

is not expected to affect plants and animals adversely or become invasive, and that the nutritional profile of the GM alfalfa is the same as non-GM varieties.[107] It was determined further that the GM alfalfa has no adverse effects on human health and worker safety, and that the overall risk presented by glyphosate herbicide use does not change with the use of GM alfalfa instead of non-GM alfalfa.[108]

The practical result of Roundup Ready's deregulation is that it may be cultivated for commercial purposes with no further federal oversight or restrictions.[109] This decision has drawn opposition, and many are worried about the potential economic and environmental impacts of GM alfalfa.[110]

In terms of economic risks, organic and non-GM farmers are worried that their sales could decrease if GM genes are detected in their crops, which could occur through cross-pollination.[111] As a cross-pollinating crop, contamination of non-GM varieties may be inevitable, which could decimate sales for non-GM alfalfa growers if contamination occurs.[112] It may also adversely affect organic or other non-GM dairy and beef producers who may find it prohibitively expensive, or impossible, to ensure that their animals are not being fed GM-contaminated alfalfa.[113] Further, exports of U.S. alfalfa may be harmed, given that consumers in many international markets demand GM-free products.[114]

The Roundup Ready alfalfa also poses some potential environmental risks. Introduction of the crop could lead to increased pesticide and herbicide use.[115] The crop may also pose unknown risks to non-target organisms that interact with the alfalfa, such as birds, insects, or other beneficial organisms.[116] Another risk is that weeds will develop glyphosate resistance. Because Roundup Ready alfalfa is designed to be used with Roundup, its adoption may promote more overall use of the herbicide, thus increasing development of glyphosate-resistant weeds.[117] This could lead to the use of additional, more harmful chemicals in order to control glyphosate-resistant weeds. Additionally, despite APHIS' finding of no risk from plant pests, it is possible that the GM alfalfa will itself become a weed or could cross-pollinate with native plants.[118]

At the end of the day, Roundup Ready alfalfa is now unregulated despite concerns that it could pose significant risks to the organic and other non-GM alfalfa markets and to the environment. No laws require adopters of the GM product to initiate protective measures such as using a buffer zone to avoid contamination, so the burden will largely be on growers who wish to remain GM-free to adopt preventative measures against contamination. USDA has stated that it is their goal to help maintain purity of non-GM seed and to lessen risk of gene flow to non-GM crops.[119] USDA has stated that it will be taking some steps to meet these goals: reestablishing two advisory committees, conducting research, promoting dialogue, and providing voluntary third-party audits and verification of industry-led stewardship initiatives.[120]

In sum, GM alfalfa provides a useful example of a GM plant that was recently approved for federal deregulation and some of the potential risks and benefits associated with such approval. The main concerns associated with GM alfalfa are generally present for other GM organisms and the deregulation decision illustrates the trend of unrestricted use of GM plants and lack of adoption of protective measures for non-pesticidal GM organisms.[121]

---

107. *Id.*
108. *Id.* EPA has determined that glyphosate-containing products present no unreasonable adverse effects on the environment and has a reasonable certainty of no harm to humans when used in accordance with labeling of registered products. *Id.*
109. *Id.* APHIS notes that it assumes growers will continue to be subject to Monsanto's contractual restrictions, including managing hay to prevent seed production. *Id.*
110. Pollack, *supra* note 94.
111. *Id.*
112. Crumb, *supra* note 96. Even if cross-pollination does not occur, contamination could still happen due to unintentional mixing of seeds or a GM alfalfa plant popping up as a "volunteer," or weed, in a field that was replanted. *Id.*
113. WORC, *supra* note 93.
114. *Id.*
115. *Id.*
116. *Id.*
117. *Id.* Several glyphosate-resistant alfalfa weeds already exist. *Id.*
118. *See* Jane Rissler & Margaret Mellon, The Ecological Risks of Engineered Crops 37 (1996) (noting that alfalfa is similar in genetic makeup to wild plants, which may mean that shared genes extend to those for weediness traits).
119. APHIS, Q&A—Roundup Ready Deregulation, *supra* note 105.
120. *Id.*; Crumb, *supra* note 96; Pollack, *supra* note 94.
121. *See* Crumb, *supra* note 96 (discussing the need to put protective measures in place before approval decisions are made).

B. GM Animals

Genetic modification of animals is a relatively new and rapidly developing technology that has raised controversy due to a lack of long-term research on the potential risks and benefits of GM animals as well as ethical concerns. This section examines potential costs and benefits associated with ongoing GM animal development and their possible introduction to the market. Additionally, this section briefly addresses the current regulatory process applied to GM animals, and includes a case study of GM salmon, the first GM animal designed for human consumption. Finally, the section concludes with a discussion of labeling issues for GM products in the context of GM salmon.

1. Background

A GM animal contains a recombinant DNA (rDNA construct) that introduces a new trait or characteristic into that animal. The genes that comprise an rDNA construct can be taken from another living species or synthesized in a laboratory.[122] The construct is then inserted into a target animal, where it will direct production of specific proteins.[123] Thus, this form of genetic engineering modifies animals at the cellular level.[124] The GM animal will differ from a non-GM counterpart by possessing an rDNA construct that, if successful, will introduce a trait or characteristic that a non-GM animal would probably lack or possess at a lesser degree.[125] Most GM animals are designed to have "heritable" traits, meaning that they will pass their engineered traits onto their progeny, who are also considered GM animals.[126]

GM animals are produced for a range of uses and purposes. Currently, there are no GM animals commercially available for consumption by humans or animals, although an application for approval of GM salmon is pending before FDA[127] and others are in development stages.[128] The only GM animals now available in the marketplace are a GM aquarium fish and GM laboratory rodents.[129] Many other species of GM animals exist or are in development but only for investigational or research purposes.[130] The largest class of such animals is intended for biopharm purposes, meaning that the animals are engineered to produce substances for use in animal or human pharmaceuticals.[131]

2. Benefits

The development and commercialization of GM animals has several potential benefits. Proponents maintain that GM animals have the potential to have improved characteristics of nutrition, growth rate, and disease resistance.[132] In terms of efficiency, modification through rDNA insertion may be beneficial because non-GM selective breeding methods are labor- and time-intensive, costly, imprecise, and only possible among closely related species.[133] In contrast, rDNA modification allows for faster results, less variability, and does not require that species be related or sexually compatible, thereby allowing for a greater range of traits to be introduced into a target animal.[134]

---

122. U.S. Food and Drug Administration, *Genetically Engineered Animals General Q&A*, http://www.fda.gov/AnimalVeterinary/DevelopmentApprovalProcess/GeneticEngineering/GeneticallyEngineeredAnimals/ucm113605.htm (accessed Nov. 28, 2012).
123. Mandel, *supra* note 1, at 2175.
124. *Id. See also* John Clark & Bruce Whitelaw, *A Future for Transgenic Livestock*, 4 Nat. Rev. Gen. 825 (2003) (discussing new methods for genetic modification of livestock).
125. U.S. Food & Drug Administration, *supra* note 122.
126. *Id.* In response to environmental concerns associated with cross-breeding with wild relatives, some GM animals may be sterilized.
127. U.S. Food & Drug Administration, *supra* note 122.
128. For instance, private industry is developing several species of transgenic fish and shellfish such as oysters and trout. Nathaniel Logar & Leslie Pollock, *Transgenic Fish: Is a New Policy Framework Necessary for a New Technology?*, 8 Env. Sci. & Pol'y 17, 18 (2005).
129. *Id.*
130. *Id.*
131. *Id.* For more information, see, e.g., Louis-Marie Houdebine, *Production of Pharmaceutical Proteins by Transgenic Animals*, 32 Comp. Immun. Microbiol. Infect. Dis. 107 (2009) (discussing the use of GM animals as a source for pharmaceutical proteins) *and* Nils Lonberg, *Human Antibodies From Transgenic Animals*, 23 Nature Biotech. 1117 (2005) (discussing use of GM mice and other animals for production of human antibodies). Additional health-related uses include organ and tissue transplant and development of human gene therapies. *Id.*; Chad West, Note, *Economics and Ethics in the Genetic Engineering of Animals*, 19 Harvard J.L. & Tech. 414, 415 (2006).
132. U.S. Food & Drug ADministration, *supra* note 122.
133. Mandel, *supra* note 1, at 2175.
134. *Id.* at 2176.

In terms of animals meant for consumption, the largest probable advantage is the potential to increase world protein supply in an efficient manner. For example, if a transgenic fish is modified to grow at a faster rate and to consume less feed than non-GM counterparts raised in aquaculture, overall inputs to produce the GM fish will be reduced,[135] lowering market price for that species and thereby making it accessible for a wider range of people. Further, pressure on already threatened wild stocks could be reduced. Potential benefits associated with livestock animals also include efficiency where animals are developed to require less food or with improved growth rates or disease resistance.[136] Other possibilities include the production of products with improved nutritional value (such as lowered cholesterol in eggs) or animals that have a lower environmental impact by producing less waste or requiring fewer inputs.[137] Ultimately, the benefits associated with a GM animal will depend on the type of characteristic that is introduced and the purpose for which the animal is used.[138]

3. Risks and Concerns

The development and commercialization of GM animals also carries with it several risks and concerns. First, some people are opposed to the genetic modification of animals for ethical or environmental reasons. Notably, public opposition appears to be stronger to GM animals than it is to GM plants,[139] which might be due to concerns about animal welfare—for example, in animal testing.[140] Some suggest that consumers are more likely to be accepting of GM animals that are used for purposes other than human consumption (such as biopharmaceutical uses).[141] In any case, it is worth noting public resistance to GM animals because it may act as a hindrance to the emergence of the industry or development of technologies.

Some opposition to the development of GM animals for human consumption stems from concerns about human health. One such concern is potential allergenicity or toxicity—in other words, the expression of new proteins may increase risk of allergic response in consumers or cause the creation or bioaccumulation of toxins in GM animal meat.[142] Additionally, use of animals for organ or tissue transplant or for human consumption raises the potential for disease or virus transfer.[143] It is also unclear how consumption of disease-resistant GM animals might affect humans or animals.[144] Another concern is that GM animals that are *not* intended for consumption could end up in the food supply unless control measures are created.[145]

In terms of environmental risk, the main concern is that GM animals might escape and interact with wild or non-GM animals. If a GM animal were to breed with a non-GM species, it would be possible for gene transfer to occur, thereby spreading the GM traits to nontarget species.[146] In theory, the spread of GM traits could lead to extinction of wild populations. Even if the GM animal is poorly adapted to survival in the wild, it may have short-term breeding advantages (such as faster growth rate) that could dilute a wild or non-GM gene pool.[147] The GM animal could also have the potential to transfer diseases or parasites to wild or other non-GM populations, especially where GM animals have been developed to be

---

135. For a more detailed discussion, see Martin Smith et al., *Genetically Modified Salmon and Full Impact Assessment*, 330 Sci. 1052.
136. Mandel, *supra* note 1, at 2188.
137. *Id.*
138. Additional benefits include the use of GM animals for human and animal health purposes. Use of GM animals for pharmaceutical purposes has the benefits of efficiency, cost reduction, and possible environmental benefits due to reduced manufacturing inputs needed to produce pharmaceutical proteins. The possibility for organ and tissue transfer is also a large benefit, given the shortage of human donors. *Id.* at 2188-89.
139. *See, e.g.*, Logar & Pollack, *supra* note 128, at 19 (noting that Americans are "far less comfortable" with GM animals than plants); Mary Clare Jalonick, *Shoppers Wary of GM Foods Find They're Everywhere*, Wash. Post, Feb. 25, 2011, at http://www.washingtonpost.com/wp-dyn/content/article/2011/02/25/AR2011022500643.html ("Most consumers are more open to modifications in fruits and vegetables than in animals"); and Clark & Whitelaw, *supra* note 124, at 832 (discussing "keenly held" ethical views about GM animals).
140. Paul Rincon, *GM Animal Tests Continue to Rise*, BBC News (Dec. 8, 2005).
141. *Id.*; C.W. Forsberg et al., *The Enviropig Physiology, Performance, and Contributions to Nutrient Management Advances in a Regulated Environment: The Leading Edge of Change in the Pork Industry*, 81 J. Anim. Sci. E68, E75 (2002).
142. Gijs Kleter & Harry Kuiper, *Considerations for the Assessment of Safety of Genetically Modified Animals Used for Human Food or Animal Feed*, 74 Livestock Prod. Sci. 275, 279 (2002); Mandel, *supra* note 1, at 2201.
143. Mandel, *supra* note 1, at at 2201.
144. *Id.*
145. *Id.* at 2202.
146. *Id.* at 2200.
147. Tillman J. Benfey, Environmental Impacts of Genetically Modified Animals 3 (2003), *available at* ftp://ftp.fao.org/es/esn/food/GMtopic5.pdf.

disease-resistant.[148] This risk can be managed, though not entirely eliminated, through bioconfinement (for instance, by sterilization of GM species) and containment (fencing in GM livestock or on-land facilities for fish). This concern is less relevant for livestock species than for species such as fish and shellfish, as there are generally no wild populations of agricultural livestock species.[149] In contrast, farmed fish and shellfish lines can breed easily with other captive lines and with wild relatives, and hybridization of closely related species is common.[150]

Beyond extinction, escape concerns include invasion hazards. These include outcompetition and displacement because GM animals may possess traits—such as a fast growth rate or cold or disease tolerance—that could provide a competitive edge over wild or other non-GM species.[151] Essentially, a GM animal could harm wild or other non-GM populations by acting as an invasive species. While some GM animals may not be able to persist in the wild, others, such as fish, may have a good chance of reestablishment in nature due to a short history of domestication.[152] Both extinction and invasion hazards could have larger, cascading ecosystem effects by disrupting other species in a community (for instance, a GM animal could, through outcompetition, eliminate a prey species and thereby harm a predator species).[153]

There are considerable environmental concerns associated with GM animals because the effect of their interaction with the natural environment is unknown. Fish species are of the most concern because they possess phenotypes that predispose them to successful survival, dispersal, and reproduction in the wild and are especially difficult to recapture.[154] GM fish species in particular may warrant high levels of safeguards and monitoring to prevent deleterious environmental impacts.

4. Regulation of GM Animals in the United States

The regulation of GM animals in the United States[155] is new and largely untested. As is the case with GM plants, the United States has not enacted specific legislation to address GM animals.[156] Instead, like plants, GM animals are regulated under preexisting laws and regulations pursuant to the Coordinated Framework.[157] FDA is the only federal agency to have asserted authority over GM animals; EPA and USDA have not done so.[158]

FDA's regulatory authority is derived from the FFDCA. FDA has interpreted the Act as providing the agency with authority over GM animals before the animals can be marketed for human or animal consumption by defining GM animals as a "new animal drug."[159] The FFDCA defines a new animal drug as "an article . . . intended to affect the structure or any function of the body of . . . animals."[160] The GM animal itself is not a "drug," but FDA considers the rDNA inserted in GM animals to be an "article" intended to affect the structure or function of the body of the animal—and thus it is the rDNA article that is technically regulated by FDA under the FFDCA.[161] Evaluations of new animal drug applications (NADA) are

---

148. *Id.*
149. Clark & Whitelaw, *supra* note 124, at 832.
150. NATIONAL RESEARCH COUNCIL, *supra* note 18, at 47.
151. Logar & Pollock, *supra* note 128, at 18; William M. Muir & Richard D. Howard, *Assessment of Possible Ecological Risks and Hazards of Transgenic Fish With Implications for Other Sexually Reproducing Organisms*, 11 TRANSGEN. RES. 101 (2002).
152. Muir & Howard, *supra* note 151, at 109. Again, this concern is diluted with respect to livestock as their long history of domestication would reduce their ability to survive in the wild.
153. *Id.* at 102.
154. Benfey, *supra* note 147, at 1; Edward Bruggemann, *Environmental Safety Issues for Genetically Modified Animals*, 71 J. ANIM. SCI. 47 (1993).
155. To date, the United States is the first country to consider approval for genetically modified animals meant for human consumption. The United States' decisions and experiences regarding the regulation of GM animals have the potential to influence the regulatory structures or guidelines of other countries or international bodies such as the World Health Organization or the FAO. Logar & Pollock, *supra* note 128, at 25.
156. Emily Marden et al., *The Policy Context and Public Consultation: A Consideration of Transgenic Salmon*, 6 INTEGRATED ASSESSMENT J. 73, 78 (2006).
157. While this section discusses federal regulation of GM animals, it is worth noting that regulatory efforts also occur at the state or local level. For instance, a number of states have adopted legislation banning the growth of transgenic fish or requiring labeling for GM animals. *Id.* at 79.
158. *Id.*
159. Logar & Pollock, *supra* note 128, at 19 (citing 21 U.S.C. §360b). *See also* FDA, GUIDANCE FOR THE INDUSTRY: REGULATION OF GENETICALLY ENGINEERED ANIMALS CONTAINING HERITABLE RECOMBINANT DNA CONSTRUCTS (Jan. 15, 2009), *available at* http://www.fda.gov/downloads/AnimalVeterinary/GuidanceComplianceEnforcement/GuidanceforIndustry/UCM113903.pdf ; FDA, Guidance for Industry: Regulation of Genetically Engineered Animals Containing Heritable rDNA Construct, 73 Fed. Reg. 54407 (Sept. 19, 2008).
160. 21 U.S.C. §321(g)(1).
161. U.S. Food and Drug Administration, *supra* note 122.

conducted by FDA's Center for Veterinary Medicine.[162] NADA is the process that companies seeking to market GM animals must utilize to gain premarket approval for their products.[163] Ultimately, the rDNA in the GM animal is treated as a drug under the FFDCA—rather than as an animal or as food—even if it is intended for human consumption.

In essence, FDA's mandate with respect to GM animals is to ensure that a GM animal is safe in the context of a new animal drug. Although "safe" is left undefined, applicable statutes use the term in reference to "health of man or animal" and "cumulative effect on man or animal."[164] The main considerations taken up by FDA under the NADA process are whether a GM animal is safe to eat, nutritionally equivalent, and safe for the environment.[165] The main way in which these considerations are evaluated is through a "materially equivalent" assessment, which compares the risks posed by a GM animal with a non-GM animal of the same species.[166] For instance, health risks are analyzed by comparing nutritional profiles and screening for toxins and allergens.[167]

On the question of environmental risks, FDA has interpreted the FFDCA provisions in such a way that some environmental risks are considered within the jurisdictional reach of FDA. Because safety is determined with "reference to the health of man or animal," FDA should include as part of its assessment environmental risks that affect the health of humans or animals either directly or indirectly.[168] As a federal agency, FDA is under a mandate to comply with NEPA by assessing environmental impacts of its actions.[169] Thus, a NADA applicant is required to submit an environmental assessment (EA) or an EIS that addresses the environmental impacts of its product (or claim an exemption).[170]

The NADA process has drawn considerable criticism. Many feel that the FFDCA is an inappropriate vehicle for GM animal regulation because, at the time the FFDCA was passed, it was not intended to cover GM animals and, as such, it cannot adequately address the particular risks and considerations posed by GM animals.[171] Further, because the term "safe" with reference to GM animals is undefined by statute, it is left unclear what the exact parameters for safety considerations must be.

The scope of environmental effects that will or may be considered in the NADA process is unknown, meaning that there may be environmental effects (for instance, aesthetic harm) posed by a NADA application that could go unconsidered.[172] Additionally, FDA's internal interpretation of the FFDCA—under which FDA can consider certain environmental effects—is not codified and could be altered.[173] The primary means by which environmental risks are presented for review is through an initial EA that must be prepared by the applicant.[174] Critics say that this presents an opportunity for self-interested applicants to misrepresent environmental risks purposely or due to inherent bias.[175] The overall impact of the EA or EIS could be limited, as NEPA does not require elimination or even minimization of risk, instead requiring only that agencies

---

162. Logar & Pollock, *supra* note 128, at 19.
163. *Id.*
164. Smith et al., *supra* note 135, at 1052, 1053 (2010) (quoting 21 U.S.C. §§321(u) and 369b(d)(2)). Generally, a new animal drug is viewed as "unsafe" until it is approved, 21 U.S.C. §360b(a)(1), but it is unclear how a determination that a GM animal is "safe" should be made. Logar & Pollock, *supra* note 128, at 20.
165. Andrew Pollack, *Genetically Altered Salmon Get Closer to the Table*, N.Y. Times, June 25, 2010, at http://www.nytimes.com/2010/06/26/business/26salmon.html. *See also* FDA, *Public Meetings on Genetically Engineered Atlantic Salmon*, http://www.fda.gov/newsevents/publichealthfocus/ucm224089.htm.
166. Smith et al., *supra* note 135, at 1052.
167. *Id.*
168. 21 U.S.C. §321(u).
169. 42 U.S.C. §4332(C).
170. 21 C.F.R. §25.15(a).
171. *See, e.g.*, Logar & Pollock, *supra* note 128, at 19 (noting objections to "fitting transgenic animals into the [FFDCA] regulatory framework" raised by scientific, environmental, and consumer groups and noting that the FFDCA was initially intended to address substances such as antibiotics for pets).
172. Mandel, *supra* note 1, at 2210.
173. Logar & Pollock, *supra* note 128, at 20. *See also* FDA, Guidance for the Industry: Regulation of Genetically Engineered Animals Containing Heritable Recombinant DNA Constructs 8 (Jan. 15, 2009), *available at* http://www.fda.gov/downloads/AnimalVeterinary/GuidanceComplianceEnforcement/GuidanceforIndustry/UCM113903.pdf (noting that FDA will consider environmental factors in its exercise of enforcement discretion, including whether there is anything about the article that will pose an environmental risk and whether the GE animal poses any more of an environmental risk than its non-GE counterpart).
174. 21 C.F.R. §25.40. It reads, in part: "(b) Generally, FDA requires an applicant to prepare an EA and make necessary corrections to it. Ultimately, FDA is responsible for the scope and content of EA's and may include additional information in environmental documents when warranted."
175. Logar & Pollock, *supra* note 128, at 20-21.

consider environmental risk.[176] There are also questions regarding FDA's ability to act upon a finding of significant impact and whether an application could be rejected based entirely on environmental concerns.[177]

Aside from environmental concerns, some feel that there are additional gaps or inadequacies in the NADA process as applied to GM animals. For instance, some argue that FDA's interpretation of its jurisdictional reach is too narrow and fails to consider ancillary risks and benefits—in other words, what effects might occur if introduction of a particular GM animal *expanded* the overall market of that species (GM and non-GM) rather than the 1:1 ratio assumed by the material equivalence assessment.[178] Additionally, it remains to be seen whether FDA has authority to regulate GM animals that are not intended for consumption (or, if such authority exists, whether enforcement will be pursued).[179]

Finally, perhaps the most commonly faulted aspect of the NADA is its lack of transparency. The NADA approval process takes place "almost entirely behind closed doors."[180] Under the FFDCA and Trade Secrets Act (2004), FDA is prohibited from sharing certain information or data about an application because the GM animal is entitled to protection as a trade secret.[181] As a result, commentators, public interest groups, and the media have criticized application of the NADA process to GM animals on the grounds that it "blocks public input" and "does not allow full assessment of the possible environmental impacts of genetically engineered animals."[182]

While these concerns and criticisms of the NADA process are not unwarranted, much of the problem stems from the Coordinated Framework directive to regulate GM products under existing regulatory schemes rather than to develop more specific new regulations. Although the NADA process under the FFDCA is not a perfect fit, it may represent "the most stringent review available" under existing schemes.[183] When compared to the regulatory process for GM crops, the NADA process may be preferable because FDA approval for GM animals is mandatory and occurs before the GM animal may be marketed. In contrast, where a GM plant is a "substantial equivalent" to non-GM food, it may be marketed without review.[184] Given the lack of developed information regarding the risks and benefits of GM animals, the NADA scheme therefore appears at least preferable to the process applied to GM plants because it considers risks prior to going to market. In the future, it may make sense for the United States to develop regulations that are specific to GM animals and address the considerations associated with those animals more adequately.

5. Case Study: AquaBounty Salmon

Genetically modified salmon are at the forefront of the debate surrounding GM animals and provide a good illustration of some of the issues pertinent to that debate. FDA is currently reviewing an application by AquaBounty Technologies to approve a GM salmon. If approved, AquaBounty's salmon (branded as AquAdvantage fish) would be the first genetically engineered animal intended for human consumption.[185]

---

176. *Id.* ("NEPA does not mandate specific outcomes; it merely requires that agencies engage in the process of considering environmental impacts of their major actions.").
177. *Id.* at 20 ("[i]t is unclear whether FDA has the express authority to act upon a finding of significant environmental impact, that is, whether it may reject an application solely on environmental grounds.").
178. *See* Smith et al., *supra* note 135, at 1052-53 (discussing what effect salmon market expansion due to introduction of GM salmon might have in terms of public and environmental health).
179. Mandel, *supra* note 1, at 2233. *See also* U.S. Food & Drug Administration, *supra* note 122 (noting that while "any animal containing rDNA" intended to alter its structure or function is subject to FDA regulation prior to commercialization, pursuant to FDA's "enforcement discretion," "based on risk, there are some animals for which the agency may not require an approval."). Indeed, the GloFish (an aquarium pet fish) was granted an exception based on risk to the NADA process and was introduced to U.S. and Canadian markets in 2003. Marden et al., *supra* note 156, at 77.
180. Logar & Pollock, *supra* note 128, at 21.
181. 18 U.S.C. §1905.
182. Pollack, *supra* note 165. *See also, e.g.*, Marris, Transgenic *Fish Go Large*, 467 Nature 259 (2010) (noting criticism leveled at the NADA process due to its capacity to "allow . . . companies to shield some details of their product from public view"); Logar & Pollock, *supra* note 128, at 21-22 (arguing that the secrecy of the NADA process forces interested parties to rely on speculation, fails to account for public values, and hinders opportunity for meaningful public input on environmental matters).
183. Marden et al., *supra* note 156, at 80.
184. David A. Taylor, *Genetically Engineered Salmon on FDA's Table*, 118 Envtl. Health Persp. A384 (2010); Logar & Pollock, *supra* note 128, at 23. *See also* U.S. Food & Drug Administration, *supra* note 122 (noting that the difference in regulatory schemes applied to GM plants and GM animals is due to differences in how U.S. law treats plants and animals generally and because animals can pose health risks to humans that are not present in plants).
185. Pollack, *supra* note 165. Some claim that unregulated GM fish are already present in the marketplace. *See, e.g.*, Stephen Nottingham, Genescapes 134 (Zed Books, 2002) (stating that GM salmon that grow "ten times faster than normal fish" has been sold in U.S. supermarkets since the late 1990s and that due to U.S. labeling policies, consumers have been unaware that they are consuming GM fish).

The AquaBounty salmon's approval process has been understandably contentious, given that this is the first time that FDA has considered approval for a GM animal-based food source, and the process is new and unfamiliar to both consumers and regulators.

AquaBounty's salmon would grow at about double the rate of non-GM salmon, taking 16-18 months to reach market size instead of the usual three years.[186] The salmon contain a growth hormone from the Chinook salmon and a genetic on-switch from the ocean pout,[187] which alter the salmon's normal growth cycle. Salmon typically produce growth hormones only in warm months, but the insertion of the pout gene makes it possible to produce growth hormones in winter months as well—resulting in a continuous, rather than seasonal, growth cycle.[188] Ultimately, the AquaBounty salmon will not achieve an overall larger size than non-GM salmon but will reach market size in about half the time that it would take for non-GM salmon.[189]

The approval process for the AquaBounty salmon is being conducted by FDA, and the salmon are being regulated under animal drug law rather than food safety laws. Although it appeared that FDA was set to make a decision in 2011 after 15 years of efforts by AquaBounty to gain approval,[190] there have been efforts by members of the U.S. Congress to block FDA approval of the salmon.[191] There has been no approval as of this writing. In response to criticism of FDA's animal drug approval process allowing testing data to be kept confidential,[192] all information relating to FDA's decision on approval of AquaBounty salmon has been posted online and deliberations have been opened to the public.[193]

The key environmental concern associated with the AquaBounty salmon—as with GM fish in general—is that it could have a deleterious impact on wild populations. Specifically, some worry that there is the potential for the GM salmon to escape into marine environments and influence native populations through either gene transfer or competition for food or mates.[194] However, in the case of the AquaBounty salmon currently under review, only sterilized female eggs will be sold to growers, who will then grow the fish in inland tanks rather than the ocean net pens typically used in fish farming, which are known to pose a risk of escape.[195] The combination of sterilization and use of land-based tanks appears to address concerns related to influence on wild relatives since escape would be unlikely and, if escape did occur, the fish would be unable to reproduce.[196] Still, despite these biological and physical

---

186. Taylor, *supra* note 184.
187. Pollack, *supra* note 165.
188. Marden, *supra* note 156, at 75.
189. Pollack, *supra* note 165. "'You don't get salmon the size of the Hindenburg.'" *Id.* (quoting Ronald Stotish, the chief executive officer of AquaBounty).
190. Molly Peterson, *This Genetically Altered Salmon Is No Fish Story*, BLOOMBERG BUSINESSWEEK, Sept. 23, 2010, at http://www.businessweek.com/magazine/content/10_40/b4197021491547.htm (noting that FDA staff agreed the fish "is safe").
191. Several members of the U.S. House of Representatives who are opposed to the introduction of GM salmon, citing environmental and public health concerns, have introduced an amendment to be attached to an agricultural spending bill. The amendment would block approval of the salmon by banning FDA from spending funds on genetically modified salmon approval in the next fiscal year. Paul Voosen, *House Moves to Ban Modified Salmon*, N.Y. TIMES GREEN BLOG, June 16, 2011, at http://www.nytimes.com/gwire/2011/06/16/16greenwire-house-moves-to-ban-modified-salmon-84165.html. It is unclear whether the amendment will pass to be included with the final spending bill. Massachusetts House members have asked FDA to continue its science-based approval process for the salmon, arguing that the review process should be carried out to completion. Andrew Seidman, *Lawmakers Ask FDA to Move Forward on Salmon Review*, L.A. TIMES, Aug. 17, 2011, at http://www.latimes.com/news/politics/la-pn-lawmakers-ask-fda-to-move-forward-on-salmon-review-20110817,0,3663620.story. The U.S. Senate has held hearings on the matter, and Senate Bill 1717 has been introduced that would ban interstate commerce of genetically engineered salmon. S.B. 1717 was referred to committee in October 2011, and no further action has been taken as of August 2012. *See* LIBRARY OF CONGRESS, BILL SUMMARY AND STATUS, available at http://thomas.loc.gov/cgi-bin/bdquery/z?d112:s.01717: (last visited Aug. 17, 2012).
192. *See, e.g.*, Peterson, *supra* note 190 (quoting Wenonah Hauter of Food & Water Watch, who suggested that the animal drug approval process was being used in a deliberate attempt to keep data confidential because of a fear of public reaction).
193. Marris, *supra* note 182, at 259. FDA's information regarding AquaBounty approval is available at http://www.fda.gov/AdvisoryCommittees/CommitteesMeetingMaterials/VeterinaryMedicineAdvisoryCommittee/ucm201810.htm.
194. Taylor, *supra* note 184, at A385.
195. *Id.* AquaBounty, which is based in Waltham, Massachusetts, will produce GM salmon eggs at a facility in Prince Edward Island. Under the current application to FDA, eggs will be sold to growers and raised in inland systems in Panama and then the market-size fish will be sold by another company to U.S. consumers. Peterson, *supra* note 190, at A385.
196. *See* Smith et al., *supra* note 135, at 1053 (noting that escape risk "appears minimal," but pointing out that expanding production to other facilities, such as sea cages, could increase risk—though it would require FDA approval because it would be outside the scope of the current NADA); NATIONAL RESEARCH COUNCIL, BIOLOGICAL CONFINEMENT OF GENETICALLY ENGINEERED ORGANISMS 137-39 (Nat'l Academies Press 2004) (discussing the effectiveness of AquaBounty's proposed confinement methods and noting that use of land-based facilities is "reliable" and solves "a host of environmental problems" associated with cage farming); Josh Ozersky, *How I Learned to Love Farmed Fish*, TIME, Sept. 1, 2010, at http://www.time.com/time/nation/article/0,8599,2015134,00.html (discussing AquaBounty salmon's improbability of escape and inability to reproduce).

safeguards, some still worry that AquaBounty salmon could interact with wild populations or otherwise harm the environment.[197]

Beyond potential escape, there are other environmental concerns associated with AquaBounty salmon and other GM fish. Examples include local pollution from waste effluents generated in the grow tanks, the effect of novel gene products contained in waste on other organisms,[198] disease, and increased pressure on wild fish stocks used as food sources because salmon are carnivorous.[199] Although AquaBounty salmon consume less feed than non-GM salmon, if the introduction of GM salmon increases overall market demand for salmon (rather than the 1:1 ratio assumed in the NADA for AquaBounty, which projects that each GM salmon will substitute for one non-GM salmon), there could be increased pressure on wild fish stocks used as food for salmon.[200] These environmental concerns, along with concerns associated with FDA's approval process, general opposition to genetic modification, ethical concerns, and desire for labeling, are fueling a backlash against AquaBounty salmon. Indeed, many environmental groups and members of the public are vocally opposed to AquaBounty's approval and are calling the salmon a "frankenfish" that would "open[] the door to federal approval of all kinds of freaks from the farm."[201]

On the other hand, introduction of AquaBounty salmon and other GM fish in the marketplace has several potential benefits. Due to declining wild stocks of fish and development of new technologies, aquaculture now produces nearly half of the total weight of fish consumed worldwide.[202] Given that AquaBounty salmon take less time to mature to market size and require less feed to grow, the salmon could lower input costs needed to produce an individual fish. Lower input costs and quicker turnaround time could cause salmon prices to decline, thus enabling low-income households better access to a fish with known health benefits such as high levels of omega-3 fatty acids.[203] The salmon also have the potential to increase overall protein production, thereby improving world food security and reducing pressure on wild fish stocks.[204] Some suggest that AquaBounty salmon could be used as a model of ecologically sustainable aquaculture for developing nations and help to avoid a "growth stage" of less eco-friendly aquaculture that relies on the use of antibiotics or other chemicals.[205]

While it remains to be seen whether AquaBounty will be approved as the first GM animal designed for food consumption, approval does seem likely.[206] Approval of the salmon could pave the way for developments in other GM fish and animals.[207] On the other hand, a way to meet global protein demands other than through the use of GM fish—or aquaculture in general—could be to produce species like tilapia that are more tolerant to captivity and require less inputs to grow than do the carnivorous salmon or to modify fishing and consumption patterns to better sustain wild stocks.[208]

---

197. Peterson, *supra* note 190, at A385 (discussing the possibility that some eggs might not be sterile and could be fertilized); April Fulton, *Biotech Battle: Are Genetically Engineered Fish Safe?*, NPR, Sept. 20, 2010, at http://www.npr.org/templates/story/story.php?storyId=129939819 (discussing fears that fish could still escape and that further research is needed).
198. Logar & Pollock, *supra* note 128, at 18 (quoting NATIONAL RESEARCH COUNCIL, ANIMAL BIOTECHNOLOGY: IDENTIFYING SCIENCE-BASED CONCERNS (2002)).
199. Smith et al., *supra* note 135, at 1053.
200. *Id.* Salmon aquaculture currently consumes 40% of world fish oil production, and there is a 3:1 ratio of kilograms of wild fish to produce kilograms of salmon. *Id.*
201. Timothy Egan, *Frankenfish Phobia*, N.Y. TIMES OPINIONATOR BLOG, Mar. 17, 2011, at http://opinionator.blogs.nytimes.com/2011/03/17/frankenfish-phobia/. *See also* Peterson, *supra* note 190 (discussing opposition to the salmon's approval and quoting Alaska Senator Mark Begich, who said the salmon are "frankenfish" and that approval would be "'unprecedented, risky, and a threat to the survival of wild species.'"); Pollack, *supra* note 165 (discussing objections to the salmon's approval); Ozersky, *supra* note 196 (stating that the salmon had "inspired hysteria" in the media and noting the use of the term "frankenfish" as an example of such hysteria).
202. *See* FOOD AND AGRICULTURE ORGANIZATION OF THE UNITED NATIONS, WORLD REVIEW OF FISHERIES AND AQUACULTURE 3 (2010), *available at* http://www.fao.org/docrep/013/i1820e/i1820e01.pdf (reporting that 46% of world food fish supply came from aquaculture in 2010).
203. Smith et al., *supra* note 135, at 1053.
204. Taylor, *supra* note 184, at A 385.
205. *Id.*
206. *See* Peterson, *supra* note 190 (noting that FDA staff have stated that meat from AquaBounty fish is "safe" and has "no biologically relevant differences" from non-GM salmon).
207. Current developments that may be up for approval soon include the Enviropig, marketed as an eco-friendly pig that produces less phosphorus in its manure. Forsberg et al., *supra* note 141. Another example is a flu-resistant chicken. Pallab Ghosh, *World's First Flu-Resistant Chickens "Created,"* BBC NEWS, Jan. 13, 2011, at http://www.bbc.co.uk/news/science-environment-12181382.
208. Bryan Walsh, *Food: Why the Debate Over GM Salmon Misses the Point*, TIME ECOCENTRIC BLOG, Sept. 21, 2010, at http://ecocentric.blogs.time.com/2010/09/21/food-why-the-debate-over-gm-salmon-misses-the-point/. *See generally* PAUL GREENBERG, FOUR FISH (Penguin Press 2010) (arguing that aquaculture operations should focus on fish that are naturally better-adapted to aquaculture rather than energy-intensive carnivorous fish such as salmon and tuna).

6. Labeling for GM Salmon?

Because labeling for GM plant foods isn't required, in the grocery store one usually cannot tell traditional products from those whose ingredients have been genetically engineered. But the potential approval of AquaBounty salmon has reopened the public labeling debate and led many consumers to demand labeling for GM animals.

It is a fair assumption that, if approved, GM salmon and other GM animal products that follow will be treated the same as food that contains genetically modified plant products. In other words, the lack of labeling will continue. But why will GM salmon not be labeled? Do consumers have a right to know?

The answer is not so easy. Although some commentators argue that the federal government simply opposes mandatory labeling of GM products,[209] that view does not consider federal authority and its limits. FDA does not have unlimited authority to require food producers to label products. Under federal law, FDA may require specific labeling where a label is misleading or where the absence of labeling would be misleading.[210] For example, if a package says "Atlantic salmon" but does not say it's GM, the salmon could be considered mislabeled. However, FDA has determined that it may not require labeling about a product's production method—such as being genetically modified—unless the resulting product is materially different.[211]

Arguably, FDA does have the authority to require labeling of GM animals. Federal law states that food is considered "misbranded" if it is "misleading in any particular,"[212] or if the labels fails to reveal "material" facts.[213] Failure to disclose that a human food animal has been genetically engineered could be considered to be a failure to disclose a material fact. Nevertheless, in applying the law to GM plants, FDA has created a "material difference" standard requiring some kind of *physical* difference between the GM product and its non-GM counterpart in order for mandatory labeling to be required.[214] Accordingly, FDA's approach does not consider consumer demand or underlying ethical or environmental justifications for labeling as a sufficient basis for mandatory labeling absent a physical material difference.

Thus far, while much scientific information remains unknown or confidential, testing has not revealed any significant physical differences between the GM salmon and non-GM salmon.[215] Accordingly, it is unlikely that FDA will require GM salmon or other GM food products to be labeled unless studies indicate that such a physical material difference exists. Of course, FDA's interpretation of what requires labeling—its "material difference" standard—is not the only possible interpretation of the FFDCA.

The statutory language should be interpreted to consider realities of 21st century food. Misbranding, defined as "misleading in any particular," could include failing to label as genetically modified (or, more narrowly, as having recombinant DNA) because these products have DNA and proteins that have never existed and which consumers would not expect in food absent labeling. Given the rapidly changing landscape of food regulation due to new technologies, it is possible that FDA could come up with a new legal view for mandatory labeling. A new interpretation could, for instance, allow for consumer demand of labeling or environmental impacts of bioengineering as a basis for mandatory labeling in the case of GM animals or other foods. Another possibility could be new federal legislation mandating such labeling.

For now, however, it appears unlikely that AquaBounty salmon, if approved, will be labeled as GM, leaving consumers to rely on voluntary labeling—by producers of bio-engineered and non-GM foods alike. Absent any voluntary labeling, consumers will be unable to tell whether the salmon on their dinner plate is GM or not.

---

209. *See, e.g.*, Mark Bittman, *Why Aren't G.M.O. Foods Labeled?*, N.Y. Times Opinionator Blog, Feb. 15, 2011, at http://opinionator.blogs.nytimes.com/2011/02/15/why-arent-g-m-o-foods-labeled/ (calling the federal "unwillingness" to label GM products "demeaning and undemocratic").
210. Section 403(a)(1) of the Federal Food, Drug & Cosmetics Act states that a food is mislabeled if "its labeling is false or misleading in any particular." 21 U.S.C. §343(a)(1).
211. 75 Fed. Reg. 52602 (Aug. 26, 2010).
212. FFDCA §403(a)(1), 21 U.S.C. §343(a)(1).
213. Section 201(n) of the FD&CA provides more information about how labeling can be misleading. 21 U.S.C. §321(n).
214. FDA has determined that the kind of "material difference" that would necessitate labeling would be one with respect to nutritional value, functional properties, or organoleptic qualities. *Id.*
215. U.S. Food and Drug Administration, Transcript for the September 20, 2010 Veterinary Medicine Advisory Committee Meeting, and transcript for the September 19, 2010, Veterinary Medicine Advisory Committee Meeting, *available at* http://www.fda.gov/newsevents/publichealthfocus/ucm224089.htm.

Conclusion

The rapid and widespread adoption of GM plants—and potentially forthcoming introduction of GM animals—has left many unsure about the potential short- and long-term effects that biotechnology poses. Of particular concern is the possibility that bioengineering could have irreversible effects on the environment—essentially, that we may not be able to undo whatever effects GM products may have.[216] The uncertainty and lack of information pertaining to the risks and benefits associated with GM products has contributed to controversy surrounding their use.

Given this uncertainty, there are several steps that could be taken to help ensure the safety of GM use. First, there must be continuing research regarding the environmental costs and benefits of GM use, which is an area that has not been emphasized sufficiently in past research. Future studies should include long-term analyses that mimic or utilize natural settings. Such research would help to understand the overall impacts that GM plants and animals may have, and could help to inform the regulatory process.

Second, the regulatory system for GM plants and animals should be improved. The current regulatory scheme that is applied to GM organisms is based on the Coordinated Framework, which directs that GM organisms be regulated under existing laws and regulations. However, it has become apparent that biotechnology presents unique considerations and risks that are not addressed adequately under available regulatory options. A comprehensive regulatory framework should be developed that specifically addresses GM organisms. A key feature of such a framework would include a meaningful environmental risk analysis that considers existing research and requires ongoing environmental monitoring. A more effective regulatory system would help to minimize the risks—and maximize the benefits—associated with GM use.

Finally, a domestic labeling system should be considered. At this point, mandatory labeling seems unlikely. There is no federal law that requires labeling of GM products, and FDA has interpreted the FFDCA to only require labeling of GM products where there is a physical "material difference" between a GM product and its non-GM counterpart.[217] Most GM products are outside the scope of this limit because measurable physical features such as nutritional content and presence of allergens are often unaltered by the use of biotechnology. Further, state efforts to require mandatory labeling may be preempted by FDA's interpretation of the FFDCA or may be considered a violation of the commercial free-speech rights of the manufacturer.[218] Because FDA's interpretation of its labeling authority for GM products under the FFDCA is unnecessarily narrow, it appears that the best option for a labeling system may be to encourage voluntary labeling efforts that are either positive (e.g., "this product contains GM ingredients") or negative (e.g., "this product is GM-free").

Although the current "material difference" standard requires some sort of physical difference between GM and non-GM varieties, FDA could take a different approach and become more responsive to new developments in food technologies. Because federal law states that food is considered misbranded if it is "misleading in any particular," or if the labels fails to reveal material facts, FDA could require labeling of GM products by recognizing that the presence of GM in a food product is a material fact, or is misleading because consumers would not typically expect the food to be GM. Future legislation could address GM labeling, or labeling could be incorporated as part of a comprehensive regulatory system tailored to address

---

216. An example of irreversible environmental costs might be a decrease in biodiversity due to gene drift, impacts on nontarget organisms, and pest resistance. Matty Demont et al., *Irreversible Costs and Benefits of Transgenic Crops: What Are They?*, in ENVIRONMENTAL COSTS AND BENEFITS OF TRANSGENIC CROPS 113 (J.H.H. Wesseler ed., 2005).

217. Doug Farquhar & Liz Meyer, *State Authority to Regulate Biotechnology Under the Federal Coordinated Framework*, 12 DRAKE J. AGRIC. L. 439, 452 (2007). Consumer demand for labeling is not considered sufficient justification for labeling without nutritional or food safety concerns. *Id*.

218. *Id*. at 470. *See also* Int'l Dairy Foods Ass'n v. Amestoy, 92 F.3d 67 (2d Cir. 1996) (striking down Vermont's mandatory labeling requirements for milk from cows that had been treated with rBST on First Amendment grounds). California, Connecticut, and Vermont are states that have attempted to pass state-specific labeling requirements for GE fish, but no state has passed legislation as of August 2012. *See* Carl Entier, *Vermont Not Alone in Pushing for GMO Labeling of Foods*, VT DIGGER, Apr. 3, 2012, at http://vtdigger.org/2012/04/03/vermont-not-alone-in-pushing-for-gmo-labeling-of-foods/ (discussing attempts by Vermont and other states to pass labeling legislation).

GM products. Ultimately, consumers want to know if they are consuming GM food,[219] and labeling would help ease consumer concern and allow for informed decisionmaking at the grocery store.[220]

In sum, the use and commercialization of GM plants and animals come with both benefits and risks. But the full scope of those benefits and risks is largely unknown, and GM products have been adopted without a full understanding of their socioeconomic or environmental impacts. The use of GM organisms is no longer a remote possibility—instead, given that GM products are an established reality on U.S. farms and in U.S. grocery stores, research and regulatory efforts need to catch up in order to effectively address their attendant risks and benefits.

---

219. *See* Mark Bittman, *supra* note 209 (calling the federal "unwillingness" to label GM products "demeaning and undemocratic" and arguing that the reason for the lack of labeling is a fear that labeling would lead to customer avoidance).

220. *See, e.g.*, Mario Teisl et al., *Labeling Genetically Modified Foods: How do U.S. Consumers Want to See It Done?*, 6 AgBioForum 48, 51 (2003) (predicting that labeling would lead to an initial decrease in sales of GM products but would eventually lead to consumer acceptance).

Chapter 7
Environmental and Climate Impacts of Food Production, Processing, Packaging, and Distribution

Jason J. Czarnezki and Elisa K. Prescott

Getting the average meal from farm to plate in the United States is no small task; a great deal of energy and material inputs goes into producing, processing, packaging, and distributing food. A commodity crop like corn, for example, could be cultivated in one state, say a midwestern state like Iowa, shipped to a processing plant in Tennessee to take on an entirely new form like high fructose corn syrup, packaged in plastic in Pennsylvania, and finally distributed as ketchup to a supermarket chain in California. The modern food system is fossil fuel-intensive and often environmentally damaging, making the true cost of our food much greater than the price we pay at the supermarket. Only recently have environmentalists and environmental law scholars focused on the ecological impacts of food choices and considered how industrial food production has been responsible for significant greenhouse gases emissions.[1]

This chapter, focusing on produce and grain,[2] discusses the environmental and climate change impacts of food production, processing, packaging, and distribution, which ultimately contribute to both economic and social costs.[3] The chapter addresses environmental energy costs in the food supply. Figure 7.1 shows, for example, the significant amount of energy used in various aspects of food production, transportation, and processing.

---

1. *See, e.g.,* James E. McWilliams, Just Food 1 (Back Bay Books 2010); Jason J. Czarnezki, Everyday Environmentalism: Law, Nature & Individual Behavior (2011).
2. Although the production of livestock is a significant contributor to the carbon footprint and environment impacts, it will not be discussed here.
3. At the outset, we must recognize the difficulty scientists have had in evaluating the environmental costs of the life-cycle of food, given the food system's geographic diffusion, the inability to disaggregate all commodity production from food (e.g., ethanol), and the complexity of import and export systems. Martin C. Heller & Gregory A. Keoleian, *Assessing the Sustainability of the U.S. Food System: A Life Cycle Assessment*, 76 Agric. Systems 1007, 1009 (2003); Karin Anderson & Thomas Ohlsson, *Life Cycle Assessment of Bread Produced on Different Scales*, 4 Int'l J. Life Cycle Assessment 25, 25 (1999).

Figure 7.1
Life-Cycle Energy Use in Supplying U.S. Food[4]

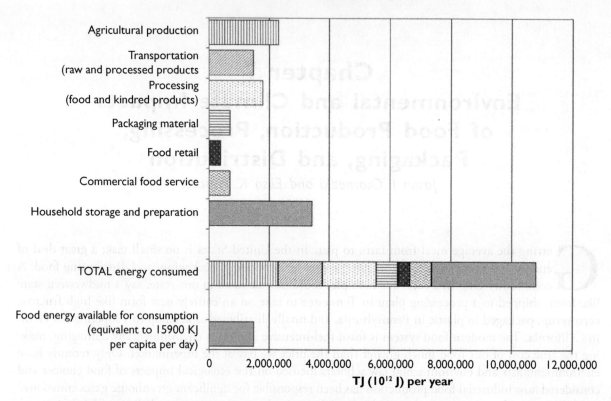

Note: KJ is a Kilojoule and TJ is a terajoule. Both are measurements of energy.

Much of this chapter's focus will be on commodity crops. Along with wheat and rice, corn and soybeans constitute the world's most popular planted and consumed crops.[5] The United States is the leading producer of corn, growing nearly 40% of the world's total,[6] with more than half of that production coming from only 20% of U.S. corn growers.[7] In 2008 over 85 million acres of corn and more than 75 million acres of soybeans were planted in the United States,[8] and the crops "have faced increasing demand in the world market over the past ten years as they are sources of both human and animal food."[9]

This chapter illustrates the environmental impacts of the food system. It opens with a discussion of the industrial food system, exploring both conventional and organic models. It then looks at the carbon footprint of food production, as well as the environmental impacts of production, specifically soil, water, and air pollution. The next section discusses food processing and associated greenhouse gas emissions and environmental impacts. The chapter then considers the environmental and climate change impacts of packaging from creation to disposal. The final section addresses food distribution and the distance food travels to get to our plates, also known as food miles.

A. Food, Agriculture, and the Environment

There is a closed-loop relationship between the current agriculture system and the climate crisis. With changing temperature and rainfall patterns, climate change may have a dramatic impact on food production and agricultural geography. Completing the circle, current agricultural practices are fossil fuel-inten-

---

4. Heller & Keoleian, *supra* note 3, at 1025.
5. Madeleine Pullman & Zhaohui Wu, Food Supply Chain: Economic, Social and Environmental Perspectives 64 (2011). *See also* Czarnezki, *supra* note 1, at 75-77 (2011).
6. Kimbrell, Andrew, *A Blow to the Breadbasket: Industrial Grain Production*, in Fatal Harvest: The Tragedy of Industrial Agriculture 100 (Andrew Kimbrell ed., 2002).
7. *Id.*
8. Darrel Good, *Weekly Outlook: Corn and Soybean Acreage*, ACES News (Jan. 20, 2009), http://www.aces.uiuc.edu/news/stories/news4630.html.
9. Pullman & Wu, *supra* note 5, at 67.

sive and contribute to greenhouse gas emissions, accelerating this change in climate and rainfall patterns.[10] In the United States, food production alone accounts for 20% of overall fossil fuel consumption.[11] The American food system is dominated by industrial agriculture, though the organic food market has seen significant gains.[12] Despite the domination of industrial agriculture, both conventional and organic food production contribute to greenhouse gas emissions and ecological harm.

The Green Revolution dramatically transformed U.S. agriculture into high-yield, fossil-fueled, mass production in the 1940s.[13] The replacement of human labor with technological innovations, mechanized farm equipment, and other fossil fuel inputs were the primary changes resulting from the Green Revolution. The Green Revolution also led to the extensive use of chemical pesticides and synthetic fertilizers, as described in Chapter 3. The high-yield goals of the Green Revolution accelerated production of commodity crops, and, at the time, was essential in addressing concerns of food security and hunger. Between the start of the Green Revolution and 2000, commodity crop production increased threefold in order to keep up with the population growth; the population grew from three billion to six billion people during that period.[14]

With increased use of chemical fertilizers and pesticides, consumers became concerned with the health and environmental impacts of conventional agriculture. Rachel Carson's *Silent Spring*, along with the growing environmental movement at the time, brought awareness and consciousness about the problems associated with agricultural chemicals, leading to consumer demand for food to be grown without harsh chemicals. Over the past 20 years, organic foods, and more recently industrial organic foods, have begun to enter the market. In response, the U.S. Congress passed the Organic Foods Production Act (OFPA) in 1990 as part of the U.S. farm bill (discussed in more detail in Chapter 13).[15] In order for a product to be called organic or carry the organic label, OFPA forbids the use of synthetic fertilizers and growth hormones and the addition of synthetic ingredients during processing, as well as the use of antibiotics in livestock.[16] In recent years, the organic industry has shifted from small, local farms to large industrial ones. The organic market has quadrupled in the last decade, and the sales of organic food have grown from $1 billion in 1990 to over $20 billion today.[17]

---

10. Mary Jane Angelo, *Corn, Carbon, and Conservation: Rethinking U.S. Agricultural Policy in a Changing Global Environment*, 17 Geo. Mason L. Rev. 593, 600 (2010).
11. Czarnezki, *supra* note 1, at 75 (citing Dan Imhoff, Paper or Plastic: Searching for Solutions to an Overpackaged World 102 (2002)).
12. *See* Jason J. Czarnezki, *Food, Law & the Environment: Informational and Structural Changes for a Sustainable Food System*, 31 Utah Envtl. L. Rev. 263 (2011).
13. *Id.* at 602.
14. Pullman & Wu, *supra* note 5, at 66.
15. *Id.*
16. Czarnezki, *supra* note 1, at 71 (citing Organic Foods Production Act of 1990, 7 U.S.C. §§6501 et seq.).
17. Czarnezki, *supra* note 1, at 80 (citing Organic Trade Ass'n Executive Summary, Organic Trade Association's 2007 Manufacturers Survey, *available at* http://www.ota.com/pics/documents/2007ExecutiveSummary.pdf (stating that organic food sales accounted for 0.8% of total food sales in 1997, and 2.8% in 2006)).

Figure 7.2
Organic Industry Structure: Acquisitions by the Top 30 Food Processors in North America[18]

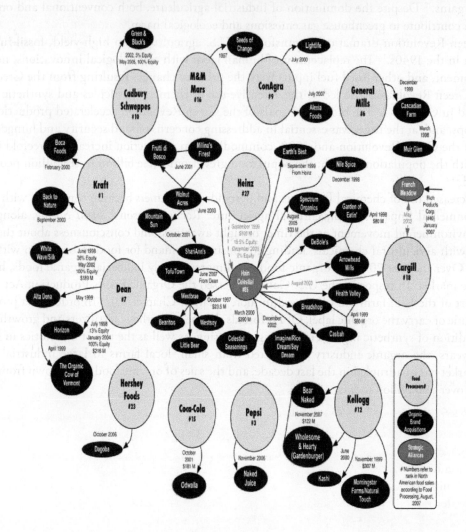

Large corporations like Kraft, Coca-Cola, and Nestle, which have traditionally processed and supplied conventional foods, have seen this increase in organic trends and jumped on board by purchasing smaller organic firms. Figure 7.2 depicts major U.S. food processing companies that now own organic brands. Large-scale organic production increases greenhouse gas emissions and may also practice questionable agricultural methods,[19] which are legally organic but not necessarily sustainable.[20]

B. Industrialized Food Production and Cultivation

The dominance of the industrialized food system drives current farm cultivation practices. The cultivation stage is when seeds are planted, soil tilled, and crops tended, watered, and harvested. Large mechanized farm equipment and irrigation systems are used to produce enormous crop yields, contributing significant greenhouse gas emissions, degrading the soil, and reducing access to water. In addition, petroleum-based

---

18. Phillip H. Howard, *Organic Processing Structure*, 5 NMC Media-N (2009), *available at* http://www.newmediacaucus.org/wp/organic-industry-structure/.
19. Czarnezki, *supra* note 12, at 275.
20. *Id.* at 275 (citing Jill Richardson, Recipes for America: Why Our Food System Is Broken and What We Can do to Fix It 11 (2009) (in terms of ecological interests, defining sustainable agriculture as "resource-conserving," "environmentally sound," and efficient use of resources)); Gill Seyfang, *Ecological Citizenship and Sustainable Consumption: Examining Local Organic Food Networks*, 22 J. Rural Stud. 383, 383 (2006) (recognizing that sustainable consumption has an elusive definition).

chemical pesticides and nitrogen fertilizers used in industrial agriculture increase its ecological footprint. Finally, both conventional and organic industrial crops are planted as monocultures, which can have grave impacts on soil health and water quality, as well as lead to a continued increase in the use of pesticides and fertilizers. It is a self-reinforcing system in which each innovation and input increases the reliance on further interventions and, ultimately, the ecological costs of modern food production.

1. Mechanized Cultivation and Irrigation

As population has increased, food production has also increased, as has the size and scope of farms. Millions of acres of the two leading crops in the United States, corn and soybeans, are planted each year.[21] The massive area it takes to grow such a large amount of commodity crops requires the use of industrial machines and large irrigation systems, which creates significant carbon emissions, and soil and water degradation.

a. Carbon Emissions

Industrial agriculture is based on high-yield crops, which, due to the immense size of farms they are grown on, are difficult to plow, till, and water in a traditional way by human labor or livestock (absent significant labor costs). According to historian Walter Prescott Webb, new technologies like the John Deere plow and mechanized harvesters (not to mention the Colt six-shooter, barbed wire, and other agricultural implements) helped farmers control the Great Plains and increase production during the rise of industrial agriculture in the 19th century.[22] As the industrial agriculture model became dominant, the energy costs of simply tilling became substantial. To till one hectare of land today with a 50 horsepower tractor, the petroleum input is 306,303 kilocalories, or 30.3 liters of gas.[23] These petroleum inputs by farm machinery are directly linked to the carbon emissions going into the atmosphere. In addition, the manufacturing of farm machinery contributes to greenhouse gas emissions. While there are negligible emissions associated with transporting the tractors from the manufacturing plant to the farms,[24] the manufacturing of tractors takes a great deal of energy. On average, roughly 12.8 kilograms carbon dioxide equivalent of greenhouse gas is emitted to manufacture just one kilogram of a tractor.[25] If calculated out, using the average weight of a 92 kilowatt tractor, it takes 187 barrels of oil consumed, or the annual greenhouse gas emissions of 15.8 passenger vehicles, to produce one tractor.[26]

In addition to the energy and carbon emissions from the manufacturing of large machinery, irrigation requires a significant amount of fossil fuel energy for pumping and delivering water to crops.[27] Generally, irrigation water is from wells or surface reservoirs located on the farm, or surface reservoirs from off the farm.[28] Fossil fuels are the primary source of energy for powering the pumps required to distribute the water.[29] Each year, pumping irrigation water accounts for an estimated 15% of the total energy expended on farms.[30]

With the increasing demand for organic products, more and more organic acreage is being put into production. Large organic farming operations share the same energy needs for cultivation and irrigation as

---

21. Good, *supra* note 8.
22. Czarnezki, *supra* note 1, at 9.
23. David Pimentel & Marcia H. Pimentel, Food, Energy, and Society 60 (3d ed. 2008) (citing David Pimentel & Marcia H. Pimentel, Food, Energy and Society (1979)).
24. *Id.* at 119.
25. *Id.* at 118-19 (citing C. Wells, Total Energy Indicators of Agricultural Sustainability: Dairy Farming Case Study (2001)).
26. The average weight of a 92 kilowatt power John Deere tractor is 6,277 kilograms, and approximately 12.8 kilograms carbon dioxide equivalent of greenhouse gas is emitted in the manufacture of one kilogram of that tractor, then roughly 80,345 kilograms carbon equivalent of greenhouse gas is emitted just to produce one tractor. *See Greenhouse Gas Equivalencies Calculator*, http://www.epa.gov/cleanenergy/energy-resources/calculator.html#results (last visited Aug. 18, 2011).
27. David Pimentel et al., *Water Resources: Agricultural and Environmental Issues*, in Food, Energy, and Society 188 (3d ed. 2008).
28. Tristram O. West & Gregg Marland, *A Synthesis of Carbon Sequestration, Carbon Emissions, and Net Carbon Flux in Agriculture: Comparing Tillage Practices in the United States*, 91 Agric., Ecosystems & Env't 217, 223 (2002), *available at* http://ornl.gov/info/ornlreview/v40_3_07/documents/article17web_West_Marland_ag_net_flux.pdf.
29. *Id.*
30. Pimentel et al., *supra* note 27, at 188 (citing David Pimentel et al., Water Resources, Agriculture, and the Environment (N.Y. State Coll. of Ag. & Life Scis. ed. 2004)). To irrigate just one hectare of corn, it requires an average of 10 million liters of irrigation water, which requires about 880 kWh/ha (kilowatt-hour per hectare) of fossil fuels. Pimentel et al., *supra* note 27, at 188 (citing J.C. Batty & J. Keller, *Energy Requirements for Irrigation*, in Handbook of Energy Utilization in Agriculture 35-44 (David Pimentel ed., 1980)).

do conventional farms. Earthbound Farm, which was originally two-and-a-half acres, is now a 40,000-acre farm.[31] Journalist Barry Estabrook describes the Earthbound spinach field:

> To step into an Earthbound Farm spinach field is to be overwhelmed by the incomprehensible vastness of it all. It looks identical to hundreds of operations that stretch across the valley floor, stopping only at the base of the faraway, hazy mountains. An area big enough to accommodate a dozen football fields is carpeted with symmetrical strips of tiny, perfect baby spinach plants with just enough space between the rows to allow for the passage of a mechanical harvester.[32]

What Estabrook describes is a sight similar to any found throughout the country on either a conventional or industrial organic farm. Like conventional fields, massive organic fields like Earthbound's spinach field cannot be cultivated or watered by hand (or not, at least, without very high cost), so they require the same fossil fuel-intensive machinery and irrigation systems as conventional agriculture.

b. Soil Degradation

The use of large farm equipment contributes to soil degradation because the heavy equipment compacts and disturbs the soil, and over-tilling of crops causes soil to erode.[33] In addition, the large amount of water applied to fields from the irrigation systems adds to the likelihood of erosion. Degraded soil has a reduced capacity to function properly.[34] Farmers rely on soil health and quality to stay in production; however, every year millions of acres of productive land are abandoned due to degradation.[35]

Industrial agriculture involves tilling by large machines, and due to the fact that organic farming does not use chemicals to kill weeds, increased tilling is often employed, potentially leading to overtillage and increased soil degradation. In addition, compaction from heavy machinery used in agriculture causes run-off, erosion, and flooding which can prevent water from infiltrating down to recharge the aquifer.[36]

After soil is degraded by overtilling and compaction from large machinary, it is more likely to erode, especially with the increase of water on fields, and the erosion diminishes essential nutrients, micronutrients, and organic matter in the soil.[37] According to a study by the National Resources Inventory of the USDA, 1,700 megatonnes (Mt=million metric tonnes) of agriculture soil in the United States was eroded in 1997. Wind accounted for 760 Mt of this erosion, and 960 Mt was due to sheet and rill (caused by water).[38] This amount of eroded soil would "fill a freight car train loaded to capacity that would encircle the *planet* about seven times."[39] Profs. Martin Heller and Gregory Keoleian calculated that at the current rate of topsoil erosion, "2.5 cm of topsoil [is] lost from all U.S. cropland every 34 years."[40] They explain that, when soil is supplemented with large amounts of fertile organic material, which does not happen in normal industrial farming practices, 2.2 cm of soil can rejuvenate in about 30 years.[41] However, it is estimated that under current farming conditions it could take anywhere between 200 and 1,000 years to regenerate 2.5 cm of soil.[42]

---

31. Barry Estabrook, *Politics of the Plate: Greens of Wrath*, GOURMET MAG., Nov. 2008, at http://www.gourmet.com/magazine/2000s/2008/11/politics-of-the-plate-contaminated-greens.
32. *Id.*
33. CZARNEZKI, *supra* note 1, at 79.
34. USDA, Natural Resources Conservation Service, *Soil Quality Concepts*, http://soils.usda.gov/sqi/concepts/concepts.html.
35. DALE ALLEN PFEIFFER, EATING FOSSIL FUELS: OIL, FOOD, AND THE COMING CRISIS IN AGRICULTURE 11 (2006) (citing R.A. Houghton, *The Worldwide Extent of Land Use Change*, 44 BIOSCIENCE 305 (1994)).
36. Laurel E. Phoenix, *Water and Land-Use Policies in the United States*, *in* CRITICAL FOOD ISSUES: PROBLEMS AND STATE OF THE ART SOLUTIONS WORLDWIDE, VOLUME 1: ENVIRONMENT, AGRICULTURE, AND HEALTH CONCERNS 3, 4 (Laurel E. Phoenix ed., 2009).
37. *Id.*
38. Heller & Keoleian, *supra* note 3, at 1019 (citing U.S. DEPT. OF AGRICULTURE, AGRICULTURE FACT BOOK (1999)). U.S. DEPT. OF AGRICULTURE, AGRICULTURE FACT BOOK 2000 (2000), *available at* http://www.usda.gov/documents/factbook2000.pdf.
39. Heller & Keoleian, *supra* note 3, at 1019.
40. *Id.*
41. *Id.* (citing WES JACKSON, NEW ROOTS FOR AGRICULTURE (1980)).
42. Heller & Keoleian, *supra* note 3, at 1019 (citing David Pimentel et al., *Environmental and Economic Costs of Soil Erosion and Conservation Benefits*, 267 SCI. 1117).

c.   Water Quality and Access

Commodity crop production impacts both the quality and quantity of U.S. waters.[43] Agricultural irrigation constitutes over one-third of the freshwater used in the United States, making it the largest use in the nation.[44] On average, it takes 1,000 tons of water to grow one ton of grain, with rice using the most water, and corn using the least.[45] The use of enormous amounts of water by large irrigation systems is particularly problematic because so few states have adequate regulations of groundwater withdrawals.[46] Because in some areas of the United States property owners risk losing their water rights if they do not use their entire allotment, "farms will use their whole water allowance even though they could easily use less or install efficient irrigation."[47] As an example, the Ogallala Aquifer is a critical water resource for agriculture in the Midwest, as it covers eight states—Wyoming, South Dakota, Nebraska, Colorado, Kansas, Oklahoma, New Mexico, and Texas.[48] The aquifer receives little recharge, and the water table drops each year due to unregulated and unsound irrigation practices.[49] Estimates suggest that within the next decade or two the Ogallala aquifer will be so low that using it for irrigation will become prohibitively expensive.[50]

As industrial farms increase across the United States, groundwater is being used unsustainably, reducing water tables to the point where other users may no longer be able to use the water.[51] In addition, commodity crops, like corn, are grown in states that do not have adequate water resources to support this sort of intense irrigation.[52] This results in diverting water from waterbodies far from the growing fields, sometimes causing water disputes.[53]

2.   Pesticides and Fertilizers

Approximately 500,000 tons of 600 different types of pesticides are used annually in the United States.[54] In fact, in 1997, about 98% of the corn acreage planted in the top 10 corn-producing states received commercial fertilizer.[55] The increased use of pesticides and nitrogen fertilizers contributes to the carbon footprint (as well as other forms of pollution) as synthetic pesticides are made from fossil fuels, and fertilizers are derived from natural gas made from fossil fuels.[56] Chemical pesticides and fertilizers also greatly impact soil, water, and air quality.

a.   Carbon

For better or worse, pesticides "prevent pest damage to potentially a third of the nation's crop, help food preservation, and result in cheaper food production."[57] Today, the growth of chemical use has been an "integral part of the technological revolution in agriculture that has generated major changes in productiv-

---

43. Angelo, *supra* note 10, at 603-604 (citing J.B. Ruhl, *Farms, Their Environmental Harms, and Environmental Law*, 27 Ecology L.Q. 263, 274 (2000)).
44. *Id.* at 604 (citing William S. Eubanks, *A Rotten System: Subsidizing Environmental Degradation and Poor Public Health With Our Nation's Tax Dollars*, 28 Stan. Envtl. L.J. 213, 253 (2009)).
45. Dale Allen Pfeiffer, Eating Fossil Fuels: Oil, food and the Coming Crisis in Agriculture 17 (2006).
46. Phoenix, *supra* note 36, at 5.
47. *Id.*
48. Leo Horrigan et al., *How Sustainable Agriculture Can Address the Environmental and Human Health Harms of Industrial Agriculture*, 110 Envtl. Health Persp. 445, 447 (2002), *available at*. http://www.ncbi.nlm.nih.gov/pmc/articles/PMC1240832/pdf/ehp0110-000445.pdf.
49. *See id.* at 447-48.
50. *Id.* at 447.
51. Phoenix, *supra* note 36, at 5.
52. Angelo, *supra* note 10, at 604 (citing Christine A. Klein, *Water Transfers: The Case Against Transbasin Diversions in the Eastern States*, 25 UCLA J. Envtl. L. & Pol'y 249, 253 (2007)).
53. *Id.* at 660 n.83 (citing Eubanks, *supra* note 44, at 254).
54. David Pimentel et al., *Environmental and Economic Costs of Pesticide Use*, 42 Bioscience 750, 750 (1992) (citing Pimentel et al., *Environmental and Economic Impacts of Reducing U.S. Agricultural Pesticide Use*, *in* Handbook on Pest Management in Agriculture (David Pimentel ed., 1991)).
55. Heller & Keoleian, *supra* note 3, at 1020 (citing U.S. Dep't. Agric., Economic Research Service, Agricultural Resources and Environmental Indicators 1996-97 (1997)).
56. Angelo, *supra* note 10, at 612 (citing William S. Eubanks II, *The Sustainable Farm Bill: A Proposal for Permanent Environmental Change*, 39 ELR 10493, 10504 (June 2009); and Peter Warshall, *Tilth and Technology: The Industrial Redesign of Our Nation's Soils, in* The Fatal Harvest Reader: The Tragedy of Industrial Agriculture, *supra* note 6, at 225).
57. Czarnezki, *supra* note 1, at 68.

ity techniques, shifts in input use, and growth in output and productivity," and has resulted in the use of more chemicals and less manual labor.[58]

Many pesticides are powerful toxins that persist in the environment.[59] The most widely used pesticides, those containing synthetic chemicals, are produced primarily from fossil fuels[60] and contain ingredients like refined oil and kerosene. Depending on the type of pesticide, it is estimated that it "takes the equivalent of a gallon of diesel fuel to make one pound of active ingredient of pesticides."[61]

In the past, animal manure and other farm refuse were used as nutrient sources, but today commercially manufactured chemical fertilizers are the major source of applied plant nutrients.[62] Nitrogen fertilizers were first introduced into industrial agriculture after World War II; their use has shifted soil fertility from "a total reliance on the energy of the sun to a new reliance on fossil fuel."[63] Fertilizers are petroleum-based, and the energy needed to fix nitrogen into fertilizers is supplied by fossils fuels.[64] California organic farmer Jason McKenny estimates that "[t]he production of nitrogenous fertilizers consumes more energy than any other aspect of the agricultural process."[65] This is a bold statement, considering all the inputs in the agricultural process; however, it takes the "energy from burning 2,200 pounds of coal to produce 5.5 pounds of usable nitrogen."[66]

b. Soil

The increased use of pesticides, including herbicides and insecticides, used on crops significantly impacts soil health. Pesticides, including ones banned years ago like DDT, can linger in the soil for decades and kill important biota that nourish and aerate soil.[67] Creatures like earthworms and microorganisms nourish the soil, and soils are more productive when these important biota are present.[68] Many herbicides and insecticides kill these biota.[69] Chemical inputs also degrade soil structure, hindering water and gas relationships between plants and soil.[70] Severely degraded soil can take decades, centuries, or millennia to fully recover.[71] The increased amount of nitrogen added to soil from synthetic fertilizers destroys the soil biodiversity by

> diminishing the role of nitrogen fixing bacteria and amplifying the role of everything that feeds on nitrogen. These feeders then speed up the decomposition of organic matter and humus. As the organic matter decreases, the physical structure of soil changes.[72]

Soil becomes less efficient at absorbing and retaining water, air, and other essential nutrients.[73] In addition, fertilizers make soil more acidic, which causes declines both in soil humus content and crop output.[74] The large amount of chemicals applied to crops today clearly hinders the soil that the crops need to flourish. As a result, more chemicals are applied because soils are degraded and lack nutrients to crop productively, creating a vicious cycle.

---

58. NOEL D. URI, AGRICULTURE AND THE ENVIRONMENT 8-9 (1999).
59. PFEIFFER, *supra* note 35, at 23.
60. *Id.*
61. Mike Duffy, *Prices on the Rise: How Will Higher Energy Costs Impact Farmers?*, LEOPOLD LETTER, Spring 2001, at 1, 4.
62. Heller & Keoleian, *supra* note 3, at 1020.
63. MICHAEL POLLAN, THE OMNIVORE'S DILEMMA: A NATURAL HISTORY OF FOUR MEALS 41, 44 (2007).
64. Jason McKenney, *Artificial Fertility: The Environmental Costs of Industrial Fertilizers, in* THE FATAL HARVEST READER: THE TRAGEDY OF INDUSTRIAL AGRICULTURE, *supra* note 6, at 121, 127 (Andrew Kimbrell ed., 2002).
65. *Id.* at 127.
66. *Id.*
67. Phoenix, *supra* note 36, at 4.
68. Wes Jackson, *Farming in Nature's Image: Natural Systems Agriculture, in* THE FATAL HARVEST READER: THE TRAGEDY OF INDUSTRIAL AGRICULTURE, *supra* note 6, at 70.
69. *Id.* at 70.
70. *Id.*
71. Catherine Badgley, *Can Agriculture and Biodiversity Coexist?, in* THE FATAL HARVEST READER: THE TRAGEDY OF INDUSTRIAL AGRICULTURE, *supra* note 6, at 203.
72. McKenney, *supra* note 64, at 125. "Humus is the basis for nutrient storage and cycling," and absorbs water, air, and nutrients. Humus is essential in healthy soil. *See* McKenney, *supra* note 69, at 122.
73. McKenney, *supra* note 64, at 125.
74. AIT-UNEP REGIONAL RESOURCE CENTRE FOR ASIA AND THE PACIFIC, DPR KOREA: STATE OF THE ENVIRONMENT 47 (2003), *available at* http://www.rrcap.unep.org/pub/soe/dprk_land.pdf.

c. Water

Farming practices harm water quality when soil, loaded with pesticides and nitrogen fertilizers, erodes into water systems due to overproduction and over-tilling. EPA "has identified siltation associated with erosion in rivers and lakes as the second leading cause of water quality impairment."[75] Agricultural chemicals easily leach into groundwater when rain or irrigation water comes into contact with farm fields.[76] In addition, surface water can become contaminated when groundwater naturally flows into surface water.[77] According to EPA, farming practices account for 70% of the pollution in the nation's rivers and streams.[78] EPA reported that "runoff of chemicals, silt, and animal waste from U.S. farmland has polluted more than 173,000 miles of waterways."[79] A 2006 U.S. Geological Survey study stated:

> Pesticides or their degradates were detected in one or more water samples from every stream sampled. One or more pesticides or degradates were detected in water more than 90 percent of the time during the year in agricultural streams, urban streams, and mixed-land-use streams. . . . Organochlorine pesticides (such as DDT) and their degradates and by-products were found in fish or bed-sediment samples from most streams in agricultural, urban, and mixed-land-use settings—and in more than half the fish samples from streams draining undeveloped watersheds.[80]

Chemical runoff from pesticides and fertilizers also results in a "build-up of excess nutrients in waterbodies that create algae blooms, which use up oxygen and essentially suffocate fish and shellfish populations."[81] One of the largest "dead zones" lies in the Gulf of Mexico, where the Mississippi River deposits high-nutrient runoff from midwestern farms.[82] Agricultural runoff can impact both local water sources as a direct point source pollutant, as well as harm waters thousands of miles away when in it flows down river. (A more detailed discussion of the water impacts from fertilizers and pesticides can be found in Chapter 3.)

Even organic farms, which often use animal manure to fertilize crops, pollute surrounding water and land. Like synthetic fertilizers, animal manure is high in nitrogen and phosphorous. A 1992 study showed that farms throughout the United States had manure-based nitrogen levels that exceeded potential plant uptake, and many areas showed excess levels of phosphorous.[83] The excess nitrogen and phosphorous levels, when not absorbed by plants, runs off into area streams and pollutes groundwater just the same as chemical fertilizers. This runoff can also contribute to the "dead zones" and excess nitrogen in water.

d. Air

Nitrogen fertilization contributes to emissions of the greenhouse gas nitrous oxide, which is a large contributor to climate change[84] by impacting the first and second layers of the atmosphere, the troposphere and stratosphere.[85] When nitrogen oxide is emitted from agricultural soil it increases tropospheric ozone, a component of smog, and impacts human health, natural ecosystems, and ironically, agriculture.[86] Nitrogen oxide from agricultural practices can be transported through the air over long distances and deposited around the world, impacting water and land ecosystems.[87] This widespread exposure to nitrogen oxide

---

75. Uri, *supra* note 58, at 139 (citing U.S Envtl. Prot. Agency, Office of Water, National Water Quality Inventory—1994 (1995)).
76. Angelo, *supra* note 10, at 605 (citing Eubanks, *supra* note 44, at 258).
77. *Id.* at 605 (citing Eubanks, *supra* note 44, at 255).
78. Horrigan et al., *supra* note 48, at 447.
79. *Id.* at 447.
80. Robert J. Gilliom et al., Pesticides in the Nation's Streams and Ground Water, 1992-2001—A Summary (U.S. Geological Survey ed., 2006), *available at* http://pubs.usgs.gov/circ/2005/1291/.
81. Czarnezki, *supra* note 1, at 78.
82. *Id.*
83. Heller & Keoleian, *supra* note 3, at 1020 (citing R.L. Kellogg & C.H. Lander, *Trends in the Potential for Nutrient Loading From Confined Livestock Operations*, in Proceedings of the State of North America's Private Land Conference (National Resources Conservation Service ed., 1999)).
84. Czarnezki, *supra* note 1, at 78.
85. David Tilman et al., *Agricultural Sustainability and Intensive Production Practices*, 418 Nature 671, 673 (2002), *available at* http://www.cedarcreek.umn.edu/biblio/fulltext/t1860.pdf (citing R.J. Cicerone & R.S. Oremland, *Biogeochemical Aspects of Atmospheric Methane*, 2 Global Biogeochem. Cycl. 299 (1988); and S.J. Hall et al., NO_x *Emission From Soil: Implications for Air Quality Modeling in Agricultural Regions*, 21 Ann. Rev. Energy & Env't 311 (1996)).
86. *Id.* at 673 (citing R. Delmas et al., *Global Inventory of NO_x Sources*. 48 Nutr. Cycl. Agroecosyst. 51 (1997)).
87. *Id.* at 673.

can cause "eutrophication, loss of diversity, dominance by weedy species and increased nitrate leaching or NOx fluxes."[88] Pesticides pollute air through runoff and airborne pesticide "drift" and can act as a means of transport, further polluting land and water.[89]

3. Monoculture

Monoculture is the planting of a single crop as opposed to a variety of diverse crops,[90] and the practice has increased dramatically worldwide.[91] It is estimated that the amount of crop diversity has decreased dramatically in recent years.[92] Today, there are over 50,000 varieties of corn grouped together into hundreds of races; however, U.S. commercial production "relies almost exclusively on the cultivation of a handful of hybrid varieties from two of these races."[93] This "sameness," as Helena Norberg-Nodge refers to monoculture, is beneficial to the transnational food corporations like Cargill, but "in the long term a homogenized planet is disastrous for us all."[94] Norberg-Nodge writes: "It is leading to a breakdown of both biological and cultural diversity, erosion of our food security, an increase in conflict and violence, and devastation of the global biosphere."[95] Traditionally, farmers grew a variety of diverse crops and continuously rotated their crops.[96] This diversity allows crops to adopt a resistance to pests, and if one crop is impacted by pests and disease, the entire year's harvest will not be lost.[97] Monoculture, on the other hand, greatly reduces the biodiversity of farmland and

> [p]art of the instability and susceptibility to pests of agroecosystems can be linked to the adoption of vast crop monocultures, which have concentrated resources for specialist crop herbivores and have increased the areas available for immigration of pests. This simplification has also reduced environmental opportunities for natural enemies. Consequently, pest outbreaks often occur when large numbers of immigrant pests, inhibited populations of beneficial insects, favorable weather and vulnerable crop stages happen simultaneously.[98]

A single monoculture crop, planted continuously and in great volume, is more susceptible to pest infestations and disease, and it impairs soil quality and accelerates soil erosion.[99] Genetically uniform crops are more vulnerable, as seen in 1996 when "the fungal disease known as Karnal Bunt swept through the U.S. wheat belt, ruining over half of that year's crop and forcing the quarantine of more than 290,000 acres."[100] The crop failed because farmers had planted only a few varieties of wheat with low resistance to the disease.[101]

Loss of diversity and increased risk of pest infestation and disease due to the planting of monocultures, in turn, requires the application of more chemical pesticides to kill the pests.[102] Monocultures also require increased application of nitrogen fertilizers because the lack of diversity impoverishes the soil and reduces its ability to naturally retain nitrogen.[103] This vulnerable soil lacks nutrients and loses its capacity to retain moisture, and therefore becomes more sensitive to drought and erosion.[104] Monocultures are perpetuated

---

88. *Id.*, at 673 (citing P.M. Vitousek et al., *Human Alteration of the Global Nitrogen Cycle: Sources and Consequences*, 7 ECOL. APPLIC. 737 (1997)).
89. Horrigan et al., *supra* note 48, at 446.
90. Jerry Mander, *Machine Logic: Industrializing Nature and Agriculture*, in THE FATAL HARVEST READER: THE TRAGEDY OF INDUSTRIAL AGRICULTURE, *supra* note 6, at 89.
91. Miguel A. Altieri, *Modern Agriculture: Ecological Impacts and the Possibilities for Truly Sustainable Farming* (University of California, Berkley, Division of Insect Biology), *available at* http://nature.berkeley.edu/~agroeco3/modern_agriculture.html (accessed Nov. 2, 2011).
92. *Id.*
93. Andrew Kimbrell, *A Blow to the Breadbasket: Industrial Grain Production*, in THE FATAL HARVEST READER: THE TRAGEDY OF INDUSTRIAL AGRICULTURE, *supra* note 6, at 102.
94. Helena Norberg-Hodge, *Global Monoculture: The Worldwide Destruction of Diversity*, in THE FATAL HARVEST READER: THE TRAGEDY OF INDUSTRIAL AGRICULTURE, *supra* note 6, at 59.
95. *Id.*
96. *Id.*
97. *Id.* at 59-60.
98. Altieri, *supra* note 91.
99. Horrigan et al., *supra* note 48, at 448. *See, e.g.*, Dan Koeppel, *Yes, We Will Have No Bananas*, N.Y. TIMES, June 18, 2008, at http://www.nytimes.com/2008/06/18/opinion/18koeppel.html.
100. Kimbrell, *A Blow to the Breadbasket*, *supra* note 93, at 102.
101. *Id.*
102. Frederick Kirschenmann, *Scale—Does It Matter?*, in THE FATAL HARVEST READER: THE TRAGEDY OF INDUSTRIAL AGRICULTURE, *supra* note 6, at 95.
103. *Id.*
104. Peter Warshall, *Tilth and Technology: The Industrial Redesign of Our Nations Soils*, in FATAL HARVEST: THE TRAGEDY OF INDUSTRIAL AGRICULTURE, *supra* note 6, at 224.

only by adding large amounts of fertilizer and pesticide. It has been shown that "rotating crops provides better weed and insect control, less disease buildup, more efficient nutrient cycling and other benefits."[105] Unfortunately, monoculture practices are encouraged by modern agroeconomics, as farmers receive greater government payments for growing high-erosion monocrops.[106]

C. Food Processing

Commodity crops like corn, soybeans, alfalfa, and wheat—which are cultivated with large machines and irrigation systems, pesticides, and nitrogen fertilizers—do not go directly to the consumer. Cultivation is only the first of many steps to get processed and packaged foods (think macaroni and cheese, and TV dinners) on American plates. The next stage in the food system, for both conventional and organic agriculture crops, is processing—a stage that also produces greenhouse gas emissions. This section discusses the emissions associated with food processing, focusing on the emissions from processing plants as well as the wastewater pollution problems associated with processing facilities.

1. High Fructose Corn Syrup, Commodity Crops, and the Farm Bill

Today, most food found in grocery stores is loaded with hydrogenated fats, salts,[107] and most commonly, high fructose corn syrup (HFCS), a substance found in virtually every processed food.[108] Most of these ingredients, like HFCS, are "basically a clever arrangement of carbohydrates and fats teased out of corn, soybeans and wheat—three of the five commodity crops that the farm bill supports."[109] Every year, about 530 million bushels of corn are turned into 17.5 billion pounds of HFCS in processing facilities.[110] Author Michael Pollen writes: "Today there are hundreds of things processors can do with corn: They can use it to make everything from chicken nuggets and Big Macs to emulsifiers and nutraceuticals."[111] Soft drinks are the most common place to find HFCS, but it also shows up everywhere at your Fourth of July cookout, from the ketchup and mustard to the hot dog and buns.[112]

Why have we chosen commodity crops to be transformed into highly processed additives and ingredients? As commodity crops, grains like corn were what the first processors focused their efforts on.[113] Author Richard Manning explains that "[n]o one took say, carrots or tangerines or broccoli to a mill and attempted alchemical transformation. Chemically, those items are too complex to lend themselves to reconstitution. They are food."[114] Unlike vegetables and fruits, grains are generally not consumed immediately as food, but need fermenting, baking, or grinding.[115]

More recently, the processing of commodity crops has been accelerated by the policies in the U.S. farm bill that encourage large-scale, high-yield commodity production (discussed in Chapters 1 and 2). Industrial farming and food processing is sustained through government subsidies on commodity grains. The farm bill's role has shifted from its original intent of supporting farmers to subsidizing industrial agriculture and supporting the growth of cheap and plentiful soybeans and corn.[116] Agricultural subsidies within the U.S. farm bill create incentives for farmers to produce large amounts of commodity grain, in particular soybeans and corn.[117] The monetary incentives to grow these crops put new fields into production whenever

---

105. John P. Reganold et al., *Sustainable Agriculture: Traditional Conservation-Minded Methods Combined With Modern Technology Can Reduce Farmers' Dependence on Possibly Dangerous Chemicals. The Rewards Are Both Environmental and Financial*, Sci. Am., June 1990, at 115, *available at* http://oregonstate.edu/instruct/bi430-fs430/Documents-2004/7B-MIN%20TILL%20AG/Sustainable%20Agr%C3%89hn%20Reganold.pdf.
106. Warshall, *Tilth and Technology: The Industrial Redesign of Our Nations Soils*, *supra* note 104, at 224.
107. Marion Nestle, What to Eat 305 (2007).
108. Angelo, *supra* note 10, at 611.
109. Michael Pollan, *You Are What You Grow*, N.Y. Times, Apr. 22, 2007, at http://simplysassysauces.com/archivedarticles/YouAreWhatYouGrow.pdf.
110. Pollan, *supra* note 63, at 103.
111. *Id.*
112. *Id.* at 104.
113. Richard Manning, Against the Grain: How Agriculture Has Hijacked Civilization 167 (2004).
114. *Id.*
115. *Id.*
116. Czarnezki, *supra* note 1, at 66-67. *See* Eubanks, *A Rotten System*, *supra* note 44.
117. Czarnezki, *supra* note 1, at 66.

possible and boost crop yields through increased use of pesticides, herbicides, and monoculture.[118] With the excessive amounts of corn and soybeans, and other commodity crops being produced, there is a need to process all this grain into new products. The government subsidies in the farm bill drive the industrial agriculture system, which in turn drives the processing of commodity crops.

2. Greenhouse Gas Emissions and Air Pollution

Air pollution, in the form of a number of greenhouse gases, is considered one of the most threatening environmental hazards associated with the industrial food system. With over 20,000 companies processing food in the United States,[119] a great deal of greenhouse gases is released into the atmosphere by processing plants. Major air pollutants involved in food processing plants include sulfur dioxide (SO_2), carbon monoxide (CO), ozone (O_3), carbon dioxide (CO_2) and nitrogen dioxide (NO_2).[120] Emissions from food processing can be classified into three categories: direct emissions, indirect emissions from purchased electricity, and other indirect emissions.[121] Direct emissions are from sources owned by processors, including boilers, heaters, cookers, vehicle fleets, and wastewater treatment.[122] Other key contributors to energy use and carbon emissions within the plant include processing equipment, like ovens, dehydrators, retorts and pasteurizers, coolers and freezers, compressed-air systems, air-handling systems, and lighting.[123] Indirect emissions, the second category, come from the use of purchased electricity.[124] And finally, the category of other indirect emissions include "emissions that occur as a result of food processing activities but from sources not owned or controlled by the manufacturer" such as "ingredients, freight, equipment manufacture, solid waste disposal, contractor, [and] employee business travel."[125]

In addition to sulfur dioxide, carbon monoxide, ozone, carbon dioxide, and nitrogen dioxide emissions from processing plants, some plants use products that contain volatile organic compounds (VOC), also a greenhouse gas.[126] VOCs are used as flavorings, dyes, inks, adhesives, and other surface coatings.[127] The greenhouse gas emissions from processing plants contribute substantially to climate change.

3. Wastewater Pollution

In addition to increased greenhouse gas emissions from the processing facilities themselves, wastewater is another environmentally damaging cost of food processing. On average, large food processing facilities produce about 1.4 billion liters of wastewater annually.[128] The wastewater from processing facilities "is high in suspended solids, and organic sugars and starches and may contain residual pesticides."[129] These solids include "organic materials from mechanical preparation processes, that is, rinds, seeds, and skins from raw materials."[130]

---

118. *Id.* at 67.
119. Steven W. Van Ginkel et al., *Biohydrogen Gas Production From Food Processing and Domestic Wastewaters*, 30 INT'L J. HYDROGEN ENERGY 1535, 1536 (2005) (citing Howard Elitzak, *Food Marketing Costs: A 1990's Retrospective*, FOOD REV., Sept.-Dec. 2000, at 27, 27-30).
120. *Id.* at 349 (citing CHARLES E. KUPCHELLA & MARGARET C. HYLAND, ENVIRONMENTAL SCIENCE: LIVING WITHIN THE SYSTEM OF NATURE (1993); PRASAD MODAK & ASIT K. BISWAS, CONDUCTING ENVIRONMENTAL IMPACT ASSESSMENT IN DEVELOPING COUNTRIES (1999); and Ogbonnaya Chukwu, Development of Predictive Models for Evaluating Environmental Impact of the Food Processing Industry: Case Studies of Nasco Foods Nigeria Limited and Cadbury Nigeria Plc (2005) (Dissertation, Dept. of Agric. Eng'g, FUT, Minna, Niger State, Nigeria)).
121. Tim Bowser, Oklahoma Cooperative Extension Service: Division of Agriculture Science and Natural Resources, *Carbon Strategy for the Food Industry*, at http://pods.dasnr.okstate.edu/docushare/dsweb/Get/Document-7196/FAPC-172web.pdf.
122. *Id.*
123. Kate B. Connolly, *Conquering the Carbon Footprint: Operating Costs and the Environment Have Processors Scrambling to Reduce Carbon Emissions*, FOOD PROCESSING, Oct. 8, 2008, at http://www.foodprocessing.com/articles/2008/372.html.
124. Bowser, *supra* note 121.
125. *Id.*
126. Michigan Department of Environmental Quality, Environmental Assistance Program, Air Quality Division, *Volatile Organic Compound (VOC) Emissions at Food Manufacturing Facilities* (2009), available at http://www.michigan.gov/documents/deq/deq-p2ca-VOCEmissionsFoodMfg-Facilities_281942_7.pdf.
127. *Id.*
128. Van Ginkel et al., *supra* note 119, at 1536 (citing McIlvaine, *Food Processing*, http://www.mcilvainecompany.com/generic_examples/food.htm).
129. McIlvaine, *supra* note 128.
130. *Id.*

D. Food Packaging

As food processors grew in size early in the 20th century, "they relied increasingly on the innovation of product packaging to keep their commodities clean and fresh."[131] Packaging "maintains the benefits of food processing after the process is complete, enabling foods to travel safely for long distances from their point of origin and still be wholesome at the time of consumption."[132] But environmental damage can result from the materials used in food packages and, arguably worse, the disposal of the packaging.

The Tellus Institute has described the three categories of costs in the production and disposal of food packaging in an equation:

> Full cost of packaging production and disposal =
> Environmental costs (pollutant costs) of packaging production +
> Conventional (monetary) costs of disposal +
> Environmental costs (pollutant costs) of disposal[133]

Applying the equation to the production of a plastic ketchup bottle, the pollutant costs of packaging production include the materials, like the chemicals and petroleum, used to make the bottle. There is the monetary cost to the consumer to dispose of the bottle, either by recycling or municipal waste. Finally, the environmental costs, such as air and water pollution associated with disposing the ketchup bottle, are added. Even without specific numbers plugged into this equation, one can see that the full cost of food packaging is a great deal more than the price at the supermarket.

Virtually all packaging materials impact the environment in some manner. Depending on the type of packing materials used, the manufacturing, use, and disposal of packaging materials may contribute to greenhouse gas emissions in the form of carbon dioxide, the release of toxin, like vinyl chloride monomer, and the scarring of landscape from the extraction raw materials used in packaging.[134]

The primary packaging materials used in food production are glass, metals (aluminum, laminate and metalized films, tinplate, tin-free steel), plastics (polyolfins, polyesters, polyvinyl chloride, polyvinylidene chloride, polystyrene, polyamide, ethylene vinyl alcohol, laminates, and co-extrusions), paper, and cardboards.[135] Plastics are by far the most common packaging material, and arguably the most harmful to the environment. Aside from e-waste, plastics are the fastest growing portion of municipal waste;[136] "Americans trash more than 40 million plastic Pepsi bottles a day."[137] Flexible plastic wrap is one of the most common packaging materials, constituting about two-thirds of all food packaging, and is made from low-density polyethylene, sometimes with a chemical abbreviated as DEHA, which is added to make the wrap more flexible and adhesive.[138] During the production stage of packaging, plastics are known to be particularly hazardous to workers and the environment.[139]

All the packaging must go somewhere. Author Heather Rogers writes, "Tossed as soon as it is empty, sometimes within minutes of purchase, packaging is garbage waiting to happen."[140] Most household solid waste ends up in landfills, and of the 250 million tons of municipal garbage produced each year, two-thirds ends up in landfills, while only one-third is recovered in recycling or composting programs.[141] Rogers explains that "[t]he rate of climate change is perhaps the broadest barometer of environmental health and is closely linked to trash; the more that gets thrown out, the more pollution-causing processes are relied on to make replacement goods."[142]

---

131. HEATHER ROGERS, GONE TOMORROW: THE HIDDEN LIFE OF GARBAGE 66 (2005).
132. Kenneth Marsh & Betty Bugusu, *Food Packaging: Roles, Materials and Environmental Issues*, 72 J. FOOD SCI. 39.
133. TRISTEN E. RAGSDALE, FOOD PACKAGING STUDY: A REPORT ON ENVIRONMENTAL IMPACT 2 (2005), *available at* http://www.ashlandfood.coop/sites/default/files/packaging-study.pdf.
134. *Id.*
135. Marsh & Bugusu, *supra* note 132, at 40-44.
136. *Id.* at 177.
137. *Id.*
138. HARVEY BLATT, AMERICA'S FOOD: WHAT YOU DON'T KNOW ABOUT WHAT YOU EAT 213 (2008).
139. ROGERS, *supra* note 131, at 6.
140. *Id.* at 7.
141. CZARNEZKI, *supra* note 1, at 39.
142. ROGERS, *supra* note 131, at 7.

Food packaging—such as soda cans, milk cartons, cardboard boxes, and ketchup bottles—represents the majority of the packaging waste going into landfills.[143] Over 50% of all paper (though not all of it packaging), which is an easily recyclable material, ends up as garbage, accounting for half of the discards of all materials in U.S. landfills.[144] Only 5% of all plastic is recycled, and two-thirds of all glass and half of aluminum beverage cans end up in the trash.[145] The recycling rate of PET plastics (polyethylene terephthalate), the most widely collected type, was 19.9% in 2002, with 3.2 billion pounds of PET bottles being buried or burned.[146] Water bottles with the No. 1 recycling code[147] are recycled less then soda bottles with the No. 1 recycling code.[148] With Americans consuming so much bottled water (13 billion liters in 2003, and only 11% being recycled[149]), a great deal of plastic is being trashed.

When materials like plastic are not recycled properly, they end up in landfills, where they cause significant environmental impacts. First, with an increase in the amount of packaging being trashed, there are more collection trucks on the roads releasing more carbon dioxide into the atmosphere.[150] Second, incinerators release large amounts of toxins into the air, which contaminate soil and water.[151] According to the United Nations Environment Programme, as of 2000, "municipal waste incinerators were responsible for creating 69 percent of worldwide dioxin emissions."[152] Even in facilities with proper filtration equipment, "dioxin cannot be destroyed or neutralized because it is generated through the very process of incineration."[153] Dioxins are formed when diverse packaging materials, like paper and plastic, are burned together.[154] When the plastics are burned in incinerators, the remaining ash can also "contain heavy metals like lead, mercury, cadmium and other toxic substances that can leach once buried in landfills."[155] The dioxins can travel through the air and be dispersed on the land and into water on a global scale.

Landfills produce what is known as "landfill gas," the emissions of decomposing waste, which consists primarily of methane, another large contributor to climate change.[156] In fact, EPA suggests that "'[m]ethane is of particular concern because it is 21 times more effective at trapping heat in the atmosphere than carbon dioxide.'"[157]

Plastic is the most environmentally hazardous packaging material when not recycled because, like all synthetics, it cannot be safely returned to the environment and will stay intact for an unknown number of years, with estimates ranging from 200 to 1,000 years.[158] The Container Recycling Institute has estimated that if the number of bottles ending up in incinerators were recycled, "an estimated 6.2 million barrels of crude oil equivalent could have been saved, and over a million tons of greenhouse gas emissions could have been avoided."[159]

Despite national recycling initiatives, few food packages get recycled and curbside recycling programs are at risk, due in part to increased costs and municipal budget cuts.[160] In addition, "[r]ecycling experts link the drop [in recycling] to the rising number of beverages consumed away from the home—in offices, parks, cars, and other places that lack a handy recycling bin."[161] Food is now made to be convenient—wrapped in a plastic, ready to be microwaved in a minute, and eaten with one hand. Convenient foods can be eaten on

---

143. Marsh & Bugusu, *supra* note 132, at 47.
144. ROGERS, *supra* note 131, at 6-7.
145. *Id.* at 7.
146. ELIZABETH ROYTE, GARBAGE LAND: ON THE SECRET TRAIL OF TRASH 177 (2006).
147. Recycling codes refer to the Resin Identification Code (RIC), which is used to identify the plastic resin used in a manufactured plastic. CZARNEZKI, *supra* note 1, at 39 (citing SPI, *Material Container Coding System, at* http://www.plasticsindustry.org/AboutPlastics/content.cfm?ItemNumber=825&navItemNumber=1124).
148. *Id.*
149. *Id.*
150. ROGERS, *supra* note 131, at 4.
151. *Id.* at 5.
152. *Id.*
153. *Id.*
154. *Id.*
155. *Id.* at 5.
156. *Id.* at 4.
157. *Id.* at 4-5 (citing U.S. EPA, *Frequent Questions: Health, Safety, and the Environment, Landfill Methane Outreach Program,* http://www.epa.gov/lmop/faq/public.html).
158. ROGERS, *supra* note 131, at 6 (citing Brian Howard, *Message in a Bottle,* E: THE ENVTL. MAG. 36 (Sept./Oct. 2003)).
159. News Release, Container Recycling Institute, Report Shows Plastic Bottle Waste Tripled Since 1995 (Sept. 15, 2003), *available at* http://www.container-recycling.org/assets/docs/Plaswaste9-15-03--PR.doc.
160. ROGERS, *supra* note 131, at 180.
161. ROYTE, *supra* note 146, at 177.

the run, and ultimately the packaging ends up in trash cans and then landfills, increasing greenhouse gas emissions and environmental impacts.

E. Food Distribution

The food supply chain has become increasingly global, with food being produced and shipped all around the United States and the world. The use of the fossil fuels in transporting food products increases the climate change impacts of the food system tremendously.

1. Food Miles

Food miles are the "distance food travels from where is it produced to where it is consumed. Food miles have increased dramatically in the last couple of decades, largely as a result of globalization."[162] On average, food travels between 1,300 and 1,500 miles before it is consumed.[163] Depending on distribution channels, for example, food moving anywhere within the United States and Canada may first travel through Los Angeles.[164] As author Dale Allen Pfeiffer explains, "[e]ven food distributed within North America is first shipped to L.A. So pears and apples from Washington, right next to the Canadian border, make a longer journey to reach Toronto than carrots from California."[165] Because most food is moved by truck, train, or plane,[166] all of which are currently fueled by fossil fuels, transportation increases carbon emissions.

Food miles may be similar regardless of whether or not the product is conventional or organic. Some may believe that organically grown means locally grown, and that local products are organic. This is not necessarily the case, because "[a] food may be Certified Organic, but it is not necessarily locally grown."[167] This can lead to consumer confusion, as it can be difficult to find the food with the least carbon footprint. Marion Nestle writes about her predicament in a New York City Whole Foods store:

> I found peaches, corn, and tomatoes from New Jersey, and apples from New York, but all were conventionally grown. I looked hard for local organic foods but found only one (some red cabbage from New York State), unless you consider organic corn and tomatoes from Vermont as 'local'. On that particular midsummer day, hardly any of the produce was grown locally, and hardy any of the local produce was Certified Organic.[168]

Many organic foods, particularly industrialized organic foods, travel the same great distances as conventional products to get to a store. With the rise in the organic market, both conventional grocery stores and high-end stores like Whole Foods are carrying more organic products from all around the world in order to meet the demand.

2. Importing Food

Generally speaking, the calculation of food miles only accounts for food traveling within the United States and does not consider imported foods, in which case, the number grows significantly. Food in the United States is increasingly grown in other countries, "including an estimated 39% of fruits, 12% of vegetables, 40% of lamb, and 78% of fish and shellfish in 2001."[169] Furthermore, "[t]he typical American prepared meal contains, on average, ingredients from at least five other countries."[170]

However, despite the large amounts of energy used in the long miles food travels, transportation is arguably the least energy-intensive step in the entire food system. Rich Pirog, who was the first to analyze food miles, has shown that transportation is actually the lowest of all fossil fuel usage in the food system,

---

162. PFEIFFER, *supra* note 35, at 24.
163. HOLLY HILL, FOOD MILES: BACKGROUND AND MARKETING 1 (National Center for Appropriate Technology 2008), *available at* http://kirikiva.com/PDF/Foodmiles.pdf.
164. PFEIFFER, *supra* note 35, at 24-25.
165. *Id.* at 25.
166. BLATT, *supra* note 138, at 216.
167. NESTLE, *supra* note 107, at 39.
168. *Id.* at 41-42.
169. PFEIFFER, *supra* note 35, at 24.
170. *Id.* at 24.

at about 11%.[171] Production and processing, Pirog suggests, account for much more—45.6% of fossil fuel use.[172] Scientists at the Landcare Institute in New Zealand explain that "localism is not always the most environmentally sound solution if more emissions are generated at other stages of the product life cycle."[173] This is not to say we should not be concerned with where our food is coming from, but that buying local alone will not solve all the environmental problems associated with the food system.

Conclusion

The current food system, including conventional and organic food, contributes to environmental degradation and climate change. The production stage relies on fossil fuel-intensive machinery and irrigation systems that harm the soil, water, and air. Harmful pesticides and fertilizers are used extensively, and the dominant monoculture practices pollute natural resources. To the extent that organic products are grown without synthetic pesticides and fertilizers, they are less harmful to the environment. But if they are produced with the same machinery, irrigation, and monoculture as conventional products, organic products present many of the same environmental issues. Processing, especially of commodity crops, increases greenhouse gas emissions and environmental impacts of the food system because processing facilities use energy derived from fossil fuels, and they pollute water systems. Packaging is a large part of both the conventional and organic system, and is environmentally damaging, in particular when food packages are not properly disposed of. Finally, distribution, although it may not be the most intensive stage of fossil fuel use, does increase the carbon footprint of the entire system.

---

171. McWilliams, *supra* note 1, at 25-26.
172. *Id.* at 25.
173. *Id.* at 26 (citing Landcare Research scientists who were quoted in *Greener by the Miles*, Daily Telegraph, Mar. 3, 2007).

Chapter 8
The Federal Insecticide, Fungicide, and Rodenticide Act

Mary Jane Angelo

Pesticides play a significant role in modern agriculture. Although the synthetic chemical pesticides that are widely used in modern industrial agriculture comprise the most obvious category of pesticides, the term *pesticide* includes a wide range of substances, including naturally occurring substances and living organisms.[1] Nevertheless, synthetic chemical pesticides are the most ubiquitous—and frequently the most harmful to the environment.

Humans have used pesticides for thousands of years.[2] But it was not until the early part of the 20th century that pesticide use became widespread in agriculture. Until then, pest control was accomplished primarily through cultural controls such as cultivation,[3] sanitation,[4] crop rotation,[5] and sowing and harvesting practices.[6] During the early years of the century, humans began to rely on metals such as arsenic, lead, and copper as important agricultural pest controls. Not until the latter half of the century did the development of synthetic chemical pesticides lead to a global explosion of pesticide use.[7] These new synthetic chemical pesticides were highly effective at controlling a wide variety of pests.

It did not take long for agriculture to become reliant on large-scale synthetic chemical pesticide use. Current estimates of global use are staggering. More than 1,600 types of pesticides are currently available.[8] More than five billion pounds of pesticides, with a value of over $40 billion, are used annually around the

---

Portions of this chapter have been adapted from, with permission, Mary Jane Angelo, *The Killing Fields: Reducing the Casualties in the Battle Between U.S. Endangered Species and Pesticide Law*, 32 Harv. Envt'l L. Rev. 96 (2008), and Mary Jane Angelo, *Embracing Uncertainty, Complexity and Change to Protect Ecological Integrity: An Eco-Pragmatic Reinvention of a First Generation Environmental Law*, 33 Ecology L.Q. 105 (2006).

1. The term "pesticide" is defined very broadly in the law to include "any substance or mixture of substances intended for preventing, destroying, repelling, or mitigation any pest . . ." 7 U.S.C. §136(u).
2. Homer described how Odysseus used burning sulfur as a fumigant to control pests. Homer, The Odyssey (Edward McCrorie transl. 2004). For a more complete description of the history of pesticide use, see Clive A. Edwards, *The Impact of Pesticides on the Environment*, in The Pesticide Question: Environment, Economics, and Ethics 281 (David Pimentel & Hugh Lehman eds., 1993).
3. Helmut F. van Emden & David B. Peakall, Beyond Silent Spring: Integrated Pest Management and Chemical Safety 115-17 (1996). Many pest insects live out at least part of their life cycle in soil, weeds, or accumulated crop debris in farm fields. Plowing the top layer of soil kills many of these pest insects. Accordingly, soil tillage historically was a critical component of agricultural pest management. It was not until relatively recently that, as a way to minimize soil erosion, tillage was abandoned in favor of zero or minimum tillage systems, which rely on herbicide usage to control weeds. The demise of tillage as a core component of modern agricultural systems has resulted in a dramatic increase in certain soil-dwelling pests. *Id.* at 115. Other cultivation pest control techniques used historically include mulching, compacting, and manuring. *Id.* at 115-17.
4. Sanitation practices are one of the most effective pest control practices used in both ancient and modern agriculture. *Id.* at 117-18. By destroying residues of crops left in fields after harvesting, many pest populations that live in such residues are destroyed. Related practices such as destruction of weed hosts and selective pruning also serve as effective pest control tools. *Id.* at 118-19.
5. Crop rotation, one of the oldest forms of pest control, is a very effective pest control technique for minimizing soil-dwelling pests. By alternating the planting of different crops in a particular field, populations of soil-dwelling insects that feed on a particular crop will not be able to build up during periods when their food crop is not present. Thus, when the crop eventually is planted, populations of the pest species generally will not be large enough to cause serious problems. *Id.* at 120-21.
6. Timing sowing and planting dates to avoid pest outbreaks or to ensure the crop plant is in a resistant growth stage when pest outbreaks are likely to occur, as well as carefully tailoring seed and planting rates and early harvesting, can be effective tools for avoiding pest damage to crops. *Id.* at 121-23.
7. Edwards, *supra* note 2, at 13.
8. These figures are based on EPA pesticide market estimates for the years 2000-2007. *See* Arthur Grube et al., Environmental Protection Agency, Pesticide Industry Sales and Usage, 2006 and 2007 Market Estimates (2011), *available at* http://www.epa.gov/opp00001/pestsales/.

world.[9] Pesticide use in the United States accounts for approximately 22% of global pesticide usage, with U.S. exports to other countries exceeding 450 million pounds of pesticides per year.[10] (A detailed discussion of the different types of pesticides currently in use and the environmental risks associated with them can be found in Chapter 3.)

Pesticides, by design, are intended to kill or disrupt living organisms. Moreover, agricultural pesticides are intentionally released into the environment, and many synthetic pesticides pose risks to human health and the environment. In the United States, these risks are addressed primarily through the regulatory process under the Federal Insecticide, Fungicide, and Rodenticide Act.

A. History and Provisions of the Federal Insecticide, Fungicide, and Rodenticide Act

Domestically, the U.S. Environmental Protection Agency (EPA) has the primary responsibility for regulating pesticides under the authority of the Federal Insecticide, Fungicide, and Rodenticide Act (FIFRA).[11] Generally, FIFRA establishes a licensing program for pesticides manufactured, distributed or sold in the United States. FIFRA contains a number of provisions designed to address the human health and environmental risks associated with pesticides.

1. History of the Act

The origins of FIFRA can be traced back to the federal Insecticide Act of 1910,[12] a consumer economic protection statute aimed at addressing false claims about the efficacy of pesticide products, many of which turned out to be useless, and the converse problem of pesticides that were too strong and thus caused crop damage.[13] This emphasis on consumer economic protection carried over into the first enactment of FIFRA in 1947. The 1947 Act contained the first requirement for pesticides (referred to by the Act as "economic poisons") to be registered prior to being marketed in interstate commerce.[14] The 1947 Act, however, did not establish any environmental standards or any significant safety standards for pesticides. A pesticide could be registered if its composition warranted the proposed claims for it and if the pesticide and its labeling complied with the requirements of FIFRA.[15] The 1947 Act remained intact until 1972.

The controversy over the pesticide DDT, which had gained attention through the publication of Rachel Carson's *Silent Spring* in 1962, became one of the primary motivators behind the establishment of EPA in 1970. The controversy also paved the way in 1972 for a major reform of FIFRA. The 1972 Amendments to FIFRA completely overhauled the statute and for the first time included provisions aimed at protecting environmental interests. The 1972 Amendments form the backbone of the current FIFRA.

2. Registration

FIFRA requires a premarket review of all pesticides to be sold, distributed, or used in the United States. Section 3(a) provides that the EPA Administrator shall register a pesticide if the Administrator determines that, when considered with any restrictions imposed, its composition warrants the proposed claims, its labeling and other submitted materials comply with the requirements of FIFRA, and, when "used in accordance with widespread and commonly recognized practice, it will not generally cause unreasonable adverse affects on the environment."[16]

The burden of providing EPA with the necessary information to determine whether the standard for registration is met rests with the registrant or applicant for registration.[17] As defined by FIFRA, "unreason-

---

9. *Id.*
10. Edwards, *supra* note 2, at 13.
11. 7 U.S.C. §§136-136(y).
12. Act of 1910, April 26, 1910, ch. 191, 36 Stat., repealed 61 Stat. 163, 172 (1947).
13. William H. Rodgers, Environmental Law 412-13 (West, 2d ed. 1994).
14. Ch. 125, 61 Stat. 163 (1947).
15. *Id.*
16. 7 U.S.C. §136a(c)(5).
17. *Id.* §136A(c)(1). *See also* http://www.epa.gov/oecaagct/lfra.html.

able adverse affects on the environment" are "(1) any unreasonable risks to man or the environment, taking into account the economic, social, and environmental costs and benefits of the use of any pesticide, or (2) a human dietary risk from residues that result from a use of a pesticide in or on any food inconsistent with the standard [under the Federal Food, Drug, and Cosmetics Act]."[18] Accordingly, when determining whether to register a pesticide, EPA must consider not only any risks the pesticide poses to man or the environment, but also the economic and social implications of using the pesticide. Noticeably, however, while the U.S. Congress did direct EPA to take into account economic factors in defining unreasonable adverse effect on the environment, it did not explicitly mandate that EPA conduct a strict cost/benefit analysis.[19] In fact, the legislative history of FIFRA suggests that adverse effects were not intended to be tolerated in absence of "overriding benefits" from the use of the pesticide.[20] Nevertheless, for more than 30 years, EPA has interpreted FIFRA to require a cost/benefit balancing except in the case of human dietary risk, for which the more stringent Federal Food, Drug and Cosmetic Act standard applies, and this interpretation has been upheld by courts.

Although the registration standard requires EPA to determine that the pesticide "will perform its intended function" without unreasonable adverse effects on the environment,[21] FIFRA expressly states that EPA shall not make any lack of essentiality a criterion for denying registration of any pesticide, and that where two pesticides meet the requirements for registration, one should not be registered in preference to the other.[22] In other words, there is no requirement to demonstrate that a pesticide is essential to obtain a registration, and the availability of alternative pesticides for the same use does not preclude registration. Moreover, FIFRA expressly authorizes EPA to waive all data requirements pertaining to efficacy, and in fact EPA has, by rule, done so.[23] Thus, as a practical matter in making registration decisions, EPA does not require any showing of the economic or social benefits to be derived from the pesticide, but instead assumes that such benefits will accrue.

3. Data Requirements

One of the most significant aspects of FIFRA is that it requires an applicant for a pesticide registration to submit data to EPA.[24] The data requirements are aimed at developing the information needed for EPA to make the "unreasonable adverse effects" determination. The vast majority of EPA's data requirements under FIFRA relate to human health effects.[25] These requirements include testing on residue chemistry to estimate human exposure to pesticides, acute human hazard, subchronic human hazard, chronic human hazard, mutagenicity, metabolism, reentry hazard, spray drift evaluation, as well as testing on oncogenicity, teratogenicity, neurotoxicity, and reproductive effects in humans. EPA's data requirements for testing for wildlife and ecological effects are extremely limited.[26] EPA does require the submission of environmental fate data to "assess the presence of widely distributed and persistent pesticides in the environment which may result in loss of usable land, surface water, ground water, and wildlife resources, and assess the potential environmental exposure of other nontarget organisms, such as fish and wildlife, to pesticides."[27] EPA's data requirements related to wildlife impacts or other ecological effects are much less ambitious, but the

---

18. Section 136(bb) defines the term "unreasonable adverse effects on the environment" as any "unreasonable risk to man or the environment, taking into account the economic, social, and environmental costs and benefits of the use of any pesticide...." *Id.* §136(bb). Human dietary risk from pesticide residues in food is addressed by the Federal Food, Drug and Cosmetic Act, as described in Chapter 13.
19. SIDNEY A. SHAPIRO & ROBERT L. GLICKSMAN, RISK REGULATION AT RISK: RESTORING A PRAGMATIC APPROACH 29, 32 (2003).
20. *See* RODGERS, *supra* note 13, at 451-53.
21. 7 U.S.C. §136a(c)(5)(B).
22. *Id.* §136a(c)(5).
23. 40 C.F.R. §158.640(b)(1).
24. 7 U.S.C. §136a. Data requirements are found at 40 C.F.R. Part 158, and provide for the submission of health and environmental effects data. The applicant for registration must bear the cost of gathering and generating the necessary data.
25. *See* 40 C.F.R. §158.202(a), (c), (e), (f), and (g) and *id.* §§158.240, 158.390, 158.440, and 158.340. *See also id.* §158.34 (providing that certain human health effects data submitted to EPA must be flagged as indicating potential adverse effects).
26. *See* Leslie W. Touart & Anthony F. Macriowski, *Information Needs for Pesticide Registration in the United States,* 7 ECOLOGICAL APPLICATIONS 1086-93 (1997) (describing and evaluating EPA's ecological risk data requirements for pesticide registration).
27. 40 C.F.R. §158.202(d)(1). These data requirements include studies to determine the rate of pesticide degradation; metabolism studies to determine the nature and availability of pesticides to rotational crops and to aid in the evaluation of the persistence of a pesticide; mobility studies pertaining to leaching, adsorption/desorption, and volatility of pesticides; dissipation studies; and accumulation studies. Environmental fate data are used to evaluate human exposure to pesticides, as well as wildlife exposure. Consequently, these data requirements appear to be fairly comprehensive. *Id.* §158.202(d)(2), (3), (4), (5), and (6). *See also id.* §158.290.

agency does require submission of some data designed to evaluate impacts to wildlife and aquatic organisms. The wildlife and aquatic organism data requirements include avian toxicity studies[28] and freshwater fish and invertebrate acute toxicity studies[29] for most pesticides intended for outdoor use. Additional data are only required on a case-by-case basis depending on the result of lower tier studies. Such conditionally required data include studies on mammal toxicity, avian reproduction, simulated and actual field testing of mammals and birds, acute toxicity to estuarine and marine organisms, fish early life stage, aquatic invertebrate life cycle, fish life-cycle and aquatic organisms accumulation, and simulated or actual field testing of aquatic organisms[30] for most outdoor uses.

With regard to wildlife, EPA's main concern is with acute toxicity testing, and EPA typically does not require data submission on the potential adverse effects of pesticides on wildlife behavior, neurology, reproduction, birth defects, or other nonacute effects. EPA's data requirements do not contain any studies, whatsoever, aimed at evaluating effects on other species such as amphibians or reptiles or other species not specifically identified in the rules.

As to organisms other than birds, mammals, and fish, EPA's requirements are even more limited. In fact, EPA rarely requires data submission related to adverse effects on nontarget insects. Although EPA does conditionally require acute toxicity testing for honey bees and other pollinators if the proposed use will result in honeybee or other pollinator exposure, EPA does not have any data requirements related to pollinator subacute feeding studies,[31] nontarget aquatic insects, or nontarget predatory or parasitic insects.[32]

In its data requirements rule, EPA identifies pollinator data requirements as "reserved pending development of test methodology" and data requirements for nontarget predatory or parasitic insects as "reserved pending further evaluation to determine what and when data should be required, and to develop appropriate test methods."[33]

4. Regulation of Pesticide Use

Once EPA evaluates submitted data, it must determine whether use restrictions are necessary to minimize risks sufficient to be outweighed by benefits, and thus to meet the registration standard. However, EPA's ability to regulate pesticide use under FIFRA is limited. Unlike many other environmental statutes, FIFRA does not establish a permitting system for pesticide use. Specifically, no EPA approval is required prior to using a pesticide, whether by permit or any other mechanism, even for large-scale usage. Consequently, the risks associated with release of pesticides in a particular geographic location at a particular time are not evaluated under FIFRA prior to release of pesticides into the environment. This is significant because risks vary greatly depending on the specific circumstances of the place and time of release. For example, applying a pesticide in a particular location during a time of year when migratory birds are present may pose a much more significant risk that would applying the same pesticide in the same place at a different time of year.

Instead, FIFRA's regulation of pesticide "use" is achieved through labeling restrictions. It is the registration applicant's responsibility to propose all labeling with the registration application.[34] All registered pesticide products must bear a label or labeling containing precautionary statements, warnings, directions

---

28. Avian oral LD50 and dietary LC50s (the concentration at which 50% of the test animals die) are required when using the preferred test animal species, the mallard and the bobwhite. *Id.* §158.490.
29. Freshwater fish LC50 studies are required, with the preferred test species being the rainbow and bluegill fish, and acute LC50 studies are required on freshwater invertebrates, with the preferred test species being Daphnia. *Id.* §158.490.
30. *Id.* §158.490. Conditionally required studies are required only on a case-by-case basis depending on the results of lower tier studies, such as acute and subacute testing, intended use pattern and environmental fate characteristics, or if certain specified criteria are met.
31. In its data requirements rule, EPA identifies this type of requirement as "reserved pending development of test methodology." *Id.* §158.590 (2005).
32. In its data requirements rule, EPA identifies these types of requirements as "reserved pending further evaluation to determine what and when data should be required, and to develop appropriate test methods." *Id.* §158.590.
33. *Id.*
34. 7 U.S.C. §136a(c)(1)(C). FIFRA defines the term "label" as the written, printed, or graphic matter on, or attached to the pesticide. "Labeling," on the other hand, is much broader and includes the label as well as all other written, printed, or graphic matter that accompanies the pesticide or to which reference is made on the label. *Id.* §136(p)(2).

for use of the product, and an ingredient statement.[35] A product with labeling that does not contain the information required by EPA, or which sets forth false or misleading information, is misbranded.[36]

The primary means by which EPA regulates pesticide use under FIFRA is by requiring users of pesticides to follow all label directions. Pesticide product labels are required to state that it shall be unlawful for any person to use any pesticide in a manner inconsistent with its labeling.[37] This is the sole obligation placed by FIFRA on users of pesticides. Accordingly, directions for use are the only mechanism to regulate user behavior to reduce risks. Unfortunately, pesticide users may not understand, or be willing to follow, the complex labeling instructions necessary to prevent environmental harms. Moreover, it is virtually impossible for EPA to know who, where, when, and how persons are using pesticides, much less to monitor each and every pesticide user in the country to assure the labeling instructions are followed.

5. Restricted-Use Pesticides

Although FIFRA authorizes EPA to classify higher-risk pesticides as "restricted-use pesticides," such a classification is of limited value. A restricted-use pesticide may be used only by or under the supervision of a certified applicator. EPA may classify a pesticide for restricted use if it would cause unreasonable adverse effects on the environment in the absence of such a restriction.[38] These products may not be purchased by the general public.[39] However, such a designation is designed primarily to protect the users themselves and not to reduce risks to wildlife or ecosystems. Over one-half of all registered agricultural pesticides are restricted-use pesticides[40] and thus can only be applied only under the supervision of a certified applicator.

Certification of applicators is primarily conducted by the state. Each state must have a certification plan that conforms to certain standards enumerated in FIFRA. The law provides that, if any state, at any time, desires to certify applicators of pesticides, the governor of such state shall submit a state plan for such purpose.[41] EPA shall approve the plan submitted by any state provided it meets certain conditions regarding the state's legal authority and funding mechanisms. Federal certification is required to be conducted by the EPA Administrator, in consultation with the governor of any such state, in which a state plan for applicator certification has not been approved by the Administrator.[42] Unfortunately, the certified applicator requirement does not mandate consideration of local ecological factors and consideration of lower risk alternatives to address ecological risks posed by the use of a specific pesticide in a specific location at a specific time. The law does not require certified applicators to directly oversee the application of pesticides. Instead, certified applicators may simply be what are referred to as "arm chair" supervisors. They do not need to be present on site, only available via telephone during the time of the pesticide application. Moreover, FIFRA does not require that certified applicators obtain any training in local ecological systems and their vulnerability to particular pesticides.[43] Finally, although FIFRA §11 requires EPA and states to make instructional materials on integrated pest management (IPM) available to certified applicators at their request, the statute expressly states that certified applicators are not required to receive instruction on IPM and are not required to be competent with respect to such techniques.

6. Other Approval Mechanisms

FIFRA gives EPA discretionary authority to grant conditional registration—to register products in certain situations even though not all data necessary to make a decision on registration have been generated. Conditional registration can be used for products with composition and proposed uses identical or substantially similar to currently registered pesticides, products with proposed new uses, or certain products with a new

---

35. The pesticide labeling requirements are codified in 40 C.F.R. §156.10.
36. *Id.* §§136(q) and 136j(a)(1)(E).
37. *Id.* §136j(a)(2)(G).
38. *Id.* §136a(d)(1).
39. *Id.*
40. RODGERS, *supra* note 13, at 458.
41. 7 U.S.C. §136i(a)(2).
42. *Id.* §136i.
43. For a description of certified applicator training programs, see RODGERS, *supra* note 13, at 462-63.

active ingredient.[44] For the first two categories, EPA must determine that, despite the lack of data, approval of the conditional registration would not significantly increase the risk of adverse effects on the environment.[45] For new active ingredients, EPA must determine that the use of the pesticide during the period of conditional registration will not cause unreasonable adverse effects on the environment and that use of the pesticide is in the public interest.[46]

FIFRA provides for several forms of pesticide approval in addition to registration under §3. First, EPA may grant an emergency exemption under FIFRA §18,[47] which provides that the Administrator has discretion to exempt any federal or state agency from any provision (normally, the registration requirement) of the Act under emergency conditions.[48] An emergency condition means an urgent, nonroutine situation and is deemed to exist under specified conditions: (1) there are no effective pesticides available that have labeled uses registered for control of the pest under the conditions of the emergency; (2) there are no economically or environmentally feasible alternative practices to provide adequate control; and (3) the situation involves the introduction or dissemination of a new pest, will present significant health risks, will present significant environmental risks, or will cause significant economic loss.[49]

In addition to federal pesticide registration, under FIFRA §24(c) states may issue registrations of pesticide products or limit uses of such products to meet special local needs.[50] A §24(c) registration may be issued to (1) allow use of a new formulation of a federally registered pesticide; (2) amend federal registration to permit use on additional crops or pests or at additional sites, or to permit use of different application techniques, rates, and equipment; (3) amend federal registration with special label directions necessary to prevent adverse effects or to ensure efficacy under local conditions; or (4) for any other purposes consistent with FIFRA. Valid state registrations are treated as federal registrations under FIFRA.[51]

FIFRA §5 authorizes EPA to issue an experimental use permit (EUP) for field testing of an unregistered pesticide.[52] The Administrator may issue an EUP if the applicant needs such a permit to accumulate information necessary to register a pesticide under §3 of FIFRA.[53] Finally, §3(a) authorizes EPA, to the extent necessary to prevent unreasonable adverse effects on the environment, to issue regulations limiting the distribution, sale, or use of any pesticide that is not registered under the Act and that is not subject to an EUP under §5 or an emergency exemption under §18.[54]

7. Continuing Duties of Registrants

Once a pesticide is registered, registrants face a number of continuing responsibilities, particularly with regard to supplying additional data. In 1978, Congress added §3(c)(2)(B) to FIFRA, giving EPA the authority to require holders of existing registrations to provide data to support the continued registration of a pesticide.[55] The penalty for failure to supply this data is suspension of the registration, which results in a prohibition on sale and distribution of the product.[56] Prior to suspension under §3(c)(2)(B), a registrant has a right to a limited adjudicatory hearing. The only issues to be considered at such a hearing are whether "the registrant has failed to take the action" that is the basis of the suspension and whether the disposition of existing stocks is consistent with the Act.[57]

In addition to information required under §3(c)(2)(B), registrants are under a continuing obligation under FIFRA §6(a)(2) to submit factual information regarding the pesticide's unreasonable adverse effects on the environment whenever the registrant has such information.[58] EPA has adopted a rule that

---

44. 7 U.S.C. §136a(c)(7).
45. *Id.* §136a(c)(7)(A).
46. *Id.*
47. *Id.* §136.
48. *Id.*
49. *Id.*
50. *Id.* §136v.
51. *Id.* §136v(c)(1).
52. *Id.* §136c.
53. *Id.*
54. *Id.* §136a(a).
55. *Id.* §136a(c)(2)(B).
56. *Id.* §136a(c)(2)(B)(iii).
57. *Id.*
58. *Id.* §136d(a)(2).

describes specifically the types of information that must be reported and the time frame for submission of these reports.[59]

For pesticides registered before the more environmentally rigorous FIFRA standards were enacted, a number of additional requirements apply. The 1972 revisions to FIFRA included a tougher standard for initial registration of pesticides and mandated that EPA reexamine previously registered pesticides.[60] This reexamination or reregistration reflects a congressional determination that previously registered pesticides ought to be as safe as newer ones and a recognition that the data EPA had for these older pesticides were not as complete or up-to-date as that for newer pesticides.

Reregistration has proven to be one of the most critical and most difficult regulatory tasks for EPA's pesticide program.[61] Because reregistration efforts were moving so slowly, in 1988, Congress enacted a new §4 of FIFRA, which prescribes specific reregistration requirements intended to dramatically change both the pace and the nature of reregistration.[62] The 1988 Amendments required EPA to complete, over a nine-year period, the reregistration review of each registered product containing any active ingredient initially registered before November 1, 1984.[63] The amendments redirected the initial burden of identifying data gaps from EPA to the affected registrants. Moreover, the amendment established a multiphase process with a number of deadlines that ensures that reregistration moves at a more accelerated pace. Failure of registrants to meet the prescribed deadlines would result in suspension or cancellation of registration.[64]

In 1996, Congress passed the Food Quality Protection Act (FQPA), which among other things required EPA to go back and reassess the tolerances (safe level of residues in food under the Federal Food, Drug and Cosmetics Act) for registered pesticides, to ensure they meet the safety standards adopted in the 1996.[65] Although the FQPA is primarily focused on risks to human health from pesticides in or on food, the process of reevaluating tolerances led to the cancellation of some pesticides—sometimes merely because it was not worth it to the manufacturer to provide the data necessary for the evaluation—or the reduction of amounts of pesticides that could be used. Thus, the FQPA indirectly affected ecological risks by eliminating some pesticides or restricting the use of others. In 2006, EPA initiated a new program, which it calls "registration review." The purpose of this process is to reevaluate all pesticides every 15 years to ensure that as the ability to assess risks improves and as policies and practices change, registered pesticides continue to meet legal requirements with regard to human health and the environment.

8. Cancellation and Suspension

EPA may cancel or suspend existing registrations based upon certain risk/benefit determinations. FIFRA §6(b), which addresses cancellation, states that EPA may issue a notice of intent to cancel if a pesticide or its labeling does not comply with FIFRA or if, when used in accordance with widespread and commonly recognized practice, the pesticide generally causes unreasonable adverse effects on the environment.[66] Under §6(b), there are two types of cancellation actions: §6(b)(1)—notice of intent to cancel or change classification; and §6(b)(2)—notice of intent to hold a hearing to determine whether registration should

---

59. 40 C.F.R. Part 159. In addition to authority to require information reporting, EPA has broad enforcement authority, which it shares with the states under FIFRA. EPA generally is responsible for manufacturer/producer enforcement, while the states have primary responsibility for user enforcement. The manufacturer/producer enforcement provisions give the Agency authority to register pesticide establishments, 7 U.S.C. §136e; to inspect and to take samples, *id.* §136g; to inspect books and records, *id.* §136f; and to issue "stop sale, use or removal" orders and to institute seizure actions, *id.* §136i-2. Pursuant to §27 of FIFRA, a state must have adequate pesticide laws and regulations and must be implementing such laws and regulations in order to maintain primary enforcement responsibility for pesticide use situations. *Id.* §136v. The Agency can respond to an emergency requiring immediate action if a state is unwilling or unable to respond. *Id.* §136w-1. Under §16(c), the Agency is authorized to seek an injunction against violations of the Act in federal district court. *Id.* §136n(c). A person who violates any provision of the Act may be subject to civil penalties under §14(a). *Id.* §136l. The amount of the penalty is determined by a consideration of the appropriateness of the penalty to the size of the business, the effect on the violator's ability to stay in business, and the gravity of the violation. *Id.* §136l(a)(4). Moreover, a person who knowingly violates any provision of the Act may be subject to criminal penalties which carry larger fines and the possibility of a prison sentences. *Id.* §136l(b).
60. *Id.* §136a-1.
61. Rodgers, *supra* note 13, at 431.
62. 7 U.S.C. §136a-1.
63. *Id.*
64. *Id.*
65. Food Quality Protection Act of 1996, Pub. L. No. 104-170, 110 Stat. 1489. A detailed discussion of the FFDCA is provided in Chapter 13 of this book.
66. 7 U.S.C. §136d(b).

be cancelled or classification changed.[67] For both sections, EPA must make a finding that the risks appear to outweigh the benefits. For §6(b)(2), however, a hearing may be held when the Administrator's judgment concerning the risks and benefits of a pesticide is only tentative.[68] Before taking final action under §6(b), the Administrator must determine whether any unreasonable risks posed by a pesticide's use can be sufficiently reduced by regulatory measures short of cancellation. Such measures include imposition of additional labeling restrictions and/or classification of the pesticide for restricted use. If the Administrator determines that adequate risk reduction cannot be achieved by such regulatory measures, the registration of the pesticide for that use must be cancelled. An EPA final order on a cancellation is reviewable in district court.[69]

FIFRA also authorizes EPA to suspend the registration of a pesticide based on certain findings. FIFRA provides for two types of suspension proceedings—"ordinary" and "emergency" suspension.[70] Ordinary suspension is issued where such action is necessary to prevent an imminent hazard during the time required for cancellation proceeding. "Imminent hazard" is defined as a substantial likelihood of serious harm occuring during the time required forcancellation proceedings.[71] The term is not limited to a concept of potential crisis. The function of a suspension action is to assess the evidence required to determine the risks and benefits for the period involved, not an ultimate resolution of the cancellation issues.[72] In an ordinary suspension, notification to the registrant of the intent to suspend and an opportunity for a hearing is required prior to effectiveness of suspension. Only a registrant may request an adjudicatory hearing. The order becomes effective either after a favorable decision following a hearing or five days after notification if no hearing is requested.[73] If no hearing is requested, the suspension order is not reviewable by a court.[74] If a hearing is requested, an expedited administrative adjudicatory hearing is held before an administrative law judge in which interested persons can intervene. The sole issue at the hearing is whether an imminent hazard exists.[75]

An emergency suspension order, which is effective immediately, may be issued if an emergency exists that does not permit even an expedited hearing before suspension takes place.[76] Registrants have five days to request an expedited hearing and the hearing must begin within five days of the agency's receipt of such a hearing request.[77] If an expedited hearing is requested, the emergency order remains in effect until the issuance of a final suspension order following the hearing.[78] No party other than the registrant and the agency may participate in the expedited hearing except for the filing of briefs.[79] An emergency suspension order is subject to immediate review in district court.[80]

B. FIFRA's Strengths and Limitations

Although FIFRA has its roots in decades-old laws that were not concerned with environmental protection, the current statute does provide EPA with the regulatory authority to protect against environmental harms. The environmental protection authority granted to EPA by FIFRA has several strengths as well as serious limitations.

---

67. *Id.* §136d.
68. There is no distinction between §136d(b)(1) and §136d(b)(2) hearings in the manner of conduct, burden of proof, or nature of initial decision by an ALJ. One issue generally considered as part of the cancellation process is whether the agency should allow the continued sale and use of existing stocks of the pesticide.
69. Of the more than 60 pesticide cancellations and suspensions, only approximately one-third have been judicially reviewed. RODGERS, *supra* note 13, at 480. EPA's refusal to initiate proceedings to cancel or suspend a registration is considered a final order reviewable in district court. *See* Environmental Defense Fund v. EPA, 465 F.2d 528 (D.C. Cir. 1972).
70. 7 U.S.C. §136d(c) (2004).
71. *Id.* §136(l).
72. *Id.* §136d(c)(1).
73. *Id.* §136d(c)(2).
74. *Id.*
75. *Id.*
76. *Id.* §136d(c)(3).
77. *Id.*
78. *Id.*
79. *Id.*
80. *Id.* §136d(c)(4).

1. Strengths

FIFRA's strength is that it allows consideration of a broad range of human health and environmental concerns prior to approval of pesticides for sale or use in the United States. Specifically, the standard for registration under FIFRA addresses a large array of ecological concerns as well as human health concerns. FIFRA's regulatory standard aims to prevent "unreasonable adverse effects on the environment."[81] The word *environment* is defined very broadly by FIFRA to include water, air, land, and all plants and man and other animals living therein, and the interrelationships which exist among these."[82]

Another strength of FIFRA is that unlike many areas of environmental law, FIFRA adopts a precautionary approach in that it requires premarket approval based on an environmental assessment.[83] FIFRA's precautionary approach also manifests in its allocation of the burden of proof. While not expressly stated in the language of FIFRA, pursuant to a series of administrative and judicial decisions, the burden of proof that a pesticide does not pose an unreasonable adverse effect of the environment remains at all times on the proponent of registration or continued registration.[84] Thus, a proponent for registration must demonstrate that it meets this burden prior to a pesticide being registered. Further, if EPA proposes cancellation of a pesticide or use of the pesticide, the burden of proof rests on the proponent for continued registration during any cancellation or suspension hearing that may ensue.[85]

Another strength is that FIFRA's consideration of environmental concerns and the duties of registrants to ensure that their products do not pose unreasonable adverse effects does not end once a product receives a registration. Many of FIFRA's provisions are specifically designed to seek new information, to adapt to new information, or to tailor the level of regulation to the level of certainty of risks based on the sufficiency of available data. For example, FIFRA establishes two different levels of registration—full registration and conditional registration. EPA may conditionally register a pesticide under certain circumstances despite the fact that sufficient data have not been generated to support full registration. As described above, such circumstances may include a new proposed use for a pesticide already registered for another use. In such a situation, sufficient data exist to support the existing use, but additional data may be required to support full registration of the newly proposed use. Under such circumstances, EPA may conditionally register the pesticide for the new use if conditional registration would not significantly increase the risk of any unreasonable adverse effects on the environment. Accordingly, through the conditional registration process, a degree of flexibility is built into FIFRA allowing products to be used in new ways prior to full data generation.[86]

Other provisions of FIFRA allowing unique circumstances and changing information to be considered include those for emergency exemption, state registration, and the experimental use permits. The emergency exemption provision of §18 of FIFRA authorizes EPA to grant an emergency exemption to any state or federal agency in emergency conditions—i.e., urgent nonroutine conditions for which no economically or environmentally feasible alternative practices that provide adequate control are available.[87] Section 18 provides flexibility to adapt to changing circumstances, which could include the outbreak and spread of a new pest or the spread of a disease that endangers public health. In such circumstances, EPA is authorized to act quickly to control the problem before the pest or

---

81. *See, e.g., id.* §§136(l), 136(x), 136(ee)(2).
82. *Id.* §136(j).
83. In contrast, new nonpesticide chemicals entering the marketplace do not require a premarket environmental review under the Toxic Substances Control Act (TSCA), 15 U.S.C. §2604. Instead, prior to manufacturing these new nonpesticide chemical substances under TSCA, all that is required is a 90-day notification to EPA. *Id.* §2604(a). During the premarket notification period, EPA conducts a cursory review of the proposed new chemical, but unless a determination is made that generation of new data is required, EPA typically does not require environmental testing. *Id.* §2604. If a non-pesticide substance is later found to pose unforeseen risks, EPA can require additional testing or impose regulations to reduce the risk from such a substance under §§4, *id.* §2603 and 6, *id.* §2606, of TSCA, respectively.
84. Environmental Def. Fund, Inc. v. EPA (heptachlor-chlordane), 548 F.2d 998, 1004 (D.C. Cir. 1976), *cert. denied,* 431 U.S. 925 (1977); Envtl. Def. Fund, Inc. v. EPA (DDT II), 439 F.2d 584 (1971); Stearns Elec. Paste Co. v. EPA, 461 F.2d 293, 304 (7th Cir. 1972).
85. Environmental Protection Agency, Federal Insecticide, Fungicide, and Rodenticide Act (FIFRA): Cancellation and Suspension of Pesticide Registrations, *available at* http://www.epa.gov/oecaagct/lfra.html (last visited Apr. 25. 2012).
86. Of course, should the new data demonstrate that the new use does not meet the standards for full registration, full registration will not be granted.
87. 7 U.S.C. §136p.

disease vector is widely disseminated and to minimize the harm without waiting for complete data to support registration.[88]

The provision for state registration in §24(c) authorizes states to issue registration to meet special local needs.[89] Accordingly, this provision allows states to consider local circumstances warranting use of pesticide products—or particular uses of those products—not generally approved under FIFRA for nationwide use. In a state where a particular pest causes more severe harm than in other states, the cost/benefit analysis for the use of the pesticide in that state may have a different result from the nationwide cost/benefit analysis and, accordingly, a special local needs registration may be granted for that state only. In this way, FIFRA's flexibility allows registrations to be tailored to the special agricultural, environmental, economic, or other needs of a state.

Finally, the provision allowing a permit for experimental use contained in FIFRA §5 is another example of flexibility in tailoring the amount of data necessary to the level of risk resulting from a particular use. FIFRA §5 authorizes EPA to issue permits for the field testing of pesticides necessary to generate data to support full registration.[90] As the risk from exposure to a pesticide increases (i.e., as it moves from lab testing, to small-scale field testing, to full-scale use), progressively greater data requirements attach to ensure that sufficient data are available to make a determination of unreasonable adverse effects for each level of use. Similarly, classification of a pesticide as either general or restricted use allows EPA to adapt the amount of regulation required to the risks associated with the particular pesticide. By classifying a pesticide as restricted use, EPA ensures that users of the pesticide will have at least some training and supervision to reduce the risks associated with the use of that pesticide.

2. Weaknesses

Despite its strengths, FIFRA has significant limitations in its ability to prevent or minimize environmental harm. A fundamental shortcoming is that FIFRA, by its very nature, is designed to approve the use of pesticides, which are, by definition, designed to be released in to the environment to kill, harm, or disrupt living organisms. In this regard, FIFRA is fundamentally different from other environmental laws, which typically seek to reduce the amount of hazardous substances released into the environment or seek to substitute less hazardous substances for more hazardous ones in the product or process, thereby minimizing environmental impacts to the extent feasible.

FIFRA does not authorize EPA to deny or cancel a registration simply because it believes a particular pesticide is not essential or that other lower risk pesticides exist. Instead, EPA engages in a cost/benefit balancing. Under such a balancing, even a high-risk pesticide will be registered provided it has significant economic or social benefits. Nothing in FIFRA directs or authorizes EPA to reduce the number of chemical pesticides registered or to reduce the overall quantity of pesticides released into the environment. Moreover, FIFRA does not even require EPA to obtain efficacy data to ensure that the pesticides it registers will actually be efficacious against pests.

Further, EPA's implementation of FIFRA is limited in the sense that EPA does not have robust data requirements to address certain nontarget wildlife and ecological concerns. EPA's data requirements do not adequately address risks to wildlife species, in particular those listed under the Endangered Species Act (ESA).[91] EPA's data requirements for testing for ecological effects are also limited. For example, wild mammal toxicity, avian reproduction, simulated and actual field testing of mammals and birds, acute toxicity to estuarine and marine organisms, fish early life stage, aquatic invertebrate life cycle, fish life cycle and

---

88. *Id.*
89. 7 U.S.C. §136v(c).
90. 7 U.S.C. §136c.
91. The minimum data requirements for registration, experimental use permits, and reregistration are set forth in 40 C.F.R. §158. More detailed standards for conducting tests, guidance on evaluation, and reporting of data and additional guidance is provided in a series of advisory documents that EPA makes available to applicants and the public. *See id.* §158.20(c). In its data requirement rules, EPA identifies some data as required and other data as "conditionally required." Conditionally required data are required only if the product's proposed pattern of use, results of other tests, or other factors meet the criteria specified in the rules. *See id.* §§158.25(a) and 158.101. EPA's rules also allow certain data requirements to be waived if they are not applicable to the particular pesticide or use. *See id.* §158.25(b) (setting forth policy on flexibility and waiver); 40 C.F.R. §158.35 (describing the flexibility in data requirements); and §158.45 (regarding waiver of data requirements). In addition, EPA's rules set forth varying data requirements for minor use of a pesticide, i.e., used on a minor crop, and biochemical and microbial pesticides. *See id.* §§158.60 and 158.65, respectively.

aquatic organisms accumulation, and simulated or actual field testing of aquatic organisms are only conditionally required[92] for most outdoor uses. As illustrated by EPA's primary focus on acute toxicity testing, EPA does not generally require data related to potential adverse effects of pesticides on wildlife behavior, neurology, reproduction, birth defects, or other nonacute effects. Moreover, EPA's data requirements do not contain any studies aimed at evaluating effects on other species such as amphibians or reptiles or other species not specifically identified in the rules.

Likewise, EPA's data requirements for nontarget insects are limited to conditionally requiring acute toxicity testing for honey bees and other pollinators only where proposed uses would result in honeybee or other pollinator exposure. There are no data requirements related to honeybee subacute feeding studies,[93] nontarget aquatic insects, or nontarget predatory or parasitic insects.[94] Moreover, EPA does not have any data requirements related to soil microorganisms, which provide critical ecological services such as decomposition and nitrogen fixation, or any data requirements designed to evaluate the effects of pesticides on any other ecological services.

Although EPA's data requirements include some studies designed to evaluate risks to fish, wildlife, aquatic organisms, and nontarget insects, the agency's primary purpose in requiring such studies is not to determine *whether* to register a pesticide product, but instead is to "provide data which determines the need for (and appropriate wording for) precautionary label statements to minimize the potential adverse effects to nontarget organisms."[95] However, label requirements do not always provide sufficient protection against the environmental harms resulting from pesticide use.

Despite all of the testing and labeling that EPA imposes, large numbers of birds, insects, amphibians, and aquatic species, including threatened and endangered species, continue to be harmed by EPA-registered pesticides.[96] Most hazard-related label restrictions that EPA imposes are aimed at protecting human users of pesticides[97] and other humans, such as children,[98] from accidental poisonings. EPA does require certain limited environmental hazard information to appear on pesticide labels. For example, if a pesticide intended for outdoor use contains an active ingredient with a specified level of acute mammalian or avian toxicity, the label must bear a precautionary statement such as "This pesticide is toxic to wildlife."[99] If either accident history or field studies demonstrate that the use of the pesticide may result in fatality to birds, fish, or mammals, the pesticide label must bear a precautionary statement such as "This pesticide is extremely toxic to wildlife (fish)."[100] Similarly, if a product intended for certain uses contains an active ingredient toxic to pollinating insects, the label must bear an appropriate label caution.[101] Finally, if a product is intended for outdoor use other than aquatic applications, the label must bear the precautionary statement "Keep out of lakes, ponds, or streams. Do not contaminate water by cleaning of equipment or disposal of wastes."[102]

Although EPA requires these precautionary statements on labels, their practical effect is unclear. For example, if a farmer intends to apply a particular pesticide to combat a particular pest and the pesticide label indicates it is toxic to wildlife, how will this information influence the farmers' behavior? It is unlikely that the farmer will choose not to apply the pesticide because virtually all of the major chemical pesticides used in agriculture today are acutely toxic to at least some nontarget organisms and thus bear label lan-

92. *Id.* §158.490. Conditionally required studies are required only on a case-by-case basis depending on the results of lower-tier studies, such as acute and subacute testing, intended use pattern, and environmental fate characteristics, or if certain specified criteria are met.
93. In its data requirements rule, EPA identifies this type of requirement as "reserved pending development of test methodology." 40 C.F.R. §158.590.
94. In its data requirements rule, EPA identifies these types of requirements as "reserved pending further evaluation to determine what and when data should be required, and to develop appropriate test methods." *Id.* §158.590.
95. *Id.* §158.202(h)(1).
96. *See* Brian Litmans & Jeff Miller, *Silent Spring Revisited: Pesticide Use and Endangered Species* (A Center for Biological Diversity Report, 2004), *available at* http://www.biologicaldiversity.org/swcbd/Programs/science/pesticides/.
97. 40 C.F.R. §§156.10 (general labeling requirements); 156.60 (human hazard and precautionary statements); 156.62 (human hazard toxicity categories); 156.64 (signal words for human hazard toxicity categories); 156.68 (first aid statement); 156.70 (precautionary statements for human hazards); 156.78 (precautionary statements of physical or chemical hazards); and §§156.200-156.212 (worker protection statements).
98. *See id.* §156.66 (child hazard warning).
99. The specified level of acute toxicity for mammals warranting such a statement is an oral LD50 of 100 mg/kg or less. The specified level of acute toxicity for fish warranting such a statement is an LC50 of 1 ppm or less. The specified level of acute toxicity for birds warranting such a statement is an oral LD50 of 100 mg/kg or less or a subacute dietary LC50 of 500 ppm or less. *Id.* §156.85(b) paragraphs (1), (2), and (3), respectively.
100. *Id.* §156.85(b)(4).
101. *Id.* §156.85(b)(5).
102. *Id.* §156.85(b)(6).

guage indicating toxicity to wildlife. It is difficult to imagine that a statement on a label indicating that a product is toxic to wildlife will have any significant influence on user behavior. Without more specific directions about when, where, or how it is appropriate to apply the pesticide to minimize risks to wildlife, the farmer is left with an essentially useless warning. The lack of more useful directions to minimize risk is likely due to the fact that, because most chemical pesticides are acutely toxic to at least some wildlife, it is impossible to release them into the environment in large amounts without creating the possibility of harm to wildlife.

Arguably the most significant shortcoming with FIFRA is its limited authority to regulate the actual use of pesticides. Currently, no federal system is in place—and only very limited state or local systems exist—to regulate the uses of pesticides registered under FIFRA other than the label restrictions on each registered pesticide. These label restrictions are generally the same nationwide; they can limit the crops the particular pesticides can be used on, the application rates and methods of application, and provide certain precautionary statements. However, the labels do not allow detailed determinations of the specific risks posed by the use of a specific pesticide in a particular place at a particular time. While a pesticide may pass the cost/benefit balancing test on a nationwide basis, the pesticide could pose unacceptable risks if used in certain locations such as those adjacent to fragile or threatened ecosystems or locations where protected species breed or nest. A label simply cannot anticipate every potential concern for every location, time and event throughout the entire country. In contrast, most other environmental laws impose permitting requirements wherein the permit issuers can make particularized decisions with regard to time, place, and method of environmental release.

The consideration of local factors in making the determination of whether or how to use a specific pesticide in a specific location is of particular import. Local factors could include the presence of threatened, endangered, or otherwise rare species, the presence of sensitive species, soil conditions, climatic conditions, proximity to environmentally sensitive lands, types of crops grown, types of farming practices used, severity of pest infestations, or other relevant site-specific factors.

A variety of potential mechanisms are available for achieving local decisionmaking regarding actual pesticide use. One such mechanism is to encourage local government regulation of pesticide use. Another mechanism is to provide better training to certified applicators in the IPM on nonchemical controls and better information regarding endangered species, ecological processes, the role of predators and parasites, and other local environmental conditions. Similarly, better training could be provided to local agricultural extension agents. A variation on this theme would be to empower local officials—whether they be local government officials or extension agent officials—to make case-by-case or season-by-season decisions on the actual use of pesticides.

For example, a local official could be required to evaluate the local conditions—including the particular pest concerns, the climatic conditions, and a wide variety of local environmental factors—before prescribing that a particular pesticide be used. This idea is similar to that of a medical doctor prescribing that a patient take a particular medication. Prior to issuing such a prescription, the doctor would consider a number of factors such as the patient's overall health, other medical conditions, other medications the patient is taking, any allergies or sensitivities the patient may have to certain types of medications, the patients age, the patient's health and lifestyle objectives, and the patient's willingness to accept certain risks to achieve such goals. Moreover, the doctor could adjust the type or amount of medication over time to fine-tune the treatment in accordance with changing circumstances or new information. Such an approach to pesticide application could similarly adjust over time after consideration of changed local conditions or new information about local environmental factors. With a pesticide prescription system in place, pesticide manufacturers will likely be able to convince decisionmakers to prescribe their pesticides. Nevertheless, such a system, if properly instituted, could result in at least some level of informed decisionmaking prior to the release of large amounts of pesticides into the environment.

The likely criticisms of such a system would be that it could entail high costs and possibly the creation of a new bureaucracy. However, relying on existing infrastructure to facilitate such a system without the need for a completely new body or significant additional personnel may be possible. Perhaps the existing agricultural extension services could be used to administer such a system. Alternatively, existing state requirements for certified applicator training and certification could be expanded to better educate applicators about local environmental factors that should be taken into account and nonchemical alternative pest control mechanisms that in many cases may be preferable to chemical approaches. Such an approach might also rely on existing extension infrastructure and resources.

C. Recent Legal Developments

The shortcomings of FIFRA have led to a number of legal controversies in recent years. Two issues have been particularly important. The first involves the relationship between the federal regulation of pesticides under FIFRA and attempts by states or local governments to impose additional or different regulations on the use of pesticides.

Although FIFRA provides a regulatory system that applies to any pesticide sold or distributed in the United States, FIFRA does not generally preempt state or local government regulation of pesticide use. In 1991, the right of a local government to regulate pesticide use was clearly established by the U.S. Supreme Court in the case of *Wisconsin Public Intervenor v. Mortier*.[103] In that case, a Wisconsin local government had adopted an ordinance that required a permit from the local government prior to certain types of pesticide use. Prior to the *Mortier* decision only a small number of states had in place laws that preempted local governments from regulating pesticide use. After *Mortier*, all but 11 states have laws preempting local regulation of pesticides.[104]

Justice Byron R. White, writing for the majority in *Mortier*, recognized the benefit of local decisionmaking for the actual use of pesticides when he wrote: "FIFRA nowhere seeks to establish an affirmative permit scheme for the actual use of pesticides. It certainly does not equate registration and labeling requirements with a general approval to apply pesticides throughout the [n]ation without regard to regional and local factors like climate, population, geography, and water supply."[105] And yet this is, in practice, what FIFRA does. Once a pesticide receives a FIFRA registration, unless a state actively seeks to further regulate such a pesticide, it can be used anywhere in the United States with the only limitation that it must be used in accordance with the FIFRA label instructions. Most states do not have detailed environmental permitting requirements for pesticide use. Although EPA attempts to impose risk-reducing measures on users through detailed labeling requirements, a set of instructions on a container that have been drafted to apply to the entire United States is a poor substitute for a site-specific and circumstance-specific decision on what pesticide to use where, when, and how.

As described in more detail in Chapter 9 of this book, another issue of local decisionmaking regarding pesticide use has arisen in another context in recent years. Beginning with *Headwaters, Inc. v. Talent Irrigation District*[106] in 2002, the courts and EPA have been grappling with whether, and under what circumstances, a National Pollutant Discharge Elimination Systems (NPDES) permit is required under the Clean Water Act (CWA) for the application of pesticides into waters of the United States.[107] Historically, EPA had not required NPDES permits for such pesticide applications. In *Talent*, the U.S. Court of Appeals for the Ninth Circuit addressed the issue of whether the application of FIFRA-compliant aquatic pesticides to irrigation canals eliminated the need for a NPDES permit under the CWA §402 program.[108] Ultimately, the Ninth Circuit held that in addition to FIFRA regulation, aquatic pesticide residues were a "chemical waste" and therefore a "pollutant" subject to the NPDES permitting program.[109] However, unaddressed

---

103. Wisconsin Public Intervenor v. Mortier, 501 U.S. 597 (1991).
104. Laura A. Haight, *Local Control of Pesticides in New York: Perspectives and Policy Recommendations*, 9 Alb. L. Envtl. Outlook J. 39 (2004). The U.S. Supreme Court recently addressed FIFRA preemption again in Bates v. Dow Agrosciences, LLC., 125 S. Ct. 1788 (2005) (holding that FIFRA does not preempt claims for defective design, defective manufacture, negligent testing, breach of express warranty, and violation of the Texas Deceptive Trade Practices Act, and remanding the issue of whether FIFRA preempts fraud and failure-to-warn claims).
105. *Wisconsin Public Intervenor*, 501 U.S. 597.
106. Headwaters, Inc. v. Tallent Irrigation Dist., 243 F.3d 526 (9th Cir. 2001) (holding that the application of an aquatic pesticide to irrigation canals in compliance with the registration and labeling requirements under FIFRA did not eliminate the need for an NPDES permit).
107. For a detailed discussion of the judicial decisions and EPA's position on the issue of requiring NPDES permits for aquatic pesticide application, see Kelly C. Connelly (case note), *Pesticides and Permits: Clean Water Act v. Federal Insecticide, Fungicide and Rodenticide Act*, 8 Great Plains Nat. Resources J. 35 (2003), and Paul Herran (case note), *Headwaters, Inc. v. Talent Irrigation District: Application of Aquatic Pesticides to Irrigation Canals, a Discharge, Which Requires a Clean Water Act Permit?*, 25 Haw. L. Rev. 629 (2003).
108. *Headwaters*, 243 F.3d 526.
109. *Id.* at 532-33.

by the *Talent* court was the issue of whether FIFRA-compliant pesticides leaving no chemical residue in the water similarly qualified as a "pollutant" under the CWA.[110] In 2005, the Ninth Circuit took up this question in *Fairhurst v. Hagener*, and ruled that FIFRA-compliant pesticides producing no residue were not "pollutants" subject to the NPDES permit program of the CWA.[111]

After *Fairhurst*, the CWA definition of "pollutant" (including both "chemical wastes" and "biological materials" within its definition) continued to create considerable controversy.[112] In 2006, EPA issued a rule that attempted to clarify and codify EPA's view prior to *Talent* that FIFRA-compliant aquatic pesticides were not CWA pollutants, and therefore did not require an NPDES permit.[113] A year later environmental and industry groups challenged EPA's pesticide rule that FIFRA-compliant aquatic pesticides were not subject to the NPDES permitting program because they were neither chemical wastes nor biological materials under the CWA definition of pollutants. Ultimately, the litigation that ensued led the U.S. Court of Appeals for the Sixth Circuit to vacate EPA's rule in *National Cotton Council et al. v. EPA*.[114] After finding that the rule was not a reasonable interpretation of the CWA, the Sixth Circuit held that NPDES permits were required for all biological and chemical pesticide applications leaving a residue in, over, or near waters of the United States.[115] In response to the *National Cotton Council* ruling, on June 2, 2010, EPA issued a proposed pesticide general permit (PGP), which would authorize certain specified discharges of pesticides to waters of the United States. Pesticide applications not covered by the PGP would continue to require individual NPDES permits. Pesticide applications that would be subject to the PGP would include mosquito and other flying insect control, aquatic weed control, and forest canopy pest control.[116] On October 31, 2011, EPA issued the final NPDES pesticide general permit.[117] A pending Congressional bill, H.R. 872, looks to codify and reinstate EPA's pre-*National Cotton* view that FIFRA-compliant aquatic pesticides should be exempt from the NPDES permit program.[118]

The final, and perhaps most intractable, area of legal controversy under FIFRA in recent years is the relationship between FIFRA and the Endangered Species Act. As described above, nationwide decisionmaking regarding pesticides can lead to disproportionate risks being placed on vulnerable populations of people, and on vulnerable species or ecosystems. While an overall cost/benefit analysis for a particular pesticide may weigh in favor of use of the pesticide, geographic or ecological "hot spots" may occur where the risks outweigh the benefits on those local areas or for those particular species.

The standard of FIFRA and the ESA are not easily reconcilable. FIFRA utilizes a cost/benefit balancing standard to approve ex ante nationwide registration for a particular pesticide. The ESA, on the other hand, utilizes a risk-based standard that applies to particular acts that take place in particular geographic locations, at particular times and under particular circumstances. This disconnect between the two statutes has led to a rash of litigation attempting to force EPA to comply with the ESA while still carrying out its duties under FIFRA.

The litigation over the impacts to protected wildlife species from pesticide use heated up in 2002, when 40 environmental groups, including the American Bird Conservancy and Defenders of Wildlife, sent EPA a Notice of Intent to Sue for Violations of the Endangered Species Act (ESA), Migratory Bird Treaty Act (MBTA), and Administrative Procedure Act Concerning the Registration of the Pesticide Fenthion due to the high risks fenthion posed to a number of bird species. Later in 2002, the U.S. Fish and Wildlife Service (FWS) recommended that EPA cancel existing registrations for fenthion immediately due to unreasonable

---

110. *See NPDES Permits Required to Spray Aquatic Pesticides*, Marten Law, *available at* http://www.martenlaw.com/newsletter/20090123-npdes-aquatic-pesticides (last visited Sept. 30, 2011).
111. Fairhurst v. Hagener, 422 F.3d 1146, 1152 (9th Cir. 2005).
112. Claudia Copeland, Cong. Research Serv., Paper No. 9, Pesticide Use and Water Quality: Are the Laws Complementary or in Conflict? (2007), *available at* http://digitalcommons.unl.edu/cgi/viewcontent.cgi?article=1028&context=crsdocs.
113. U.S. EPA, *Background Information on EPA's Pesticide General Permit*, *available at* http://cfpub.epa.gov/npdes/home.cfm?program_id=414#decision (last visited Sept. 30, 2011).
114. National Cotton Council v. EPA, 553 F.3d 927 (6th Cir. 2009).
115. *Id.* at 940.
116. *Id.*
117. 76 Fed. Reg. 68750 (Oct. 31, 2011).
118. *See* Ashlie Rodriguez, *Pesticide Spraying Near Streams to Expand Under Congressional Bill*, L.A. Times, June 21, 2011, at http://latimesblogs.latimes.com/greenspace/2011/06/house-bill-senate-agriculture-committee-pesticide-clean-water-act-epa.html.

adverse effects it posed to avian species protection under ESA[119] and MBTA.[120] EPA failed to take action to reduce the risks as requested by the plaintiffs and as recommended by FWS. Consequently, in October 2002, Defenders of Wildlife, the American Bird Conservancy, and the Florida Wildlife Federation filed suit against EPA in federal district court alleging that EPA had violated the ESA and MBTA. The suit was rendered moot, however, when in 2003, the manufacturer voluntarily canceled its registration of fenthion.[121]

In September 2004, environmentalists won a significant victory when the Ninth Circuit issued a decision affirming a January 2004 order by the U.S. District Court for the Western District of Washington that found that EPA had violated the ESA because it had failed to take steps to ensure that the registration of 54 pesticides would not jeopardize the survival of listed salmon species. The court's ruling upheld the district court's injunction, which imposed detailed buffer zones restricting the use of more than 30 pesticides along listed salmon-supporting waters in California, Oregon, and Washington State.[122] EPA argued, among other things, that since it had already granted a FIFRA license, any action that would result in a cancellation or modification of that license must be according to the statutory requirements of FIFRA.[123] Furthermore, EPA claimed that FIFRA, when read in conjunction with the ESA, already took into account any concerns that registration might affect listed species.[124]

The Ninth Circuit, relying on the U.S. Court of Appeals for the Eighth Circuit's logic in the 1989 case *Defenders of Wildlife v. EPA*,[125] in which the court held that the EPA action had caused the deaths of endangered species and as a result an illegal taking had occurred,[126] concluded that FIFRA does not allow EPA to exempt itself from the requirements of the ESA, and that EPA must comply with the ESA if its registration of pesticides will affect listed species.[127] The court held that, while the statutes have different purposes and different calculations, EPA could not avoid its duties under the ESA simply "because it is bound to comply with another statute that has consistent, complementary objectives."[128] The court explained that under FIFRA, EPA utilizes a cost/benefit analysis to measure the risk to people or the environment from the pesticide's use. The ESA, on the other hand, provides a virtual blanket prohibition against the takings of endangered species. The court then summarily dismissed EPA's argument that it lacked discretion to cancel registration except under the statutory requirements of FIFRA.[129] The court ultimately upheld the district court's injunctive relief, noting that because it was the "maintenance of the 'status quo' that [was] alleged to be harming the endangered species,"[130] the injunction was appropriate pending EPA compliance with the ESA. Furthermore, the court placed the burden of proof on EPA to show that its action did not jeopardize the listed species, finding that such burden shifting was appropriate under the ESA for agency actions that have violated §7(a)(2).[131]

After its dramatic loss in *Washington Toxics Coalition*, EPA stopped litigating suits brought to force the agency to comply with §7 of the ESA in the FIFRA registration process and pursued a policy of settling these cases.[132] For example, one post-*Washington Toxics* settlement occurred in 2005, when EPA agreed to make "effects determinations" for six pesticides harmful to the Barton Springs salamander within speci-

---

119. 16 U.S.C. §§1532-1544 (2004).
120. 16 U.S.C. §§703-711 (2004).
121. *See* Fenthion: Product Registrations Cancellation Order, 68 Fed. Reg. 55609-55611 (Sept. 26, 2003). EPA approved the manufacturer's request to cancel Fenthion in May 2003. *See* 68 Fed. Reg. 32495-32497 (May 30, 2003).
122. *See* Washington Toxics Coalition et al. v. EPA et al., Case No. C01-013132C, Order issued Jan. 22, 2004. This order was the third in a series of orders granting injunctive relief to the environmental plaintiffs in this matter. *See* Wash. Toxics Coalition et al. v. EPA et al., Case No. C01-013132C, Orders issued July 16, 2003 and Aug. 8, 2003. All of these orders are available on EPA's website at http://www.epa.gov/espp/litstatus/wtc/index.htm.
123. *Id.* at 14.
124. *Id.* at 15.
125. Defenders of Wildlife v. EPA, 882 F.2d 1294 (8th Cir. 1989).
126. *Id.*
127. *Washington Toxics Coalition*, 413 F.3d at 1032.
128. *Id.*
129. *Id.* at 1032-33.
130. *Id.* at 1035.
131. *Id.* Some settlements actually occurred prior to the court decision in Washington Toxics Coalition. In *Californians for Alternatives to Toxics v. EPA*, No. C00-3150 CW (N.D. Cal.), EPA agreed to make "effects determinations" for approximately 20 pesticides harmful to dozens of plant and salmon species by specified deadlines. 83 Fed. Reg. 21232 (Apr. 30, 2002), *available at* http://www.epa.gov/fedrgstr/EPA-PEST/2002/April/Day-30/p10725.htm.
132. *See* Californians for Alternatives to Toxics website at http://www.alternatives2toxics.org.

fied time frames in response to a January 26, 2004, lawsuit against the agency brought by the Center for Biological Diversity and the Save Our Springs Alliance. The suit, brought in the U.S. Court of Appeals for the District of Columbia (D.C.) Circuit, alleged that EPA violated the anti-take provisions of the ESA when it registered six pesticides without reviewing the potential negative effects on the Barton Springs salamander.[133] The pesticides in question were atrazine, diazinon, carabaryl, prometon, metolachlor, and simazine.[134] The plaintiffs specifically charged that EPA had failed to comply with §§7(a)(1) and 7(a)(2) of the Endangered Species Act,[135] which require federal agencies to consult with the services to guarantee that agency action will not jeopardize the continued existence of any listed endangered or threatened species.[136] Under the terms of the settlement agreement, EPA agreed to make "effects determinations" relating to the Barton Springs salamander for the six pesticides at issues and would initiate consultation under ESA §7 for any of the pesticides found to "likely adversely affect" the listed species.[137]

Other recent settlements include EPA's 2006 agreement with the Natural Resources Defense Council to make "effects determinations" for atrazine's effect on 21 threatened and endangered species within specified time frames,[138] and EPA's 2006 agreement with the Center for Biological Diversity in which the agency agreed to make "effects determinations" for 66 pesticides harmful to the California red-legged frog within specified time frames.[139] More recently, EPA agreed to a stipulated injunction to resolve a lawsuit brought by the Center for Biological Diversity, which establishes a schedule for EPA to review the effects of 75 pesticide active ingredients on 11 federally listed threatened or endangered species.[140]

The most recent litigation over EPA's failure to comply with ESA's §7 consultation requirements when making regulatory decisions on pesticides under FIFRA is a lawsuit filed on January 20, 2011, by the Center for Biological Diversity and the Pesticide Action Network North America.[141] In this lawsuit the plaintiffs assert that EPA has illegally failed to consult under §7 of the ESA "regarding the impacts of hundreds of pesticides known to be harmful to more than 200 endangered and threatened species." This lawsuit, known as the "mega lawsuit" because it involves hundreds of pesticides, could have significant impacts on farmers and the agricultural industry. Accordingly, the lawsuit has caught the attention of many industry groups which, arguing it could have devastating effect, have brought it to the attention of Congress. Congress has since held hearings on the topic and a National Academies, National Research Council committee has been established to look into the risk assessment process used by EPA in making its "effects determinations." On June 6, 2011, the court denied motions filed by numerous chemical industry and farming stakeholder groups to intervene in the lawsuit, thereby barring them from becoming parties to the suit.

D. Encouraging Reduced-Risk Pesticides

Despite the many limitations of FIFRA, EPA has taken some steps to attempt to encourage the development of lower-risk pesticides. Although there are no specific laws directing EPA to conduct comparative risk analyses or to promote one type of pesticide over another, the agency has taken certain policy initiatives to encourage the use of reduced-risk pesticide. In 1992, EPA issued a *Federal Register* notice entitled "Incentives for Development and Registration of Reduced Risk Pesticides," which sought comment on potential approaches to encourage the development, registration, and use of pesticides or pest management practices that present lower risks to human health and the environment. Subsequently, EPA took steps to exempt from FIFRA registration requirements more than 30 naturally occurring pesticides that EPA found to pose little or no risk.[142] EPA has also taken steps to promote the use of lower-risk biological pesticides

---

133. U.S. EPA, *EPA Signs Settlement Agreement Regarding Endangered Species*, http://www.epa.gov/oppfead1/cb/csb_page/updates/es-settlement.htm (last viewed Sept. 27, 2005).
134. *Id.*
135. 16 U.S.C. §§1536(a)(1) and 1536(a)(2).
136. Center for Biological Diversity vs. Johnson, No. 1:04-cv-00126-CKK, Settlement Agreement, at 2-3 (D.C. Cir. 2005), *available at* http://www.epa.gov/oppfead1/cb/csb_page/updates/bartonsprings-agreemt.pdf.
137. *Id.* at 5-6.
138. *See* http://www.epa.gov/espp/litstatus/es-settlement-atrazine.pdf.
139. *See* http://www.epa.gov/espp/litstatus/stipulated-injunction.pdf.
140. Order Approving Stipulated Injunction and Order, Center for Biological Diversity v. EPA, No. 07-2794-JCS (May 17, 2010).
141. Center for Biological Diversity et al. v. EPA et al., No. 3:11-cv-00293 (Jan. 20, 2011).
142. 40 C.F.R. §152.25(f) (2010).

and to streamline the registration process for pesticides that EPA classifies as posing reduced risk. Although important, these steps do not go far enough to accomplish the difficult task of reducing environmental harms from pesticides while continuing to provide adequate pest control mechanisms to protect our agricultural system.

Conclusion

Widespread use of synthetic chemical pesticides has, in some cases, significantly impacted ecological resources by killing nontarget organisms, including protected species; contaminating water and soil; and disrupting and destabilizing the natural processes that tend to suppress pest populations. FIFRA provides a premarket review process for evaluating risks posed by pesticides. However, it does not establish a regulatory program under which the use of particular pesticides in specific locales and under specific conditions can be comprehensively evaluated. The limited authority for regulating the use of pesticides, coupled with the nationwide cost/benefit balancing standard of FIFRA, is not adequate to provide significant reduction in the environmental risks posed by agricultural pesticides. As is highlighted by the recent rash of litigation over the relationship between FIFRA and the ESA, the current system is failing and the time is ripe to consider new approaches to address the environmental risks of chemical pesticides.

Chapter 9
Agriculture and the Clean Water Act
Mary Jane Angelo and James F. Choate

Agriculture has the potential to impact significantly both water quality and water quantity. The primary federal authority for addressing the environmental impacts to water resources from agriculture, as well as impacts from industrial and domestic sources of water pollution, is the federal Clean Water Act (CWA). The CWA includes both regulatory and nonregulatory provisions designed to protect the integrity of the nation's waters. Its focus is on protecting water quality through a range of permitting and incentive-based programs.

This chapter covers the role of the federal CWA in addressing water resource impacts associated with agricultural activities. It is important to note, however, that in addition to water quality impacts, agricultural activities also cause water quantity impacts. Enormous volumes of groundwater and surface water are withdrawn to irrigate crops.[1] In 2008, for example, farmers in the United States irrigated nearly 55 million acres of agricultural land.[2] These water withdrawals can severely impact existing natural systems, as well as the availability of water for other human uses, such as public water supply. However, water quantity issues are not directly addressed by federal law and the CWA does not include provisions specifically designed to address water quality concerns. Water use and its associated water quality impacts are regulated on a state-by-state basis in accordance with state law.[3] (A detailed discussion of the environmental impacts of agricultural irrigation is provided in Chapter 4.)

A. A Brief Overview

As one of the most significant features of the CWA, the National Pollutant Discharge Elimination System (NPDES) permit program requires a permit for any addition[4] of a "pollutant" from a "point source" into "navigable waters" (i.e., "waters of the United States").[5] Either the U.S. Environmental Protection Agency (EPA) or a state to which EPA has delegated the authority administers the NPDES permit program.[6] Currently, 46 states administer the NPDES permit program on behalf of EPA.[7] Idaho, Massachusetts, New Hampshire, and New Mexico are the states that do not administer the NPDES permit program.[8]

The NPDES permit program has been relatively successful at reducing the amount of pollutants discharged from point sources, defined under the CWA as "any discernable, confined and discrete convey-

---

James Choate's contributions to this chapter are expressed in his individual capacity and do not necessarily reflect those of the U.S. Department of Defense, the U.S. Army, or the U.S. Army Corps of Engineers, which have indicated neither approval nor disapproval of the positions Mr. Choate takes in this chapter.

1. J.B. Ruhl, *Farms, Their Environmental Harms, and Environmental Law*, 27 ECOLOGY L.Q. 263, 279 (2000).
2. U.S. Dep't of Agric., National Agric. Stat. Serv., 2008 Census of Agriculture, Farm and Ranch Irrigation Survey, *available at* http://www.agcensus.usda.gov/Publications/2007/Online_Highlights/Farm_and_Ranch_Irrigation_Survey/index.php (last visited July 29, 2012).
3. Christine A. Klein et al., *Modernizing Water Law: The Example of Florida*, 61 FLA. L. REV. 403 (2009).
4. "Addition" is undefined within the CWA, but is found in 33 U.S.C. §1362(12)(a) (defining "discharge of a pollutant" as "any addition of any pollutant to navigable waters from any point source").
5. 33 U.S.C. §1311; 33 U.S.C. §1342. *See also* 33 U.S.C. §1362(7) (defining "navigable waters"). For the definition of "waters of the United States," see 40 C.F.R. §230.3(s).
6. *See* U.S. EPA, *National Pollutant Discharge Elimination System (NPDES): State Program Status*, http://cfpub.epa.gov/npdes/statestats.cfm (last visited July 29, 2012).
7. *Id.*
8. *Id.*

ance." One of the greatest failures of the program, however, is that it does not apply to nonpoint source (NPS) discharges, including agricultural runoff.[9] Most of the water quality impacts from agriculture result from stormwater runoff—i.e., when rainwater picks up sediments, fertilizers, or pesticides from the soil and carries them into water bodies. In fact, agricultural runoff is one of the leading sources of water pollution in the United States.[10] Common agricultural pollutants include sedimentation, nutrients, pathogens, pesticides, metals, and salts, with sediment runoff itself serving as the primary vehicle for the transport of these pollutants.[11] In recent years, NPS pollution,[12] which includes polluted runoff from urban, suburban, and mining areas, as well as from agriculture and silviculture, has been the primary source of water quality problems in the United States.[13] A 2000 report by the National Water Quality Inventory identified agricultural NPS pollution as "the leading source of water quality impacts on surveyed rivers and lakes, the second largest source of impairments to wetlands, and a major contributor to contamination of surveyed estuaries and ground water."[14]

Sediment, fertilizer, and pesticide runoff from farmed fields are major contributors to water quality problems.[15] Rain-induced sediment runoff from agricultural fields decreases water clarity in receiving waters, while the fertilizers, pesticides, and heavy metals attached to the transported soil particulate create contaminated algal blooms, and deplete oxygen levels in the nearby lakes, rivers, wetlands, and receiving waters.[16] Natural and manmade contaminants often attach to sediment runoff in-route, only further contaminating the receiving waters (e.g., rivers, lakes, wetlands, coastal waters, groundwaters).[17]

In addition to the farm field runoff concerns, other agriculture activities—including animal feeding operations, aquaculture, and wetland conversion/agricultural development—also contribute to agriculturally related NPS pollution.[18] According to EPA, the agricultural activities primarily responsible for NPS pollution are "poorly located or managed animal feeding operations; overgrazing; plowing too often or at the wrong time; and improper, excessive or poorly timed application of pesticides, irrigation water and fertilizer."[19]

A wide range of normal agricultural practices contribute to water quality degradation. When land is cleared, tilled, or plowed for planting, the bare disturbed land is vulnerable to wind or rainfall driven erosion.[20] Sediments from the erosions are washed off by rain or carried off by wind and eventually find their way into water bodies.[21] One of the most significant agricultural contributors to water quality problems is the use of fertilizers.[22] Fertilizers are comprised of nutrients, primarily nitrogen and phosphorous, which are needed for plant growth.[23] Just as fertilizers can result in rapid and hardy crop plant growth, they also can cause rapid and hardy algal growth in water bodies.[24] Thus, when fertilizers are applied to land and subsequently carried out in stormwater sheet flows into water bodies, they can cause the high nutrient levels in the water bodies, which ultimately can result in excessive algal growth, or what is known as eutrophication.[25] Eutrophication in water bodies can have a number of significant consequences: oxygen depletion, loss of cold/deeper water fish and other animals, algal blooms and a shift in algal species to toxin-producing

---

9. For an explanation of "point source" versus "nonpoint source" pollution, see *infra* Section B.1.a.
10. *See* U.S. EPA, Protecting Water Quality From Agricultural Runoff, *available at* http://water.epa.gov/polwaste/nps/upload/2005_4_29_nps_Ag_Runoff_Fact_Sheet.pdf (asserting that agricultural runoff is the leading source of pollution to water bodies surveyed by EPA).
11. U.S. EPA, Protecting Water Quality From Agricultural Runoff (2005), *available at* http://water.epa.gov/polwaste/nps/upload/2005_4_29_nps_Ag_Runoff_Fact_Sheet.pdf.
12. For the definition of "nonpoint source" pollution, see *infra* note 43 and accompanying text.
13. U.S. EPA, *Polluted Runoff (Nonpoint Source Pollution): Basic Information*, http://www.epa.gov/owow_keep/NPS/whatis.html (last visited Sept. 30, 2011).
14. U.S. EPA, *Agriculture*, http://water.epa.gov/polwaste/nps/agriculture.cfm (last visited July 29, 2012). *See also* U.S. EPA, *2000 National Water Quality Inventory*, http://water.epa.gov/lawsregs/guidance/cwa/305b/2000report_index.cfm (last visited July 29, 2012).
15. Ruhl, *supra* note 1, at 284-85.
16. *Id.*
17. U.S. EPA, *What Is Nonpoint Source Pollution?*, http://water.epa.gov/polwaste/nps/whatis.cfm (last visited July 29, 2012).
18. *Agriculture, supra* note 14.
19. *Id.*
20. Ruhl, *supra* note 1, at 277-78.
21. *Id.* at 278.
22. *Id.* at 284.
23. *Id.*
24. *Id.* at 285.
25. *See* U.S. EPA, *Water: Nutrients*, http://water.epa.gov/scitech/swguidance/standards/criteria/nutrients/problem.cfm#eutrophication (last visited July 29, 2011).

cyanobacteria, increases in low-oxygen tolerant "trash" fish, loss of shallow water vegetation through shading and other effects,[26] taste and odor problems, and loss of aesthetic and recreational value.[27] Fertilizers also can end up in groundwater where they can contaminate drinking water wells.[28] Another significant water quality impact from agriculture is a result of the aerial or land-based application of pesticides, which are carried by stormwater runoff into water bodies. Many agricultural pesticides are harmful to aquatic life (as described in Chapter 3).

Finally, agriculture can impact water bodies when agricultural activities take place in the water bodies (including wetlands) themselves. Central to the ongoing NPS pollution problem is the continual degradation of the nation's wetlands, which provide vital water quality protection by operating as a natural filter for pollutants.[29] Over the course of the last four centuries, land management and development have converted over half of the United States' wetlands to other uses (including agriculture).[30] Although the conversion of wetlands for agricultural use has slowed in recent decades, many remaining wetlands are, in large part, continually degraded by NPS pollution.[31] As a direct result of NPS pollution, wetlands are less able to provide water quality protection, filtration, and floodwater storage, and also become less suitable as fish and wildlife habitat.[32]

B. The Clean Water Act

The origin of today's CWA dates back to the middle of the 20th century. In 1948, the U.S. Congress enacted the Federal Water Pollution Control Act (FWPCA)[33] as the first major U.S. law to address water pollution.[34] For several decades, this law primarily provided technical and financial support to states to help address water pollution. It was not until the late 1960s, when water pollution concerns became a central focus of the environmental movement, that efforts were made to amend the Act to contain more aggressive measures to reduce water pollution. In 1972, Congress adopted extensive amendments in the 1972 Federal Water Pollution Control Act, which for the first time established a comprehensive water pollution regulatory program.[35] In 1977, Congress further amended the FWPCA, leading to what is commonly referred to today as the Clean Water Act.[36] With a stated objective "to restore and maintain the chemical, physical, and biological integrity of the Nation's waters,"[37] the CWA is the primary federal regulatory authority for addressing water pollution. Significant amendments to the CWA were adopted in 1987, but the basic structure of the regulatory programs from the 1972 Act remains intact.[38]

1. The National Pollutant Discharge Elimination System Permitting Program

The primary regulatory program under the CWA for addressing water pollution is the National Pollutant Discharge Elimination (NPDES) permit program, which is found in CWA §402.[39] Section 301 of the CWA explicitly prohibits "the discharge of any pollutant" to navigable waters from point sources unless

---

26. S.R. Carpenter et al., *Nonpoint Pollution of Surface Waters With Phosphorus and Nitrogen*, 8 Ecological Application 559-61 (1998).
27. *See* U.S. EPA, An Urgent Call to Action: Report of the State-EPA Nutrient Innovations Task Group (2009), *available at* http://water.epa.gov/scitech/swguidance/standards/criteria/nutrients/upload/2009_08_27_criteria_nutrient_nitgreport.pdf. *See also* U.S. EPA, Hypoxia in the Northern Gulf of Mexico: An Update by EPA Science Advisory Board, Doc. No. EPA-SAB-08-003, (2007), *available at* http://yosemite.epa.gov/sab/sabproduct.nsf/95eac6037dbee075852573a00075f732/c3d2f27094e03f90852573b800601d93!OpenDocument; Donna M. Schiffer, *Hydrology of Central Florida Lakes—A Primer*, U.S. Geological Survey Circular 1137, 1998, *available at* http://www.sjrwmd.com/minimumflowsandlevels/pdfs/USGS_hydrology_centfla_lakes.pdf.
28. Erik Lichtenberg & Lisa K. Shapiro, *Agriculture and Nitrate Concentrations in Maryland Community Water System Wells*, 26 J. Envtl. Quality 145, 145-47 (1997).
29. U.S. EPA, Threats to Wetlands (2001), EPA 843-F-01-002d, *available at* http://www.epa.gov/owow/wetlands/pdf/threats.pdf.
30. *Id.*
31. *Id.*
32. *Id.*
33. Water Pollution Control Act of 1948, Pub. L. No. 80-845, 62 Stat. 1155.
34. U.S. EPA, *History of the Clean Water Act*, http://www.epa.gov/lawsregs/laws/cwahistory.html (last visited July 29, 2012).
35. Federal Water Pollution Control Act Amendments of 1972, Pub. L. No. 92-500, 86 Stat. 816; U.S. EPA, *supra* note 34.
36. Clean Water Act of 1977, Pub. L. 95-217, 91 Stat. 1566; U.S. EPA Region 6, *Clean Water Act*, http://www.epa.gov/region6/6en/w/cwa.htm.
37. 33 U.S.C. §1251(a).
38. Water Quality Act of 1987, Pub. L. No. 100-4, 101 Stat. 7. *See* U.S. EPA, *Clean Water Act*, http://www.epa.gov/agriculture/lcwa.html (last visited July 29, 2012).
39. 33 U.S.C. §1342.

the discharge is in accordance with an NPDES permit under §402.[40] In other words, an NPDES permit is required for (1) any discharge (2) of a pollutant (3) from a point source (4) into navigable waters. The interpretation of each step in this four-part test for determining when an NPDES permit is required has been the subject of debate and litigation. Of particular significance to agricultural discharges are the issues of what types of discharges are considered to be point source discharges, what constitutes a pollutant, and what the reach of federal jurisdiction is (i.e., what are considered navigable waters).

a. Point Source Discharges Versus Nonpoint Source Discharges

Whether a discharge is considered a point source is of major concern because NPDES permits are required only for point sources; nonpoint sources are not directly regulated at the federal level. Many agricultural discharges do not fall within the definition of point sources and thus are outside of the jurisdiction of the NPDES permitting program. The CWA defines the term "point source" as:

> [A]ny discernable, confined and discrete conveyance, including but not limited to any pipe, ditch, channel, tunnel, conduit, well, discrete fissure, container, rolling stock, concentrated animal feeding operation, or vessel or other floating craft, from which pollutants are or may be discharged. *This term does not include agricultural stormwater discharges and return flows from irrigated agriculture* (emphasis added).[41]

Congress clearly intended to exclude normal agricultural runoff from permitting requirements by explicitly excluding agricultural stormwater discharges and irrigation return flows from the definition of point source.[42] According to EPA, "nonpoint source" is understood to mean "any source of water pollution that does not meet the legal definition of 'point source' in §502(14) of the Clean Water Act."[43] Thus, any source that is not a point source is a "nonpoint source."

In contrast to regulatory controls over point source pollution (through the NPDES permit program), the CWA does not require any regulatory control of nonpoint source discharges. NPS pollution is often addressed by the states through nonregulatory means. CWA §208 calls for states to adopt "areawide waste management plans," which can include state controls on NPS water pollution.[44] However, this section does not require any form of NPS pollution control. In addition, the 1987 Amendments to the CWA established the §319 Nonpoint Source Management Program, which looked to curb NPS pollution by requiring states to prepare NPS state assessment reports, and to establish state management programs.[45] Unfortunately, measures to reduce NPS pollution, primarily recommending best-management practices, have proven largely unsuccessful at scaling back widespread NPS pollution.[46]

In 1987, Congress amended the CWA to include certain stormwater discharges in the NPDES permit program. However, Congress once again chose to continue to exclude agricultural stormwater runoff from the definition of point source.[47] Consequently, most of the current significant quality problems with the nation's waters are caused by these unregulated nonpoint source discharges.[48] The two greatest contributors to NPS pollution are runoff from agriculture and runoff from urban and suburban land uses.[49]

---

40. *Id.* §1311(a).
41. *Id.* §1362(14).
42. *Id.*
43. *What Is Nonpoint Source Pollution?, supra* note 17.
44. Clean Water Act §208, 33 U.S.C. §1288.
45. *See* L. Allan James, *Non-Point Source Pollution and the Clean Water Act: Policy Problems and Professional Prospects*, 126 J. Contemp. Water Res. & Educ. 60, 60 (2003), *available at* http://opensiuc.lib.siu.edu/jcwre/vol126/iss1/10. Section 319 also provides for grants to states to implement management programs, which supports "a wide variety of activities including . . . technical assistance, financial assistance, education, training, technology transfer, demonstration projects, and monitoring to assess the success of specific NPS implementation projects." U.S. EPA, *Questions and Answers on the Relationship Between the Sec. 319 Nonpoint Source Program and the Sec. 314 Clean Lakes Program*, http://www.epa.gov/owow/NPS/Section319/qa.html (last visited July 29, 2012); *see also* U.S. EPA, *Polluted Runoff (Nonpoint Source Pollution): Clean Water Act Section 319*, http://water.epa.gov/polwaste/nps/cwact.cfm (last visited Nov. 29, 2012).
46. *See* James, *supra* note 45, at 60.
47. 33 U.S.C. §1342(p)(2).
48. *See* Jackson B. Battle & Maxine I. Lipeles, Water Pollution 537 (3d ed. 1998).
49. *See id.* at 535-36.

b. Pollutant

The scope of the definition of the term "pollutant" is also significant in determining whether certain agricultural practices are subject to NPDES permitting. This issue is of particular significance with regard to the application of pesticides. When rainwater picks up pesticide residues from farm fields and carries them into water bodies, it is considered an NPS discharge, which is not subject to NPDES permitting. This is not as clear, however, for pesticides applied directly to water bodies or directly adjacent to water bodies. For example, the application of pesticides to control aquatic weeds in streams, ditches, and ponds and wetlands, whether on or off the farm, can contribute to water quality degradation and harm to nontarget aquatic organisms.

Over the last 10 years, the CWA definition of "pollutant" has led the courts and EPA to wrestle with the issue of whether and under what circumstances EPA should require an NPDES permit for aquatic pesticides already registered and labeled under the Federal Insecticide, Fungicide, and Rodenticide Act (FIFRA) (discussed extensively in Chapter 8).[50] The issue with regard to NPDES permitting is whether applying a pesticide to water bodies to exert their pesticidal effect in the water body constitutes a discharge of a pollutant. CWA §502 defines "pollutant" as:

> dredged spoil, solid waste, incinerator residue, sewage, garbage, sewage sludge, munitions, *chemical wastes, biological materials*, radioactive materials, heat, wrecked or discarded equipment, rock, sand, cellar dirt and industrial, municipal, and agricultural waste discharged into water (emphasis added).[51]

This issue is whether a pesticide applied to a water body for a particular purpose (i.e., to control aquatic weeds) is considered to be a chemical waste. The issue of whether the pesticide is a waste is important because for biological pollutants, Congress chose to use the term "biological materials" rather than "biological wastes." Thus, it must be assumed that Congress made a conscious decision to use the word "waste" when referring to chemicals and "materials" when referring to biological matter. The first example of the difficulties both EPA and the courts have faced in understanding the intersection of the CWA definition of "pollutant" and FIFRA is the decision in 2001 by the U.S. Court of Appeals for the Ninth Circuit in *Headwaters, Inc. v. Talent Irrigation District*. In *Talent*, the Ninth Circuit addressed the issue of whether the application of aquatic pesticides to irrigation canals in compliance with the requirements of FIFRA still required an NPDES permit under the CWA §402 program.[52] The court interpreted the term "chemical waste" in the CWA to include aquatic pesticide residues. Accordingly, the court held that the aquatic pesticide residues were a pollutant subject to the NPDES permitting program.[53] The court did not, however, address the issue of whether FIFRA-compliant pesticides leaving no chemical residue in the water similarly qualified as a chemical waste and therefore a pollutant subject to NPDES permitting.[54] In 2005, the Ninth Circuit addressed this issue in *Fairhurst v. Hagener*, in which the court ruled that FIFRA-compliant pesticides producing no residue were not chemical wastes and therefore not pollutants subject to the NPDES permit program.[55]

The *Talent* and *Fairhurst* cases created significant confusion and controversy. As a factual matter, it was unclear how an aquatic pesticide could leave no residue whatsoever. As a practical matter, if all or virtually all applications of aquatic pesticides fell under NPDES, thousands of farmers and local, state, and federal agencies would have the burden of obtaining permits every time they applied an aquatic pesticide, and EPA and state agencies would have the burden of issuing the permits.[56] In 2006, EPA, through rulemaking, attempted to clarify and codify its pre-*Talent* view that FIFRA-compliant aquatic pesticides were not CWA

---

50. For a detailed discussion of the judicial decisions and EPA's position on the issue of requiring NPDES permits for aquatic pesticide application, see Kelly C. Connelly, Case Note, *Pesticides and Permits: Clean Water Act v. Federal Insecticide, Fungicide and Rodenticide Act*, 8 Great Plains Nat. Resources J. 35 (2003) and Paul Herran, Case Note, *Headwaters, Inc. v. Talent Irrigation District: Application of Aquatic Pesticides to Irrigation Canals, a Discharge, Which Requires a Clean Water Act Permit?*, 25 Haw. L. Rev. 629 (2003).
51. Clean Water Act §502, 33 U.S.C. §1362(6).
52. Headwaters, Inc. v. Talent Irrigation Dist., 243 F.3d 526 (9th Cir. 2001).
53. *Id.* at 532-33.
54. *See NPDES Permits Required to Spray Aquatic Pesticides*, Marten Law, available at http://www.martenlaw.com/newsletter/20090123-npdes-aquatic-pesticides (last visited July 29, 2012).
55. 422 F.3d 1146, 1152 (9th Cir. 2005).
56. Claudia Copeland, Cong. Research Serv., Paper No. 9, Pesticide Use and Water Quality: Are the Laws Complementary or in Conflict?, (2007), available at http://digitalcommons.unl.edu/cgi/viewcontent.cgi?article=1028&context=crsdocs.

pollutants, and did not require an NPDES permit.[57] Under the rule, EPA construed FIFRA-compliant aquatic pesticides as "products that EPA has evaluated and registered for the purpose of controlling target organisms . . . designed, purchased, and applied to perform that purpose" to justify that such pesticides were not chemical wastes within the CWA definition of pollutant.[58] The rule also provided that FIFRA-compliant aquatic pesticides were also not biological materials within the CWA definition of pollutant, to prevent an inconsistency between biological and chemical pesticide regulation pursuant to the CWA definition of pollutant.[59]

In 2007, environmental and industry groups initiated a judicial challenge to the rule. In 2009, the U.S. Court of Appeals for the Sixth Circuit in *National Cotton Council et al. v. EPA* vacated EPA's rule, finding that it was not a reasonable interpretation of the CWA.[60] The court concluded that NPDES permits were required for all biological and chemical pesticide applications leaving a residue in, over, or near waters of the United States, regardless of whether the pesticides were applied in compliance with FIFRA.[61] The court looked to the plain meaning of the word "waste," and found that the term "chemical wastes" included "excess chemicals."[62] Although the Sixth Circuit agreed with the Ninth Circuit's analysis in the *Fairhurst* case that FIFRA-compliant aquatic pesticides leaving no residue were not a chemical waste, the Sixth Circuit also ruled that excess pesticide and pesticide residue could be pollutants under the CWA.[63] As to biological pesticides, the Sixth Circuit concluded that such pesticides and associated residues, when discharged into water, were subject to the NPDES permitting program as biological materials within the definition of pollutants.[64] The court observed that the NPDES requirement applies in situations where pesticides are applied on land near water bodies and their residues subsequently are carried into jurisdictional waters, as well as in situations where pesticides are directly applied to water bodies.

EPA estimated that the *National Cotton* decision to vacate the 2006 rule would subject approximately 365,000 pesticide applicators that perform 5.6 million pesticide applications annually to NPDES permitting requirements.[65] Accordingly, EPA sought a two-year stay of the issuance of the court's mandate in order to provide time to develop, propose, and issue a final NPDES general permit for pesticide applications and to conduct outreach and education to the regulated community.[66] On June 8, 2009, the Sixth Circuit granted the two-year stay of the mandate, delaying the effective date of the court's ruling until April 9, 2011.[67]

On June 2, 2010, EPA issued a proposed pesticide general permit (PGP), which seeks to bring pesticides within the boundaries of the NPDES permit program.[68] The PGP, as proposed, authorizes certain specified discharges of pesticides to waters of the United States. However, pesticide applications not covered by the PGP would still be required to obtain an individual NPDES permit, unless covered by an alternative NPDES general permit. The PGP would cover discharges to waters of the United States from the application of biological pesticides and chemical pesticides that leave a residue.[69] The use patterns covered by the PGP include mosquito and other flying insect control, aquatic weed control, and forest canopy pest control.[70] As proposed, the PGP would not cover discharge of pesticides or their degradates to waters already impaired by these specific pesticides or degradates or discharges to outstanding national resource waters. On October 31, 2011, EPA issued the final NPDES pesticide general permit.[71] Interestingly, a pending

---

57. U.S. EPA, *Background Information on EPA's Pesticide General Permit*, http://cfpub.epa.gov/npdes/home.cfm?program_id=414#decision (last visited July 29, 2012).
58. EPA Pesticide Rule, 71 Fed. Reg. 68486 (Nov. 27, 2006).
59. *Id.*
60. National Cotton Council v. EPA, 553 F.3d 927, 940 (6th Cir. 2009).
61. *Id.*
62. *Id.* at 938.
63. *Id.* at 936.
64. *Id.* at 937-38.
65. EPA Pesticide Rule, *supra* note 58.
66. Respondent United States Environmental Protection Agency's Motion for Stay of Mandate, National Cotton Council et al. v. EPA, Case No. 06-4630, Doc. 00615475372 (6th Cir. filed Apr. 9, 2009).
67. Order Granting Respondent United States Environmental Protection Agency's Motion for Stay of Mandate, National Cotton Council et al. v. EPA, Case No. 06-4630, Doc. 00615559373 (6th Cir. filed June 9, 2009).
68. *See also* Meline MacCurdy, *EPA Releases Draft General Permit for Pesticide Applications*, MARTEN LAW, June 16, 2010, *available at* http://www.martenlaw.com/newsletter/20100616-pesticide-applications-draft-permit.
69. *Id.*
70. *Id.*
71. 76 Fed. Reg. 68750 (Oct. 31, 2011).

congressional bill, H.R. 872, looks to codify and reinstate EPA's pre-*National Cotton* view that FIFRA-compliant aquatic pesticides should be exempt from the NPDES permit program.[72]

c. Navigable Waters

In addition to the point source and pollutant elements of the test to determine whether an NPDES permit is required for a particular discharge, the discharge must be to a water body within the jurisdictional reach of the CWA. The CWA defines the term "discharge of a pollutant" to mean any addition of any pollutant into navigable waters from any point source.[73] The term "navigable waters" is defined as the "waters of the United States, including the territorial seas."[74] Due to Congress' failure to include specific definitions or guidance regarding exactly what waters are meant to be considered navigable waters or waters of the United States, there has been much debate about how far the jurisdiction of the CWA extends. This is particularly true with regard to waters that are not clearly "navigable in fact"—i.e., waters such as wetlands, streams, and tributaries that cannot be navigated by boats,. Historically, the regulatory agencies responsible for implementing the CWA—EPA and the U.S. Army Corps of Engineers (the Corps)—have interpreted the term "waters of the United States" very broadly to include not only navigable-in-fact waters, but also to include, among other things, all interstate waters and "[a]ll other waters such as intrastate lakes, rivers, streams (including intermittent streams), mudflats, sandflats, wetlands, sloughs, prairie potholes, wet meadows, playa lakes, or natural ponds, the use, degradation or destruction of which could affect interstate or foreign commerce."[75] This broad interpretation of the statute has been subject to a number of legal challenges.

The seminal decision upholding a broad interpretation of the term "navigable waters" is the 1985 U.S. Supreme Court case *Riverside Bayview Homes*.[76] In that case, the Court, citing *Chevron U.S.A., Inc. v. Natural Resources Defense Council, Inc.*,[77] articulated that EPA's interpretation of the CWA was entitled to deference provided it was reasonable. Consequently, the Court held that the Corps' interpretation of the term "waters of the U.S." as including "adjacent wetlands" was reasonable. After this decision, EPA and the Corps continued to assert broad jurisdiction under the CWA, even in some cases over wetlands without an obvious physical connection or proximity to navigable-in-fact waters. For example, in 1986, EPA and the Corps adopted what is commonly referred to as the "migratory bird rule,"[78] which asserted federal jurisdiction over isolated *and* intrastate wetlands pursuant to the CWA, provided the wetlands served as a habitat for migratory birds.[79] In 2001, the Supreme Court, in *Solid Waste Agency of Northern Cook County (SWANCC) v. U.S. Army Corps of Engineers*,[80] rejected this broad interpretation and struck down the migratory bird rule as exceeding the scope of jurisdiction under the CWA.[81] The *SWANCC* Court declined to expand the CWA's definition of "navigable waters" to include "isolated ponds, some only seasonal, . . . [just] because they serve as habitat for migratory birds."[82] In rejecting the broad interpretation of CWA jurisdiction, the Court emphasized the "'independent significance'" of the term "navigable,"[83] stating that "[t]he term 'navigable' has at least the import of showing us what Congress had in mind as its authority for

---

72. H.R. 872: Reducing Regulatory Burdens Act of 2011, *available at* http://www.govtrack.us/congress/bills/112/hr872. *See* Ashlie Rodriguez, *Pesticide Spraying Near Streams to Expand Under Congressional Bill*, L.A. Times, June 21, 2011, at http://latimesblogs.latimes.com/greenspace/2011/06/house-bill-senate-agriculture-committee-pesticide-clean-water-act-epa.html.
73. 33 U.S.C. §1362(12).
74. 33 U.S.C. §1362(7).
75. 33 C.F.R. §328.3(a)(3).
76. United States v. Riverside Bayview Homes, Inc., 474 U.S. 121 (1985).
77. Chevron U.S.A., Inc. v. Natural Resources Defense Council, Inc. (1984). The *Chevron* case established a two-part test for judicial review of an agency's interpretation of a statute it is charged with implementing or enforcing. Part 1 requires that the court look to see if Congress has directly spoken to the precise question at issues. If the answer is yes, the court must give effect to Congress' unambiguously expressed intent and may not engage in any further analysis. If Congress has not directly spoken to an issue, either because the statute is silent on that issue or because the language of the statute is ambiguous, the court must defer to the agency interpretation provided that interpretation is a permissible construction of the statute. The court is not free to impose its own interpretation.
78. Migratory Bird Rule, 51 Fed. Reg. 41216 (Nov. 13, 1986).
79. *Id.*
80. Solid Waste Agency of N. Cook County v. U.S. Army Corps of Eng'rs, 531 U.S. 159 (2001).
81. *Id.* at 171-72.
82. *Id.*
83. *Id.* at 172 (quoting Transcript of Oral Argument at 28, Solid Waste Agency of N. Cook County (SWANCC) v. U.S. Army Corps of Eng'rs, 531 U.S. 159 (2001) (No. 99-1178)).

enacting the CWA: its traditional jurisdiction over *waters that were or had been navigable in fact or which could reasonably be so made*" (emphasis added).[84] The Court further stated that "Congress intended that the [CWA's] jurisdiction be limited to navigable waters and non-navigable waters that have a 'significant nexus' to navigable waters, such as wetlands adjacent to navigable waters"[85]

The most recent Supreme Court case addressing the proper interpretation of "navigable waters" was the 2006 case of *Rapanos v. United States*,[86] in which the Court granted certiorari on the question of whether the CWA's jurisdiction extends to wetlands that are adjacent to tributaries of navigable-in-fact water bodies.[87] A majority of the court could not agree on the proper interpretation of CWA jurisdiction, and the case resulted in a 4-1-4 vote. The four dissenting justices would have upheld the Corps' broad exercise of jurisdiction over the wetlands at issue, finding that the interpretation should be afforded *Chevron* deference.[88] The four justices joining in the plurality opinion, as well as Justice Anthony M. Kennedy, who wrote his own concurring opinion, concluded, however, that the case should be remanded for "proper" evaluation of whether the wetlands at issue are "waters of the United States."[89] The plurality and Justice Kennedy provided different reasoning for reaching that conclusion. The plurality set forth a two-part test for determining whether the wetlands are subject to CWA jurisdiction: (1) the channel adjacent to the wetland must be a relatively permanent body of water connected to traditional interstate navigable waters; and (2) the wetland adjacent to the channel must have a continuous surface connection with that water, making it difficult to determine where the water ends and the wetland begins.[90]

Justice Kennedy, in his concurring opinion, set forth a different test. He concluded that the proper evaluation of whether wetlands are waters of the United States depends on whether there is a "significant nexus" with traditional navigable waters.[91] Justice Kennedy explained that in his view, a significant nexus includes the ability to affect "the chemical, physical, and biological integrity" of the traditional navigable water.[92] He also stated that adjacent wetlands do not necessarily need a direct surface water connection and that the term "adjacent wetlands" does not necessarily extend to wetlands adjacent to tributaries.[93] Instead, he asserted, such determinations must be made on a case-by-case basis.[94]

Since the *Rapanos* decision there has been considerable confusion about whether the plurality test or Justice Kennedy's test governs and, if Justice Kennedy's test governs, how the case-by-case determinations of jurisdiction should be made. In 2007, in an attempt to clear up some of the confusion, the Corps and EPA issued a guidance document stating that the agencies believe that because there was no majority opinion in *Rapanos*, CWA jurisdiction over a water body exists if either the plurality's or Justice Kennedy's standard is met.[95] The agencies indicated that they would consider waters to fall within CWA jurisdiction if the waters would be considered jurisdictional under the reasoning of a majority, meaning five or more, of the justices.[96] Accordingly, in the view of the agencies, CWA jurisdiction exists in any of the following situations: (1) under the reasoning of *Riverside Bayview* for traditional navigable waters and their adjacent wetlands[97]; (2) under the *Rapanos* plurality test, for relatively permanent nonnavigable tributaries of traditional navigable waters and adjacent wetlands that have a continuous surface connection to relatively permanent nonnavigable tributaries; and (3) under the *Rapanos* Justice Kennedy concurrence on a case-by-case basis under the significant nexus standard.[98] The courts have also struggled with this issue, with some finding

---

84. *Id.* at 172 (emphasis added).
85. Bradford C. Mank, *The Murky Future of the Clean Water Act After SWANCC: Using A Hydrological Connection Approach to Saving the Clean Water Act*, 30 ECOLOGY L.Q. 811, 813 (2003) (quoting *SWANCC*, 531 U.S. at 167).
86. 547 U.S. 715 (2006).
87. *Id.* at 740.
88. *Id.* at 788 (Stevens, J., dissenting).
89. *Id.* at 724, 757, 759 (Kennedy, J., concurring).
90. *Id.* at 732-42.
91. *Id.* at 779 (Kennedy, J., concurring).
92. *Id.*
93. *Id.* at 784-85 (Kennedy, J., concurring).
94. *Id.*
95. U.S. EPA & ACOE, *Clean Water Act Jurisdiction Following the U.S. Supreme Court's Decision in* Rapanos v. United States *and* Carabell v. United States, U.S. EPA, June 5, 2007, *available at* http://www.epa.gov/owow/wetlands/pdf/RapanosGuidance6507.pdf.
96. *Id.*
97. Adjacency does not require a continuous surface connection between the wetland and traditional navigable water.
98. U.S. EPA & ACOE, *supra* note 95.

that CWA exists only where the Justice Kennedy test has been met, and others finding jurisdiction where either the Justice Kennedy or plurality test has been met.[99]

In 2011, EPA and the Corps released the "Draft Guidance on Identifying Waters Protected by the Clean Water Act"[100] to clarify how the agencies will interpret CWA jurisdiction in light of *Riverside Bayview*, *SWANCC*, and *Rapanos*.[101] In the document, the agencies explain that they:

> continue to believe, as expressed in previous guidance, that it is most consistent with the *Rapanos* decision to assert jurisdiction over waters that satisfy either the plurality or the Justice Kennedy standard, since a majority of justices would support jurisdiction under either standard. However, after careful review of these opinions, the agencies concluded that previous guidance did not make full use of the authority provided by the CWA to include waters within the scope of the Act, as interpreted by the Court. This draft guidance provides a more complete discussion of the agencies' interpretation, including of how waters with a "significant nexus" to traditional navigable waters or interstate waters are protected by the CWA. In addition, this guidance explains the legal basis for coverage of waters by the CWA in cases that were not addressed by the previous guidance (for example, interstate waters).[102]

Because most agricultural discharges are considered to be NPS and thus do not require NPDES permits, the reach of CWA jurisdiction is not generally an important issue when considering agricultural water pollution. However, as described in more detail below, the reach of CWA jurisdiction becomes very significant for agricultural discharges, such as those from concentrated animal feeding operations (CAFOs) that do require NPDES permits and for certain agricultural activities that require CWA §404 permits for discharge of dredged or fill material to wetlands.

2. Standards for Effluent Limitations and Water Quality

Once it is determined that an NPDES permit is required for a particular discharge, permits must contain limitations that both comply with specified technology-based standards and assure compliance with state water-quality standards. The NPDES permitting agency, either EPA or a state to which EPA has delegated the authority to implement the program,[103] must ensure that permitted discharges meet both types of standards.[104] Technology-based standards are established on an industry-wide basis to ensure that polluters are treating discharges to the extent feasible.[105] There are different types of technology-based standards that apply with varying levels of stringency and differences in the extent to which economic considerations are taken into account. The technology-based standard that applies to a particular discharge will depend on a number of factors, including the type of pollutants being discharged and whether the source is considered a new source or whether it existed prior to the implementation of the CWA.

Water quality standards are generally established by a state for each water body within its jurisdiction. Water quality standards are comprised of (1) determination of designated use; (2) water quality criteria; and (3) antidegradation standards.[106] Each state determines the designated use of each water body within the state.[107] For example, states may determine that a particular water body should be designated for drinking water, for shellfish harvesting, for fishing and swimming, for agricultural use, or for industrial use.

---

99. *See, e.g.*, United States v. Gerke, 464 F.3d 723 (7th Cir. 2006) (explaining the import of the various opinions in the *Rapanos* case and the likelihood of their significance in future cases). For a summary of CWA jurisdiction including the lower court cases decided since *Rapanos*, see Mark A. Chertok, *Federal Regulation of Wetlands*, SS042 ALI-ABA 965 (2011).
100. U.S. EPA, *Draft Guidance on Identifying Waters Protected by the Clean Water Act*, available at http://www.epa.gov/indian/pdf/wous_guidance_4-2011.pdf. *See also* U.S. EPA and Army Corps of Engineers Guidance Regarding Identification of Waters Protected by the Clean Water Act, 76 Fed. Reg. 24479-02 (May 2, 2011).
101. *Draft Guidance on Identifying Waters Protected by the Clean Water Act*, *supra* note 100.
102. *Id.*
103. *See supra* note 6 and accompanying text.
104. Ruhl, *supra* note 1, at 294. *See also* 33 U.S.C. §§1311-17, 1342.
105. Three different technology-based standards exist under the CWA. The applicable technology is determined based on the type of pollutant discharged and whether the discharging source is new or existing. BATTLE & LIPELES, *supra* note 48, at 167. "Best available technology" is the technology-based standard applied to existing sources of nonconventional and toxic pollutants. 33 U.S.C. §1311(b)(2)(A). "Best conventional technology" is applied to existing sources of conventional pollutants. *Id.* "Best available demonstrated control technology" is the technology-based standard applied to new sources of water pollutants. *See* 33 U.S.C. §1316(a)(1); *see also* BP Exploration & Oil, Inc. v. EPA, 66 F.3d 784, 789-90 (6th Cir. 1995) (explaining the different standards EPA uses to control pollutants under the CWA). For an extensive overview of water law, see BATTLE & LIPELES, *supra* note 48.
106. BATTLE & LIPELES, *supra* note 48, at 182.
107. *See id.* at 182-83.

Then, numerical, or in some cases narrative, criteria are established for particular pollutants to protect such uses.[108] In theory, all NPDES permits must ensure that these water quality criteria are met and the designated uses are protected. Water quality standards were intended to serve as a backstop to technology-based standards as a means of protecting designated uses in situations where technology-based standards were not sufficient to protect a designated use of a particular water body.[109]

An important, but until fairly recently long-ignored, standard for implementing water quality standards is the total maximum daily load (TMDL).[110] The CWA defines a TMDL as the sum of allocated loads of pollutants set at a level necessary to implement the applicable water quality standards, including waste-load allocations from point sources, load allocations from nonpoint sources, and natural background conditions.[111] The CWA further provides that a TMDL must contain a margin of safety and a consideration of "seasonal variations."[112] In other words, a TMDL can be described as the amount of a particular pollutant that a particular water body can assimilate without resulting in a violation of a water quality standard.

Once TMDLs are established by states and approved by EPA, the next challenge is the allocation of TMDLs among all point and nonpoint source dischargers, and the implementation of the TMDLs. For point source discharges, TMDLs are allocated and implemented through the NPDES permit program and may require pollution reductions beyond what would be required using only technology-based standards.[113] For nonpoint sources,[114] which include agricultural runoff as well as urban and suburban runoff, and which are not addressed by the NPDES permit program, the allocation and implementation of TMDLs is much more challenging. In most places, it is likely that a multifaceted, watershed-based approach will be needed. Components of such an approach would most likely have to include, among other things, some or all of the following pollution reduction approaches: state regulation of urban, suburban, and agricultural runoff; adoption of best management practices to reduce pollutant loadings in stormwater and agricultural discharges; retrofitting existing urban areas to treat stormwater; land acquisition programs to protect riparian areas that provide the function of filtering pollutants from runoff; wetland and water body restoration programs; and public education.[115] To meet TMDLs in areas with significant agriculture, a variety of regulatory and nonregulatory mechanisms will need to be imposed. These could include voluntary, incentive-based, or mandatory best-management practices, state permitting programs to limit agricultural discharges, farmer education and technical support, and preservation of riparian vegetated buffers along water bodies running through or near agricultural lands.

3.   Animal Feeding Operations and Concentrated Animal Feeding Operations

Animal Feeding Operations (AFOs) are a major component of the problems associated with addressing agricultural runoff. In general, AFOs are agricultural operations involving the farming of animals in confined situations or small land areas, where feed is brought to the animals in lieu of pasture grazing.[116] By definition, an AFO is "a lot or facility (other than an aquatic animal production facility)" that (1) confines and feeds animals for 45 or more days per year, and (2) does not have sustained vegetation during the growing season over any portion of the facility.[117] In contrast to most agricultural runoff, AFOs are not always exempt from the NPDES permit program. Specifically, EPA designates certain AFOs as "concentrated" Animal Feeding Operations (CAFOs), which allows for potential NPDES regulation, as CAFOs are expressly listed within the definition of "point source" in the CWA.[118]

---

108. *See id.* at 183-84.
109. *See id.* at 181-82.
110. For an historical perspective on TMDLs, see Oliver Houck, Clean Water Act TMDL Program: Law, Policy, and Implementation (Envt. L. Inst. 2002).
111. *See* 33 U.S.C. §1313(d)(1)(C); *see also* Ruhl, *supra* note 1, at 300-05 (discussing TMDLs).
112. 33 U.S.C. §1313(d)(1)(C).
113. *See* Battles & Lipeles, *supra* note 48, at 184.
114. The term "nonpoint source" is defined to mean any source of water pollution that does not meet the legal definition of "point source" as defined in §502(14) of the Clean Water Act. *See What Is Nonpoint Source Pollution?*, *supra* note 17. For the definition of "point source," see *supra* note 41 and accompanying text.
115. *See* Oliver A. Houck, *TMDLs III: A New Framework for the Clean Water Act's Ambient Standards Program*, 28 ELR 10415, 10423 (Aug. 1998).
116. 40 C.F.R. §122.23(b); U.S. EPA, *National Pollutant Discharge Elimination System (NPDES): Animal Feeding Operations*, http://cfpub.epa.gov/npdes/home.cfm?program_id=7 (last visited July 29, 2012).
117. *See* 40 C.F.R. §122.23(b).
118. 33 U.S.C. §1362(14); 40 C.F.R. §401.11(d).

As recognized by EPA, CAFO-generated manure and wastewater have the potential to contribute "pollutants, such as nitrogen and phosphorus, organic matter, sediments, pathogens, heavy metals, hormones and ammonia, to the environment."[119] Accordingly, such CAFO discharges (e.g., manure, litter, or process wastewater) do not qualify for an agricultural stormwater discharge exemption to point source regulation.[120] As regulated by EPA, an AFO becomes a CAFO based upon the actual number and type of animals at the operation.[121]

EPA, or an authorized state authority, also reserves the right to "designate any AFO as a CAFO upon determining that [the AFO] is a significant contributor of pollutants to waters of the United States."[122] Of the approximately 450,000 AFOs in the United States,[123] roughly 20% are considered to be CAFOs.[124] AFOs falling outside of the NPDES permit program (i.e., non-CAFOs) may still be regulated by state programs.[125]

The Natural Resources Conservation Service (NRCS) of the U.S. Department of Agriculture (USDA) and EPA encourage both AFO and CAFO owners and operators to take voluntary actions to minimize impacts.[126] Conservation plans known as comprehensive nutrient management plans (CNMPs) are required for CAFOs and encouraged for AFOs, and are designed to help provide AFO operators with a plan to control soil erosion and eliminate polluted runoff by managing manure and organic by-products through conservation practices and management activities.[127]

In 2003, EPA revised the CWA regulations for CAFOs by expanding the number of CAFOs needing NPDES permits to operate in compliance with the statute, and by adding requirements to the land application of manure by CAFOs.[128] Following legal challenges to the rule, the U.S. Court of Appeals for the Second Circuit in 2005 issued *Waterkeeper Alliance v. EPA*,[129] which directed EPA to (1) lift the requirement that all CAFOs apply for NDPES permits, and to (2) require nutrient management plans (NMPs) for CAFO permit applications.[130] In response to the *Waterkeeper* decision, EPA issued a 2008 rule that now requires only discharging CAFOs, as well as CAFOs proposing to discharge, to apply for NPDES permits.[131] Specifically, the 2008 rule calls upon the CAFO owner or operator to decide whether the CAFO does or will discharge from its production area or land application area.[132] According to EPA, this case-by-case analysis is to be "based on an objective assessment of the CAFO's design, construction, operation, and maintenance."[133]

Another substantive change in the 2008 rule is that it adds new NMP requirements for permitted CAFOs.[134] In addition to the requirement for CAFOs to develop and implement NMPs under the 2003 rule, the 2008 rule requires them to submit NMPs with the NPDES permit application so that such NMPs become enforceable elements of the permit (for both individual and general permits).[135] The application review also allows an opportunity for public review and comment on the proposed plans.[136] In assessing the 2008 rule's impacts to CAFOs and the environment, EPA claims that the rule effects only minor changes

---

119. U.S. EPA, *Region 7 Concentrated Animal Feeding Operations (CAFOs): How Do CAFOs Impact the Environment*, http://www.epa.gov/region7/water/cafo/cafo_impact_environment.htm (last visited July 29, 2012).
120. 40 C.F.R. §122.23(e); U.S. EPA, *Implementation Guidance on CAFO Regulations—CAFOs That Discharge or Are Proposing to Discharge*, 1-2 (2010), EPA-833-R-10-006, *available at* http://www.epa.gov/npdes/pubs/cafo_implementation_guidance.pdf.
121. 40 C.F.R. §122.23. For a brief summary of how EPA regulates CAFOs by size, see U.S. EPA, *Regulatory Definitions of Large CAFOs, Medium CAFOs, and Small CAFOs*, http://www.epa.gov/npdes/pubs/sector_table.pdf.
122. 40 C.F.R. §122.23(c).
123. U.S. EPA, *Agriculture: Animal Feeding Operations*, http://www.epa.gov/agriculture/anafoidx.html (last visited July 29, 2012).
124. U.S. EPA, *Region 7 Concentrated Animal Feeding Operations (CAFOs): What Is a CAFO?*, http://www.epa.gov/region7/water/cafo/ (last visited Jan. 29, 2013).
125. U.S. EPA, Agricultural Counselor Office of the Administrator, *Summary of Major Existing EPA Laws and Programs That Could Affect Agricultural Producers*, June 2007, at 5, *available at* http://www.epa.gov/agriculture/agmatrix.pdf.
126. NRCS, *Animal Feeding Operations (AFO) and Confined Animal Feeding Operations (CAFO)*, http://www.nrcs.usda.gov/wps/portal/nrcs/detailfull/national/programs/?&cid=nrcsdev11_000330 (last visited Nov. 29, 2012).
127. *Id.*
128. U.S. EPA, *Concentrated Animal Feeding Operations Final Rulemaking—Fact Sheet* (2008), *available at* http://www.epa.gov/npdes/pubs/cafo_final_rule2008_fs.pdf [hereinafter *CAFO Fact Sheet*].
129. Waterkeeper Alliance v. EPA, 399 F.3d 486 (2d Cir. 2005).
130. *CAFO Fact Sheet, supra* note 128, at 1.
131. *Id.*
132. *Id.*
133. *Id.*
134. *Id.*
135. *Id.* at 1-2.
136. *Id* at 2.

in the administrative burden for discharging CAFOs, and preserves the beneficial environmental benefits of the 2003 rule by leaving the technical requirements for discharging CAFOs intact.[137]

Several agricultural interests challenged EPA's 2008 rule, and on March 15, 2011 the U.S. Court of Appeals for the Fifth Circuit issued *National Pork Producers Council v. EPA*, granting the agricultural interests' petitions in part, denying them in part, and dismissing them in part.[138] Of greatest significance, the court held that EPA had authority to impose the duty to apply for NPDES permits on CAFOs that were discharging pollutants, but lacked the authority to issue a regulation requiring CAFOs that merely proposed to discharge pollutants to apply for an NPDES permit. In response to the decision in *National Pork Producers Council*, on July 30, 2012, EPA promulgated a final rule amending its regulations to eliminate the requirements that and owner or operator of a CAFO "proposing to discharge" must apply for an NPDES permit.[139]

4. Aquaculture

Discharges into aquaculture projects are subject to the NPDES permit program[140] as point source pollutants under §318 of the CWA.[141] An aquaculture project is a "defined managed water area which uses discharges of pollutants into that designated area for the maintenance or production of harvestable freshwater, estuarine, or marine plants or animals."[142] Fish farms and hatcheries, as another form of aquaculture regulation within the CWA, are also often subject to the NPDES permit program as a point source discharge if they fall within the definition of a "concentrated aquatic animal production facility" (CAAP).[143] EPA defines a CAAP as an aquatic feeding operation that either (1) produces more than 9,090 harvest weight kilograms of cold water fish (e.g., trout, salmon), or (2) produces more than 45,454 harvest weight kilograms (about 100,000 pounds) of warm water fish (e.g., catfish, sunfish, minnows).[144]

Similar to CAFO regulations, EPA reserves the right to proceed on a case-by-case basis in designating CAAPs. Specifically, EPA has the authority to designate both warm and cold water aquatic animal production facilities as CAAPs if they are determined to be a "significant contributor of pollution to waters of the United States."[145] In 2004, EPA promulgated effluent limitation guidelines for CAAPs in order to help reduce discharges of conventional (e.g., suspended solid) and nonconventional (e.g., nutrient) pollution.[146]

5. Wetlands Regulation Under CWA §404

Section 404 of the CWA, jointly administered by EPA and the Corps, requires individuals to obtain a permit from the Corps prior to discharging dredged or fill material into the waters of the United States. The Corps defines "a discharge of dredged or fill material" as involving "the physical placement of soil, sand, gravel, dredged material or other such materials into the waters of the United States."[147] Thus, wetlands that fall within the definition of "waters of the United States" are regulated under the §404 program.[148] The CWA regulations define "wetlands" as:

---

137. *Id.*
138. 635 F.3d 738 (2011).
139. National Pollutant Discharge Elimination System Permit Regulation for Concentrated Animal Feeding operations: Removal of Vacated Elements in Response to 2011 Court Decision, 77 Fed. Reg. 44494 (July 30, 2012).
140. 40 C.F.R. §122.25; *see also* 40 C.F.R. §125(B).
141. 33 U.S.C. §1328; 40 C.F.R. §122.3(e).
142. 40 C.F.R. §122.25(b)(1).
143. 40 C.F.R. §122.3(e).
144. 40 C.F.R. §122.24; 40 C.F.R. §122 app. C.
145. 40 C.F.R. §122.24(c)(1). In making this determination, EPA examines the capacity of the facility, "[t]he location and quality of the receiving waters of the United States," and "[t]he quantity and nature of the pollutants reaching the waters of the United States." *Id.*
146. U.S. EPA, *Agriculture: Aquaculture Operations—Laws, Regulations, Policies, and Guidance*, http://www.epa.gov/agriculture/anaqulaw.html (last visited July 29, 2012).
147. U.S. EPA & U.S. Army Corps of Engineers, Memorandum: Clean Water Act Section 404 Regulatory Program and Agricultural Activities (May 3, 1990), *available at* http://water.epa.gov/lawsregs/guidance/wetlands/cwaag.cfm.
148. U.S. EPA, *Agriculture: Clean Water Act (CWA)*, http://www.epa.gov/agriculture/lcwa.html#Wetlands and Agriculture (last visited July 29, 2012).

those areas that are inundated or saturated by surface or ground water at a frequency and duration sufficient to support, and that under normal circumstances do support, a prevalence of vegetation typically adapted for life in saturated soil conditions. Wetlands generally include swamps, marshes, bogs, and similar areas.[149]

Thus, a §404 permit generally is required for any discharge of dredged or fill material into wetlands that fall within the above definition and fall within the jurisdiction of the CWA.

In reality, most agricultural operations do not require a §404 permit (either because "they do *not* occur in wetlands or other waters of the United States or do *not* involve dredged or fill material . . . .").[150] Even where an agricultural operation does occur on wetlands, certain agricultural exemptions may apply.

Section 404 authorizes the Corps to issue general permits for certain activities involving the discharge of dredged or fill material.[151] However, because general permits are issued on a state, regional, or nationwide basis, the list of such permits often varies by geographic location.[152] According to the Corps, "[d]ischarges authorized under a General Permit may proceed without applying to the Corps for an individual permit."[153] However, conditions associated with a general permit often require notice to the Corps prior to the discharge.[154]

As it relates to agriculture, §404(f) of the CWA provides, inter alia, certain agricultural exemptions for discharges of dredged or fill material associated with "normal farming and harvesting activities that are part of established, *ongoing* farming or forestry operations."[155] Generally, ongoing farming activities such as plowing, seeding, and cultivating are exempt from the §404 program, while wetlands converted to a "new use" are not.[156] Thus, an activity involving a discharge of dredged or fill material, and constituting a "new use" of the wetland so as to impair regulated waters is not exempt.[157] An example of an agricultural activity that would not qualify for the agricultural exemption is the "deep ripping" that was the subject of *Borden Ranch v. U.S. Army Corps of Eng'rs* in the Ninth Circuit.[158] As explained by the Ninth Circuit, deep ripping is a "procedure . . . in which four-to seven-foot long metal prongs are dragged through the soil behind a tractor or a bulldozer. The ripper gouges through the restrictive layer, disgorging soil that is then dragged behind the ripper."[159] In addressing the authority of EPA and the Corps over this form of agricultural activity when it occurs in wetlands, the court held that deep ripping brings the land "into a use to which it was not previously subject,"[160] and is consequently disqualified as a "normal farming" activity under the agricultural exemptions of the CWA.[161]

Recognizing the potential confusion regarding CWA §404(f) exemptions, EPA and the Corps stressed the limitations of the exemptions:

> It should be emphasized that the use of Section 404(f) exemptions does not affect Section 404 jurisdiction. For example, the fact that an activity in wetlands is exempted as normal farming practices does not authorize the filling of the wetland for the construction of buildings without a Section 404 permit. Similarly, a Section 404 permit would be required for the discharge of dredged or fill material associated with draining a wetland and converting it to dry land.[162]

Further, the following examples, as proposed by EPA, provide contextual understanding of ongoing operations (exempt) versus new uses (nonexempt) within the §404 program:

> Activities that bring a wetland into farm production where the wetland has not previously been used for farming are not considered part of an established operation, and therefore *require a permit*.

---

149. 33 C.F.R. §328.3(b).
150. U.S. EPA, *Section 404 and Swampbuster: Wetlands on Agricultural Lands*, http://www.epa.gov/owow/wetlands/facts/fact19.html (last visited July 29, 2012) (emphasis in original).
151. Memorandum: Clean Water Act Section 404 Regulatory Program and Agricultural Activities, *supra* note 147.
152. *Id.*
153. *Id.*
154. *Id.*
155. *Agriculture: Clean Water Act (CWA)*, *supra* note 148 (emphasis added).
156. *Id.*
157. *Id.*
158. Borden Ranch v. U.S. Army Corps of Eng'rs, 261 F.3d 810 (9th Cir. 2001), *aff'd* 537 U.S. 99, 123 S. Ct. 599, 154 L. Ed. 2d 508 (2002).
159. *Id.* at 812.
160. 33 U.S.C. §1344(f)(2).
161. *Borden Ranch*, 261 F.3d at 815.
162. Memorandum: Clean Water Act Section 404 Regulatory Program and Agricultural Activities, *supra* note 147.

Introduction of a new cultivation technique such as disking between crop rows for weed control may be a new farming activity, but because the farm operation is ongoing, the activity is *exempt* from permit requirements under Section 404.

Planting different crops as part of an established rotation, such as soybeans to rice, is *exempt*.

Discharges associated with ongoing rotations of rice and crawfish production are also *exempt*.[163]

Although the §404(f) exemptions prolong many of the negative effects of agricultural runoff, the §404 program has at least resulted in a reduction of the rate in which wetlands are converted for agricultural purposes.[164] Some of this success can be directly attributed to the Swampbuster provisions of the Food Security Act of 1985[165] (FSA), which was designed to be consistent with the §404 program.

While EPA and the Corps jointly administer the §404 program, the NRCS oversees the Swampbuster program and determines whether agricultural lands are subject to the FSA.[166] Similar to the §404 program, the Swampbuster program "generally allows the continuation of most farming practices so long as wetlands are not converted or wetland drainage increased."[167] This incentive-based program functions by encouraging participants with wetlands subject to the Swampbuster program to comply with certain wetland conservation provisions in exchange for federal farm program benefits.[168] Therefore, farmers who convert or modify (e.g., drain, dredge, level,) wetlands to make agricultural commodity production possible are ineligible for program benefits.[169]

Prior to 2005 an existing memorandum of agreement (MOA) between USDA, the U.S. Department of the Interior, the Corps, and EPA simplified wetland delineations for farmers by promoting consistency between the CWA and the FSA jurisdictional determinations for CWA §404 and the FSA Swampbuster.[170] However, due to inconsistencies between the two programs resulting from amendments to the FSA in 1996 and 2002, the Corps and USDA withdrew from the MOA in 2005.[171] That same year, the NRCS and the Corps promulgated new joint guidance concerning wetland determinations in administering the CWA §404 program and the FSA Swampbuster program. Within the guidance, the agencies provided the following explanation:

> Because of the differences now existing between CWA and FSA on the jurisdictional status of certain wetlands (e.g., prior converted or isolated wetlands may be regulated by one agency but not the other), it is frequently impossible for one lead agency to make determinations that are valid for the administration of both laws.[172]

According to the guidance, although each agency is now responsible for its own wetland determinations (e.g., NRCS conducts certified wetland determinations in administering the FSA; the Corps conducts jurisdictional determinations for the §404 program),[173] the agencies have agreed to continue to work together to provide consistent wetland determinations where possible.[174]

In light of the jurisdictional differences over wetland regulation between the CWA and the FSA, "prior converted croplands" (PCs)[175] have become a topic for debate and evolving policy. In general, PCs, as wet-

---

163. U.S. EPA, *Exemptions to Permit Requirements*, http://www.epa.gov/owow/wetlands/facts/fact20.html (last visited July 29, 2012).
164. *Section 404 and Swampbuster: Wetlands on Agricultural Lands*, *supra* note 150.
165. 7 U.S.C. §1631.
166. NRCS, *Wetland Conservation Program (SWAMPBUSTER)* (Mar. 27, 2003), *available at* http://www.in.nrcs.usda.gov/programs/Training%20Guide/Cpgl%20GLCI/Wetland.pdf.
167. *Section 404 and Swampbuster: Wetlands on Agricultural Lands*, *supra* note 150.
168. *Wetland Conservation Program (SWAMPBUSTER)*, *supra* note 166.
169. *Agriculture: Clean Water Act (CWA)*, *supra* note 148.
170. Memorandum of Agreement Among the Department of Agriculture, the Environmental Protection Agency, the Department of the Interior, and the Department of the Army Concerning the Delineation of Wetlands for Purposes of Section 404 of the Clean Water Act and Subtitle B of the Food Security Act (1994), *available at* http://www.lb5.uscourts.gov/ArchivedURLs/Files/06-30917(2).pdf.
171. NRCS, Guidance on Conducting Wetland Determinations for the Food Security Act and Section 404 of the Clean Water Act: Key Points (Feb. 28, 2005), *available at* http://www.nrcs.usda.gov/Internet/FSE_DOCUMENTS/nrcs143_007868.pdf.
172. Joint Guidance From the Natural Resources Conservation Service (NRCS) and the Army Corps of Engineers (COE) Concerning Wetland Determinations for the Clean Water Act and the Food Security Act of 1985, *available at* http://www.nrcs.usda.gov/Internet/FSE_DOCUMENTS/nrcs143_007869.pdf. For the original 1994 MOA document (e.g., Memorandum of Agreement Among the Department of the Agriculture, the Environmental Protection Agency, the Department of the Interior, and the Department of the Army Concerning the Delineation of Wetlands for Purposes of Section 404 of the Clean Water Act and Subtitle B of the Food Security Act), see *supra* note 170.
173. Guidance on Conducting Wetland Determinations for the Food Security Act and Section 404 of the Clean Water Act, *supra* note 171.
174. *Id.*
175. EPA and the Corps define "prior converted croplands" as "areas that, prior to December 23, 1985, were drained or otherwise manipulated for the purpose, or having the effect, of making production of a commodity crop possible." Clean Water Act Regulatory Programs, 58 Fed. Reg.

lands converted to agriculture prior to the passage of the FSA in 1985, are exempt from regulation under both the §404 program and the Swampbuster program.[176] However, although most PCs have been extensively drained and altered, it is possible that a PC may lose CWA exemption where the PC is abandoned (i.e., cropland production ceases and the land reverts to a wetland state).[177]

Until recently, only abandonment of a PC would result in the loss of the CWA exemption (and compel the Corps to reassert jurisdiction). It now appears, however, that the Corps is moving towards a second exception (in addition to abandonment) by asserting jurisdiction over PCs that are converted to nonagricultural use. This was the issue in 2010 in *New Hope Power Company v. U.S. Army Corps of Engineers*[178] after the Jacksonville office of the Corps announced that "PCs converted to non-agricultural use would become subject to Corps regulation." According to the *New Hope* court, this announcement, known as the Stockton rules, changed the longstanding Corps policy that PCs converted to nonagricultural use were exempt from Corps jurisdiction.[179] Ultimately, the *New Hope* court struck down the Stockton rules, finding them "procedurally improper because no notice-and-comment procedures were used" (as required by the Administrative Procedure Act) prior to implementing the policy.[180] It is uncertain whether the Corps will undergo notice-and-comment rulemaking to formally promulgate this new policy. Even so, the NRCS is advising that landowners intending to develop their PC or discontinue agricultural production should contact the Corps for a CWA jurisdictional determination.[181]

Conclusion

The Clean Water Act provides a comprehensive regulatory scheme for many discharges of pollutants to waters of the United States. Through the primarily regulatory NPDES permitting program, significant improvements have been made to the quality of the country's water bodies. However, the NPDES permitting program only applies to point sources discharges, thus most agricultural discharges are not subject to permitting or other federal regulatory control. Nonpoint sources, including those from agriculture, remain the most significant water quality challenge facing the nation.[182] Moreover, the CWA's exemption from §404 permitting for normal farming practices continues to allow many wetlands to be degraded by agricultural activities.

Because the CWA does not provide direct federal authority for regulating many agricultural sources of water pollution and wetlands degradation, the responsibility for addressing water quality degradation from agricultural activities has fallen largely to the states. To date, most programs designed to address agricultural water pollution have been voluntary or incentive-based programs designed to encourage farmers to implement best-management practices. These programs have been only minimally successful, and agricultural pollution continues to be one of the most significant sources of water quality degradation in the United States, meaning that there is a need for a more comprehensive regulatory system to address the water impacts of farming.

45008-01, 45031 (Aug. 25, 1993). The NRCS defines "prior converted cropland" as "a converted wetland where the conversion occurred prior to December 23, 1985, an agricultural commodity had been produced at least once before December 23, 1985, and as of December 23, 1985, the converted wetland did not support woody vegetation . . . ." 7 C.F.R. §12.2.

176. *See* 33 C.F.R. §328.3(a)(8) ("Waters of the United States do not include prior converted cropland. Notwithstanding the determination of an area's status as prior converted cropland by any other Federal agency, for the purposes of the Clean Water Act, the final authority regarding Clean Water Act jurisdiction remains with EPA."); NRCS, *Fact Sheet—Certified Wetland Determinations, available at* http://www.nrcs.usda.gov/Internet/FSE_DOCUMENTS/stelprdb1043614.pdf.

177. *See* 33 C.F.R. §328.3; Guidance on Conducting Wetland Determinations for the Food Security Act and Section 404 of the Clean Water Act, *supra* note 171.

178. New Hope Power Co., v. U.S. Army Corps of Eng'rs, No. 10-2277, 2010 U.S. Dist. LEXIS 103231 (S.D. Fla. Sept. 29, 2010).

179. *Id.* at *24.

180. *Id.* at *29.

181. *Wetland Conservation Program (SWAMPBUSTER), supra* note 166.

182. *See* Ruhl, *supra* note 1, at 288 (stating that nonpoint source pollution is the most significant form of pollution affecting water bodies in 33 states).

Chapter 10
Agriculture and the Clean Air Act
Teresa B. Clemmer

As livestock production has become more concentrated, factory farming has generated vast quantities of air pollution and caused severe human health and environmental impacts in surrounding communities.[1] Factory farms emit several different types of toxic and conventional air pollution, and these are far more dangerous than many people may realize.[2] Despite the serious health and environmental impacts resulting from factory farm air pollution and despite the clear applicability of federal air pollution control and release-reporting obligations, the U.S. Environmental Protection Agency (EPA) has shown a reluctance to hold the factory farm industry to the same standards as other industries regulated under the Clean Air Act and other federal statutes.[3]

EPA has even taken the unusual step of entering into an industrywide consent agreement effectively excusing the industry from liability associated with its longstanding noncompliance with federal permitting and release reporting requirements.[4] EPA's industry-friendly approach has discouraged many states from vigorously applying and enforcing the Clean Air Act and other laws in the factory farm context as well.[5] Some states, however, have recognized the air pollution threats associated with factory farms, and they have taken some important first steps to address the most egregious problems.[6] Moreover, citizen groups frustrated with the lack of regulation for this highly polluting industry have repeatedly called upon EPA to implement and enforce existing air pollution permitting and release reporting requirements and to initiate rulemakings that would establish appropriate regulatory tools.[7]

EPA's decision not to apply traditional Clean Air Act permitting requirements to factory farms is now converging with the ongoing controversy over the applicability of these same permitting requirements to greenhouse gases, such as methane and nitrous oxide, which are emitted in large quantities by factory farms.[8] Here again, some states are taking steps to address greenhouse gas emissions from factory farms,[9] but EPA leadership in implementing and enforcing applicable Clean Air Act requirements will be important in broadening the scope of these efforts and expediting the effort needed to meet the challenges ahead.

---

1. *See supra* Chapter 5.
2. *See infra* Section A.1. (discussing health and environmental impacts of particulate matter (PM) and ozone pollution); and Sections A.4. and B. (discussing health hazards posed by ammonia and hydrogen sulfide pollution).
3. The factory farm industry has lobbied aggressively against governmental efforts to protect public health and welfare through air pollution regulation. *See, e.g.,* PEW COMMISSION ON INDUSTRIAL FARM ANIMAL PRODUCTION, PUTTING MEAT ON THE TABLE: INDUSTRIAL FARM ANIMAL PRODUCTION IN AMERICA viii, *available at* http://www.ncifap.org/bin/e/j/PCIFAPFin.pdf; MICHELE M. MERKEL, EPA AND STATE FAILURES TO REGULATE CAFOS UNDER FEDERAL ENVIRONMENTAL LAWS 5-6 (2006), *available at* http://www.environmentalintegrity.org/pdf/publications/EPA_State_Failures_Regulate_CAFO.pdf.
4. *See infra* Section C.2.
5. *See infra* Section D.2.
6. *See infra* Sections D.2-D.4.
7. *See infra* Section A.1. (discussing citizen petition filed in 2011 urging EPA to regulate ammonia as a criteria air pollutant under the Clean Air Act); Sections A.3. and E.2. (discussing citizen petition filed in 2009 urging EPA to regulate factory farm air pollution on an industrywide basis under the NSPS program of the Clean Air Act); Section A.4. (discussing citizen petition filed in 2009 urging EPA to regulate hydrogen sulfide as a hazardous air pollutant under the Clean Air Act); Section C.2. (discussing citizen group comments and subsequent lawsuit objecting to EPA's entry into an industrywide consent agreement in 2005); and Section E.2. (discussing citizen petition filed in 2008 and subsequent lawsuit seeking to compel EPA to regulate greenhouse gas emissions from non-road engines, including those used in agricultural operations).
8. *See infra* Section E.1.
9. *See infra* Section E.3.

A. Clean Air Act

The purpose of the Clean Air Act is "to protect and enhance the quality of the Nation's air resources so as to promote the public health and welfare and the productive capacity of its population."[10] Because the Clean Air Act represents the increasing involvement of the federal government in areas traditionally regulated by the states—public health, nuisance, industrial facility siting, transportation, and land use—the U.S. Congress has relied on the principle of "cooperative federalism" to preserve state sovereignty as much as possible and thereby minimize state opposition to federal participation in air pollution control.[11] As a result, the concept of a federal-state partnership has been at the heart of the modern Clean Air Act since it was enacted in 1970.[12] With this in mind, Congress designed the Clean Air Act in a bifurcated fashion: Congress assigned the federal government the duty to set ambient air quality standards,[13] and it gave states the responsibility to develop laws and regulations sufficient to achieve and maintain these overarching standards.[14] Both EPA and states have been granted responsibility for enforcing applicable air quality requirements.[15]

1. National Ambient Air Quality Standards and State Implementation Plans

Under the Clean Air Act, EPA must designate as "criteria" pollutants those ubiquitous air pollutants generated by "numerous or diverse mobile or stationary sources" that, in EPA's judgment, "may reasonably be anticipated to endanger public health or welfare."[16] To date, EPA has designated six criteria air pollutants: ozone, particulate matter (PM), nitrogen oxides (NO_x), sulfur dioxide, carbon monoxide, and lead.[17] This designation, in turn, triggered a duty for EPA to establish a national ambient air quality standard (NAAQS) for each criteria pollutant, and to review and, if appropriate, revise such standards at least once every five years.[18] Primary NAAQS are set at levels that EPA deems "requisite to protect the public health" with an "adequate margin of safety."[19] Secondary NAAQS are set at levels that EPA deems "requisite to protect the public welfare," including environmental, recreational, economic, aesthetic, and other considerations, "from any known or anticipated adverse effects associated with the presence of such air pollutant in the ambient air."[20] After the promulgation or revision of a NAAQS, each state must prepare and submit to EPA for approval a state implementation plan (SIP) that "provides for implementation, maintenance, and enforcement" of each NAAQS within each air quality control region of the state.[21] The Clean Air Act sets forth detailed requirements for SIPs, and EPA relies on these criteria to assess their adequacy.[22]

Presently, the NAAQS standards most relevant for agricultural operations are those for fine and coarse PM ($PM_{2.5}$ and PM_{10}), as well as ozone.[23] Scientific studies have shown that PM pollution can cause and contribute to a variety of human health problems, including decreased lung function, asthma, chronic bronchitis, irregular heartbeat, heart attacks, and premature death.[24] PM is also responsible for haze and visibility impairment in many areas, and it contributes to acid rain.[25] Ozone is the primary constituent of

---

10. 42 U.S.C. §7401(b)(1).
11. *See generally* Robert Glicksman, *From Cooperative to Inoperative Federalism: The Perverse Mutation of Environmental Law and Policy*, 41 Wake Forest L. Rev. 719 (Jan. 2011).
12. *See generally id.*
13. *See* 42 U.S.C. §7409.
14. *See id.* §7410.
15. *See id.* §§7410, 7413.
16. *Id.* §7408(a)(1).
17. 40 C.F.R. §§50.1-50.17.
18. 42 U.S.C. §7409(a), (b), (d)(2)(B).
19. *Id.* §7409(b)(1).
20. *Id.* §7409(b)(2).
21. *Id.* §7410(a)(1). For purposes of developing and implementing SIPs, EPA designates one or more "air quality control regions" within each state. *See id.* §7407. Commonly, these coincide with major metropolitan areas or county boundaries, but there are examples of smaller and larger air quality control regions throughout the nation.
22. *Id.* §7410(a)(2)-(6) et seq.
23. 40 C.F.R. §50.6 (PM_{10}), §50.7 ($PM_{2.5}$), §50.9 (one-hour ozone standard), §50.10 (eight-hour ozone standard), §50.13 ($PM_{2.5}$), §50.15 (eight-hour ozone standard).
24. *See, e.g.,* U.S. EPA, Policy Assessment for the Review of the Particulate Matter National Ambient Air Quality Standards (Apr. 2011) [hereinafter 2011 Policy Assessment for PM NAAQS], *available at* http://www.epa.gov/ttn/naaqs/standards/pm/data/20110419pmpafinal.pdf; National Ambient Air Quality Standards for Particulate Matter, Final Rule, 71 Fed. Reg. 61144 (Oct. 17, 2006) [hereinafter 2006 PM NAAQS].
25. *See* 2011 Policy Assessment for PM NAAQS and 2006 PM NAAQS, *supra* note 24.

smog.[26] It is a lung-searing chemical that can cause and contribute to a variety of health problems, including chest pain, difficulty breathing, bronchitis, emphysema, asthma, inflammation and scarring of lung tissue, and premature death.[27] Ozone also damages vegetation, leading to decreased agricultural yields and degradation of forested areas.[28] EPA has recently increased the stringency of the NAAQS for both PM and ozone, and it is in the process of reviewing them to determine whether they should be strengthened even further.[29] Very few states include controls on PM or ozone emissions from factory farms in their SIPs.[30] Instead, most states choose to place the burden primarily on other industries to control these emissions.[31] If EPA does strengthen the NAAQS for PM or ozone, these new NAAQS standards will not directly impose any new requirements on agricultural facilities. Nevertheless, some states may find it necessary to start regulating, or to strengthen their requirements for, agricultural operations in their SIPs. This would help reduce the serious health effects of PM and ozone across the nation, particularly in rural areas.

EPA has not yet designated as criteria pollutants either of the other two key factory farm pollutants—hydrogen sulfide and ammonia.[32] In 2011, however, citizen groups petitioned EPA to designate ammonia as a new criteria pollutant, based largely on its emissions from agricultural operations.[33] If EPA does so, it would then be required to establish a NAAQS for ammonia as well. This would represent an important first step in protecting public health, particularly in rural communities that are increasingly becoming inundated with factory farm air pollution. An ammonia NAAQS would lead to the designation of attainment and nonattainment areas and the installation of ambient air monitoring equipment, which would improve our understanding of how many communities are breathing unhealthy air and to what degree. This enhanced understanding would, in turn, provide a basis for the promulgation of SIP provisions designed to achieve and maintain the NAAQS for ammonia, thereby protecting communities from the serious health effects of this pollutant.[34]

New and strengthened NAAQS and SIP provisions will be important in addressing the ongoing public health threats from factory farm air pollution, but it is unlikely that they alone will be sufficient. Despite Congress' lofty objectives, the division of responsibility between the federal government and the states in the Clean Air Act has proven challenging and frustrating in day-to-day implementation. States have consistently failed to attain NAAQS standards by the mandatory deadlines, and the program has faced long delays in EPA's review and revision of NAAQS standards, state revisions of SIPs, and protracted litigation brought by industry and citizen groups.[35] Recognizing these problems, Congress amended the Clean Air Act in 1977 and 1990 to shore up the NAAQS-SIP program.[36] In doing so, Congress has employed two main strategies: (1) requiring states to implement permitting programs designed to enhance compliance

---

26. *See, e.g.,* National Ambient Air Quality Standards for Ozone, Proposed Rule 75 Fed. Reg. 2938 (Jan. 19, 2010) [hereinafter 2010 Proposed Ozone NAAQS]; National Ambient Air Quality Standards for Ozone, Final Rule, 73 Fed. Reg. 16436 (Mar. 27, 2008) [hereinafter 2008 Ozone NAAQS].
27. *See* 2010 Proposed Ozone NAAQS and 2008 Ozone NAAQS, *supra* note 26.
28. *See id.*
29. For more information about the status of the PM NAAQS, see 2011 Policy Assessment for PM NAAQS, *supra* note 24; 2006 PM NAAQS, *supra* note 24; and U.S. EPA, *Coarse Particulate Matter (PM_{10}) Standards and Agriculture Fact Sheet*, http://www.epa.gov/air/particlepollution/agriculture.html (last visited Nov. 28, 2012). For more information about the status of the ozone NAAQS, see 2010 Proposed Ozone NAAQS, *supra* note 26; 2008 Ozone NAAQS, *supra* note 26; U.S. EPA, *Fact Sheet: EPA to Reconsider Ozone Pollution Standards,* http://www.epa.gov/airquality/ozonepollution/pdfs/O3_Reconsideration_FACT%20SHEET_091609.pdf (last visited Oct. 12, 2011); U.S. EPA, *Clean Air Scientific Advisory Committee (CASAC) Response to Charge Questions on the Reconsideration of the 2008 Ozone National Ambient Air Quality Standards* (Mar. 30, 2011), *available at* http://yosemite.epa.gov/sab/sabproduct.nsf/0/F08BEB48C1139E2A8525785E006909AC/$File/EPA-CASAC-11-004-unsigned+.pdf; and Statement by the President on the Ozone National Ambient Air Quality Standards (Sept. 2, 2011), *available at* http://www.whitehouse.gov/the-press-office/2011/09/02/statement-president-ozone-national-ambient-air-quality-standards (last visited Nov. 28, 2012).
30. *See generally* Jody M. Endres & Margaret Rosso Grossman, *Air Emissions From Animal Feeding Operations: Can State Rules Help?*, 13 Penn St. Envtl. L. Rev. 1 (2004) (showing that state SIP rules focus primarily on hydrogen sulfide and odors).
31. *See, e.g.,* EPA, *Coarse Particulate Matter (PM_{10}) Standards and Agriculture Fact Sheet* (June 14, 2012), http://www.epa.gov/air/particlepollution/agriculture.html ("The vast majority of states have not required the agriculture industry to take any actions that require PM_{10} emission reductions; focusing their efforts to reduce PM_{10} on sources such as industrial processes, and construction and demolition.").
32. The health hazards associated with these pollutants will be discussed in Sections A.4. and B. below.
33. Environmental Integrity Project (EIP) et al., Petition for the Regulation of Ammonia as a Criteria Pollutant Under Clean Air Act §§108 and 109, at 1 (Apr. 2011), *available at* http://www.environmentalintegrity.org/documents/PetitiontoListAmmoniaasaCleanAirActCriteriaPollutant.pdf [hereinafter Ammonia Petition].
34. *See infra* Section A.4.
35. *See generally* Julie R. Domike & Alec C. Zacaroli, The Clean Air Act Handbook, ch. 3 (3d ed. 2011).
36. *See generally id.* chs. 3, 6, 9, and 15.

with and enforcement of requirements applicable to stationary sources, and (2) creating new direct federal standard-setting authorities as overlays to the SIP approach.

2. New Source Review and Title V Permits

The Clean Air Act requires two key types of permits—preconstruction permits and operating permits. Before commencing construction, an entity planning to construct a new "major source" or make a "major modification" to an existing major source must apply to the relevant permitting authority (usually the state environmental agency) for a preconstruction permit.[37] In areas that are already achieving the NAAQS, these are commonly referred to as PSD permits because they are part of a program designed to "prevent significant deterioration" of air quality in attainment areas.[38] Major sources include (1) certain listed types of sources that emit 100 tons per year or more of a pollutant subject to regulation under the Clean Air Act, and (2) all other nonlisted sources that emit 250 tons per year or more of a regulated pollutant.[39]

To obtain a PSD permit, the major source must undergo a comprehensive new source review (NSR) process. The NSR process results in, among other things, the establishment of a technology-based emission standard known as the "best available control technology" (BACT) standard.[40] The NSR program is not limited to criteria pollutants. A preconstruction permit must include BACT standards for all pollutants subject to regulation under the Clean Air Act, i.e., those which have been defined as "regulated NSR pollutants" under EPA's regulations.[41] These include "designated" pollutants regulated under the New Source Performance Standard (NSPS) program, such as hydrogen sulfide,[42] as well as pollutants regulated under the mobile source program, such as methane, nitrous oxide, and other greenhouse gases.[43] It also includes ammonia, since it is regulated as a precursor to $PM_{2.5}$, as well as volatile organic compounds (VOCs), since they are regulated as precursors to ozone.[44]

Operating permits are similarly required for "major stationary sources" of air pollution, including facilities emitting 100 tons per year or more of "any air pollutant."[45] Since these permits are required under Title V of the Clean Air Act, they are commonly referred to as Title V permits. Each Title V permit must incorporate the BACT standards established for a facility through the preconstruction process, as well as requirements derived from other programs of the Clean Air Act.[46]

---

37. 42 U.S.C. §§7475(a)(1), 7479(1), 7602(j).
38. *Id.* §7470(4) (Clean Air Act, Title I, Part C, "Prevention of Significant Deterioration of Air Quality"). Areas that are not achieving the NAAQS are subject to a parallel preconstruction permitting program known as Nonattainment New Source Review (NNSR). *See id.* §7503.
39. *See id.* §§7479(1), 7602(j); 40 C.F.R. §51.166(b)(1)(i).
40. 42 U.S.C. §7475(a)(4). If the facility is proposed to be constructed in an air pollution control region that is in attainment for the relevant pollutant, BACT will apply. BACT also applies as the default for facilities emitting non-criteria pollutants (such as designated pollutants or greenhouse gases) where there is no attainment or nonattainment designation. If the facility is proposed to be constructed in a nonattainment area, the more stringent "lowest achievable emission rate" (LAER) standard will apply, along with other requirements. *Id.* §§7501(3), 7503(a)(2).
41. *See* 40 C.F.R. §52.21(b)(49) (defining the term "subject to regulation" as used in the provision of the Clean Air Act requiring BACT for new and modified facilities in attainment areas, 42 U.S.C. §7475(a)(4), to include "any air pollutant" that is subject to "actual control of the quantity of emissions of that pollutant" under a nationally-applicable EPA regulation for air programs), 40 C.F.R. §52.21(b)(50) (definition of "regulated NSR pollutant"), §52.21(b)(1)(i) (definition of "major stationary source"), §52.21(b)(2)(i) (definition of "major modification").
42. *See* 40 C.F.R. §52.21(b)(50)(ii) (defining the term "regulated NSR pollutant" as used in EPA's regulatory definitions of "major stationary source" and "major modification," which trigger the applicability of NSR requirements, as including "[a]ny pollutant that is subject to any standard promulgated under §111 of the Act," i.e., standards within the NSPS program); 40 C.F.R. Part 60, Subpart Ja, §60.102a(f)(1)(ii) and (2)(ii) (regulating hydrogen sulfide emissions from sulfur recovery plants within petroleum refining operations). If hydrogen sulfide is ever designated as a HAP, it would shift to a MACT standard, and would no longer be subject to the BACT requirement. HAPs are not considered "regulated NSR pollutants" for purposes of triggering NSR or applying BACT requirements (*see* 40 C.F.R. §52.21(b)(50)(v)), but the question whether a source "will comply" with applicable MACT standards generally must be considered as part of the overall new source review process (42 U.S.C. §7412(i)(1)).
43. *See* 40 C.F.R. §52.21(b)(49), (50) (describing in detail when and to what extent greenhouse gas emissions will be considered "subject to regulation" and "regulated NSR pollutants" for purposes of BACT and other Clean Air Act requirements).
44. 40 C.F.R. §52.21(b)(50)(i)(a) (VOCs are precursors to ozone) and (50)(vi) (PM_{10} and $PM_{2.5}$ emissions include gaseous emissions that condense to form PM at ambient temperatures). U.S. EPA, *Final Rule on the Implementation of the New Source Review Provisions for Particulate Matter Less Than 2.5 Microns ($PM_{2.5}$) Fact Sheet* (May 8, 2008), *available at* http://www.epa.gov/NSR/documents/20080508_fs.pdf. Ammonia is regulated if the state demonstrates that ammonia emissions are a significant contributor to the formation of $PM_{2.5}$ for an air quality control region in the state. *See id.*
45. 42 U.S.C. §7661(2)(B) (referencing §7602, containing a definition of "major stationary source" at §7602(j)). The "major source" threshold for Title V permitting may be lower in nonattainment areas with respect to the pollutant for which the area is not in attainment. *See* 42 U.S.C. §7661(2)(B) (referencing part D of subchapter I of the Clean Air Act, which applies to nonattainment areas).
46. *See* 42 U.S.C. §7661c(a); 40 C.F.R. §70.2.

Many large factory farms exceed the major source thresholds for PM, VOCs, ammonia, hydrogen sulfide, methane, and other regulated NSR pollutants, and there is nothing in the Clean Air Act exempting them from PSD and Title V permitting requirements.[47] Nevertheless, EPA and states have been reluctant to implement or enforce these permitting requirements at factory farms.

3. NSPS Program

The NSPS program is a federal overlay to the NAAQS-SIP approach, and it authorizes EPA to establish technology-based regulations on an industry-by-industry basis.[48] The program generally applies to all sources in each industrial category regardless of whether each source qualifies as a major source for purposes of other Clean Air Act programs. EPA must establish a performance standard for each industrial category known as the "best demonstrated technology" (BDT) standard.[49] Under the NSPS program, EPA regulates emissions of criteria pollutants and designated pollutants.[50] Much like the PSD permitting program, the main emphasis of the NSPS program is on new and modified sources. However, the statute also requires EPA to help develop standards for existing sources through coordination with states.[51] Using EPA guidance, states must adopt and implement performance standards for existing sources that would otherwise be regulated by EPA if they were new sources.[52]

Some agriculture-related facilities, including manufacturers of fertilizers, soil amendments, and pesticides, as well as grain elevators, are presently regulated under the NSPS program based on their emissions of NO_x, PM, fluoride, and VOCs.[53] Factory farms have not yet been regulated under the NSPS program, but in 2009, nine citizen groups filed a petition urging EPA to designate factory farms as a new NSPS industrial category and to regulate their emissions of hydrogen sulfide, ammonia, PM, VOCs, and greenhouse gases under this program.[54]

More generally, EPA has failed to take full advantage of its authority under the NSPS program. EPA has only established NSPS standards for approximately 70 industrial categories, and many of these standards are based on out-of-date technologies because EPA has not kept up with its obligation to periodically review and revise the standards. In many instances, the NSR-derived standards for particular facilities are much more stringent than the NSPS standards.

Nevertheless, the NSPS program could play an important role in the future in regulating factory farms. A key benefit of the NSPS program is that EPA has the ability to consider and address multiple types of air pollution at once. This is particularly important in the factory farm context because many of the same operational activities—e.g., storage and handling of manure—are responsible for emissions of toxic, conventional, and greenhouse gas pollution. EPA could address a substantial portion of the problems resulting from factory farm pollution by establishing a single, carefully crafted BDT standard relating to manure management.[55] Technologies and work practices for better management of manure are already available and in use in many locations, including waterproof covers for storage lagoons, anaerobic digesters, aeration,

---

47. 42 U.S.C. §§7401-7515, 7661-7661f. *See, e.g.,* EPA Finding of Substantial Inadequacy of Implementation Plan; Call for California State Implementation Plan Revisions, 68 Fed. Reg. 7327, 7328 (Feb. 13, 2003) (explaining that the Clean Air Act does not provide for the exemption of agricultural sources from the obligation to comply with NSR permitting requirements).
48. *See generally* 42 U.S.C. §7411(b), (d), (f).
49. *Id.* §7411(a)(1), (b)(1)(B).
50. *See id.* §7411(b), (d); 40 C.F.R. §60.22(a).
51. *See id.* §7411(d); 40 C.F.R. Part 60, Subpart B.
52. *See id.* §7411(d); 40 C.F.R. Part 60, Subpart B.
53. *See* 40 C.F.R. Part 60, Subpart G (NO_x standard for nitric acid plants); Subpart T through Subpart X (fluoride standards for various types of phosphate fertilizer plants); Subpart DD (PM standard for grain elevators); Subpart HH (PM standard for lime manufacturing plants); Subpart NN (PM standard for phosphate rock plants); Subpart PP (PM standard for ammonium sulfate plants); and Subpart VV, Subpart VVA, Subpart III, Subpart NNN, and Subpart RRR (VOC standards for plants producing the synthetic organic chemicals that may be used in pesticides and other agriculture-related applications).
54. *See* The Humane Society of the United States et al., Petition to List Concentrated Animal Feeding Operations Under Clean Air Act §111(b)(1)(A) of the Clean Air Act, and to Promulgate Standards of Performance Under Clean Air Act §§111(b)(1)(B) and 111(d) (Sept. 21, 2009) [hereinafter CAFO NSPS Petition], *available at* http://foe.org/sites/default/files/HSUS_et_al_v_EPA_CAFO_CAA_Petition.pdf.
55. *See, e.g.,* 2011 Swine Research Review—Manure Management (Dec. 22, 2011), *available at* http://nationalhogfarmer.com/environment/2011-swine-research-review-manure-management (explaining that certain types of biofilter media "were found to be acceptable alternatives for reducing hydrogen sulfide, ammonia and greenhouse gas (methane and nitrous oxide) emissions from swine manure and storage pits").

and other measures.⁵⁶ Another advantage of the NSPS program is that it generally applies to all facilities in an industry, rather than just major sources and, as noted above, it includes a mechanism for addressing existing facilities within an industry, rather than just new and modified sources.⁵⁷ Moreover, because the NSPS program establishes nationwide standards, it has the potential to achieve a more fair and consistent industrywide approach than through reliance on state-by-state regulation under SIPs or facility-by-facility permitting under the PSD and Title V programs.

4. Hazardous Air Pollutant Program

Another exception to the cooperative federalism theme in the Clean Air Act is the hazardous air pollutant (HAP) program, in which the standard-setting function is carried out by EPA rather than states.⁵⁸ The focus of the HAP program is primarily on pollutants that present very serious health effects, such as those which are known or reasonably anticipated to be "carcinogenic, mutagenic, teratogenic, neurotoxic, which cause reproductive dysfunction, or which are acutely or chronically toxic."⁵⁹ Once a pollutant is listed as a HAP, either by statute or through EPA designation,⁶⁰ EPA must promulgate maximum achievable control technology (MACT) standards for new and existing major sources of those HAPs in various industrial categories.⁶¹ MACT standards are technology-based, rather than risk-based, and they tend to be more stringent and less flexible than most of the other standards in the Clean Air Act.⁶² The "major source" thresholds for HAPs are also much lower than for conventional pollutants.⁶³ EPA also has authority to regulate nonmajor sources of HAPs where their emissions are sufficient to warrant listing and regulation under the Clean Air Act's "area source" provisions.⁶⁴

EPA has promulgated MACT standards addressing the HAP emissions of several agriculture-related industries, including manufacturers of fertilizers, pesticides, cellulose products, and soil amendments.⁶⁵ Factory farms have not yet been regulated under the HAP program, even though the toxicity of hydrogen sulfide, ammonia, and other emissions from factory farms is comparable to that of other pollutants already regulated as HAPs. For instance, the health effects of hydrogen sulfide have been described as similar to hydrogen cyanide, which is currently regulated as a HAP.⁶⁶ Moreover, both hydrogen sulfide and ammonia exhibit acute and chronic health effects, including adverse respiratory, cardiovascular, cytotoxic, and neurotoxic effects.⁶⁷ These effects have been severe enough to cause death in factory farm workers, brain

---

56. *See* Jeff El-Hajj, *Confined Animal Feeding Operations in California: Current Regulatory Schemes and What Must Be Done to Improve Them*, 15 Hastings W.-Nw. J. Envtl. L. & Pol'y 349, 357, 366-67 (Summer 2009).
57. *See* 42 U.S.C. §7411(d); 40 C.F.R. Part 60, Subpart B.
58. States do still play a role in the HAP program as they are usually delegated authority to implement the Title V program within their borders. Since Title V permits incorporate HAP requirements, states thereby play a role in implementing, monitoring, and enforcing HAP requirements.
59. 42 U.S.C. §7412(a)(6), (b)(2).
60. The current list of HAPs includes all those listed by Congress in 1990, except for a few that have since been delisted or added by EPA. *See* U.S. EPA, *Modifications to the 112(b)(1) Hazardous Air Pollutants*, http://www.epa.gov/ttn/atw/pollutants/atwsmod.html (last visited Nov. 28, 2012).
61. These standards are codified in EPA's regulations as National Emission Standards for Hazardous Air Pollutants (NESHAPs), located at 40 C.F.R. Parts 61 and 63.
62. For new sources, the MACT standard for a given category "shall not be less stringent than the emission control that is achieved in practice by the best controlled similar source." 42 U.S.C. §7412(d)(3). For existing sources, the MACT standard for a given category "shall not be less stringent" than either "the average emission limitation achieved by the best performing 12 percent of the existing sources" where the category has at least 30 or more sources in it, or "the average emission limitation achieved by the best performing 5 sources" where the category has fewer than 30 sources. *Id.* §7412(d)(3)(A), (B).
63. A facility is considered a major source for purposes of the HAP program if it emits 10 tons per year or more of any single HAP, or an aggregate of 25 tons per year or more of any combination of HAPs. *Id.* §7412(a)(1).
64. *Id.* §7412(a)(2), (c)(3), (k).
65. *See* 40 C.F.R. Part 63, Subpart F and Subpart G (regulating multiple HAPs from synthetic organic chemical manufacturing plants); Subpart BB (regulating hydrogen fluoride HAP from phosphate fertilizers manufacturing plants); Subpart MMM (regulating multiple organic HAPs from pesticide active ingredient manufacturing plants); Subpart UUUU (regulating carbon disulfide, toluene, organic HAPs, and other HAP emissions from cellulose products manufacturing plants); and Subpart AAAAA (regulating HAPs associated with PM emissions from lime manufacturing plants).
66. *See* 42 U.S.C. §7412(b)(1) (listing cyanide compounds as a category of HAPs); 40 C.F.R. §63.1103(g); Letter from Neil J. Carman, Sierra Club, to EPA Administrator Lisa Jackson, requesting that hydrogen sulfide be listed as a hazardous air pollutant under §112 of the Clean Air Act, at 4 (Mar. 30, 2009) [hereinafter Hydrogen Sulfide Petition], *available at* http://www.texas.sierraclub.org/press/newsreleases/H2S-LetterToEPA.pdf; Sujal Mandavia, MD, *Hydrogen Sulfide Toxicity* (updated May 3, 2011), *available at* http://emedicine.medscape.com/article/815139-overview#showall.
67. *See* Michele Merkel, Raising a Stink: Air Emissions From Factory Farms, 3 n.10 (2002) [hereinafter Raising a Stink], *available at* http://www.environmentalintegrity.org/news_reports/Report_Raising_Stink.php (citing Agency for Toxic Substances and Disease Registry, Minimal Risk Levels for Hazardous Substances, http://www.atsdr.cdc.gov/mrls.html); EPA, Integrated Risk Information System, www.cpa.gov/iris/subst.

damage in members of neighboring communities, increased rates of infant mortality, and other serious consequences (as discussed above in Chapter 5). In addition, both hydrogen sulfide and ammonia are designated as "regulated substances" subject to requirements relating to the prevention of accidental releases and risk management planning under §112(r) of the Clean Air Act.[68] Thus these pollutants are sufficiently hazardous to be regulated as HAPs, and many large factory farms emit them in quantities far exceeding the major source thresholds for the HAP program.[69]

Hydrogen sulfide was included in Congress' original list of 189 HAPs in 1990, but it was removed a few months later—under pressure from industry[70]—through a legislative amendment.[71] In 2009, citizen groups filed a petition asking EPA to reinstate hydrogen sulfide on the list of HAPs.[72] In their petition the groups provided extensive evidence showing that hydrogen sulfide is at least as dangerous as other pollutants already regulated as HAPs and that it is being emitted in high concentrations at large factory farms in many regions of the country.[73] If EPA responds favorably to the petition,[74] the HAP program would provide a powerful tool for reducing the unhealthy levels of hydrogen sulfide emissions from factory farms in many parts of the country.

B. Reporting Requirements Under EPCRA and CERCLA

In addition to Clean Air Act requirements, agricultural facilities are subject to requirements for reporting releases of hazardous substances that exceed specified quantities pursuant to the Emergency Planning and Community Right-to-Know Act (EPCRA) and the Comprehensive Environmental Response, Compensation, and Liability Act (CERCLA).[75]

The EPCRA statute was enacted in response to the growing threat to communities posed by the storage and handling of hazardous materials at industrial sites. In 1984, a Union Carbide plant near Bhopal, India, released large quantities of highly toxic methyl isocyanate gas into the air,[76] killing several thousand people and injuring tens of thousands.[77] Moreover, in a 1985 report, EPA determined that 6,928 chemical accidents had occurred in the United States during the previous five years.[78] In response, Congress enacted EPCRA, which is "based on the principle that the more you and your neighbors know about hazardous chemicals in your community, the better prepared your community will be to manage these potential hazards and to improve public safety and health as well as environmental quality."[79] The main purposes of EPCRA are to encourage emergency planning for releases of hazardous materials, and to inform citi-

---

html; Mich. Dep. Env. Qual., Concentrated Animal Feedlot Operations (CAFOs) Chemicals Associated With Air Emissions 4-7 (May 2006), *available at* http://www.michigan.gov/documents/CAFOs-Chemicals_Associated_with_Air_Emissions_5-10-06_158862_7.pdf.

68. *See* 42 U.S.C. §7412(r); 40 C.F.R. §68.130.
69. *See* Sierra Club, Hydrogen Sulfide Petition, *supra* note 66.
70. *See* Amy Mall et al., Protecting Western Communities From the Health and Environmental Effects of Oil and Gas Production, Natural Resources Defense Council 13 (2007), *available at* http://www.nrdc.org/land/use/down/down.pdf (citing Jim Morris, *Lost Opportunity; EPA Had Its Chance to Regulate Hydrogen Sulfide*, Houston Chron. (Nov. 9, 1997) ("[c]ompanies in Texas were very successful in removing [hydrogen sulfide] from the list because of its presence in the extraction of oil")).
71. EPA's official position is that "[a] clerical error led to the inadvertent addition of hydrogen sulfide to the §112(b) list of Hazardous Air Pollutants," and that this was rectified by a joint resolution to remove hydrogen sulfide from the list. *See* EPA, Modifications To The 112(b)(1) Hazardous Air Pollutants, *available at* http://www.epa.gov/ttn/atw/pollutants/atwsmod.html (last visited Oct. 27, 2011).
72. *See* Sierra Club, Hydrogen Sulfide Petition, *supra* note 66. As noted above, citizen groups also asked EPA to list ammonia as a criteria pollutant in 2011, but so far they have not sought the listing of ammonia as a HAP. Ammonia's status as a criteria pollutant would not preclude it from being designated as a HAP as well. For instance, lead is regulated as a criteria pollutant, and lead compounds are regulated as HAPs. 42 U.S.C. §7412(b)(1); 40 C.F.R. §§50.12, 50.16. Similarly, many VOCs are regulated based on their contribution to the formation of ozone, a criteria pollutant, as well as based on their toxicity as HAPs. For example, benzene, ethylbenzene, toluene, and xylene are regulated as VOCs and as HAPs. *See* 42 U.S.C. §7412(b)(1); 40 C.F.R. §51.100(s).
73. *See* Sierra Club, Hydrogen Sulfide Petition, *supra* note 66.
74. At the time of this writing, EPA still has not responded to the hydrogen sulfide petition. In 2011, the Sierra Club and other groups threatened to sue the agency to compel it to acknowledge and respond to the petition, but it appears no litigation has commenced. *See* Sierra Club, Press Release, Sierra Club Wants Answers From EPA on Toxic Hydrogen Sulfide Gas: Coalition Threatens Legal Action (June 22, 2011), *available at* http://action.sierraclub.org/site/MessageViewer?em_id=209544.0 (last visited Sept. 1, 2012).
75. 42 U.S.C. §11023(g), (h) (toxic chemical release form under EPCRA); 42 U.S.C. §9603(a) (release notification requirement under CERCLA).
76. U.S. EPA, *Methyl Isocyanate*, http://www.epa.gov/ttnatw01/hlthef/methylis.html (last visited Oct. 13, 2011).
77. *Id.*
78. Stuart Diamond, *U.S. Toxic Mishaps in Chemicals Put at 6,298 in 5 Years*, N.Y. Times, Oct. 3, 1985, at http://www.nytimes.com/1985/10/03/us/us-toxic-mishaps-in-chemicals-put-at-6298-in-5 years.html?scp=1&sq=The%20New%20York%20Times%20said%20at%20least%206,928%20chemical%20accidents%20occurred%20since%201980&st=cse.
79. U.S. EPA, Office of Solid Waste and Emergency Response, EPA 550-K-99-001, Chemicals in Your Community 33 (Dec. 1999), *available at* http://www.epa.gov/osweroe1/docs/chem/chem-in-comm.pdf.

zens and local officials about potential hazards in their communities.[80] The release-reporting provisions of CERCLA serve similar objectives.

Under EPCRA, both hydrogen sulfide and ammonia are listed as "extremely hazardous substances" and have a "reportable quantity" (RQ) threshold of 100 pounds released.[81] Similarly, under CERCLA, both hydrogen sulfide and ammonia are listed as "hazardous substances," and they each have an RQ of 100 pounds.[82] Hydrogen sulfide and ammonia are also considered to be toxic chemicals for which cumulative releases must be reported annually as part of EPCRA's toxic release inventory (TRI) program.[83]

Ammonia reporting under the TRI program has been very limited due to lack of EPA enforcement at factory farms resulting from EPA's Consent Agreement, which will be discussed further below. The end result is that the TRI data compiled for the purpose of informing the public about health risks has dramatically understated actual emissions. Georgia, for instance, is the nation's top producer of broiler chickens. In 2007, factory farms in Georgia emitted an estimated 97.6 million pounds of ammonia, yet the TRI data showed ammonia emissions from all industrial sources in the state amounted to only 11.9 million pounds.[84] Similarly, Iowa is the nation's top producer of eggs. In 2007, layer hen factory farms in Iowa emitted an estimated 53 million pounds of ammonia, while the TRI data showed a total of only 9.4 million pounds from all industrial sources.[85] These estimates were made by the nonprofit organization Environmental Integrity Project after analyzing and extrapolating from data compiled by the U.S. Department of Agriculture, studies of poultry farming operations conducted by Purdue University (as part of the monitoring activities initiated by EPA under the Consent Agreement), and studies conducted by Iowa State University and the University of Kentucky.[86]

Factory farms also have not been reporting their hydrogen sulfide emissions under the TRI program. Although hydrogen sulfide was added to the TRI list of toxic chemicals by EPA in December 1993,[87] in response to pressure from industry, EPA issued an administrative stay in August 1994 and decided to conduct further evaluation of the effects of hydrogen sulfide on human health and the environment.[88] In February 2010, EPA published notice of its intent to consider lifting the administrative stay, explaining that its "technical evaluation of hydrogen sulfide shows that it can reasonably be anticipated to cause chronic health effects in humans" at low concentrations and, "because of its toxicity, significant adverse effects in aquatic organisms."[89] Accordingly, EPA stated that "there is no basis for continuing the Administrative Stay of the reporting requirements for hydrogen sulfide," that "the Administrative Stay should therefore be lifted," and that its findings "clearly demonstrate the correctness of the Agency's final decision in December 1993 to list hydrogen sulfide on the EPCRA §313 toxic chemicals list."[90] Therefore, factory farms that are subject to annual reporting obligations will have to file reports concerning hydrogen sulfide by July 1, 2013, for the 2012 reporting period.[91] EPA's failure to enforce the reporting requirements from 1994 through 2011, however, means that it has lost the opportunity to collect 18 years' worth of data regarding releases of hydrogen sulfide into the environment from factory farms and other industrial facilities.

---

80. *Id.*
81. 40 C.F.R. Part 355, app. A.
82. *Id.* §302.4 (tbl.).
83. *Id.* §372.65(a).
84. *See* EIP, Ammonia Petition, *supra* note 33, at 35 n.142 (citing EIP, *A Holiday Gift for Big Poultry: Bush Administration Rushes Emissions Reporting Exemption* (corrected Dec. 2009), *available at* http://environmentalintegrity.org/news_reports/documents/RTKEXEMPTIONFINAL_12_14_09.pdf [hereinafter EIP Report]).
85. *Id.* at 35 n.143 (citing EIP Report, *supra* note 84).
86. *See* EIP Report, at 3 n. ii, vii, and viii (citing Purdue Univ., National Air Emissions Monitoring Study: Data From Layer Site IN2H (May 12 to June 30, 2007)), 3 n. v (citing U.S. Dept. of Agriculture, Poultry—Production and Value 2007 Summary (released Apr. 2008)), 3 n. vi (citing U.S. Dept. of Agriculture, Chicken and Eggs (released Nov. 21, 2008)), and 3 n. viii (citing Iowa State Univ. and Univ. of Kentucky, Tyson Broiler Ammonia Emission Monitoring Project: Final Report (released May 1, 2007)).
87. U.S. EPA, Hydrogen Sulfide: Intent to Consider Lifting Administrative Stay; Opportunity for Public Comment, 75 Fed. Reg. 8889, 8889 (Feb. 26, 2010).
88. *See* U.S. EPA, Hydrogen Sulfide; Methyl Mercaptan; Toxic Chemicals Release Reporting; Community Right-to-Know; Stay of Reporting Requirements, 59 Fed. Reg. 43048, 43049 (Aug. 22, 1994). EPA has repeatedly acknowledged that the administrative stay was granted because of concerns raised by the regulated community regarding the adequacy of the scientific basis for the agency's final decision to include hydrogen sulfide on the TRI list. *See id.* at 43049; 75 Fed. Reg. at 8889; and U.S. EPA, Hydrogen Sulfide: Lifting of Administrative Stay, http://www.epa.gov/tri/lawsandregs/hydrogensulfide/indexf.html (last visited Nov. 28, 2012).
89. 75 Fed. Reg. at 8893.
90. *Id.*
91. U.S. EPA, Hydrogen Sulfide: Community Right-to-Know Toxic Chemical Release Reporting, 76 Fed. Reg. 64022, 64022 (Oct. 17, 2011).

Additionally, future reporting will not capture all emissions of hydrogen sulfide and ammonia from factory farms because in December 2008, EPA promulgated a regulation creating a permanent exemption from CERCLA and EPCRA reporting requirements for a large segment of the livestock industry.[92] The exemption encompasses "[a]ny release to the air of a hazardous substance from animal waste," including hydrogen sulfide and ammonia, for small and medium-sized factory farms that "stable or confine" livestock—including dairy cows, veal calves, other cattle, swine, horses, sheep, lambs, turkeys, chickens, and ducks—below specified numbers of animals, regardless of whether their emissions exceed the reporting thresholds.[93] The rule also exempts all farms at which livestock are not "stabled or confined," such as farms where the animals graze in pastures, without reference to air pollutant emission levels.[94]

For all these reasons, the information available to the public concerning releases of hydrogen sulfide, ammonia, and other air pollutants from factory farms in their communities is woefully inadequate, yet recent trends heighten the importance of release-reporting obligations. As discussed above, factory farms are growing larger and more concentrated, and they are housing more and more animals at individual sites, with corresponding increases in air emissions. At the same time, population growth and expanding residential development is bringing human populations into closer proximity with factory farms.[95] In light of the serious health effects associated with hydrogen sulfide and ammonia, these trends suggest there is a growing threat of tragic accidents for residents near factory farms as well as increasing risks of severe health impacts. The EPCRA and CERCLA reporting requirements are intended to address these types of threats through greater transparency and public awareness. If EPA takes steps to improve and expand release reporting at factory farms, these efforts would help in the achievement of these goals.

C. Enforcement Issues

Prior to 2000, very little attention was paid by industry, government, or citizen groups to the air pollution requirements applicable to the agriculture industry under the Clean Air Act or the release-reporting requirements under CERCLA and EPCRA.[96] However, as the scale and concentration of animal confinement facilities grew larger, and as the seriousness of the associated air pollution problems started becoming more apparent, federal officials and citizen groups began to bring enforcement actions in order to compel factory farms to come into compliance. This prompted industry to exert pressure on EPA for some form of relief, and EPA ultimately entered into a broad consent agreement with thousands of livestock producers in which EPA agreed not to enforce applicable air pollution standards and release reporting requirements for an indefinite period of time. As a result, EPA enforcement has been nonexistent since 2005.

1. Initial Enforcement Actions

By the early 2000s, EPA had begun pursuing enforcement actions against factory farms based on violations of the Clean Air Act, EPCRA, and CERCLA. For instance, in October 1999, EPA intervened in a citizen suit initially brought to enforce Clean Water Act violations occurring at factory farms operated by Premium Standard Farms, Inc., the second largest pork producer in the nation.[97] EPA later amended its

---

92. U.S. EPA, CERCLA/EPCRA Administrative Reporting Exemption for Air Releases of Hazardous Substances From Animal Waste at Farms, 73 Fed. Reg. 76948 (Dec. 18, 2008).
93. 40 C.F.R. §355.31(g). In this regard, the reporting exemption is broader than the Consent Agreement, which did not cover farms that raise veal calves, other non-dairy cattle, ducks, sheep, lambs, or horses. *See* Consent Agreement, 70 Fed. Reg. 4958, 4963, ¶ 8 (defining "Agricultural Livestock" or "Livestock" to include "dairy cattle, swine and/or poultry among others").
94. 40 C.F.R. §355.31(h).
95. *See generally* USDA Economic Research Service, Rural Economy and Population, *Population and Migration*, http://www.ers.usda.gov/Briefing/Population/Rural.htm (last visited Nov. 28, 2012).
96. The modern Clean Air Act was enacted in 1970. During the 1970s, livestock production was not as concentrated as it is now and did not pose as much of an air pollution threat. During the 1980s and 1990s, the factory farm industry was growing and becoming more concentrated, but the Reagan and Bush administrations favored industry interests and implemented a laissez faire and deregulatory approach in many arenas, including environmental protection. By 2000 the seriousness of the threat posed by factory farm air pollution had become widely recognized, and the incoming Clinton administration showed a greater willingness to implement and enforce environmental regulations, at least initially, as discussed in Section C.1. below. The Clinton-era EPA later reversed course, however, as evidenced by its entry into the Consent Agreement, discussed in Section C.2. below.
97. U.S. EPA, Premium Standard Farms, Inc. and Continental Grain Company, Inc. Civil Settlement Fact Sheet (Nov. 19, 2001), *available at* http://www.epa.gov/compliance/resources/cases/civil/mm/psf.html.

pleadings to assert violations of minor source preconstruction permitting requirements under the Clean Air Act, as well as release-reporting violations under EPCRA and CERCLA, against both Premium Standard Farms and Continental Grain Company, Inc., which had bought a controlling interest in Premium Standard Farms.[98] At the time, Premium Standard was permitted to house more than 900,000 hogs at its factory farms in Missouri, where it produced 2 million hogs annually, and stored and applied more than 750 million gallons of animal waste annually.[99] In November 2001, the parties entered into a consent decree settling the claims. The consent decree provided for (1) injunctive relief valued at $50 million, which required improvements to manure management practices and monitoring of PM, hydrogen sulfide, ammonia, and VOCs from selected barns and lagoons; (2) payment of a $350,000 penalty (in addition to a $650,000 penalty paid to the state of Missouri under a separate consent judgment); and (3) implementation of a supplemental environmental project (SEP) involving the testing and installation of an experimental oil sprinkling system to control PM and odor emissions, at a cost of up to $400,000.[100]

During roughly the same time period, in January 2001, EPA issued a notice of violation to Ohio-based Buckeye Egg Farm, LP and ordered it to start collecting data on its PM emissions.[101] At that time, Buckeye housed 12 million chickens in over 100 barns and produced 2.6 billion eggs annually, representing 4% of the nation's total egg production.[102] Because Buckeye had failed to test some of its confinement buildings, EPA issued a second compliance order in December 2001.[103] EPA testing showed that three Buckeye facilities in Marseilles, Croton, and Mt. Victory, Ohio, were each emitting 550 to 700 tons per year of PM, as well as 275 to over 800 tons per year of ammonia.[104] Ambient air monitoring also showed "significantly elevated levels of ammonia and particulates in residential areas over one kilometer away from Buckeye's largest facility."[105] Since the PM emissions were well over the 100 and 250 ton per year "major source" thresholds in the Clean Air Act, Buckeye was clearly obligated to obtain and comply with PSD and Title V permits.[106] EPA initially brought an enforcement action against Buckeye for failing to comply with the emissions testing order, and later EPA added claims alleging Clean Air Act violations based on Buckeye's failure to obtain PSD permits.[107] In February 2004, EPA and Buckeye settled the claims through a consent decree under which Buckeye agreed to (1) injunctive relief requiring it to install and test innovative pollution controls for PM and ammonia at a cost of $1.6 million; and (2) payment of a civil penalty of $880,598.[108]

EPA also brought an enforcement action against Seaboard Foods, LP, a large pork producer with more than 200 farms in Oklahoma, Kansas, Texas, and Colorado. In April 2002, EPA ordered Seaboard to test for emissions of hydrogen sulfide, PM, and VOCs from the facilities' confinement buildings and lagoons,[109] but Seaboard failed to fully comply with the EPA order. In September 2006, the parties entered into a settlement based on alleged violations of the Clean Air Act monitoring order, violations of release reporting obligations under EPCRA and CERCLA, and violations of the Clean Water Act.[110] Under the consent

---

98. *Id.* at 3.
99. *Id.*
100. *Id.* at 1-2.
101. Michele Merkel, Raising a Stink, *supra* note 67, at 10.
102. U.S. Department of Justice, *Significant Environmental Enforcement Section Cases: U.S. v. Buckeye Egg Farms, L.P. (N.D. Ohio)*, http://justice.gov/enrd/4467.htm (last visited Nov. 28, 2012).
103. *See* Letter to Bill Glass from Kevin Vuilleumier Regarding June 4-8 Emission Testing (Dec. 11, 2001).
104. U.S. Department of Justice, *Significant Environmental Enforcement Section Cases: U.S. v. Buckeye Egg Farms, L.P. (N.D. Ohio)*, http://justice.gov/enrd/4467.htm (last visited Nov. 28, 2012).
105. *Id.*
106. *See* 42 U.S.C. §§7470-7492 (Prevention of Significant Deterioration requirements at Subchapter I, Part C of the Clean Air Act), 42 U.S.C. §§7661-7661(f) (permitting requirements of Subchapter V of the Clean Air Act).
107. U.S. Department of Justice, *Significant Environmental Enforcement Section Cases: U.S. v. Buckeye Egg Farms, L.P. (N.D. Ohio)*, http://justice.gov/enrd/4467.htm (last visited Nov. 28, 2012).
108. *Id.* Buckeye's successor, Ohio Fresh Eggs, became subject to the requirements of the consent decree upon its purchase of the company. *See* U.S. Department of Justice, *Ohio's Largest Egg Producer Agrees to Dramatic Air Pollution Reductions From Three Giant Facilities*, http://www.justice.gov/opa/pr/2004/February/04_enrd_105.htm (last visited Nov. 28, 2012).
109. U.S. Department of Justice, Notice of Lodging of Consent Decree Between the United States and Seaboard Foods LP, 71 Fed. Reg. 56553, 56554 (Sept. 27, 2006).
110. U.S. Department of Justice, *Government Reaches Settlements With Seaboard Foods and PIC USA* (Sept. 15, 2006), http://www.justice.gov/opa/pr/2006/September/06_crm_625.html.

decree, Seaboard agreed to pay a civil penalty of $205,000, and it was credited with $100,000 toward this amount based on its payment of a civil penalty under a separate air compliance agreement with EPA.[111]

Environmental advocates also started initiating their own citizen enforcement actions. In 2000, the Sierra Club initiated a lawsuit against Seaboard Farms, Inc., asserting that it had violated CERCLA by failing to report ammonia releases from the Dorman Farm, one of its factory farms in Beaver County, Oklahoma.[112] The Dorman Farm actually consisted of two farms on adjacent parcels of land, each with eight buildings and a common waste management system, including several lagoons, barns, and land application areas.[113] Together, these facilities housed approximately 25,000 swine.[114] The parties disputed the meaning of the term "facility" for purposes of the release-reporting provision of CERCLA.[115] Seaboard viewed each barn, lagoon, and land application area as a separate facility and argued that it was only subject to reporting obligations if releases from any one of these individual sources exceeded the reporting threshold. On appeal, the U.S. Court of Appeals for the Tenth Circuit held that CERCLA's definition of facility was "unambiguous and unequivocal" and that it "encompasses the entire Dorman Farm site."[116] Thus, the Tenth Circuit established a broad interpretation of the term "facility" that requires factory farms to comply with release-reporting requirements if, in the aggregate, the emissions from the various barns, lagoons, and other facilities at a particular site exceed the reporting threshold.

In reaching this conclusion, the Tenth Circuit noted that a federal district court within the Sixth Circuit had reached the same conclusion in 2003 in a case brought by the Sierra Club and local residents against Tyson Foods, Inc.[117] That case involved allegations of ammonia reporting violations at several poultry farms, each with multiple barns housing thousands of chickens. The district court agreed with the plaintiffs' contention that under both CERCLA and EPCRA "the whole farm site" constituted the relevant facility for reporting purposes, not the individual poultry barns.[118] In addition, the court rejected several other arguments presented by the defendants and thereby clarified the applicability of federal release-reporting obligations to factory farms.[119]

2. EPA's Consent Agreement

Alarmed by this increased scrutiny and liability exposure, thousands of owners and operators of factory farms began negotiating a voluntary agreement with EPA that they hoped would delay their compliance obligations and shield them from further enforcement actions.[120] Despite vehement objections from environmental and public health advocates,[121] EPA entered into a sweeping consent agreement with thousands of factory farm operators in 2005.[122]

In the Consent Agreement, EPA granted to participating factory farms in the "egg, broiler chicken, turkey, dairy and swine industries" broad releases and covenants not to sue for all civil violations of:

- Preconstruction permit requirements in the NSR program;

- Operating permit requirements in the Title V program;

- All other federally enforceable SIP requirements "based on quantities, rates, or concentrations of air emissions of" hydrogen sulfide, ammonia, PM, and VOCs; and

---

111. *Id.*
112. Sierra Club v. Seaboard Farms, Inc., Third Amended Complaint, 2002 WL 34338408 (Mar. 6, 2002).
113. Sierra Club v. Seaboard Farms Inc., 387 F.3d 1167 (10th Cir. 2004).
114. *Id.*
115. *Id.* at 1169-76 (discussing CERCLA §103(a)).
116. *Id.* at 1176.
117. *See id.* at 1172 (referring to 299 F. Supp. 2d 693, 708).
118. Sierra Club v. Tyson Foods, Inc., 299 F. Supp. 2d 693, 708-11 (W.D. Ky. 2003).
119. *Id.* at 711-24.
120. Leonard H. Dougal, *EPA Issues Consent Agreement Regarding Air Emissions From Animal Feeding Operations and CAFOs*, http://images.jw.com/com/publications/455.pdf (explaining that "[t]he Consent Agreement resulted from nearly three years of negotiations between agricultural representatives and EPA seeking to address the recent lawsuits and how to deal with quantifying and reporting air emissions from AFOs").
121. *See* Letter from Brent Newell, Center on Race, Poverty & the Environment et al. to Christine Todd Whitman, EPA Administrator (May 5, 2003); Comments of the Association of Irritated Residents, Center on Race, Poverty & the Environment, Environmental Defense, Environmental Integrity Project, and Sierra Club, on Animal Feeding Operations Consent Agreement and Final Order (Mar. 1, 2005).
122. *See* U.S. EPA, Notice of Consent Agreement and Final Order, and Request for Public Comment, 70 Fed. Reg. 4958 (Jan. 31, 2005).

- Release reporting requirements under CERCLA and EPCRA for hydrogen sulfide and ammonia (except for accidental releases caused by explosions, fires, or other abnormal occurrences).[123]

EPA's releases and covenants not to sue covered all violations involving animal waste at emission units,[124] and the term "emission units" was broadly defined to include both livestock confinement houses and animal waste storage and treatment facilities.[125]

In exchange for these very broad releases and covenants not to sue, participating factory farms agreed to allow their facilities to be monitored as part of an EPA study of air pollution emissions and monitoring methodologies, pay nominal penalties, eventually apply for coverage under the necessary Clean Air Act permits, and eventually start submitting release reports, as required under CERCLA and EPCRA.[126]

EPA anticipated that the monitoring would begin in 2005 and continue for two years. Then, "[w]ithin 18 months after the conclusion of the nationwide emissions monitoring study," EPA planned to start publishing "on a rolling basis as the work is completed, the methodologies for estimating emissions" for the types of factory farms included in the studies.[127] Factory farm owners and operators agreed that, within 120 days after EPA published the emission-estimating methodologies applicable to the relevant emission units, factory farms would be expected to apply for preconstruction and operating permits under the Clean Air Act, and to report all qualifying releases of hydrogen sulfide and ammonia under CERCLA and EPCRA, using the methodologies developed by EPA.[128] In other words, EPA agreed not to enforce any of the specified types of air pollution violations that had occurred prior to the factory farms' entry into the Consent Agreement, or any future violations for a minimum of approximately four years thereafter.

Under the terms of the Consent Agreement, operators with only one farm below the "large" farm threshold were assessed a penalty of $200,[129] and most other operators were assessed a penalty of $500 per farm.[130] Farms containing more than 10 times the number of animals that define a "large" farm were assessed a penalty of $1,000 per farm.[131] The overall penalty to be paid by any single operator was also capped at $10,000 to $100,000 depending on the number of farms involved.[132] It appears, however, that EPA did not fully recover even these token penalties. EPA ultimately ratified agreements with a total of 2,588 operators covering a total of 13,843 factory farms,[133] and EPA has collected a total of $2,786,700 in civil penalties under the Consent Agreement.[134] This means the penalties collected by EPA average approximately $1,077 per operator and $201 per factory farm.[135] These penalties are far lower than the penalties and injunctive relief imposed by EPA in its initial enforcement actions against Premium Standard Farms and other factory farms in the early 2000s, which ranged from hundreds of thousands to millions of dollars. Moreover, EPA has taken much longer to complete the monitoring work than originally planned. The monitoring study did not even commence until spring 2007.[136] In March 2012, EPA published draft documents describing its proposed emission-estimating methodologies for broiler chicken, swine, and dairy animal feeding operations, and the opportunity for submitting public comments ended in June 2012.[137] At the time of this writing, EPA has not issued any final emission-estimating methodolo-

---

123. *Id.* at 4959, 4963, ¶ 26.
124. *Id.* at 4963, ¶ 27(a).
125. *Id.* at 4963, ¶ 11.
126. *Id.* at 4963-67, ¶¶ 28-38, 48-63.
127. *Id.* at 4960.
128. *Id.* at 4964, ¶ 28(C)(i)(a), (ii).
129. *Id.* at 4966, ¶ 48(A).
130. *Id.* at 4966, ¶ 48(B).
131. *Id.*
132. *Id.*
133. *See* U.S. EPA, *Animal Feeding Operations Air Quality Compliance Agreement Information Sheet* (Mar. 11, 2009), http://www.epa.gov/compliance/resources/agreements/caa/cafo-infosht-0309.html. U.S. EPA, *Summary of the AFO Air Compliance Agreement Participants* (Feb. 23, 2009), *available at* http://www.epa.gov/compliance/resources/agreements/caa/caforespondentlist-022309.pdf.
134. *See* U.S. EPA, *Animal Feeding Operations Air Quality Compliance Agreement Information Sheet* (Mar. 11, 2009), http://www.epa.gov/compliance/resources/agreements/caa/cafo-infosht-0309.html.
135. Operators were also expected to pay a pro rata share of the cost of the monitoring study up to a maximum of $2,500 per farm. *See* 70 Fed. Reg. 4958, 4966, ¶ 53(A). It is unclear whether, or to what extent, these funds have been collected or what the total cost of the monitoring study is.
136. *See* U.S. EPA, *Animal Feeding Operations Air Quality Compliance Agreement Information Sheet* (Mar. 11, 2009), http://www.epa.gov/compliance/resources/agreements/caa/cafo-infosht-0309.html.
137. *See* U.S. EPA, Notice of Availability: Draft Documents Related to the Development of Emissions Estimating Methodologies for Broiler Animal Feeding Operations and Lagoons and Basins for Swine and Dairy Animal Feeding Operations, 77 Fed. Reg. 14716 (Mar. 13, 2012).

gies for factory farms. As a result, the nearly 14,000 factory farms covered by the Consent Agreement are still protected by its nonenforcement provisions.

EPA justified its decision to enter into the Consent Agreement by emphasizing the importance of conducting more studies and developing better methods of estimating air pollutant emissions from factory farms.[138] EPA had no need, however, to provide generous incentives and covenants not to sue in order to encourage factory farms to participate in its air emissions study and to bear the costs of these studies. EPA already had broad legal authority to require the owner or operator of any facility that may have information relevant to Clean Air Act implementation to install monitoring equipment at its facility, sample air emissions emanating from the facility, and use other methods and procedures for data gathering specified by EPA "on a one-time, periodic or continuous basis" and submit this information to EPA.[139] EPA also has a "right of entry to, upon, or through any premises" in order to inspect records, inspect monitoring equipment and methods, and sample emissions that the facility operator has been required to monitor.[140] Furthermore, while the Consent Agreement covered nearly 14,000 factory farms, EPA's ongoing monitoring activities and studies are actually taking place at only 21 farms.[141]

The Consent Agreement also represents a substantial departure from EPA's normal enforcement practices for several reasons. First, EPA usually enters into consent decrees with one or at most a few violators a time.[142] In comparison, the scope of the Consent Agreement is breathtaking in that it encompasses roughly 2,600 operators and 14,000 factory farms, comprising the vast majority of the U.S. livestock industry.

Second, EPA typically enters into consent decrees to resolve known past violations. By contrast, under the Consent Agreement EPA has resolved (1) all past violations that had ever occurred prior to parties' entry into the Consent Agreement, including known and unknown violations dating back five or 10 years or more, (2) all violations present or ongoing at the time the Consent Agreement was finalized, and (3) all future violations that might occur prior to EPA's completion of monitoring studies and issuance of emission factors, which has been eight years so far and counting.[143]

Third, before entering into any consent decree in an enforcement context, EPA generally makes determinations concerning (1) the specific provisions of laws or regulations that have been violated; (2) the factual circumstances that form the basis of EPA's allegations that violations have occurred; and (3) the factors that influenced its decision regarding the amount of the penalty, such as the severity of the violations, the resulting harm to human health and the environment, the violator's efforts to come into compliance, and the economic benefit to the violator of its noncompliance.[144] Under the Consent Agreement, however, EPA did not make any of these types of determinations, which is not surprising given the incredible breadth of the violations resolved.[145] Finally, while EPA has often been accused of lax enforcement against a particular industry or under a particular statute,[146] this is usually because of inaction, rather than an affirmative decision not to enforce through a legally binding document that ties EPA's hands and precludes it from enforcing for an indefinite period of time into the future. Public health and environmental advocates challenged the Consent Agreement in federal court. They characterized the decision as a rulemaking subject to judicial review, rather than an enforcement action not subject to judicial review. In particular, the petitioners argued that EPA's action satisfied the statutory definition of a "rule" because it was an "agency statement of general or particular applicability and future effect designed to implement, interpret, or prescribe, law or policy."[147] They argued that, despite EPA's characterization of the Consent Agreement as an enforcement

---

138. *See* 70 Fed. Reg. 4958, 4961 ("EPA believes that the Air Compliance Agreement will be the quickest and most effective way to address the current uncertainties regarding air emissions from AFOs and to bring the entire AFO industry into compliance with the CAA, §103 of CERCLA, and §304 of EPCRA."), 4960 (discussion of purpose of the monitoring study.).
139. 42 U.S.C. §7414(a)(iii)(1).
140. *Id.* §7414(a)(iii)(2).
141. *See* U.S. EPA, *Monitored AFOs*, http://www.epa.gov/airquality/agmonitoring/data.html (last visited Nov. 28, 2012).
142. *See* U.S. EPA OECA, *Cases and Settlements*, http://cfpub.epa.gov/compliance/cases/ (last visited Jan. 30, 2012).
143. *See generally* U.S. EPA, *Agriculture, Animal Feeding Operations—Air Programs*, http://www.epa.gov/agriculture/anafoair.html (last visited Nov. 28, 2012).
144. *See, e.g.*, U.S. EPA OECA, *Air Products LLC Settlement*, http://www.epa.gov/compliance/resources/cases/civil/rcra/airproducts.html (last visited Nov. 28, 2012).
145. *See* Letter from Brent Newell, Center on Race, Poverty & the Environment, to Gerardo Rios, U.S. EPA Region IX (Mar. 1, 2010); Comments of the Association of Irritated Residents et al., on Animal Feeding Operations Consent Agreement and Final Order (Mar. 1, 2005).
146. *See, e.g.*, Seth Borenstein, *EPA Inspector Says Agency Failing to Monitor Refinery Emissions*, SEATTLE TIMES, June 26, 2004, at http://seattletimes.nwsource.com/html/nationworld/ 2001966112_refineries26.html.
147. Petitioner's Opening Brief, Association of Irritated Residents v. EPA, 2006 WL 3622127 at *15 (citing 5 U.S.C. §551(4)).

action, it did not qualify as such because EPA had not developed any factual basis for determining whether or to what extent violations had occurred at any particular facility, and because EPA had never filed any complaint or otherwise made any allegations identifying specific violations of statutory or regulatory provisions.[148] The petitioners also argued that the Consent Agreement qualified as a rulemaking because it amounted to a deferral or temporary exemption of factory farms from various statutory requirements, applied generally to a wide array of facilities, and would have a prospective effect rather than simply resolving past violations.[149] The U.S. Court of Appeals for the District of Columbia (D.C.) Circuit disagreed with the petitioners and accepted EPA's argument that the Consent Agreement was an enforcement action rather than a rulemaking.[150] The court therefore concluded that it lacked jurisdiction to consider the merits of the case.[151] As a result, the Consent Agreement remains in place, and EPA is still not enforcing air pollution requirements at the nearly 14,000 factory farms covered by the Agreement.

D. State Efforts

Given EPA's reluctance to regulate air emissions from factory farms, many have looked to states to address the ongoing public health threats posed by hydrogen sulfide, ammonia, PM, and other pollutants. Some states have taken steps to protect public health and the environment from these emissions. Most of these efforts have been limited in scope, however, and have not addressed the full range of air pollutants emitted by factory farms.

1. State Authority

Under the Clean Air Act, states have the obligation to include laws and regulations in their federally approved SIPs that are sufficient to attain the NAAQS and to avoid any significant deterioration of air quality in areas already in attainment.[152] States also have a duty to control emissions of PM and other criteria pollutants from factory farms through preconstruction and operating permits regardless of the status of EPA's enforcement moratorium under the Consent Agreement.[153] Additionally, states are required to issue preconstruction permits and establish BACT standards for all regulated NSR pollutants, including those emitted by factory farms, such as hydrogen sulfide, ammonia, PM, VOCs, methane, and nitrous oxide.[154]

Moreover, while many programs in the Clean Air Act—such as the NSPS, NSR, Title V, and HAP programs—are limited to facilities that meet the definition of "stationary sources,"[155] states have much more flexibility in developing their SIPs. In an effort to reduce PM emissions, for example, a SIP may include provisions designed to reduce dust (PM) associated with the spraying of manure from factory farms onto crop fields as fertilizer even though these provisions are unlikely to qualify as controls on emissions from "stationary sources."[156]

States also retain broad authority to enact their own laws and regulations governing air pollution above and beyond what the Clean Air Act and other federal statutes require.[157] They may adopt emission controls and reporting requirements for toxic air pollutants associated with factory farm operations, such as hydrogen sulfide and ammonia, regardless of whether these pollutants have been regulated under the Clean Air

---

148. *Id.* at **20-21.
149. *Id.* at **18-19.
150. Association of Irritated Residents v. EPA, 494 F.3d 1027, 1028-37 (D.C. Cir. 2007).
151. *Id.* at 1037.
152. 42 U.S.C. §§7410(a)(2)-(6), 7471.
153. *See id.* §§7410(a)(2)(C), 7475, 7503.
154. *Id.* §7475(a)(4).
155. *Id.* §7411(a)(2)-(6) (NSPS); §7475(a) (New Source Review); §7602(j) (general definitions); §§7661, 7661a(a), 7661b (Title V); §7412(a) (HAP).
156. *See id* §7410(a)(2)(A) ("Each implementation plan . . . shall . . . include enforceable emission limitations and other control measures, means, or techniques . . . , as well as schedules and timetables for compliance, as may be necessary or appropriate to meet the applicable requirements of this chapter."). For nonattainment areas, the statute requires reasonably available control measures and reasonably available control technology, without limiting these to particular sources. *See id.* §7502(c).
157. *See* 70 Fed. Reg. 4958 et seq. (explaining that the Consent Agreement does not address criminal violations, imminent and substantial endangerment, state enforcement actions, independent state laws, and citizen enforcement actions). *See generally* 42 U.S.C. §7416 (retention of state authority to adopt or enforce more stringent requirements under the Clean Air Act) and §11041 (nothing in EPCRA shall preempt any state or local law).

Act, CERCLA, or EPCRA. States are likewise free to regulate other air pollutants—such as PM, VOCs, and greenhouse gases—more stringently than the federal government or to fill in gaps in federal regulation where it is lacking.

States therefore have extensive authority to regulate air pollution from factory farms and other agricultural operations. As explained below, some states have shown a willingness to exercise this authority to varying degrees, while others have remained reluctant to do so.

2. State Permitting

EPA's Consent Agreement with the factory farm industry included explicit language clarifying that it would not have any effect on state permitting activities under the Clean Air Act.[158] In practice, however, EPA's reluctance to implement and enforce Clean Air Act permitting requirements has encouraged states to drag their feet as well. As discussed above, EPA has decided to forego any enforcement until it completes monitoring studies and develops emission factors for factory farms. This presents a hurdle for state enforcement because it suggests to a reviewing court that it is impossible to determine factory farm emissions with a reasonable level of accuracy in the absence of such emission factors.[159] In addition, the lack of enforcement at the federal level tends to enhance industry's sense that it is justified in failing to comply with applicable laws and this, in turn, tends to compound the political difficulties for states in bringing aggressive enforcement actions.[160]

California is an example of a state that has been reluctant to regulate agricultural air emissions and has been embroiled in controversy relating to this permitting issue for the past several years. The San Joaquin Valley in California is a heavily agricultural region, and it has some of the worst air quality in the nation largely because of emissions from dairy farms and other agricultural activities. Nevertheless, in 1998, the San Joaquin Valley Unified Air Pollution Control District (Air District) adopted a rule exempting agricultural equipment from its preconstruction permitting rules.[161] In 2000, EPA disapproved of the exemption because it conflicted with Clean Air Act requirements.[162] The Air District removed the exemption in its entirety in 2002, and EPA approved the revised exemption-free rules in 2004.[163]

In addition to the San Joaquin Valley rule, a longstanding statewide California law exempted "any equipment used in agricultural operations in the growing of crops or the raising of fowl or animals" from the obligation to obtain an operating permit.[164] Environmental advocates sued EPA in 2002, alleging that its approval of this Title V exemption violated the Clean Air Act.[165] EPA settled the litigation and promulgated a notice of deficiency in 2002, which explained that the statute's exemption of agricultural equipment was illegal because it conflicted with the Clean Air Act's Title V operating permit requirements.[166] Furthermore, in 2003, EPA determined that the California SIP was substantially inadequate based on the agricultural exemption and called upon California to revise its SIP.[167] In response, the California Legisla-

---

158. 70 Fed. Reg. 4958, 4959 ("In addition, the Agreement will not affect the ability of States or citizens to enforce compliance with nonfederally enforceable State laws, existing or future, that are applicable to AFOs.").
159. It is, in fact, possible to estimate emissions from many factory farms with reasonable certainty, as demonstrated by the EPA enforcement actions in the early 2000s. Nevertheless, EPA's current approach could create confusion and impede state enforcement efforts.
160. *See, e.g.,* Claudia Copeland, Cong. Research Serv., Air Quality Issues and Animal Agriculture: EPA's Air Compliance Agreement 6-7 (updated Apr. 9, 2008), *available at* http://www.nationalaglawcenter.org/assets/crs/RL32947.pdf ("Many among those who support the [Consent Agreement] believe that livestock operations should be entirely exempt from CERCLA and EPCRA reporting requirements because, in their view, Congress did not intend for these laws to apply to animal agriculture.").
161. U.S. EPA, Approving Implementation Plans; California State Implementation Plan Revision, San Joaquin Valley Unified Air Pollution Control District, 65 Fed. Reg. 58252, 58253-54 (proposed Sept. 28, 2000) (asserting that Rule 2020 was not approvable because it contained an exemption for "any equipment used in agricultural operations in the growing of crops or the raising of fowl or animals," which could apply to major sources subject to NSR under the Clean Air Act).
162. U.S. EPA, Final Approval and Promulgation of Implementation Plans; California State Implementation Plan Revision, San Joaquin Valley Unified Air Pollution Control District, 66 Fed. Reg. 37587, 37589-90 (July 19, 2001) (requiring that the district remove the agricultural exemption).
163. In 2003, EPA proposed approval of the agriculture exemption-free Rule 2020. U.S. EPA, Revisions to the California State Implementation Plan, San Joaquin Valley Unified Air Pollution Control District, 68 Fed. Reg. 7330 (proposed Feb. 13, 2003). Subsequently, EPA approved the December 19, 2002 version of Rule 2020 as part of the State Implementation Plan. U.S. EPA, Revisions to the California State Implementation Plan; San Joaquin Valley Unified Air Pollution Control District, 69 Fed. Reg. 27837, 27838 (May 17, 2004).
164. Cal. Health & Safety Code §42310(e) (2003).
165. *See* Settlement Agreement, *Association of Irritated Residents v. EPA,* Consolidated Case Nos. 02-70160, 02-70177, 02-70191 (9th Cir., May 22, 2002), *available at* http://www.epa.gov/region9/air/ca/titlevsettlement0502.pdf.
166. *Id.* U.S. EPA, Notice of Deficiency for 34 Clean Air Act Operating Permits Programs in California, 67 Fed. Reg. 35990 (May 22, 2002).
167. U.S. EPA, Finding of Substantial Inadequacy of Implementation Plan; Call for California State Implementation Plan Revision, 68 Fed. Reg. 37746, 37747 (June 25, 2003).

ture passed U.S. Senate Bill 700, which removed the agricultural exemption, but to a lesser extent than the Air District rules had done because SB 700 continued to exempt certain sources.[168]

In 2005, environmental advocates filed three citizen suits in federal court seeking to enforce the newly revised San Joaquin Valley Air District permitting requirements against the owners and operators of several dairy farms in the heavily polluted San Joaquin Valley.[169] The first of the cases reached a decision in 2007, and the court held that the dairy farm defendants had violated the Air District's permitting rules by not obtaining a permit prior to construction and not installing BACT.[170] At that point, California was poised to be one of the first states to start issuing permits to factory farms under the Clean Air Act.

The new court precedent, however, prompted the agriculture industry to pressure the state of California and EPA for leniency. In 2008 and 2009, California submitted three proposed amendments to its rules for the San Joaquin Valley Air District, including amendments that would incorporate by reference the agricultural exemptions set forth in SB 700.[171] Moreover, in January 2010, EPA proposed to revise its prior approval of the Air District rules in the SIP to correct its purported "error" in approving these rules (which completely abolished the agricultural exemption) because they were in conflict with SB 700 (which allowed some facilities to remain exempt).[172] EPA thus proposed to revise its SIP approval to give effect to the agricultural exemptions in SB 700 and thereby reinstate the same kind of agricultural exemption it had deemed illegal back in 2002.[173] In March 2010, environmental advocates submitted comments urging EPA not to finalize its proposed action.[174]

In May 2010, EPA deferred its proposed action to correct its "error," explaining that it would be seeking a legal opinion from the state of California regarding the extent of the Air District's authority to regulate agricultural sources of air pollution under California law.[175] At the same time, however, EPA issued a limited approval and limited disapproval of the proposed rule amendments that the Air District had submitted in 2008 and 2009. With respect to the proposed agricultural exemptions, EPA reasoned that "approval of the amended Rules 2020 and 2201 would be consistent with regional planning efforts to attain and maintain the NAAQS" and that "the exemption for minor agricultural sources from the offset requirements is consistent with federal requirements."[176] On this basis, EPA approved the amended rules and allowed them to become part of the SIP.[177] Nevertheless, because of the ambiguity created by the rules' cross-reference to SB 700, which had not been approved by EPA as part of the SIP, EPA deemed the rules deficient and required California to submit revised rules.[178] California did so through two submittals in May and September 2011,[179] and EPA issued a proposed approval of these rules containing the revived agricultural exemptions in December 2011.[180]

Based on a survey of public records, the Center for Race, Poverty and the Environment has determined that there are 132 dairy facilities in the San Joaquin Valley that should be subject to preconstruction permitting under the Clean Air Act.[181] The Air District has determined, however, that all of these dairies are minor sources subject only to the less burdensome state permitting requirements, and it has issued state

---

168. Cal. Health & Safety Code §42310(b); SB 700 §11, 2003-2004 Sess. (Cal. 2003), *available at* http://www.arb.ca.gov/ag/sb700/sb700.pdf.
169. *See* Association of Irritated Residents v. C&R Vanderham Dairy et al., No. 05-01593 (E.D. Cal.); Association of Irritated Residents v. Fred Schakel Dairy et al., No. 05-00707 (E.D. Cal.); Association of Irritated Residents v. Foster Farms, Inc. et al., No. 06-01648 (E.D. Cal.).
170. *See* Association of Irritated Residents v. C&R Vanderham Dairy, 2007 WL 2815038 at *29 (E.D. Cal., Sept. 25, 2007) (*Vanderham*).
171. *See* U.S. EPA, Approval and Promulgation of Implementation Plans, State of California, San Joaquin Valley Unified Air Pollution Control District, New Source Review, Proposed Rule, 75 Fed. Reg. 26102, 26102 (May 11, 2010).
172. *See* U.S. EPA, Approval and Promulgation of Implementation Plans, State of California, San Joaquin Valley Unified Air Pollution Control District, New Source Review, Proposed Rule, 75 Fed. Reg. 4745 (Jan. 29, 2010).
173. *See* Comments on Proposed Rulemaking (EPA Docket Nos. EPA-R09-OAR-2010-0062 and EPA-R09-OAR-2009-0269), submitted by Brent Newell, Center for Race Poverty, and the Environment et al., to Gerardo Rios, EPA Region IX (Mar. 1, 2010) (citing EPA Proposed Rule, 75 Fed. Reg. 4745).
174. *See id.*
175. *See* 75 Fed. Reg. at 26103.
176. *Id.* at 26105.
177. *See id.* at 26111.
178. *See id.*
179. *See* Letter from R. Fletcher, California Air Resources Board, to J. Blumenfeld, EPA-Region 9 (May 19, 2011) (enclosing SIP revision materials for San Joaquin Valley Air District Rule 2201); and Letter from R. Fletcher, California Air Resources Board, to J. Blumenfeld, EPA-Region 9 (Sept. 28, 2011) (enclosing SIP revision materials for San Joaquin Valley Air District Rule 2020).
180. *See* U.S. EPA, Approval and Promulgation of Implementation Plans, State of California, San Joaquin Valley Unified Air Pollution Control District, New Source Review, 76 Fed. Reg. 76112 (Dec. 6, 2011) (proposed rule); U.S. EPA, Interim Final Determination to Defer Sanctions, San Joaquin Valley Unified Air Pollution Control District, 76 Fed. Reg. 76046 (Dec. 6, 2011).
181. *See* Comments on Proposed Rulemaking, *supra* note 173.

permits to only a few of these sources.[182] EPA's recent proposal to approve California's amended rules reinstating the agricultural exemptions would, if finalized, reinforce the validity of the state's approach.

Idaho has been involved in similar litigation and controversy concerning permitting of factory farms under the provisions of its EPA-approved SIP, but it has emerged with a better outcome than California has to date. In 2004, the Idaho Conservation League filed a lawsuit in federal court opposing Adrian Boer's proposal to construct a factory farm, K&W Dairy, in Gooding County, Idaho that would house 5,750 milking cows and 840 nonmilking cows.[183] The League alleged that Mr. Boer was unlawfully planning to construct a dairy farm without obtaining a "permit to construct" from the state of Idaho before beginning construction.[184] The League alleged that this obligation would be triggered because the lagoons, barns, and milking parlors at the dairy farm had the capacity to release emissions of ammonia, hydrogen sulfide, and PM_{10} in quantities exceeding 100 tons per year.[185] In September 2004, the federal district court denied the defendant's motion for summary judgment, rejecting the defendant's argument that ammonia and hydrogen sulfide were not subject to regulation under the Clean Air Act, and that emissions from different units at a facility should not be aggregated in making the threshold determination.[186]

Following this ruling, the case was settled and, as part of the settlement, the parties agreed to ask the Idaho Department of Environmental Quality to undertake a negotiated rulemaking to establish a process for permitting dairies.[187] In 2006, Idaho finalized a rule that requires dairies to obtain air quality permits through a permit-by-rule process if they emit 100 tons or more of ammonia per year.[188] This rule made Idaho the first state to regulate ammonia emissions from factory farms.[189] Approximately 53 dairy facilities are currently operating under Idaho's specialized permit-by-rule for dairy farms.[190] After registering for coverage under the permit-by-rule, facilities must comply with best management practices for waste storage and treatment systems; vegetative and wooded buffers; manure management in barns, open lots, and corrals; animal nutrition; composting practices; and land application of manure.[191]

3. State Air Quality Standards

Although only a few states are implementing air permitting requirements for factory farms, a majority of states have adopted air quality standards for hydrogen sulfide.[192] California, for instance, has a risk-based management program, with an acute reference exposure level (REL) of 42 mg/m^3 for one hour and a chronic exposure REL of 10 mg/m^3 (8 ppb, or 0.008 ppm).[193] California requires facilities to model both cancer and noncancer risks for all listed toxic substances, including hydrogen sulfide. Facilities having a noncancer risk below a hazard quotient (HQ) of 1 for hydrogen sulfide do not have to conduct any further assessments or implement control measures. If the HQ is greater than 1, the air district conducts a risk analysis and, if the risk is determined to be significant, the facility must undergo a public notification process and implement a risk-reduction plan.[194]

---

182. *See id.*
183. Complaint at ¶¶ 1, 29, *Idaho Conservation League v. Adrian Boer dba K&W Dairy*, 362 F. Supp. 2d 1211 (D. Idaho May 26, 2004).
184. *Id.*, ¶¶ 37-38.
185. *Id.*, ¶¶ 32, 36.
186. *Idaho Conservation League*, 362 F. Supp. 2d 1211, 1214-18 (D. Idaho 2004).
187. Stipulation for Dismissal, *Idaho Conservation League*, 362 F. Supp. 2d 1211 (executed Mar. 3, 2005).
188. *See* IDAHO ADMIN. CODE r. 58.01.01.760 -.764 (2011); Idaho Department of Environmental Quality, *Permit by Rule for Dairies*, http://www.deq.idaho.gov/permitting/air-quality-permitting/permit-by-rule/dairies.aspx; Idaho Department of Environmental Quality, Rules for the Control of Ammonia From Dairy Farms, *available at* http://www.deq.idaho.gov/media/635657-58_0101_0502_fact_sheet3.pdf. In Idaho, a permit-by-rule is a process in which a facility registers with the Idaho Department of Environmental Quality and complies with a set of standardized air pollution control requirements set forth in the rule that are applicable to all members of the same industry, rather than obtaining and complying with an individual permit. *See* Idaho Department of Environmental Quality, *Permit by Rule*, http://www.deq.idaho.gov/permitting/air-quality-permitting/permit-by-rule.aspx (last visited Sept. 2, 2012).
189. CLAUDIA COPELAND, CONGRESSIONAL RESEARCH SERVICE, AIR QUALITY ISSUES AND ANIMAL AGRICULTURE: A PRIMER 11-12 (Dec. 15, 2010), *available at* http://www.nationalaglawcenter.org/assets/crs/RL32948.pdf.
190. *See* Current Permits Issued by DEQ, sorted by Air-PBR-Dairy, *at* http://www.deq.idaho.gov/permitting/issued-permits.aspx?records=all&type=Air+-+PBR+-+Dairy&sort=nameAscending (last visited Nov. 28, 2012).
191. *See* IDAHO ADMIN. CODE r. 58.01.01.764.
192. Sierra Club, Hydrogen Sulfide Petition, *supra* note 66, at 5.
193. *Id.* at 10 (citing http://www.oehha.org).
194. *Id.*

Other states have established more traditional air quality standards for hydrogen sulfide, seen in Table 10.1.

Table 10.1
Selected State Air Quality Standards for Hydrogen Sulfide

State	Hydrogen Sulfide Standards	Averaging Period
New York	14 ppb (0.01 ppm)	One hour, average concentration[1]
Pennsylvania	5 ppb (0.005 ppm)	24 hours, average concentration[2]
	100 ppb (0.1 ppm)	One hour, average concentration[3]
Iowa	30 ppb (0.03 ppm)	One hour (health effects value)[4]
	30 ppb (0.03 ppm)	One hour daily maximum, exceeded more than seven times per year (health effects standard)[5]
Minnesota	30 ppb (0.03 ppm)	Half-hour average not to be exceeded over 2 times in any 5 consecutive days[6]
	50 ppb (0.05 ppm)	Half-hour average not to be exceeded over 2 times per year[7]
Missouri	30 ppb (0.03 ppm)	Half-hour average not to be exceeded over 2 times in any 5 consecutive days[8]
	50 ppb (0.05 ppm)	Half-hour average not to be exceeded over 2 times per year[9]
Texas	80 ppb (0.08 ppm)	30 minutes (residential, business, or commercial receiving property)[10]
	120 ppb (0.12 ppm)	30 minutes (all other properties, including industrial property and vacant tracts and range lands not normally occupied by people)[11]

1. N.Y. COMP. CODES R. & REGS. tit. 6, §257-10.3 (2011).
2. 25 PA. CODE §131.3 (2011). See CLAUDIA COPELAND, AIR QUALITY ISSUES AND ANIMAL AGRICULTURE: A PRIMER, *supra* note 190, at 16.
3. *Id.*
4. IOWA ADMIN. CODE r. 567-32.3 (455B) (2011); Endres and Grossman, *Air Emissions From Animal Feeding Operations*, *supra* note 31, at 17 nn. 104-05.
5. IOWA ADMIN. CODE r. 567-32.4 (455B).
6. MINN. R. 7009.0080 (2011). See Endres & Grossman, *Air Emissions From Animal Feeding Operations*, *supra* note 31, at 10-11 nn.51-53.
7. *Id.*
8. MO. CODE REGS. Ann. tit. 10, §10-6.010 (2011). See Endres & Grossman, *Air Emissions From Animal Feeding Operations*, *supra* note 31, at 21 n.135.
9. *Id.*
10. 30 TEX. ADMIN. CODE §112.31 (2011). See Endres & Grossman, *Air Emissions From Animal Feeding Operations*, *supra* note 31, at 29 nn.196-98.
11. 30 TEX. ADMIN. CODE §112.32.

In addition, a handful of states, particularly the top pork-producing states, have adopted numerical nuisance-based limits for odors, as listed in Table 10.2.

Table 10.2
Selected State Odor Standards

State	Odor Standard
Colorado	No odor shall be detected at any off-site receptor after odorous air is diluted with two or more volumes of odor-free air.[1]
Missouri	No odor can be perceived after odorous air taken at a location outside of the property boundary is diluted with seven volumes of odor-free air.[2]
Illinois	No odor can be detectable, after odorous air taken from residential property or adjacent property is diluted with eight volumes of odor-free air.[3]

1. 5 COLO. CODE REGS. §1001-4:B.III.B.1 (2011). See Endres & Grossman, *Air Emissions From Animal Feeding Operations*, *supra* note 31, at 36 nn. 257-58.
2. MO. CODE REGS. Ann. tit. 10, §10-6.165(3).
3. ILL. ADMIN. CODE tit. 35, §245.121(a) (2011). See Endres & Grossman, *Air Emissions From Animal Feeding Operations*, *supra* note 31, at 34 nn.234-36.

Measures designed to address unpleasant odors can potentially have significant public health benefits because the main pollutants causing the odors are hydrogen sulfide and ammonia,[195] which pose serious threats to human health, as mentioned above.

4. Other State Requirements

A majority of states have adopted a variety of other planning and operational requirements designed to minimize air pollutant emissions and odors from factory farms.[196] These measures include setback distances, manure-handling training, best-management practices, plans for controlling odor and air emissions, and public notice and comment periods.[197] Some of the most important air pollution reduction measures, developed by a few states, establish requirements for the design, construction, and operation of manure lagoons and other waste storage facilities. For instance, Iowa requires the use of lagoon aeration techniques to convert anaerobic bacterial processes to aerobic processes in order to reduce the generation of hydrogen sulfide, ammonia, and other pollutants.[198] Colorado requires anaerobic waste impoundments to be covered, with emissions vented to a treatment facility rather than the ambient air.[199] Texas and Illinois regulations similarly include design, construction, and operational requirements.[200]

E. Greenhouse Gas Controls

The struggle to apply PSD and Title V permitting requirements to factory farms is now converging with the debate over the applicability of these permitting requirements to greenhouse gas emissions. After years of controversy concerning whether EPA has authority to regulate greenhouse gas emissions at all under the Clean Air Act, the U.S. Supreme Court settled the question in a landmark 2007 ruling. In *Massachusetts v. U.S. Environmental Protection Agency*, the Court held that "greenhouse gases fit well within the Clean Air Act's capacious definition of 'air pollutant.'"[201] The Court also held that, in determining whether and how to proceed with the regulation of greenhouse gases, EPA must make a reasoned decision based on scientific evidence concerning whether greenhouse gases cause "may reasonably be anticipated to endanger public health or welfare."[202] The Court emphasized that EPA may not base its decision on policy judgments that it may be "unwise" to regulate greenhouse gases or that voluntary programs will be sufficient to address global warming.[203]

Two years after the Supreme Court's ruling, in December 2009, EPA issued a determination that greenhouse gases endanger public health and welfare.[204] In particular, EPA found that elevated concentrations of six greenhouse gases in the atmosphere "may reasonably be anticipated both to endanger public health and to endanger public welfare" for current and future generations,[205] explaining that "the body of scien-

---

195. Hydrogen sulfide has a characteristic odor of rotten eggs. *See, e.g.,* Mich. Dep. Envtl. Qual., Concentrated Animal Feedlot Operations (CAFOs) Chemicals Associated With Air Emissions 12-13 (May 2006), *available at* http://www.michigan.gov/documents/CAFOs-Chemicals_Associated_with_Air_Emissions_5-10-06_158862_7.pdf. Ammonia also has a sharp odor, which is familiar to people because it is commonly used in household cleaning agents and smelling salts. *See id.* at 3-4.
196. Gretchen Vander Wal, National Hog Farmer, *44 States Regulate Odors on Hog Farms* (Mar. 15, 2001), *available at* http://nationalhogfarmer.com/mag/farming_states_regulate_odors/ (citing Jarah Redwine, a graduate research assistant at Texas A&M University, who has researched agricultural odor regulations and has found a wide range of standards and rules across the country).
197. *Id.* Copeland, Air Quality Issues and Animal Agriculture: A Primer, *supra* note 189, at 15-16.
198. Iowa Code Ann. §459.206(1), §459.207(1) ("airborne pollutant" includes hydrogen sulfide, ammonia, and odor) (2011).
199. 5 Colo. Code Regs. §1001-4:B.IV. *See* Endres & Grossman, *Air Emissions From Animal Feeding Operations, supra* note 30, at 34 n.239.
200. 30 Tex. Admin. Code §321.37 (construction), §321.38 (design), §321.39 (operation). Endres & Grossman, *Air Emissions From Animal Feeding Operations, supra* note 30, at 27-28 nn.186-92. Ill. Admin. Code tit. 8, §§900.501 et seq. (livestock waste handling facilities other than lagoons), §§900.601 et seq. (lagoon livestock waste handling facilities), §§900.801 et seq. (livestock waste management plan). Endres & Grossman, *Air Emissions From Animal Feeding Operations, supra* note 30, at 32 nn.218-21.
201. Massachusetts v. E.P.A., 549 U.S. 497, 532, 127 S. Ct. 1438, 1462 (2007).
202. *Id.* at 532-33, 127 S. Ct. at 1462, citing §7521(a)(1).
203. *Id.* at 532-34, 127 S. Ct. at 1462-63.
204. *See* U.S. EPA, Endangerment and Cause or Contribute Findings for Greenhouse Gases Under Section 202(a) of the Clean Air Act, 74 Fed. Reg. 66496, 66497-98 (Dec. 15, 2009).
205. *Id.* at 66497.

tific evidence compellingly supports this finding."[206] These greenhouse gases include methane (CH_4) and nitrous oxide (N_2O), which are emitted in large quantities by factory farms.[207]

1. PSD and Title V Permitting

The vast majority of factory farms surpass the statutory "major source" thresholds of 100 or 250 tons per year based on their emissions of methane, nitrous oxide, and other greenhouse gases. Now that it is clear that greenhouse gases are subject to regulation under the Clean Air Act, the statute imposes a direct obligation on those factory farms to obtain PSD and Title V permits and implement BACT to control their emissions.

Indeed, factory farms and other industrial facilities tend to emit greenhouse gases in quantities several orders of magnitude greater than conventional air pollutants. In order to avoid overwhelming regulators with a tremendous increase in the number of applications for federal air permits, EPA issued a rule in June 2010 temporarily raising the applicability thresholds to 100,000 tons per year for new facilities and 75,000 tons per year for major modifications of existing facilities.[208] Many large factory farms will exceed even these thresholds. EPA has estimated that under the new thresholds approximately 37,351 agricultural facilities will require Title V permits, and that 299 agricultural facilities will require PSD permits.[209] EPA also explicitly provided that agricultural facilities would become subject to federal permitting requirements starting in January 2011 where their greenhouse gas emissions exceed the thresholds established in the rule.

There is nothing in the Consent Agreement that protects factory farms from enforcement if they fail to comply with federal permitting requirements based on their greenhouse gas emissions. EPA's agreement not to enforce permitting requirements was limited to those relating to the pollutants to be studied under the Consent Agreement, and EPA has explicitly noted that "[g]reenhouse gas emissions were not measured as part of this study."[210]

The urgency of the climate crisis may prompt EPA and states to take a fresh look at the importance of ensuring that factory farms obtain and comply with PSD and Title V permits, and this may lead them to resume the enforcement efforts initiated in the early 2000s to bring about compliance with these requirements by the factory farm industry. If so, the establishment of BACT standards for methane, nitrous oxide, and other greenhouse gases—e.g., requiring the use of anaerobic digesters, aeration, lagoon covers, or some combination of technology and best-management practices—would likely have significant side benefits since many of these measures would simultaneously reduce conventional and toxic pollutant emissions as well.

2. NSPS and Mobile Source Rules

As discussed above, there is also potential for greenhouse gas emissions from factory farms to be regulated directly by EPA under the NSPS program. EPA's first step would be to list factory farms as an industrial category, and then it would promulgate a BDT standard addressing their greenhouse gas emissions, as urged by environmental group petitioners in 2009.[211] Those petitioners emphasized the significant contribution of factory farms to U.S. greenhouse gas emissions. For instance, the petition highlights the fact that enteric fermentation and manure management accounted for "over 16 percent of United States nitrous oxide emissions, more than all energy-related nitrous oxide emissions combined," as well as "27 percent of all United States methane emissions."[212] The petition also notes that in 2006 factory farms were responsible for emissions of "almost 9 million tons of methane, or almost 185 million tons of carbon dioxide

---

206. *Id.*
207. *See* U.S. EPA, Inventory of U.S. Greenhouse Emissions and Sinks: 1990-2010, EPA 430-R-12-001, at 6-1 to 6-3 (Apr. 15, 2012), *available at* http://epa.gov/climatechange/emissions/usinventoryreport.html (last visited May 3, 2012).
208. U.S. EPA, Prevention of Significant Deterioration and Title V Greenhouse Gas Tailoring Rule, 75 Fed. Reg. 31514 (June 3, 2010). *See* Coalition for Responsible Regulation v. EPA, No. 09-1322 (D.C. Cir. June 26, 2012), *petition for reh'g denied* (D.C. Cir. Dec. 20, 2012) (upholding the Tailoring Rule in its entirety).
209. 75 Fed. Reg. 31514, 31597 (Table VI-1—Estimated Number of Affected Sources Experiencing Regulatory Relief).
210. U.S. EPA, *Fact Sheet—The National Air Emissions Monitoring Study: Data Availability and Call for Additional Information*, http://www.epa.gov/airquality/agmonitoring/fs20110113.html.
211. *See* The Humane Society of the United States et al., CAFO NSPS Petition, *supra* note 54.
212. *Id.*

equivalent," and that these emissions are growing as livestock production continues to move toward more concentrated and confined feeding operations, which tend to rely on liquid manure management systems that produce more greenhouse gas emissions.[213]

Agricultural activities also generate substantial emissions of carbon dioxide and other greenhouse gases through their use of farm equipment and machinery. These emissions have the potential to be regulated under EPA's rules for nonroad engines as part of the Clean Air Act's mobile source program.[214] In 2008, a group of six states and a coalition of several environmental groups filed separate petitions urging EPA to regulate greenhouse gases from nonroad engines, including those used in agricultural operations.[215] The environmental groups later sued to compel EPA to respond to their petition and make a determination that nonroad engines contribute significantly to greenhouse gas pollution.[216] In July 2011 the U.S. District Court for the District of Columbia dismissed the plaintiffs' claim that EPA has a nondiscretionary duty to make a cause-or-contribute finding for nonroad engines, but the court allowed them to proceed with their claim that EPA had unreasonably delayed in responding to their petition.[217] In June 2012, however, EPA announced that it would not be issuing rules for nonroad engines "in the near or medium term," explaining that it lacked adequate resources to address nonroad engines and would be prioritizing other sources.[218]

3. State and Regional Measures

In the months and years to come, factory farms could also become subject to greenhouse gas regulations under states' broad residual authority to regulate air pollution. Already, some state and regional initiatives have begun to incorporate measures to address greenhouse gas emissions from factory farms. For instance, in the scoping plan developed to implement California's Global Warming Solutions Act of 2006, the "dominant strategy for reducing agricultural GHG emissions" is to "promote the voluntary adoption of manure digester technology,"[219] and the effectiveness of this voluntary approach will be reevaluated in 2013 to "determine if the program should be made mandatory by 2020."[220] Similarly, the Northeast's Regional Greenhouse Gas Initiative allows states to receive carbon dioxide offset allowances for projects that "capture and destroy methane from animal manure and organic food waste using anaerobic digesters."[221]

Conclusion

In 1970, belching black smoke from industrial facilities captured the nation's attention, and the public demanded that their government restore air quality to healthy levels. As a result, air pollution from most industries is now being addressed under one or more programs of the Clean Air Act. The equally unhealthy

---

213. *Id.* at 29-30.
214. *See* 42 U.S.C. §7547; 40 C.F.R. Parts 89, 90, 1039, 1048, 1054, 1060; U.S. EPA, *Nonroad Diesel Equipment*, http://www.epa.gov/nonroad-diesel/ (last visited Nov. 28, 2012).
215. *See, e.g.,* Petition for Rulemaking Seeking the Regulation of Greenhouse Gas Emissions From Nonroad Vehicles and Engines, From California and Five Other States et al., to U.S. EPA, at 8-15 (Jan. 29, 2008), *available at* http://ag.ca.gov/cms_attachments/press/pdfs/n1522_finaldraft-nonroadpetition3.pdf#xml=http://search.doj.ca.gov:8004/AGSearch/isysquery/5714cc7f-1003-40e0-945b-1cad3b22ec18/1/hilite/.
216. The lawsuit was filed in the U.S. District Court for the District of Columbia on June 11, 2010, by Earthjustice and the Western Environmental Law Center on behalf of the Center for Biological Diversity, the Center for Food Safety, Friends of the Earth, the International Center for Technology Assessment, and Oceana. *See* Complaint at 1, 2, Center for Biological Diversity et al. v. U.S. EPA (D.D.C. June 11, 2010), *available at* http://earthjustice.org/sites/default/files/library/legal_docs/mobile-source-ghg-petitions-complaint-10-06-11-final.pdf.
217. Center for Biological Diversity v. U.S. EPA, __ F. Supp. 2d __, 2011 WL 2620995 (D.D.C. 2011); Memorandum Opinion and Order, Henry H. Kennedy Jr., U.S. District Judge (July 5, 2011), *available at* http://earthjustice.org/sites/default/files/11-07-05.mms_.opinion.pdf.
218. U.S. EPA, Memorandum in Response to Petitions Regarding Greenhouse Gas and Other Emissions From Marine Vessels and Nonroad Engines and Vehicles (June 2012), *available at* http://www.eenews.net/assets/2012/06/18/document_pm_06.pdf.
219. California Environmental Protection Agency, Air Resources Board, http://www.arb.ca.gov/cc/scopingplan/agriculture-sp/agriculture-sp.htm (last visited Nov. 28, 2012). *See* Assembly Bill No. 32, *available at* http://www.arb.ca.gov/cc/docs/ab32text.pdf; THE CALIFORNIA AIR RESOURCES BOARD FOR THE STATE OF CALIFORNIA, CLIMATE CHANGE SCOPING PLAN: A FRAMEWORK FOR CHANGE (December 2008), *available at* http://www.arb.ca.gov/cc/scopingplan/document/adopted_scoping_plan.pdf. Efforts to suspend and delay the implementation of AB32 have not been successful. *See* Margot Roosevelt, *Prop. 23 Battle Marks New Era in Environmental Politics*, L.A. TIMES, Nov. 4, 2010, at http://articles.latimes.com/2010/nov/04/local/la-me-global-warming-20101104.
220. California Environmental Protection *Agency, Air Resources Board*, http://www.arb.ca.gov/ag/manuremgmt/manuremgmt.htm (last visited Nov. 28, 2012).
221. Regional Greenhouse Gas Initiative, *Manure Management*, http://www.rggi.org/market/offsets/categories/manure_management (last visited Nov. 28, 2012). *See also* RON RAUSCH, NYS DEPARTMENT OF AGRICULTURE AND MARKETS, NYS AGRICULTURAL NONPOINT SOURCE ABATEMENT AND CONTROL PROGRAM, at 14-24, *available at* http://www.nyserda.org/innovationsinagriculture/presentations/session2_april17/ron_rausch_04172007.pdf.

air pollution from the factory farm industry, however, remains largely uncontrolled more than 40 years later. With the lives and health of millions of Americans threatened by agricultural air pollution, and as we struggle to meet the challenge of global climate change, many public health and environmental advocates are urging EPA and states to make better use of the tools available to them to rein in factory farm air pollution. In today's turbulent political climate, however, this is likely to be a challenging endeavor.

Chapter 11
Agriculture and the Endangered Species Act
William S. Eubanks II

In many respects, the Endangered Species Act (ESA)[1] is the federal environmental statute with the strongest substantive and procedural mandates to protect the natural environment, particularly wildlife species threatened with the risk of extinction.[2] This mandate is especially forceful in the agricultural context because, in stark contrast to other environmental laws (see Chapters 9 and 10), the U.S. Congress has never carved out an exemption to the ESA for any agricultural activity.[3] Therefore, while seen as draconian by some farmers (and other landowners) subject to its provisions,[4] the ESA stands alone among environmental laws in imposing a uniform mandate that serves our nation's paramount interests in ecosystem protection and wildlife conservation.[5]

A. The ESA: Statutory and Regulatory Framework

The ESA is a complex statute and has been the subject of entire treatises devoted to the intricacies of the law and its implementing regulations.[6] This section provides a primer on pertinent ESA provisions and regulations, with a focus on the issues where the statute has traditionally been, and will likely in the future be, applied to private individuals, corporations, and governmental entities in the agriculture and food realm.[7]

1. Background

Congress enacted the ESA in 1973 "to provide a means whereby the ecosystems upon which endangered species and threatened species depend may be conserved, [and] to provide a program for the conservation of such endangered species and threatened species."[8] As the U.S. Supreme Court explained in its landmark ruling in *Tennessee Valley Authority v. Hill*, the ESA "[a]s it was finally passed . . . represent[s] the most comprehensive legislation for the preservation of endangered species ever enacted by any nation."[9] Congress recognized that such a statute was necessary to protect at-risk wildlife and plant species, which are in some cases "rendered extinct as a consequence of economic growth and development untempered by adequate

---

1. 16 U.S.C. §§1531-1544.
2. *See, e.g.*, J.B. Ruhl & James Salzman, *Climate Change, Dead Zones, and Massive Problems in the Administrative State: A Guide for Whittling Away*, 98 Cal. L. Rev. 59, 59 (2010) (calling the ESA "one of the most potent environmental laws"); Sandra B. Zellmer & Scott A. Johnson, *Biodiversity in McElligot's Pool*, 38 Idaho L. Rev. 473, 480 (2002) (calling the ESA the 'pitbull' of environmental laws").
3. *See, e.g.*, Ved P. Nanda, *Agriculture and the Polluter Pays Principle*, 54 Am. J. Comp. L. 317, 335 (2006) (explaining that "[f]arms are not exempted under th[e] [Endangered Species] Act").
4. *E.g.*, Pacific Legal Foundation, Endangered Species Act blog, http://blog.pacificlegal.org/tag/endangered-species-act/.
5. *E.g.*, J.B. Ruhl, *Agriculture and the Environment: Three Myths, Three Themes, Three Directions*, 25 Environs Envtl. L. & Pol'y J. 101, 107 (2002) (opining that the ESA is "unyielding and uncompromising," and arguing that "[r]esolving agri-environmental policy through the ESA is not pretty, and does not usually lead to inventive solutions").
6. *See, e.g.*, Endangered Species Act: Law, Policy, and Perspectives (Donald C. Baur & W. Robert Irvin eds., 2d ed. 2010); Endangered Species Deskbook (Lawrence Liebesman & Rafe Petersen eds., 2d ed. 2010).
7. It should be noted that the ESA's nondiscretionary wildlife protection mandates are separate and distinct from voluntary Farm Bill programs aimed at incentivizing farmers to conserve wildlife and wildlife habitat. The most notable example of such a program is the Wildlife Habitat Incentives Program, which "pays up to 75 percent of the cost to private land owners of enhancing wildlife habitat on their land" and which resulted in enhanced protection on nearly 813,000 acres of land in 2009. Nat'l Wildlife Fed'n, *Background on Farm Bill and Wildlife*, http://www.nwf.org/Wildlife/Policy/Farm-Bill/Farm-Bill-Background.aspx#WHIP.
8. 16 U.S.C. §1531(b).
9. Tennessee Valley Auth. v. Hill, 437 U.S. 154, 180 (1978).

concern and conservation."[10] Indeed, as the Supreme Court has opined in clarifying the broad reach of the ESA, the statute "reveals a conscious decision by Congress to give endangered species priority over the 'primary missions' of federal agencies" and further "shows clearly that Congress viewed the value of endangered species as 'incalculable'" and thus deemed their protection of a higher priority than economic activities, including agriculture, food production, and food distribution.[11] The purpose of this policy—which is referred to as an "institutionalization of caution" with respect to endangered and threatened species conservation—is "to halt and reverse the trend toward species extinction, whatever the cost."[12]

This institutionalization of caution is embodied by various substantive and procedural mandates in the ESA. As the Supreme Court noted, it is "reflected not only in the stated policies of the Act, *but in literally every section of the statute.*"[13] Certain mandates are triggered even before a species is listed, during which time the lead agency—the U.S. Fish and Wildlife Service (FWS) for terrestrial and freshwater species (including plant species)[14] or the National Marine Fisheries Service (NMFS) for marine and anadromous species—is considering relevant biological evidence to determine whether listing is warranted, and the statutory mandates continue to protect a species once it is listed until a species has recovered to the point where listing is no longer necessary or until a species has gone extinct.[15]

2. Section 4—The Listing Process

The listing process is governed by §4 of the ESA.[16] FWS or NMFS is required within certain statutorily imposed deadlines to consider and respond to a petition submitted by any "interested person" seeking to add a species to the list of endangered or threatened species.[17] Listing determinations under the ESA must be made "solely on the basis of the best scientific and commercial data available,"[18] which has been interpreted, to the chagrin of many private landowners and legal commentators, to preclude the agency from "consider[ing] economic or social impacts as part of the listing determination."[19] FWS or NMFS *must* list a species under the ESA if the respective agency finds, based on the best available scientific evidence, that a species is "endangered" or "threatened," due to any one of five factors:

(A) the present or threatened destruction, modification, or curtailment of its habitat or range;

(B) overutilization for commercial, recreational, scientific, or educational purposes;

(C) disease or predation;

(D) the inadequacy of existing regulatory regimes;

(E) other natural or manmade factors affecting its continued existence.[20]

At the end of the listing process, FWS or NMFS must determine whether listing is warranted. If it is, the agency must then list the species as "endangered" or "threatened."[21] An endangered status is given to a species "which is in danger of extinction throughout all or a significant portion of its range,"[22] whereas

---

10. 16 U.S.C. §1531(a)(1).
11. *Hill*, 437 U.S. at 184, 187-88.
12. *Id.* at 184, 194; *see also id.* at 174 (explaining that an "examination of the language, history, and structure of the legislation under review here indicates beyond doubt that Congress intended endangered species to be afforded the highest of priorities"); *id.* at 194 ("Congress has spoken in the plainest of words, making it abundantly clear that the balance has been struck in favor of affording endangered species the highest of priorities, thereby adopting a policy which it described as 'institutionalized caution.'").
13. *Id.* at 184 (emphasis added).
14. 16 U.S.C. §1532(14) (defining "plant" under the ESA as "any member of the plant kingdom, including seeds, roots and other parts thereof").
15. *See generally id.* §1536.
16. For more detailed information on the ESA listing process, see J.B. Ruhl, *Listing Endangered and Threatened Species*, in Endangered Species Act: Law, Policy, and Perspectives, *supra* note 6, at 16-39; Oliver A. Houck, *The Endangered Species Act and Its Implementation by the Departments of Interior and Commerce*, 64 Colo. L. Rev. 277 (1993).
17. 16 U.S.C. §1533(b)(ii)(3)(A).
18. *Id.* §1533(b)(1)(A).
19. Mary Jane Angelo & Mark T. Brown, *Incorporating Emergy Synthesis Into Environmental Law: An Integration of Ecology, Economics, and Law*, 37 Envtl. L. 963, 984 (2007). For more discussion on implementation of the ESA's best available science standard, see Holly Doremus, *The Purposes, Effects, and Future of the Endangered Species Act's Best Available Science Mandate*, 34 Envtl. L. 397, 419-26 (2004); J.B. Ruhl, *The Battle Over Endangered Species Act Methodology*, 34 Envtl. L. 555 (2004).
20. 16 U.S.C. §§1533(a)(1)(A)-(E).
21. *Id.* §1533(b)(3)(B).
22. *Id.* §1532(6).

a threatened status is ascribed to a species "which is likely to become an endangered species within the foreseeable future throughout all or a significant portion of its range."[23] Although an endangered status connotes a more urgent risk of extinction, the protections afforded by the two categories are essentially identical unless FWS or NMFS promulgates a special rule under §4(d) that in some way circumscribes the protections afforded a threatened species—a restriction the agency is statutorily precluded from imposing on species listed as endangered.[24]

When a species is listed as endangered (or threatened, assuming no special rule was promulgated under §4(d) of the ESA circumscribing protections for the threatened species), various threshold mandates are triggered that afford stringent protections. For example, §4 requires FWS or NMFS to "concurrently with making a [listing] determination . . . designate any habitat of such species which is then considered to be critical habitat."[25] Congress defined "critical habitat" as "the specific areas within the geographical area occupied by the species . . . essential to the conservation of the species."[26] To further achieve the ESA's primary goal of species recovery, §4 mandates that the respective agency create a "recovery plan" for each listed species, which must include "a description of such site-specific management actions as may be necessary to achieve the plan's goal for the conservation and survival of the species," and which must incorporate "objective, measurable criteria which, when met, would result in a determination, in accordance with the provisions of this section, that the species be removed from the list."[27] Finally, §4 requires FWS and NMFS, in cooperation with the states, to monitor listed species through a five-year review process, in order to determine which species have successfully recovered to the point where listing and the protections of the ESA are no longer necessary.[28]

3. Section 9—The Take Prohibition

The listing of a species immediately triggers several substantive and procedural protections designed to safeguard the newly listed species from further risk of extinction and to begin the arduous recovery process that few species have achieved to date.[29] The most notable protection afforded listed species is the categorical prohibition on "take" of listed wildlife species, which is found in §9 of the ESA.[30] Congress defined "take" very broadly to include any action that would "harass, harm, pursue, hunt, shoot, wound, kill, trap, capture, or collect" any member of a listed wildlife species—or even any attempt to engage in any of these prohibited acts—and made it unlawful for *any person* to commit any such acts.[31] FWS has, by regulation, further defined the "harm" and "harass[ment]" forms of prohibited take to generally include any intentional or negligent activity that annoys, or otherwise actually injures wildlife, by disrupting or impairing normal behavioral patterns such as breeding, feeding, and sheltering.[32]

Importantly, in contrast to listed wildlife species, listed plant species are *not* subject to the take prohibition in §9, but instead are subject to less stringent prohibitions on the import, transport, or sale of endangered or threatened plant species and on removal of listed plants from federal lands.[33]

---

23. *Id.* §1532(20).
24. *Id.* §1533(d).
25. *Id.* §1533(a)(3)(A)(i).
26. *Id.* §1532(5)(A)(i).
27. *Id.* §1533(f)(1)(B); *see also* Houck, *supra* note 16, at 296-315 (discussing the critical habitat designation process and its effectiveness); Federico Cheever, *Critical Habitat*, in Endangered Species Act: Law, Policy, and Perspectives, *supra* note 6, at 40-69.
28. 16 U.S.C. §1533(g); *see* Houck, *supra* note 16, at 344-51 (discussing recovery plans and questioning the effectiveness of implementation of certain plans); Dale D. Goble, *Recovery*, in Endangered Species Act: Law, Policy, and Perspectives, *supra* note 6, at 70-103.
29. While the ESA is undoubtedly effective in enhancing protection for listed species, relatively few species have actually achieved full recovery sufficient to warrant delisting. *See, e.g.*, Michael J. Bean, *Historical Background of the Endangered Species Act*, in Endangered Species Act: Law, Policy, and Perspectives, *supra* note 6, at 13 (explaining that of the "nearly 2,000 species . . . in peril of extinction . . . [only] [a] few have been safely pulled back from the brink").
30. 16 U.S.C. §1538(a)(1)(B). For more detailed discussion on the ESA's take prohibition, see Patrick A. Parenteau, *The Take Prohibition*, in Endangered Species Act: Law, Policy, and Perspectives, *supra* note 6, at 146-59.
31. *Id.*; 16 U.S.C. §1532(19); *see also* Babbitt v. Sweet Home Chapter of Cmtys. for a Great Or., 515 U.S. 687, 704-05 (1994) (discussing breadth of take prohibition); Forest Conservation Council v. Rosboro Lumber Co., 50 F.3d 781, 784 (9th Cir. 1995) (explaining that the ESA defines "take" in the "broadest possible manner to include every conceivable way in which a person can 'take' or attempt to 'take' any fish or wildlife") (citation omitted).
32. 50 C.F.R. §17.3.
33. 16 U.S.C. §1538(a)(2); *see also* Zellmer & Johnson, *supra* note 2, at 481 (explaining, as a shortcoming of the ESA, the fact that "the statute fails to protect plant species on private lands"). For more detailed discussion on the ESA's disparate treatment for plants and wildlife, see Holly Wheeler, *Plants*, in Endangered Species Act: Law, Policy, and Perspectives, *supra* note 6, at 249-51.

4. Sections 7 and 10—Interagency Consultation and Incidental Take Permits

The only narrow ways in which one can lawfully engage in an otherwise unlawful take of a listed species is with direct authorization from FWS or NMFS pursuant to §§7 or 10 of the ESA.[34] Without such authorization, or in excess of such authorization, *any* take is per se unlawful and subject to the ESA's civil and criminal penalties, contained in §11, which can result in up to $25,000 worth of civil penalties per violation, up to $50,000 worth of criminal penalties, and up to one year of imprisonment.[35]

In the event that the federal government does not prosecute an ongoing or likely future unauthorized take of a listed species, private citizens or organizations may commence a civil suit, called a "citizen suit," "to enjoin any person, including the [federal government], who is alleged to be in violation of any provision of [the ESA] or regulation issued under the authority thereof."[36] Congress' inclusion of this sweeping citizen suit provision in the ESA reinforces its understanding that citizen involvement would prove critical to species survival and recovery. Indeed, the Supreme Court has noted in various opinions that "[c]itizen involvement was encouraged by the Act,"[37] particularly by the citizen suit provision that serves as "an authorization of remarkable breadth when compared with the language Congress ordinarily uses," even as compared to other environmental statutes such as the Clean Water Act, and that "the obvious purpose of the [ESA citizen suit] provision . . . is to encourage enforcement by so-called 'private attorneys general'" to supplement the federal government's prosecution efforts that are often severely limited due to resource, personnel, and political constraints.[38]

As noted above, a private individual, company, or nonfederal actor can shield itself from §9 liability by obtaining and subsequently complying with limited take authorization under §10 of the ESA, when certain conditions are satisfied.[39] In this section, Congress created three narrow exemptions from the take prohibition—an incidental take permit, a scientific permit, and a hardship exemption.[40] Most pertinent here is an incidental take permit, which was designed by Congress to authorize certain otherwise lawful activities by private landowners that have, as an unintentional consequence, a likelihood of taking members of a listed species, but which can only be granted if certain rigorous conditions are met, including the development of a habitat conservation plan that analyzes alternative actions and contains measures to minimize and mitigate the take of listed species, and after a determination by FWS or NMFS that the amount of take authorized "will not appreciably reduce the likelihood of the survival and recovery of the species in the wild."[41]

Likewise, actions taken by federal agencies—or activities by nonfederal actors who are permitted, licensed, or authorized by federal agencies—that will result in otherwise unlawful takes of listed species are strictly prohibited unless authorization is obtained from FWS or NMFS to undertake the action at issue, although this authorization comes via §7 of the ESA instead of §10.[42]

Along with the §9 take prohibition, §7 imposes several of the ESA's most critical substantive mandates. First, all federal agencies are required to "carry out programs for the conservation of endangered species and threatened species."[43] Second, the agencies must consult with FWS and/or NMFS before taking any action that might take listed species. These two restrictions "insure that any action authorized, funded, or carried out by such action . . . is not likely to jeopardize the continued existence of any [listed species] or result in the destruction or adverse modification of [critical] habitat."[44] As the Supreme Court explained in

---

34. 16 U.S.C. §§1536, 1539.
35. *Id.* §1540(a)-(b).
36. *Id.* §1540(g)(1)(A). For a more detailed discussion on the ESA's citizen suit provision, see Eric R. Glitzenstein, *Citizen Suits*, in Endangered Species Act: Law, Policy, and Perspectives, *supra* note 6, at 260-91.
37. *Tennessee Valley Auth.*, 437 U.S. at 180-81.
38. Bennett v. Spear, 520 U.S. 154, 164-65 (1997).
39. 16 U.S.C. §1539.
40. *Id.* §§1539(a)-(b).
41. *Id.* §1539(a)(1)(B). For a more detailed discussion on the interaction of §§9 and 10 of the ESA, see Steven P. Quarles & Thomas R. Lundquist, *Land Use Activities and the Section 9 Take Prohibition*, in Endangered Species Act: Law, Policy, and Perspectives, *supra* note 6, at 160-91; Sam Kalen & Adam Pan, *Exceptions to the Take Prohibition*, in Endangered Species Act: Law, Policy, and Perspectives, *supra* note 6, at 192-205; Douglas P. Wheeler & Ryan M. Rowberry, *Habitat Conservation Plans and the Endangered Species Act*, in Endangered Species Act: Law, Policy, and Perspectives, *supra* note 6, at 220-45; Patrick Duggan, *Incidental Extinction: How the Endangered Species Act's Incidental Take Permits Fail to Account for Population Loss*, 41 ELR 10628 (July 2011).
42. 16 U.S.C. §1536.
43. *Id.* §1536(a)(1).
44. *Id.* §1536(a)(2).

Hill, "[o]ne would be hard pressed to find a statutory provision whose terms were any plainer than those in [§]7 of the Endangered Species Act."[45]

To satisfy this mandate, the agency seeking FWS' authorization to proceed with a project that might adversely affect listed species (the "action agency") engages with FWS or NMFS in either informal or formal consultation, depending on whether such impacts are possible or likely.[46] In formal consultation, where impacts to listed species are likely, the action agency consults with FWS and/or NMFS to determine the foreseeable effects expected to result to listed species, in order to insure that jeopardy will not occur from the action.[47] To accomplish this goal, the action agency generally begins the consultation process by preparing for FWS or NMFS a biological assessment, "for the purpose of identifying any endangered species or threatened species which is likely to be affected by such action."[48] After receiving the biological assessment, FWS or NMFS must review the best available scientific evidence on the species and render a biological opinion.[49]

If the biological opinion determines that jeopardy is likely and there exist no reasonable and prudent alternatives to the action that can avoid jeopardy, the action may not proceed (unless the action agency opts to proceed at its own peril without take authorization from FWS or NMFS).[50] If, however, the opinion concludes that jeopardy is likely but can be avoided through specific alternatives deemed by FWS or NMFS to be "reasonable and prudent," the action may proceed subject to those conditions should the action agency accept those parameters in exchange for incidental take coverage enumerated in the opinion.[51] Finally, if the opinion concludes that jeopardy is unlikely, but that incidental take is nonetheless likely to occur, FWS or NMFS typically includes nondiscretionary "reasonable and prudent measures" that the action agency or its licensee/permittee must undertake in order to insulate itself from §9 liability via an incidental take statement.[52]

A biological opinion that authorizes the action at issue must include an incidental take statement, assuming any take is anticipated, that serves three important functions: (1) providing an accounting of the amount of take authorized by a particular agency action that will necessarily inform FWS' or NMFS' decisions as to other agency decisions affecting the same species in the future; (2) minimizing and mitigating take to the extent possible to protect affected species; and (3) creating a precise and numeric take threshold where possible, which serves as a trigger for "reinitiation of consultation" if the amount of take authorized in an opinion is exceeded, or other new and unforeseen circumstances warrant reinitiation.[53] As with the §4 listing process, "[i]n fulfilling the requirements of [§7 consultation] each agency shall use the best scientific and commercial data available."[54]

To further promote and enforce the ESA's take prohibition unless and until take authorization is granted pursuant to either §7 or §10 of the Act, §7 precludes "any irreversible or irretrievable commitment of resources with respect to the agency action which has the effect of foreclosing the formulation or implementation of any reasonable and prudent alternative measures" by "the Federal agency and the permit or license applicant."[55] This provision was designed so that actions cannot proceed with impacts to listed species before FWS or NMFS has had an opportunity to review the best available science with respect to the

---

45. *Hill*, 437 U.S. at 173.
46. 16 U.S.C. §§1536(a)(2)-(3); *see also* 50 C.F.R. §402.02 (defining informal consultation as "an optional process that includes all discussions, correspondence, etc., between the Service and the Federal agency or the designated non-Federal representative prior to formal consultation," and formal consultation as "a process between the Service and the Federal agency that commences with the Federal agency's written request for consultation under §7(a)(2) of the Act and concludes with the Service's issuance of the biological opinion under §7(b)(3) of the Act").
47. 16 U.S.C. §1536(a)(2); *see also* 50 C.F.R. §402.14.
48. 16 U.S.C. §1536(c)(1); *see also* 50 C.F.R. §402.12.
49. 16 U.S.C. §1536(b)(1).
50. *See id.* §1536(b)(4).
51. *Id.* For discussion about the coercive effect of biological opinions, see *Bennett v. Spear*, 520 U.S. at 168-70 (rejecting the notion that "the 'action agency' . . . retains ultimate responsibility for determining whether and how a proposed action shall go forward" because of "the virtually determinative effect of . . . biological opinions . . . [which] alter[] the legal regime to which the action agency is subject" by imposing "nondiscretionary" measures).
52. 16 U.S.C. §1536(b)(4)(c)(ii).
53. 50 C.F.R. §402.14(g)(7); 50 C.F.R. §402.16.
54. 16 U.S.C. §1536(a)(2); *see also* 50 C.F.R. §§402.14(d)-(g). For a more detailed discussion on interagency consultation under §7 of the ESA, see Patrick W. Ryan & Erika E. Malmen, *Interagency Consultation Under Section 7*, in ENDANGERED SPECIES ACT: LAW, POLICY, AND PERSPECTIVES, *supra* note 6, at 104-25.
55. 16 U.S.C. §1536(d); 50 C.F.R. §402.09.

project and to render a biological opinion pursuant to §7 in order to ensure that jeopardy is not likely to result from the action and that unauthorized takes will not result in the interim.

5. ESA Litigation

To understand how the ESA mandates apply in practice generally—and thus how they have been applied in the past, and how they likely will be applied in the future to domestic agricultural and food activities—it is important to understand how alleged violations of the ESA are challenged in the judicial system, by whom, and how such challenges differ depending on the provision of the ESA upon which the alleged violation originates. Two questions need to be answered.

First, who can bring suits under the ESA? Although most of the ESA lawsuits filed since the statute's enactment in 1973 have been instituted by individuals and organizations seeking to *enhance* environmental protections by enforcing provisions of the Act against alleged violators, the Supreme Court expressly held in a seminal 1997 ruling that any person—even those "seeking to *prevent application of environmental restrictions* rather than to implement them"—can invoke the ESA citizen suit provision under §11.[56] Therefore, in contexts where the ESA's citizen suit provision provides judicial review for an alleged violation, the Court held that a plaintiff need not show that it meets a zone-of-interest test (also called the "prudential standing test") to establish Article III standing to bring such a case, which is a threshold jurisdictional requirement that any plaintiff must meet before a federal court will review the merits of its legal challenge. Rather, a plaintiff in an ESA citizen suit must simply prove injury in fact—that he or she is injured or will be injured in a concrete, particularized, and legally cognizable manner—from the alleged violation (even a purely economic injury suffices), a traceable causation from the defendant's allegedly unlawful conduct to the plaintiff's asserted injury, and that a favorable decision can redress the plaintiff's asserted injury.[57] However, where the citizen suit provision does not provide judicial review—meaning that review must instead occur under the more generalized catch-all judicial review provision of the Administrative Procedure Act (APA)[58]—an ESA plaintiff must still satisfy the zone-of-interest test to establish that the plaintiff's allegedly injured interests are the types of "interests sought to be protected by the statutory provision whose violation forms the legal basis for his complaint."[59] The APA paradigm is not unique to ESA suits that fall outside of the ESA's citizen suit provision; the APA provides for judicial review in any context where "[a] person suffer[s] [a] legal wrong because of agency action, or [is] adversely affected or aggrieved by agency action within the meaning of a relevant statute."[60]

Second, when does the citizen suit provision govern judicial review of ESA lawsuits and when does the APA govern instead? In §11, the ESA's citizen suit provision "expressly vests federal district courts with jurisdiction to order the Secretary [of the Interior or Commerce] 'to perform any act or duty' arising under §4, as well as to 'enforce' any 'provision or regulation' with which 'any person' covered by the Act must comply."[61] In that section, the ESA allows citizen suits—suits directly reviewable by federal district courts that, as described above, have relaxed standing requirements (i.e., the "zone of interest" test does not apply)—"to enjoin any person, including the United States and any other governmental instrumentality or agency . . . alleged to be in violation of any provision of th[e] [ESA] or regulation issued under the [ESA]" or "against the Secretary [of Interior or Commerce] where there is alleged a failure of the Secretary to perform any act or duty under section [§4] which is not discretionary."[62] In *Bennett v. Spear,* and subsequent cases, the Supreme Court has explained that citizen suits under the ESA are proper not only for challenging §9 take violations against any party but also for challenging FWS' or NMFS' implementation of any nondiscretionary listing or critical habitat duties under §4 of the ESA (i.e., that FWS or NMFS

---

56. Bennett v. Spear, 520 U.S. 154, 166 (1997) (emphasis added).
57. *Id.* at 164-67.
58. 5 U.S.C. §§500-912.
59. *Bennett*, 520 U.S. at 176 (quotation mark omitted; emphasis deleted).
60. 5 U.S.C. §702.
61. Eric R. Glitzenstein, *Citizen Suits*, in ENDANGERED SPECIES ACT: LAW, POLICY, AND PERSPECTIVES, *supra* note 6, at 261.
62. 16 U.S.C. §1540(g)(1).

misapplied or ignored §4 altogether in making the decision at issue), as well as any action agency's reliance on an unlawful biological opinion.[63]

Outside of these specific contexts, the ESA's citizen suit provision does not provide judicial review; however, the APA authorizes judicial review by federal district courts where "there is no other adequate remedy in a court."[64] Accordingly, in such situations where the ESA citizen suit provision is not applicable—for example, challenges to emergency listing decisions, challenges to any decision document issued by FWS or NMFS under §7 (including biological opinions), or challenges to permits issued by FWS or NMFS under §10—plaintiffs must plead their case under the APA.[65]

The distinction between a challenge under the ESA or the APA is important for several reasons. First, as explained above, the requirements for standing are more relaxed where a challenge arises solely under the ESA because, unlike in an APA case, the zone-of-interest test need not be satisfied. Second, while the arbitrary and capricious standard of review in an ESA case is the same as in an APA case if an underlying agency decision is the crux of the challenge,[66] which is deferential to the agency's expertise in most circumstances, the standard of review in certain types of ESA citizen suits—namely §9 suits alleging unlawful take of listed species—is a more lenient *de novo* standard of review conducted not on an administrative record compiled by an agency but rather on evidence proffered by the parties either on summary judgment or in an evidentiary hearing. Third, because the expense of litigation can be a major hurdle for both environmental organizations and the regulated community seeking to challenge allegedly unlawful action under the ESA, the type of suit is an important consideration for potential litigants, because citizen suits under the ESA generally offer a more generous calculation of attorneys' fees to reimburse successful parties for vindicating their ESA claims than do suits under the APA, where successful plaintiffs are reimbursed at generally reduced rates under the Equal Access to Justice Act.[67] Fourth, an APA suit does not require plaintiffs to provide advance notice of violations before filing a lawsuit; however, an ESA citizen suit requires, as a jurisdictional matter, that a plaintiff provide at least 60 days of notice to any alleged violator of the ESA that is a target of a suit *and* the appropriate secretary (Commerce or Interior), depending on the species affected.[68] These and other important considerations guide the propriety of ESA litigation in various situations, and conversely shape the conduct of private parties and agencies that may or may not be routinely involved in lawsuits challenging their compliance with the Act's mandates.

The ESA is a complex statute imposing various mandates on every individual, corporation, and entity in the United States. While questions abound concerning potential major overhauls to the ESA in future years due to an increasingly "sharp partisan and ideological conflict" that persists in Congress today,[69] until such modifications occur, the current structure and broad reach of the ESA will inevitably require interests on all sides of the American agricultural and food sector to factor the ESA's mandates into internal decisionmaking whether in the business, governmental, or environmental realm.

B. Application of the ESA to Crop-Based Agricultural Inputs

When poorly practiced, agriculture can have serious ecological impacts.[70] To achieve ever-growing crop yields, farmers often manipulate large amounts of four key inputs—high-yield seeds, fertilizers, water, and pesticides. This combination of ingredients ushered in by the Green Revolution of the 1960s continues today, but has had the unintended result of severe and extensive resource degradation, including to our

---

63. Eric R. Glitzenstein, *Citizen Suits*, in Endangered Species Act: Law, Policy, and Perspectives, *supra* note 6, at 265-70; *Bennett*, 520 U.S. at 172-74.
64. 5 U.S.C. §704.
65. Eric R. Glitzenstein, *Citizen Suits*, in Endangered Species Act: Law, Policy, and Perspectives, *supra* note 6, at 266-67.
66. *Id.* at 270-71.
67. 16 U.S.C. §1540(g)(4).
68. *Id.* §1540(g)(2); Eric R. Glitzenstein, *Citizen Suits*, in Endangered Species Act: Law, Policy, and Perspectives, *supra* note 6, at 273.
69. Michael J. Bean, *Historical Background of the Endangered Species Act*, *in* Endangered Species Act: Law, Policy, and Perspectives, *supra* note 6, at 13-14; *see also* K. Mollie Smith, *Abuse of the Warranted But Precluded Designation: A Real or Imagined Purgatory*, 19 SE Envtl. L.J. 119, 120 (2011) (noting that, as of 2009, there were only 1,897 ESA-listed species, despite the fact that 10,306 species are estimated to be at risk of extinction).
70. *See supra* Chapters 1-7.

nation's biodiversity. Indeed, some of the cumulative and permanent impacts to our ecology are likely no longer capable of restoration to their natural state.[71] As two legal scholars recently explained:

> The agricultural sector continues to rely primarily on heavy inputs of chemical fertilizers and pesticides, resulting in polluted runoff and serious degradation of aquatic habitats. Whereas in past centuries ecosystems could recover from localized disturbances of terrestrial and aquatic systems, the size of human populations and the scale of economic activity and resource consumption today make ecosystem restoration much more difficult.[72]

This fundamental shift in cultivation practices starting with the Green Revolution is central to understanding how the current agricultural model encouraged by the successive farm bills and implemented by the vast majority of American farmers comes into regular conflict with the ESA. While the four key inputs described above have undoubtedly become more innovative since the introduction of this agricultural framework in the 1960s, and its adherents undoubtedly more technologically savvy to keep pace, our nation's natural resources, and particularly our imperiled species, have not seen a correlative reduction in the level of degradation caused by agriculture or the risks posed by farming inputs.[73] Indeed, approximately "84% of the plants and animals listed as endangered or threatened were listed in part due to agricultural activities," and there appears to be no change in sight.[74]

Accordingly, with litigation and other ESA conflicts between farmers, environmentalists, and the federal government inevitably on the rise, there are several agricultural scenarios in which the ESA has already created, or is anticipated to create, conflicts between divergent interests—the resolution of which will necessarily depend on the statutory and regulatory mandates described above, the willingness of diverse stakeholders to find creative and wildlife-protective solutions short of litigation, and, as a matter of last resort, pertinent ESA jurisprudence if compromise proves impossible. Because each key agricultural input is unique in the ESA issues that its use implicates, a life-cycle analysis from seed to final plant crop will explore the application of the ESA as to each input.

1.   Crop Seeds

Arguably the most important ingredient in any farmer's arsenal, the seed is the origin of life from which all crops, and thus all foods, sprout. The importance of selecting the best seed in any given year to withstand myriad climatic and environmental variances cannot be underestimated, although technology has gradually removed much of the guesswork that previously plagued farmers during the seed selection process.

Less than two decades ago there were no genetically engineered (GE) seeds planted in the United States.[75] Today, however, a handful of multinational corporations have effectively cornered the national (and world) seed market in the name of making seeds more efficient, meaning that the seeds have been genetically engineered in a laboratory to resist drought or higher-than-normal precipitation, or to tolerate certain pesticides or herbicides that can be doused on the seeds in order to kill weeds competing for space or insects that serve as common agricultural pests.[76] This heavily concentrated market of GE seed producers has fundamentally changed the composition of our nation's seeds and with it the way in which we cultivate food.[77] As of 2011, 94% of soybeans, 75% of cotton, and 72% of corn derive from GE seeds.[78]

---

71. *E.g.*, William S. Eubanks II, *A Rotten System: Subsidizing Environmental Degradation and Poor Public Health With Our Nation's Tax Dollars*, 28 STAN. ENVTL. L.J. 213, 222 n.38, 251-73 (2009) (providing history of the Green Revolution and analyzing the environmental impacts of the Green Revolution).
72. John Kostyack & Dan Rohlf, *Conserving Endangered Species in an Era of Global Warming*, SR021 ALI-ABA 147 (2009).
73. *See, e.g.*, Eubanks, *A Rotten System*, *supra* note 71, at 251-73.
74. Defenders of Wildlife, Comments for the Development of USDA Recommendations for the 2007 Farm Bill (70 Fed. Reg. 35221 (June 17, 2005)), at 1, *available at* http://familyfarmer.org/sections/pdf/farmbillforum.pdf.
75. Madison Smith, *Who Owns Your Dinner? A Discussion of America's Patented Genetically Engineered Food Sources, and Why Reform Is Necessary*, 23 LOY. CONSUMER L. REV. 182, 186 (2010).
76. Mary Jane Angelo, *Regulating Evolution for Sale: An Evolutionary Biology Model for Regulating the Unnatural Selection of Genetically Modified Organisms*, 42 WAKE FOREST L. REV. 93, 95 (2007) ("Many products have been modified by genetic engineering to possess traits that increase their ability to reproduce and survive in the environment. Such traits include insect resistance, viral infection resistance, drought tolerance, and temperature tolerance in crop plants."); *see also* Smith, *supra* note 75.
77. Smith, *supra* note 75.
78. U.S. Dept. of Agriculture, Economic Research Service, ADOPTION OF GENETICALLY ENGINEERED CROPS IN THE U.S. (2011), *available at* http://ers.usda.gov/Data/BiotechCrops/.

Other chapters of this book explore certain regulatory implications of GE foods and their impact on the environment and public health.[79] GE seeds, by their nature, also have profound effects on biodiversity, particularly endangered and threatened wildlife and plant species. Under various scenarios those impacts can invoke specific provisions of the ESA.

First, studies have found that certain wildlife and insect species, including pollinating species such as butterflies, exhibit negative symptomology and higher mortality rates when consuming pollen from GE crops that have genetically inserted *Bacillus thuringiensis* (Bt) than similarly situated individuals consuming only non-Bt crop pollen, although that finding must be balanced against the fact that use of Bt crops often leads to a reduction in use of chemical pesticides that can have even more dramatic effects on pollinators than consumption of Bt pollen in isolation.[80] Moreover, both commercial non-GE crops and wild plants are highly susceptible to cross-pollination, also known as "gene flow," when GE crop pollen is blown in the wind or carried by pollinators from field to field causing unintended and unwanted pollination.[81]

Second, particularly with herbicide-tolerant and pesticide-tolerant forms of GE seeds, there is evidence suggesting that farmers are substantially increasing the amount of toxic herbicides and pesticides sprayed on fields because crop loss from misapplied pesticides is no longer a concern due to the seeds' newfound resistance to these potent chemicals.[82] This marked increase in pesticides has correlative severe negative impacts on aquatic and terrestrial species (of both the wildlife and plant variety), as discussed in more detail in Chapter 3 and as further described in the ESA context below.[83]

Third, the use of GE seeds, and thus the overuse of certain herbicides and pesticides to which the seeds are genetically resistant, is likely to create invasive "superweeds" or "superpests" that gradually develop immunity to the chemicals and exhibit ultracompetitive and nonevolutionary traits compared to their wild, unaltered counterparts.[84] Because these superweeds and superpests have been shown to invade and displace native plants and animals, and thus rapidly disrupt the balance of the natural ecology,[85] the impacts of these entirely new and extra-competitive species must be evaluated under the ESA to ensure that the effects do not tip the balance towards extinction for any particular species.

FWS or NMFS is required during the ESA's §4 listing process to analyze the effects of GE seeds on any plant or animal species under consideration for listing, assuming that there is a plausible concern that GE seeds or their derivative pollens are impacting the species under review via predation or alternatively by serving as a manmade factor affecting the species' continued existence.[86] Moreover, when the U.S. Department of Agriculture (USDA), through its Animal and Plant Health Inspection Service (APHIS), reviews an application or petition from a seed company for a particular GE crop seed for the purpose of ultimately determining whether the crop satisfies the statutory criteria for "deregulation" under the Plant Protection Act (PPA), the ESA §7 process is triggered. If, in the PPA process, APHIS grants the petition, that variety of GE crop seed moves from "regulated" status where it cannot be planted commercially to "deregulated" status, enabling it to be planted commercially because the agency has made certain determinations required under the PPA. The deregulation of that GE crop seed, however, cannot proceed unless APHIS and FWS/NMFS have met their statutory obligations, if any, under the ESA.

The first juncture during the PPA deregulation review process at which §7 of the ESA is triggered is before APHIS allows even experimental field testing of seeds on a limited basis. While the PPA and its implementing regulations allow APHIS the discretion to grant a preliminary field testing permit if the agency makes certain findings required by regulation,[87] courts have made clear that APHIS cannot grant such a permit and authorize field testing if listed species will be impacted, until APHIS consults with

---

79. *See* Chapters 6 and 16.
80. John E. Losey et al., *Transgenic Pollen Harms Monarch Larvae*, 399 Nature 214 (1999); *see also* Angelo, *Regulating Evolution*, *supra* note 76, at 101.
81. *See generally* Angelo, *Regulating Evolution*, *supra* note 76, at 101 & n.24.
82. *E.g.*, Friends of the Earth International, Who Benefits From GM Crops: The Rise in Pesticide Use 7-12 (2008), *available at* http://www.centerforfoodsafety.org/pubs/FoE%20I%20Who%20Benefits%202008%20-%20Full%20Report%20FINAL%202-6-08.pdf.
83. *See* Eubanks, *A Rotten System*, *supra* note 71, at 258-59, 263-66.
84. Angelo, *Regulating Evolution*, *supra* note 76, at 101; *see also* Margaret R. Grossman, *Anticipatory Nuisance and the Prevention of Environmental Harm and Economic Loss From GMOs in the United States*, 18 J. Env. L. & Prac. 107, 144-45 (2008).
85. Angelo, *Regulating Evolution*, *supra* note 76, at 101; *see also id.* at 102, 107-08.
86. 16 U.S.C. §§1533(a)(1)(C), (E).
87. 7 C.F.R. §340.4.

FWS and/or NMFS as appropriate under the circumstances and satisfies any obligations flowing from that consultation.[88] For example, when APHIS granted permits to several GE seed companies authorizing field testing in Hawaii of GE sugarcane and GE corn without any compliance with the ESA, the court found that APHIS violated the ESA because "APHIS skipped the initial, mandatory step of obtaining information about listed species and critical habitats from FWS and NMFS" embodied by §7(c)(1) of the ESA.[89] The court went on to explain that APHIS's granting of these permits "without fulfilling its congressionally mandated duty to obtain information from FWS and NMFS regarding endangered species, threatened species, and critical habitats . . . . constitutes an unequivocal violation of a clear congressional mandate."[90] In rendering its opinion, the court articulated a key point that §7 imposes certain categorical procedural obligations on APHIS, which the agency violates regardless of whether listed species are ultimately harmed by the field testing permits, opining that "[a]n agency violates the ESA when it fails to follow the procedures mandated by Congress, and an agency will not escape scrutiny based on the fortunate outcome that no listed plant, animal, or habitat was harmed."[91] Therefore, in considering and granting field testing permit applications, APHIS is statutorily required to comply with the procedural mandates of §7 by consulting with FWS and/or NMFS about the existence of any listed species to be affected by the permitted activity, and in turn with any additional mandates depending on whether such species exist in the area of concern.

The second, and, equally if not more important, juncture at which §7 obligations attach is at the conclusion of APHIS' deregulation review process, in the event that the agency ultimately deregulates the GE crop at issue. Although various cases have raised ESA claims with respect to APHIS' compliance with §7 in deregulating particular crops, all such cases are either currently pending in the federal court system or have been decided on other non-ESA grounds.[92] However, considering that at least one court has held that APHIS must comply with §7 as to limited field tests, logic compels that compliance with the requirements of §7 will also be required at the deregulation stage before the newly deregulated GE seed is authorized to be planted on a commercial scale with potentially wide-ranging extensive impacts to listed species.[93]

Because GE seeds are heavily regulated pursuant to the PPA via permits and deregulation decisions that themselves require §7 compliance, the ESA's obligations in this realm run to the federal agency—in this case APHIS, and of course FWS and NMFS in their consulting roles—meaning that individual seed scientists or farmers planting such seeds in compliance with the conditions of any field testing permit or any ultimate deregulation decision by APHIS should not incur liability under the ESA. Nor will §10 permits be required, since APHIS' §7 compliance, if lawfully conducted, will protect compliant users of GE seeds from individualized ESA challenges. Accordingly, the key areas where the ESA will inevitably intersect with the newly emerging technology of GE crop seeds are in the §4 listing process and in the §7 process with respect to several crucial points in the deregulation review process.

2. Fertilizers

No matter how resilient a seed's genes are, the ultimate health of a crop is highly dependent on the nutrients it receives. Synthetic fertilizers have become increasingly complex mixtures of chemical nutrients, including high concentrations of nitrogen, phosphorus, potassium, and other key ingredients, although some farmers still rely on natural fertilizers from animal manure.

Chapter 3 provides a thorough overview of the ecological impacts of the use, overuse, and misapplication of fertilizers. Notably, high volumes of fertilizer applied to agricultural fields end up as runoff, which is leached into streams and rivers and pollutes water bodies and harms aquatic species and fishing commu-

---

88. *E.g.*, Center for Food Safety v. Johanns, 451 F. Supp. 2d 1165 (D. Haw. 2005).
89. *Johanns*, 451 F. Supp. 2d at 1182.
90. *Id.*
91. *Id.*
92. *E.g.*, Center for Food Safety v. Vilsack, 844 F. Supp. 2d 1006 (N.D. Cal. 2012) (finding that APHIS is not required to consult over effects to listed species as part of deregulation because EPA, not APHIS, regulates pesticide use (decision has been appealed, *see* Center for Food Safety v. Vilsack, Civ. No. 12-15052 (9th Cir.))); Monsanto v. Geertson Seed Farms, 130 S. Ct. 2743, 2751 (2010) (noting that Plaintiffs' complaint to GE glyphosate-resistant alfalfa included ESA claims, but the district court only ruled on National Environmental Policy Act grounds); *cf.* Center for Biological Diversity, Civ. No. 10-14175 (S.D. Fla.) (challenge to APHIS' deregulation of non-food GE eucalyptus trees to be used for cosmetics and other products for human consumptive use was rejected on non-ESA grounds).
93. The GE alfalfa appeal currently pending before the U.S. Court of Appeals for the Ninth Circuit, *see* Center for Food Safety v. Vilsack, Civ. No. 12-15052 (9th Cir.), will be the first decision squarely addressing this question.

nities.⁹⁴ In several locales this confluence of nutrient-rich fertilizers has resulted in severe eutrophication, whereby the overabundance of the chemical nutrients leads to large algal blooms and deprives the water and resident aquatic species of oxygen, resulting in "dead zones" in the Gulf of Mexico and elsewhere.⁹⁵ As journalist Michael Pollan has noted, "[b]y fertilizing the world, we alter the planet's composition of species and shrink its biodiversity."⁹⁶ Due to the exponential rate at which fertilizer application is adversely impacting at-risk species and shrinking our biodiversity, the importance of the ESA in providing a reciprocal safeguard for those species is climbing proportionately.

As with other threats affecting an imperiled species, FWS or NMFS must analyze any potential impact of fertilizers during the ESA's §4 listing process in the event that the species at issue is adversely impacted by fertilizers as a manmade factor affecting a species' continued existence.⁹⁷ Unlike GE seeds, which are a relatively new concern for species under consideration for listing, fertilizers have long contributed to the decline of wildlife and plant species for which the public seeks ESA protection. In 2011, for example, FWS made a 90-day finding on a listing petition under §4 of the ESA, analyzing as a potential listing factor "high application rates of ammonium-based nitrogen fertilizers within the . . . watershed . . . . [because] [i]f these fertilizers get into the water, the high ammonia concentrations and other nutrient inputs can lead to excess algae growth, can cause oxygen depletion due to the growth and decomposition cycle of algae, and can cause increased biochemical oxygen demand as ammonia is transformed to nitrate-nitrogen."⁹⁸ Although FWS ultimately determined that listing was not warranted in that particular case, the example provides insight into the role that fertilizers play in affecting biodiversity, and thus the ways in which FWS or NMFS must take such impacts into account during listing.

Indeed, at least one lawsuit has been brought under the APA challenging FWS' negative 90-day finding (meaning that in FWS' view listing is not warranted based on the evidence presented in the listing petition) for a particular species on the grounds that the harm to the species from fertilizers was not accorded appropriate weight by FWS in rendering its finding. The court ultimately ruled in favor of FWS, finding that "it was not enough for the plaintiffs to allege that fertilizer applications have increased [because] FWS properly expected the plaintiffs to document the extent to which, and the quantities in which, ammonia-based fertilizers are being applied . . . [and therefore] FWS reasonably refused to infer that fertilizer applications in the Palouse bioregion presently threaten the [species'] habitat."⁹⁹

In addition to the ESA's application to fertilizers in the listing process, because of the vast harm that fertilizers cause aquatic species in particular as a result of nutrient loads and eutrophication, fertilizers pose a unique concern under §9 of the ESA. Unlike GE seeds and pesticides, which are strictly regulated under the PPA and the Federal Insecticide Fungicide and Rodenticide Act (FIFRA), respectively, fertilizers are not heavily regulated by the federal government. Although certain fertilizers are listed on the Toxic Substances Inventory pursuant to §8 of the Toxic Substances Control Act, there is no process whereby the U.S. Environmental Protection Agency (EPA) regulates or otherwise authorizes fertilizer use subject to consultation with FWS or NMFS, meaning that the burden of complying with the ESA's mandates generally falls to individual users of fertilizers pursuant to §10 of the ESA if take is anticipated.¹⁰⁰ The one rare exception to this rule, of course, is where a federal agency's authorization or permitting of an activity directly increases the concentration of nutrient-rich fertilizers in a particular location or water body with foreseeable impacts to the listed species residing therein, thus requiring the action agency to consult with FWS or NMFS.¹⁰¹

---

94. Eubanks, *A Rotten System, supra* note 71, at 255.
95. *Id.* at 255-56.
96. Michael Pollan, The Omnivore's Dilemma: A Natural History of Four Meals 47 (2006).
97. 16 U.S.C. §§1533(a)(1)(C), (E).
98. FWS, 90-Day Finding on a Petition to List the Straight Snowfly and Idaho Snowfly as Endangered, 76 Fed. Reg. 46238, 46243 (Aug. 2, 2011); *see also* NMFS, Final Endangered Status for a Distinct Population Segment of Anadromous Atlantic Salmon in the Gulf of Maine, 65 Fed. Reg. 69459, 69479 (Nov. 17, 2000) (listing a distinct population segment of Atlantic salmon due in part to "[d]ischarging . . . or dumping . . . fertilizers . . . into waters supporting the DPS").
99. Palouse Prairie Found. v. Salazar, Civ. No. 08-032, 2009 WL 415596, at *5 (E.D. Wash. Feb. 12, 2009).
100. EPA does not authorize fertilizer discharges under the Clean Water Act via permits because that statute specifically exempts most agricultural discharges, including fertilizers, as nonpoint source pollution not subject to the CWA's National Pollutant Discharge Elimination System.
101. *E.g.*, Forest Serv Empl. for Envtl. Ethics v. U.S. Forest Serv., 726 F. Supp. 2d 1195 (D. Mont. 2010) (requiring formal consultation with FWS and NMFS under §7 before U.S. Forest Service could use chemical retardants composed primarily of water and fertilizer). Another scenario

To date, there has never been a §9 case brought by the government or citizens against a farm or other fertilizer applicator for the impacts of fertilizer use on listed species. Thus, on the surface, farmers—even those who apply substantial amounts of fertilizer that they know to run off into nearby rivers or lakes where listed species reside—could easily disregard the ESA's take prohibition by failing to obtain a §10 incidental take permit from FWS and/or NMFS, despite the fact that §10 was designed for precisely that type of otherwise lawful activity resulting in incidental take of listed species.[102] However, the failure of a farm to obtain a §10 permit could ultimately prove to be a very arduous and expensive process in defending such a lawsuit, and could result in serious legal obligations under §10 at the conclusion of a successful case by citizen plaintiffs (or the government, although a government-initiated suit is less likely in this context).

The critical question in any §9 suit against a fertilizer applicator would come down to whether, through expert testimony or other evidence, the plaintiff can establish causation between the defendant's application of a fertilizer (and its subsequent runoff into a nearby water body) and the unlawful take of listed species that the plaintiff is challenging as an injury to the plaintiff's interests in the species and its habitat. Importantly, as discussed above, a successful §9 plaintiff need not demonstrate lethal take of a listed species, but rather must simply prove that any ongoing unlawful takes are occurring as a result of the defendants' activities. The ESA's broad take prohibition includes nonlethal takes in the form of harassment, which can include disruption or impairment of a species' normal behavioral patterns such as breeding, feeding, and sheltering—all of which are frequent impacts to listed aquatic species resulting from eutrophication caused by nutrient-rich fertilizers.[103]

Despite the breadth of the take prohibition, two serious causation problems would necessarily have to be addressed by a plaintiff challenging a defendants' application of fertilizers under §9. First, the plaintiff would have to prove proximate causation, meaning that the injurious take complained of by the plaintiff would have to be a foreseeable result of the defendant's application of fertilizer to fields in light of his or her proximity to the water body where listed species reside and the typical precipitation patterns in the bioregion. However, an intervening event that is not foreseeable can break the chain of causation, thereby defeating proximate causation and ultimately the plaintiff's case.

Indeed, in *Babbitt v. Sweet Home Chapter of Communities for a Greater Oregon*, the Supreme Court expressly considered proximate causation in the context of assessing the legality of an FWS regulation, which defined "harm" in the take definition to encompass adverse habitat modification. In that case, the justices came to dramatically different conclusions about the outer limits of foreseeability. In a strongly worded dissent, Justice Antonin Scalia explained his view that the regulation was unlawful because in the ESA, and particularly in the "harm" form of take, Congress did not intend §9 liability where, for example, "a farmer who tills his field and causes erosion that makes silt run into a nearby river which depletes oxygen and thereby impairs the breeding of protected fish has taken or attempted to take the fish."[104] But the majority upheld the regulation, explaining instead that "proximate cause principles inject a foreseeability element into the statute, and hence the regulation, that would appear to alleviate some of the problems noted by the dissent."[105] Justice Sandra Day O'Connor, concurring with the majority opinion, stated:

> The farmer whose fertilizer is lifted by a tornado from tilled fields and deposited miles away in a wildlife refuge cannot, by any stretch of the term, be considered the proximate cause of death or injury to protected species occasioned thereby. At the same time, the landowner who drains a pond on his property, killing endangered fish in the process, would likely satisfy any formulation of the principle.[106]

Accordingly, and especially under the Supreme Court's rationale for sustaining the regulation, a citizen plaintiff challenging fertilizer impacts to listed species must show that such effects are a foreseeable result of the activity complained of, and were not instead caused by intervening events outside the realm of reasonable foreseeability (e.g., tornadoes or other uncommon natural disasters in that particular locale).

---

in which §7 would be implicated relative to fertilizers would be if the Corps of Engineers authorized a plan for changing the water flow from a localized set of levees, the result of which would lead to higher concentrations of fertilizer nutrients affecting listed species in the water body.
102. 16 U.S.C. §1539(a)(1)(B).
103. 50 C.F.R. §17.3.
104. Babbitt v. Sweet Home Chapter of Cmtys. for a Great Or., 515 U.S. 687, 719 (1995) (Scalia, J., dissenting) (quotation marks and citations omitted).
105. *Id.* at 713 (O'Connor, J., concurring).
106. *Id.*

In addition, a plaintiff must clear a second causation hurdle in order to prevail under §9 of the ESA against a fertilizer applicator. Assuming the foreseeability prong is satisfied, a plaintiff still has the burden to establish that the defendant's application of fertilizer is ultimately causing the unlawful take of listed species alleged by the plaintiff. This is a complex inquiry because as at least one scholar has commented, the majority's examples in *Sweet Home* suggest that causation under the ESA includes not only "notions of foreseeability . . . [but also] spatial and temporal limitations on the potential scope of liability."[107] The import of these limitations is that, to prevail, a plaintiff must establish that the unlawful take complained of (e.g., death, injury, or disruption of biological functions of any member of listed species) is, at least in part, the result of the defendant's activity and not the result of other human-induced or even natural events.[108] As legal scholars Holly Doremus and Dan Tarlock have explained, determining this type of causation is an inherently difficult task:

> Fertilizers . . . and manure from livestock operations wash into the rivers and lakes and cause eutrophication. There is no dispute that Upper Klamath Lake is nutrient-rich and that the impaired water quality puts the endangered fish at increased risk. The NRC Interim Report agreed that changes in the water quality of Upper Klamath Lake have increased mass mortality among adult [members of a listed fish species]. . . . It is difficult, however, to assign responsibility for the basin's water quality problems. Only about 16-40 percent of the nutrient loading is anthropogenic; the rest comes from natural background conditions. To further complicate matters, other non-anthropogenic events such as hot, relatively calm weather can cause or increase the risk of fish kills, as illustrated by the massive fish kill of September 2002.[109]

This causation problem is most acute (as is the problem of eutrophication and impacts to listed species) in the Gulf of Mexico:

> The sources and causes of Gulf hypoxia are well understood, but the problematic multiscalar spatial and temporal discontinuities create a [legal] morass. First, many agricultural sources spread throughout the vast Mississippi River watershed, applying fertilizers and manure to croplands. Next, runoff carries fertilizer and manure nutrients from the land, building up in the nested watershed structure. Then, spring flooding sends a pulse of nutrients down the system and into the Gulf in one big shot far from the sources. Finally, the hypoxic event—the result of the bloom, death, and decomposition of algae feeding on the abundant nutrients—happens well after and far from where the fertilizer and manure was applied. The story sounds simple, but the scale of the aggregation and its spatial and temporal discontinuities have made hypoxia in estuarine systems a massive problem [with few answers].[110]

In the end, the success or failure of a §9 lawsuit brought against a fertilizer applicator will most likely hinge on whether the plaintiff satisfies for the reviewing court that the take complained of is not so far attenuated from the defendants' activities in geographical or temporal scope as to render the take outside of the scope of §9's liability scheme due to a failure to establish a reasonably traceable connection between the defendant's conduct and the unlawful take of a listed species in violation of §9.

In sum, although fertilizer impacts have rarely been addressed by ESA litigation, the growing size of hypoxic dead zones and the increasing impacts to listed species from eutrophication will inevitably lead to more ESA suits over fertilizer use. Because this is an area of little federal regulation, it appears that the majority of legal conflicts will fall to the §4 listing process and the §9 take prohibition as applied to fertilizer applicators, subject to the inherently difficult questions about causation.

3. Irrigation

Once a seed is planted and fertilizer applied, a crop must receive adequate amounts of water (and sun) to thrive in the photosynthesis process. As described in detail in Chapter 4, water is generally brought to crops by complex irrigation processes in today's modern farming system, which results in serious adverse

---

107. Albert C. Lin, *Erosive Interpretation of Environmental Law in the Supreme Court's 2003-2004 Term*, 42 Houston L. Rev. 565, 617 (2005).
108. Under common principles of tort law, a plaintiff need not show that the defendant's actions are the *sole* cause of the unlawful take, and instead must simply show that the defendant's actions contribute to the plaintiff's injury (and therefore to the listed species' injury) in some measurable way.
109. Holly Doremus & A. Dan Tarlock, *Fish, Farms, and the Clash of Cultures in the Klamath Basin*, 30 Ecology L.Q. 279, 345-46 (2003) (quotation marks omitted).
110. Ruhl & Salzman, *supra* note 2, at 87-88.

impacts to both water quantity and water quality, and, in turn, to the species that depend on such water for survival.[111] This conflict is growing ever more acute as climate change and other factors continue to lead to declining availability of water resources in several regions of the United States, meaning that conflicts between agricultural interests and the ESA will inevitably increase.[112]

As with GE seeds and fertilizers, FWS and NMFS are required in the listing process to consider the impacts of irrigation on any species being considered for listing.[113] However, most of the ESA litigation to date with respect to irrigation has addressed various issues through §§7 and 9, related to federal agencies' regulation of water flows (and thus the use of water by farming interests subject to the action agency's regulation) as balanced against the competing requirements of the ESA to protect imperiled species. At their core, most of the challenges have addressed an action agency's authority to reduce water use by permit holders (e.g., farmers) during times of low flow to protect listed species, or alternatively an agency's authority to refuse excess water (and instead flood local creeks and tributaries, and thus farmland) during times of high flow.[114] Several notable cases illustrate the issues.

First, as a threshold matter, the U.S. Court of Appeals for the Ninth Circuit was asked in the irrigation context to determine whether §§7 and 9 of the ESA are unconstitutional under the Commerce Clause as applied to species that are endemic to one state and thus do not constitute interstate species.[115] In line with other courts that have reviewed similar constitutional challenges to the ESA in other contexts, the reviewing court held that, even where purely endemic species are at issue, the ESA's mandates are nevertheless constitutional because "when a statute is challenged under the Commerce Clause, courts must evaluate the aggregate effect of the statute (rather than an isolated application) in determining whether the statute relates to commerce or any sort of economic enterprise."[116] Accordingly, water users (or other affected agricultural interests) cannot avoid the ESA's reach simply because the species affected by a particular activity is found only in one state.

Second, as to substantive ESA challenges under §7 (and, relatedly, §9), some of the most intriguing and highly publicized irrigation projects in the United States are the Upper Klamath River Project in Oregon and northeastern California, the Missouri River Mainstem System regulating water in various states from Montana to Missouri, and the Central Valley Project in central and northern California.

One of the most contentious scenarios in the history of the ESA played out in the Upper Klamath River Basin throughout the 1990s and 2000s, ultimately reaching the U.S. Supreme Court. The Upper Klamath River is "home to one of the earliest federal irrigation projects, a tight socioeconomic community founded on family farming."[117] The Upper Klamath is also home to two freshwater sucker species that were listed as endangered in 1988.[118] The action agency, the Bureau of Reclamation (BOR), consulted with FWS over the impacts of the irrigation project's operation, whereby in 1992, FWS issued a biological opinion requiring BOR to maintain minimum water levels at all times to avoid jeopardy to the species.[119] In compliance with that opinion, BOR began making moderate reductions to irrigation deliveries during droughts,[120] which led a coalition of farmers, ranchers, and irrigation districts to sue FWS for issuing the biological opinion in violation of §§4 and 7 of the ESA.[121] After the case was dismissed for lack of jurisdiction by the district

---

111. *E.g.*, Eubanks, *A Rotten System*, *supra* note 71, at 252-54.
112. *See id.*; Arlene J. Kwasniak, *Water Scarcity and Aquatic Sustainability: Moving Beyond Policy Limitations*, 13 Denv. Water L. Rev. 321 (2010).
113. 16 U.S.C. §1533(a); *see, e.g.*, Listing of Six Foreign Birds as Endangered Throughout Their Range, 76 Fed. Reg. 50052, 50066 (Aug. 11, 2011) (analyzing irrigation and finding that "[t]he irrigation required to sustain agricultural activities will likely further fragment any remaining suitable habitat" for the species being considered).
114. *E.g.*, County of Okanogan v. NMFS, 347 F.3d 1081 (9th Cir. 2003) (holding that the U.S. Forest Service and NMFS possess the authority under the ESA to restrict water use for permitholders during times of low flow to ensure levels necessary to protect steelhead trout, chinook salmon, and bull trout). Indeed, in at least one case, a state has sued a federal agency for taking species within that state by authorizing upstream water use practices in neighboring states that deprive the downstream states of water for species protection and other uses. Alabama v. U.S. Army Corps of Eng'rs, 441 F. Supp. 2d 1123 (N.D. Ala. 2006).
115. San Luis & Delta-Mendota Water Auth. v. Salazar, 638 F.3d 1163 (9th Cir. 2011).
116. *Id.* at 1175 (quotation marks and citation omitted). On October 31, 2011, the Supreme Court denied certiorari, meaning that the Ninth Circuit's ruling is final and cannot be appealed. *See* U.S. Supreme Court, Oct. 31, 2011 Order, *available at* http://www.supremecourt.gov/orders/courtorders/103111zor.pdf.
117. Holly Doremus & A. Dan Tarlock, Water War in the Klamath Basin: Macho Law, Combat Biology, and Dirty Politics xvi (1st ed. 2008).
118. *Id.*
119. Bennett v. Spear, 520 U.S. 154, 159 (1997).
120. Doremus & Tarlock, Water War, *supra* note 117.
121. *Bennett*, 520 U.S. at 159-60.

court and dismissal was affirmed by the Ninth Circuit, the Supreme Court issued a unanimous ruling reversing the dismissal on jurisdictional grounds and finding that the coalition had standing to pursue their claims, and further delineating which claims are properly reviewed under the ESA's citizen suit provision and which are properly reviewed under the APA.[122]

The heated controversy over the Klamath River project became even more complex in 1997 after the Supreme Court's decision, when NMFS listed coho salmon in the Klamath River under the ESA. Whereas listed suckers require minimum water levels in lakes in the water system, listed coho salmon require minimum water levels in rivers in the same system.[123] Two scholars described this as a recipe for disaster because maintaining both required sets of water levels would leave almost no water for irrigation withdrawals for farmers and other users.[124] The situation came to a head in 2000, as BOR moved forward with an annual operations plan for irrigation deliveries without obtaining ESA authorization for its plan from FWS and NMFS.[125] Environmental organizations immediately brought suit to address the harm that would otherwise result to the listed species, and the court enjoined BOR from "sending irrigation deliveries from Klamath Project whenever Klamath River flows . . . drop below the minimum flows recommended" until an operations plan has been lawfully authorized by FWS and NMFS.[126] In April 2001, pursuant to new biological opinions from FWS and NMFS, BOR announced that, for the first time in our nation's history with respect to any irrigation system, it would deliver no water to irrigation users during the summer of 2001 due to critically dry conditions and the need to protect the listed fish species—a determination upheld by a federal court.[127] Although none of the events since has been as drastic for farmers and environmentalists as during the 2001 drought, at the time of writing, litigation continues over the Klamath project as BOR attempts to strike a difficult balance to suit all of the Klamath's stakeholders, while simultaneously maintaining compliance with the ESA.[128]

Although one commenter stated that "the ESA will rarely affect other water projects as dramatically as it did the Klamath in 2001,"[129] a similar controversy unfolded over the Missouri River Mainstem System after FWS issued a biological opinion to the U.S. Army Corps of Engineers (the Corps) in 2000, requiring certain reasonable and prudent alternatives in order for the Corps' operation of the Missouri River Mainstem System to avoid jeopardizing several species residing in the Missouri River.[130] In particular, the opinion required water levels to meet certain standards in order to ensure appropriate breeding conditions for listed species while simultaneously avoiding flooding of nests and essential habitat during periods of heavy precipitation.[131] In 2003, FWS abruptly reversed position and removed certain conditions from the 2000 biological opinion. In response, environmental organizations filed suit, challenging the Corps' failure to comply with the terms of the 2000 opinion, as well as FWS' unexplained reversal of its position with respect to the terms and conditions of its biological opinions.[132] The court ultimately enjoined the Corps from taking any action contrary to the terms and conditions of the 2000 biological opinion,[133] much to the dismay of water users, including farmers, who saw the injunction in favor of species protection as one adverse to the interests of water users.[134] The Missouri River saga, however, did not end there. After the district court held the Corps in contempt and imposed sanctions on it for failing to comply with the court's earlier injunction,[135] the ESA litigation was consolidated with other Missouri River litigation.

---

122. *Id.* at 179.
123. Doremus & Tarlock, *Fish, Farms, and the Clash of Cultures*, supra note 109, at 317-18.
124. *Id.* at 318.
125. *Id.*
126. Pacific Coast Fed'n of Fishermen's Ass'ns v. U.S. Bureau of Reclamation, 138 F. Supp. 2d 1228, 1247-50 (N.D. Cal. 2001).
127. Doremus & Tarlock, Water War, supra note 117, at 1; Doremus & Tarlock, *Fish, Farms, and the Clash of Cultures*, supra note 109, at 320-21; Kandra v. United States, 145 F. Supp. 2d 1192 (D. Or. 2001); *see also* Reed D. Benson, *Giving Suckers (and Salmon) an Even Break: Klamath Basin Water and the Endangered Species Act*, 15 Tulane Envtl. L.J. 197 (2002).
128. *E.g.*, Pacific Coast Fed'n of Fishermen's Ass'ns v. U.S. Bureau of Reclamation, 426 F.3d 1082 (9th Cir. 2005) (enjoining BOR from relying on NMFS' biological opinion for operation of a project that focused only on species' long-term needs without ensuring that short-term survival and recovery would be achieved by the operation plan).
129. Benson, *Giving Suckers (and Salmon) an Even Break*, supra note 127, at 229.
130. American Rivers v. U.S. Army Corps of Eng'rs, 271 F. Supp. 2d 230, 237 (D.D.C. 2003).
131. *Id.*
132. *Id.*
133. *Id.* at 262-63.
134. *See generally* John R. Seeronen, *Judicial Challenges to Missouri River Mainstem Regulation*, 16 Missouri Envtl. L & Pol'y Rev. 59 (2009); Brook A. Spear, *The Missouri River: Law, Politics, and Creatures Caught in the Conflicts*, 18 Buffalo Envtl. L.J. 75 (2010).
135. American Rivers v. U.S. Army Corps of Eng'rs, 274 F. Supp. 2d 62, 71 (D.D.C. 2003).

The U.S. Court of Appeals for the Eighth Circuit ultimately held that the ESA claims were moot because by 2005, when the ruling was issued, the Corps was no longer engaging in water flow schemes that would take listed species, having instead mechanically constructed shallow water habitat to protect the species.[136]

An ongoing conflict just a few hundred miles south of the Upper Klamath River compels the conclusion that history is repeating itself yet again, and will continue to do so with increasing regularity as water scarcity continues to rise throughout the nation. This more recent conflict is over the Central Valley Project in California, which diverts water from several rivers in northern and central California to the much more arid regions of central and southern California, and which is also home to the ESA-listed Delta smelt.[137] The smelt are adversely impacted in various ways by the diversions, including lethal takes when caught in intake pumps, disorientation from the water pumping which leads to their predation by striped bass, and foraging impacts due to increasing salinity in the rivers.[138] In 2005, FWS issued a no-jeopardy biological opinion authorizing pumping to continue, but the opinion was later vacated by a court for violating §7's standard requiring reliance on the best available science.[139] FWS then issued a new biological opinion in 2008, setting out mitigation measures (reduced pumping and regular releases of fresh water to accommodate the smelt) to avoid jeopardy to the species.[140]

As in 2001 in the Klamath, drought-like conditions in the late 2000s caused tensions to rise even higher.[141] Additional lawsuits were filed challenging the 2008 revised opinion, which the court enjoined and ultimately remanded to FWS without vacatur due to several §7 failures, meaning that the deficient biological opinion stayed in operation while FWS complied with the court's order.[142] While any appeals play out, the district court enjoined implementation of parts of the remanded biological opinion in response to injunction requests from agricultural users who stand to be harmed by the current operation plan.[143]

Outside of the §7 context where most of the ESA irrigation lawsuits have been brought, irrigators must also be aware of direct §9 challenges where water is diverted in such a way as to directly or indirectly take listed species. At the time of writing, there have been two challenges against irrigators for unlawfully taking members of listed species without authorization under either §7 or §10 of the ESA. In the first, NMFS brought an enforcement action against an irrigation district for ongoing takes of threatened winter-run Chinook salmon in the Sacramento River as a result of water pumping operations for agricultural and other purposes.[144] At summary judgment, the district court determined that the operations were killing salmon without lawful authorization pursuant to a §10 incidental take permit, and issued a permanent injunction halting all water diversions during the four-month period when salmon spawn and swim through the river.[145]

Likewise, citizen plaintiffs more recently filed a §9 enforcement action against a municipal irrigation diverter whose pumping provides water for the municipality's farmers—an economic activity that generates hundreds of millions of dollars in revenue annually in the community.[146] The case revolved around a tidegate that the county built on the Skagit River, which would have adverse impacts on threatened Puget Sound chinook salmon.[147] The court found that ongoing takes were occurring as a result of the installation of the tidegate and entered judgment for plaintiffs as a matter of law.[148]

As water scarcity continues to increase and become more dire, ESA conflicts will inevitably be on the rise. Based on the relatively voluminous body of ESA precedent with respect to irrigation, it seems that courts take the ESA's §7 and §9 mandates very seriously, even where there are economic and other legiti-

---

136. In re Operation of Mo. River Sys. Litig., 421 F.3d 618 (8th Cir. 2005).
137. Eric M. Yuknis, *Would a "God Squad" Exemption Under the Endangered Species Act Solve the California Water Crisis?*, 38 B.C. ENVTL. AFF. L. REV. 567, 584 (2011).
138. *Id.* at 585.
139. Natural Res. Def. Council v. Kempthorne, Civ. No. 05-1207, 2007 WL 4462395 (E.D. Cal. Dec. 14, 2007).
140. *Yuknis, supra* note 137, at 585-86.
141. *See id.*
142. Consolidated Delta Smelt Cases, 717 F. Supp. 2d 1021 (E.D. Cal. 2010); San Luis & Delta-Mendota Water Auth. v. Salazar, 760 F. Supp. 2d 855 (E.D. Cal. 2010).
143. In re Consol. Delta Smelt Cases, Civ. No. 09-892, 2011 WL 3875512 (E.D. Cal. Aug. 31, 2011).
144. United States v. Glen-Colusia, 788 F. Supp. 1126, 1128 (E.D. Cal. 1992).
145. *Id.* at 1135-36.
146. Swinomish Indian Tribal Cmty. v. Skagit Cnty. Dike Dist., 618 F. Supp. 2d 1262, 1263 (W.D. Wash. 2008).
147. *Id.*
148. *Id.* at 1270-71.

mate interests that stand to be affected. Although a shortage of water is likely to ensue in our nation's future as climate change continues unabated, a shortage of contentious ESA-related irrigation lawsuits seems rather unlikely.

4. Pesticides

To protect against weeds, insects, and other pests, the last input needed so that a fertilized and well-watered seed can have the best possibility of becoming a healthy crop is a mix of pesticides (insecticides and herbicides). The environmental and public health impacts of pesticides are examined in Chapter 3.[149] Surprisingly, their environmental effects have been mostly overlooked in recent decades. As a legal scholar opined:

> One of the great ironies of environmental law is that the ecological consequences of pesticide use, which fueled the environmental movement of the late 1960s and early 1970s, largely have been ignored for the past thirty years. Only recently has interest renewed in the ecological (as opposed to human health) risks posed by pesticides.[150]

Indeed, the ESA has been virtually the only legal basis on which environmentalists have pressured agencies and pesticide users to ensure that pesticides meet environmental and public health standards. Looking forward, in light of new technologies, like GE seeds, that arguably lead to increased rates of pesticide use, common sense says that ESA conflicts will increase unless a technological fix can be found that eliminates pesticides and their serious impacts.

As with other agricultural inputs, pesticides must be evaluated during the §4 listing process for any species being reviewed by FWS or NMFS.[151] Indeed, nearly 84% of all listed species were listed in part due to agricultural activities, with pesticides being the main reason for that statistic.[152] However, most ESA pesticide litigation to date has not been brought in response to the propriety of listings, but rather to challenge compliance by a federal agency, primarily EPA, with §7 consultation duties and related §9 duties in registering pesticides under FIFRA. (See Chapter 8 for a discussion of FIFRA.) In light of those rulings, it is clear "that EPA is not only obligated to comply with the consultation requirements under the ESA when making decisions regarding pesticide registration, but the agency is also obligated to ensure that the permitted use of a registered pesticide will not result in an unauthorized take of a listed species."[153]

Under FIFRA, EPA can only register a pesticide (i.e., authorize its use for commercial, personal, or any other purpose) if it is not expected to result in "unreasonable adverse effects on the environment."[154] As a federal agency action that authorizes third parties to use active chemical ingredients potentially harmful to listed species, EPA must consult with FWS and/or NMFS before it renders a decision on a particular pesticide registration. Much of the notable ESA litigation concerning pesticides has addressed pesticide registrations where EPA failed to consult with FWS and/or NMFS as required by §7. One highly publicized case is the *Washington Toxics Coalition* case, where environmental organizations successfully sued EPA for failing to consult about the impacts to 25 listed salmon species when registering 54 active chemical pesticides before authorizing their use under FIFRA.[155] On appeal, the Ninth Circuit held that nothing in FIFRA's language relieved EPA of its §7 consultation duties under the ESA,[156] and

---

149. For more detailed discussion of the ecological impacts of pesticides, see Mary Jane Angelo, *Embracing Uncertainty, Complexity, and Change: An Eco-Pragmatic Reinvention of a First-Generation Environmental Law*, 33 Ecol. L.Q. 105 (2006); Mary Jane Angelo, *The Killing Fields: Reducing the Casualties in the Battle Between U.S. Species Protection Law and U.S. Pesticide Law*, 32 Harv. Envtl. L. Rev. 95 (2008); Brian Litmans & Jeff Miller, Center for Biological Diversity, Silent Spring Revisited: Pesticide Use and Endangered Species (2004), *available at* http://www.biologicaldiversity.org/publications/papers/Silent_Spring_revisited.pdf.
150. Angelo, *Embracing Uncertainty, supra* note 149, at 106.
151. *E.g.*, NMFS, Final Endangered Status for a Distinct Population Segment of Anadromous Atlantic Salmon in the Gulf of Maine, 65 Fed. Reg. 69459, 69479 (Nov. 17, 2000) (listing a distinct population segment of Atlantic salmon due in part to "chronic exposure to insecticides, herbicides, fungicides, and pesticides (in particular, those used to control spruce budworm)").
152. Eubanks, *A Rotten System, supra* note 71, at 263-64 (quoting Defenders of Wildlife, Comments for the Development of USDA Recommendations for the 2007 Farm Bill (70 Fed. Reg. 35221 (June 17, 2005), at 1, *available at* http://familyfarmer.org/sections/pdf/farmbillforum.pdf))..
153. Angelo, *Embracing Uncertainty, supra* note 149, at 187.
154. 7 U.S.C. §136a(c)(5).
155. Washington Toxics Coal. v. EPA, 413 F.3d 1024, 1028 (9th Cir. 2005).
156. *Id.* at 1031-34.

further ruled that an injunction against third-party use of the pesticide was appropriate under the ESA's citizen suit provision unless EPA registered the pesticides under FIFRA in compliance with the ESA's requirements.[157]

Other challenges have similarly been brought in response to EPA's failure to consult on particular pesticides under §7, most of which EPA has settled. In 2000, for example, environmental organizations sued EPA for failing to consult concerning 33 pesticides expected to affect several salmon and plant species. Under the pressure of litigation, EPA ultimately settled the case and set deadlines for reviewing 18 of the pesticides under the consultation requirements of the ESA.[158] In addition, environmentalists brought suit in 2002 for failing to consult on 200 pesticides affecting the listed California red-legged frog. The suit was also settled, with EPA agreeing to review the effects of 66 of the challenged pesticides on the frog within three years of the date of settlement.[159] In 2002, environmental groups filed suit against EPA for registering fenthion, which is highly toxic to birds listed under the ESA and the Migratory Bird Treaty Act, and the manufacturer of fenthion ultimately canceled the registration of its chemical voluntarily, thereby mooting the litigation.[160]

In 2006, EPA settled a suit brought by environmental groups challenging EPA's failure to consult on the effects of the potent chemical atrazine, leading to deadlines being set for EPA to review atrazine's impacts to the listed species enumerated in the complaint.[161] In May 2010, EPA settled another lawsuit brought by environmentalists, agreeing to review the effects of 75 pesticides on listed species including foxes, mice, butterflies, salamanders, and shrimp.[162] In 2010, yet another lawsuit was filed to address EPA's failure to comply with its §7 mandates when registering pesticides under FIFRA, seeking redress for EPA's alleged failure to consult with FWS and NMFS on 394 pesticides potentially affecting 887 listed terrestrial and aquatic species (nearly 50% of all listed species). At the time of writing, this case is pending.[163] While environmentalists are technically succeeding in these suits via settlement or judicial ruling, the frequency with which the suits must be brought (and the regularity with which EPA is losing in court or apparently conceding defeat in settlement) illustrates a fundamental tension between the ESA and FIFRA, as well as EPA's systemic neglect of its ESA duties, that puts both pesticide users (farmers) and environmentalists in a precarious and uncertain position in attempting to understand the wildlife and ecological effects of pesticides registered by EPA.[164]

A seminal case brought by Defenders of Wildlife and other conservation organizations established an independent and distinct liability against EPA under §9 for its "continued authorization" of a pesticide, the use of which resulted in take of a listed species.[165] In that case, with farming groups intervening in support of EPA, environmental groups challenged EPA's longstanding and continued registration under FIFRA of pesticides containing strychnine, which "is highly toxic and kills both target and nontarget [listed] species."[166] After the district court enjoined EPA from authorizing registration of pesticides containing strychnine until EPA could ensure that the registration would not result in unlawful takes of listed species, EPA finally obtained a biological opinion from FWS and an accompanying incidental take statement as required by §7.[167] However, the Eighth Circuit rejected EPA's late attempt to retroactively obtain FWS

---

157. *Id.* at 1034-36.
158. Center for Biological Diversity, *supra* note 149, at 55.
159. *Id.* at 56; U.S. EPA, Court Issues Stipulated Injunction Regarding Pesticides and the California Red-legged Frog, http://www.epa.gov/espp/litstatus/redleg-frog/rlf.htm.
160. Center for Biological Diversity, *supra* note 149, at 56.
161. U.S. EPA, Settlement Agreement in Natural Res. Def. Council v. EPA, Civ. No. 03-2444 (D. Md. Mar. 28, 2006), *available at* http://www.epa.gov/espp/litstatus/es-settlement-atrazine.pdf.
162. Endangered Species Law and Policy, Stipulated Injunction in Ctr. for Biological Diversity v. EPA, Civ. No. 07-2794, 2010 WL 2143658 (N.D. Cal. May 17, 2010), *available at* http://www.endangeredspecieslawandpolicy.com/uploads/file/Injunction.pdf.
163. Center for Biological Diversity, Lawsuit Initiated to Protect Hundreds of Endangered Species From Pesticide Impacts, http://www.biological-diversity.org/campaigns/pesticides_reduction/action_timeline.html.
164. *See generally* Angelo, *The Killing Fields*, *supra* note 149; *see also* Nick Allen, *EPA "Oversight?" How the Environmental Protection Agency Has Failed to Protect Pacific Salmon From Dangerous Toxins*, 10 Ocean & Costal L.J. 51 (2005) (analyzing EPA's systemic failure under FIFRA to protect listed salmon as required by the ESA). Also, it should be noted that EPA has preemptively mooted ESA claims arising in some litigation by initiating consultation with FWS and/or NMFS during the course of litigation before a final merits ruling can be issued. *See, e.g.,* Defenders of Wildlife v. Jackson, Civ. No. 09-1814, 2011 WL 2321882 (D.D.C. June 14, 2011).
165. Defenders of Wildlife v. EPA, 882 F.2d 1294 (8th Cir. 1989).
166. *Id.* at 1296.
167. *Id.* at 1301.

authorization for the many unlawful takes that had already occurred,[168] and upheld the district court's injunction against EPA registration of all pesticides containing strychnine until EPA complied with the ESA by either ensuring that no unauthorized takes would result prospectively or by obtaining appropriate authorization from FWS pursuant to §7.[169] This case was very important in helping to clarify the reach of the ESA as applied to EPA's FIFRA duties because it made clear that EPA not only has an upfront duty to carry out the ESA's mandates when registering pesticides under FIFRA, but also has a continuing duty to ensure that its ongoing regulation of pesticides does not have unanticipated effects on, or cause unauthorized takes of, listed species.[170]

As illustrated by the number of ESA lawsuits in the pesticide arena, EPA's historic noncompliance with various sections of the ESA has led to a consistent pattern of litigation, with serious consequences for farmers, environmentalists, and ultimately listed species hanging in the balance. However, unlike with other agricultural inputs described above, pesticides present a refreshing distinction since there exists a genuine possibility of decreased litigation if EPA actively takes steps to satisfy its ESA mandates. With respect to pesticides, the need for litigation rests entirely in the hands of EPA and on whether the agency will conform to the statutory and judicial mandates to which it is subject. Only time will tell whether EPA will continue with its historical approach, or whether EPA will adopt a more transparent approach to regulating pesticides in harmony with species protection laws.[171]

C. Application of the ESA to Animal Agriculture and Nonplant GE Foods

In contrast to crop agriculture—which has been, and will continue to be, a regular source of ESA conflicts—there are two notable non-crop agricultural and food scenarios where ESA litigation has not yet emerged but is almost certain to in the very near future. Both stem from how we produce food in the United States and the consequent impacts to imperiled species. There is no case law yet in these areas, but it is possible to analyze the ESA issues that are likely to emerge from the production of factory-farmed livestock and from genetic engineering of animals for human consumption.

1. Concentrated Animal Feeding Operations

In less than one century, almost all of our nation's meat (beef, poultry, and pork) has undergone a fundamental transformation: livestock production has gone from using traditional methods (e.g., pasture feeding, no use of antibiotics or growth hormones) to a system of concentrated animal feeding operations (CAFOs), in which livestock are literally farmed as if they were crops, row after row, in poor conditions where diseases abound and an incredible amount of excrement and waste is created.[172] The most marked impact to listed species resulting from this historic change is that today's industrial livestock system produces far more nutrient-rich waste than can be safely applied to fields, in striking contrast to the much more sustainable, closed-loop agricultural system that predominated until the past half-century.[173] Inevitably, much of this manure—which itself constitutes a serious pollution problem, but is exacerbated by the antibiotics and growth hormones suspended in the waste—runs off into water bodies adjacent to factory farms and leads to the same eutrophication and dire lack of oxygen that plagues listed aquatic species when fertilizers erode into waterways. In addition, the overload of antibiotics and growth hormones in water bodies, which do *not* occur naturally in the

---

168. *Id.*
169. *Id.*
170. *See* Angelo, *The Killing Fields*, *supra* note 149, at 111-12 (analyzing the *Defenders* ruling).
171. *See generally* Angelo, *The Killing Fields*, *supra* note 149. At the time of writing, one other ongoing case of note is Dow AgroSciences LLC v. NMFS, 637 F.3d 259 (4th Cir. 2011), where in 2011 the U.S. Court of Appeals for the Fourth Circuit agreed with pesticide manufacturers and farmers that a §7 challenge to a biological opinion from FWS or NMFS concerning impacts to listed species from particular EPA pesticide registrations is judicially reviewable even before EPA renders an ultimate decision on the registrations under FIFRA since biological opinions themselves are final agency actions reviewable by courts.
172. *See* Chapter 5, for a more in-depth discussion of the environmental impacts of meat, dairy, and related production.
173. *See generally* Chapter 1, for a discussion on the rapid changes in American agriculture in response to the farm bill's policies.

environment, is another source of harm to listed aquatic species that ingest these toxic substances once they enter a water body.[174]

It is important to note that EPA (or a delegated state agency) regulates discharges from CAFOs through the Clean Water Act's National Pollutant Discharge Elimination System (NPDES) (as described in Chapter 9). Therefore, assuming that a CAFO has obtained a NPDES permit from EPA and that permit has gone through the ESA's §7 review and received take authorization, there will likely be no grounds for a §9 lawsuit against the CAFO discharging waste in compliance with the permit's terms and conditions. But if a CAFO has not obtained a NPDES permit, or alternatively has obtained a NPDES permit from EPA or a delegated state agency that has not gone through the §7 process, it could be subject to a §9 lawsuit for discharging manure (and the chemicals therein). While it is true that many CAFOs have been found in Clean Water Act litigation to be discharging pollutants without a permit,[175] meaning that there is a realistic possibility that many CAFOs would be vulnerable to a §9 suit, the same obstacles of proving causation described in the fertilizer cases above would apply here. A plaintiff would have to prove causation by establishing that the §9 takes complained of are geographically and temporally traceable to the defendant CAFO's conduct.

A more likely scenario in which ESA conflicts involving an individual CAFO will arise is during the NPDES permitting process as EPA evaluates the effects of any permit it might issue for a particular CAFO. Under §7, EPA is required to engage in consultation with FWS and/or NMFS if the anticipated discharges of pollutants (in this case manure, antibiotics, and growth hormones) authorized by the permit will affect any listed species. If EPA fails to engage in consultation over such a permit, or if other §7 violations occur during the consultation process, such a violation provides a relatively clear hook for environmental organizations to step in and protect listed species in the manner Congress envisioned in the ESA.[176] Moreover, because the cumulative effect of CAFOs is resulting in a more substantial effect on listed species than ever before, FWS and NMFS must ensure that all listing processes under §4 incorporate, consider, and analyze the best available science with respect to CAFO discharges, particularly in light of the assumption that the rate of aquatic (and terrestrial) species in need of ESA protection will continue to rise as CAFO discharges displace, kill, and otherwise harm species and decimate their populations and habitats.

2. Nonplant GE Foods

Similar to factory-farmed livestock production, fish are increasingly being produced in controlled aquaculture environments. This is a serious concern under the ESA because it inevitably disrupts the ecological balance of species when farmed fish escape from their aquaculture facilities and cause reproductive, foraging, and other harm to listed fish species occurring naturally in the environment.

Adding to this already complex problem, the U.S. Food and Drug Administration (FDA) is on the verge of approving the nation's first-ever nonplant GE food source: a salmon, which environmental organizations are gearing up to challenge on ESA and other grounds. The idea of a "transgenic" fish—meaning a fish inserted with genes from another species to enhance certain traits—has existed for many decades, and the design of a transgenic salmon has been around for well more than a decade.[177] At the time of writing, FDA is nearing a final rule on GE salmon, which is an Atlantic salmon that has been given genetic material from Chinook salmon and ocean pout to double or triple its growth rates, meaning that GE salmon can reach market weight in 16 months, as compared to three years for non-GE

---

174. It should be noted that because factory farms frequently maintain large piles of manure on their property, listed terrestrial species might also be affected by eating manure containing harmful antibiotics and/or growth hormones, or through bioaccumulation as they prey on smaller species made ill by the manure. This seems less likely to occur, however, and more difficult to prove causation than with aquatic species residing in a water body adjacent to a factory farm.

175. *E.g.*, Community Ass'n for Restoration of the Env't v. Henry Bosma Dairy, 305 F.3d 943, 954 (9th Cir. 2002); Community Ass'n for Restoration of the Env't v. Henry Bosma Dairy, 65 F. Supp. 2d 1129, 1133, 1147 (E.D. Wash. 1999); Community Ass'n for Restoration of the Env't v. Southview Farms, 34 F.3d 114, 121 (2d Cir. 1994); Idaho Rural Council v. Bosma, 143 F. Supp. 2d 1169 (D. Idaho 2001).

176. With respect to cattle grazing on federal lands, courts have upheld challenges by ranchers and farmers to faulty biological opinions where such opinions create binding terms and conditions on the ranchers and farmers that exceed FWS' authority under §7. *See, e.g.*, Arizona Cattle Growers' Ass'n v. U.S. Fish & Wildlife Serv., 273 F.3d 1229 (9th Cir. 2001).

177. Blake Hood, *Transgenic Salmon and the Definition of "Species" Under the Endangered Species Act*, 18 J. Land Use & Envtl. L. 75, 100 (2002).

farm-raised salmon.[178] AquaBounty Technologies, the company seeking approval for its GE salmon, says that its salmon "will be grown as sterile, all-female populations in land-based facilities with redundant biological and physical containment [and] [a]s a result, AquAdvantage® Salmon cannot escape or reproduce in the wild and pose no threat to wild salmon populations."[179] Despite these assurances (from a company with an obvious vested interest in having FDA approve its GE salmon design), serious environmental and public health questions abound, including the impact to natural stocks of federally protected salmon species in the United States.

Scholars have written at length about the potential dangers of approving GE salmon in light of the fact that 100% containment and sterility are nearly impossible. The major threats include adversely modifying adaptive gene pools, altering natural selection, changing behavioral and feeding patterns, and overcompetition by GE salmon ultimately leading to destruction of wild and natural stocks (and thus to the loss of their genes altogether).[180] Accordingly, it appears that GE salmon and other nonplant GE foods to follow will have serious questions to answer under the ESA. FDA will necessarily have to engage in consultation with FWS and NMFS prior to approving GE salmon, much in the same way that EPA must comply with §7 before registering a new pesticide under FIFRA or before issuing a NPDES permit under the Clean Water Act.[181]

The questions that remain to be seen, however, are whether FDA (and NMFS and FWS) will consider and analyze the best available scientific information in rendering this landmark approval for a nonplant GE food item, what terms and conditions NMFS and FWS will require of FDA (and therefore the company seeking approval) in any biological opinion to avoid jeopardy, and how the approval of GE salmon and other nonplant foods will affect the future listing processes for new species under consideration for ESA protection. Another important question is whether ESA compliance will be necessary in future situations where a nonplant GE species is not a variant of a listed species—unlike with GE salmon, where several salmon populations are already listed under the ESA. In sum, while the ideas of GE foods are not new, FDA's emerging decisionmaking in that field will inevitably lead to diverse legal challenges to the potentially sweeping environmental consequences of the new technology. As usual, we will find that the ESA will almost certainly be squarely in the middle of these conflicts.

Conclusion

Some legal scholars have argued that, despite the fact that Congress has enacted no agricultural exemptions to the ESA, the statute nonetheless "falls only lightly on the shoulders of American farmers" because of the inherent problems of proving causation during prosecution and enforcement. The result is a strong statutory framework with weak on-the-ground results.[182] It is true that the federal government only prosecutes ESA violations in the rarest of circumstances because of the thorny causation issue.[183] As citizen enforcement of the ESA continues to increase in the agricultural and food sector, however, the statute will likely play an increasingly important role. Especially in light of the rapidly growing trend towards agricultural and food transparency, coupled with related calls for sustainability in our nation's cultivation, processing, and distribution of food,[184] the ESA has the potential to serve as a key tool in not only indirectly ensuring

---

178. David A. Taylor, *Laws, Regulations, and Policy: Genetically Engineered Salmon on the FDA's Table*, ENVTL. HEALTH PERSP. 118.9 (Sept. 2010), available at http://www.ncbi.nlm.nih.gov/pmc/articles/PMC2944109/. For an explanation of non-GE farm-raised salmon, see U.S. Pub. Interest Research Grp. v. Alt. Salmon of Me., LLC, 339 F.3d 23, 26-27 (1st Cir. 2003); U.S. Pub. Interest Research Grp. v. Heritage Salmon, Inc., Civ. No. 00-150, 2002 WL 240440 (D. Me. Feb. 19, 2002).
179. AquaBounty Technologies, AquAdvantage Fish, http://www.aquabounty.com/products/aquadvantage-295.aspx.
180. Hood, *supra* note 177, at 103-05.
181. *Cf.* Wild Fish Conservancy v. U.S. EPA, Civ. No. 08-0156, 2010 WL 1734850 (W.D. Wash. Apr. 28, 2010) (finding that, even in a non-GE aquaculture context, EPA was required to consult with FWS and NMFS, and in that process all of the agencies must rely on the best available science pursuant to §7).
182. *E.g.*, Zellmer & Johnson, *supra* note 2, at 481.
183. *See supra* notes 99-105.
184. *See, e.g.*, Eubanks, *A Rotten System*, supra note 71, at 213-310; William S. Eubanks II, *The Sustainable Farm Bill: A Proposal for Permanent Environmental Change*, 39 ELR 10493 (June 2009); William S. Eubanks II, *Paying the Farm Bill: How One Statute Has Radically Degraded the Natural Environment and How a Newfound Emphasis on Sustainability Is the Key to Reviving the Ecosystem*, 27 ENVTL. F. 56 (2010); Mary Jane Angelo, *Corn, Carbon, and Conservation: Rethinking U.S. Agricultural Policy in a Changing Global Environment*, 17 GEO. MASON L. REV. 593 (2010); Jason J. Czarnezki, *The Future of Food Eco-Labeling: Organic, Carbon Footprint, and Environmental Life-Cycle Analysis*, 30 STAN. ENVTL. L.J. 3 (2011); Kathryn A. Peters, *Creating a Sustainable Urban Agriculture Revolution*, 25 J. ENVTL. L. & LITIG. 203 (2010); Mary Jane

a safe, nutritious, and transparent food system for the public, but also directly achieving a thriving and harmonious ecological balance within the parameters of that evolving food system.

Angelo et al., *Small, Slow, and Local: Essays on Building a More Sustainable and Local Food System*, 12 Vt. J. Envtl. L. 353 (2011); Susan A. Schneider, *A Reconsideration of Agricultural Law: A Call for the Law of Food, Farming, and Sustainability*, 34 William & Mary Envtl. Law & Pol'y Rev. 935 (2010).

Chapter 12
Agriculture, Food, and the National Environmental Policy Act
William S. Eubanks II

The National Environmental Policy Act (NEPA)[1] is distinct from its statutory counterparts because it imposes only procedural requirements. Nonetheless, NEPA constitutes a comprehensive regulatory scheme that "promotes its sweeping commitment to 'prevent or eliminate damage to the environment . . .' by focusing Government and public attention on the environmental effects of proposed agency action."[2] For more than four decades, environmentalists have invoked NEPA to ensure public involvement in federal agency decisionmaking and ultimately to guarantee that well-informed decisions are made before their impacts permanently alter the status quo.[3] Unlike many of the environmental statutes discussed in Chapters 8-11, NEPA is not directly applicable to private action because its statutory obligations pertain only to federal agencies.

However, as this chapter explains, the activities of private actors, including food cultivation and distribution, frequently intersect with the NEPA process through programmatic or site-specific actions authorized by federal agencies. This chapter provides an overview of NEPA, its implementing regulations, and the diverse ways in which NEPA is implicated by agriculture and the food system.

A. NEPA's Statutory and Regulatory Framework

Despite being a procedural statute, NEPA is complex, as illustrated by the litany of cases where the U.S. Supreme Court and lower courts have attempted to interpret and apply its requirements to fact-specific scenarios.[4] Accordingly, the following is an explanation of the framework in which federal decisions subject to NEPA operate, and the mandates that apply in such contexts.

1. Background

When it enacted NEPA in 1969, the U.S. Congress intended the statute to serve as "our basic national charter for protection of the environment."[5] This lofty mandate is carried out through rigorous procedural requirements described in NEPA itself or in the statute's implementing regulations promulgated by the Council on Environmental Quality (CEQ). The explicit purpose of NEPA and the procedures it triggers is to "insure that environmental information is available to public officials and citizens before decisions are made and before actions are taken."[6] As the CEQ explains in its regulations, "NEPA's purpose is not to generate paperwork—even excellent paperwork—but to foster excellent action. . . . [by] help[ing] public

---

1. 42 U.S.C. §§4321-4370f.
2. Marsh v. Oregon Natural Res. Council, 490 U.S. 360, 371 (1989) (quoting 42 U.S.C. §4321).
3. William S. Eubanks II, *Damage Done? The Status of NEPA After* Winter v. NRDC *and Answers to Lingering Questions Left Open by the Court*, 33 Vt. L. Rev. 649, 649 (2009).
4. For a more detailed discussion of NEPA case law, see Daniel R. Mandelker, NEPA Law and Litigation (2d ed. 2011).
5. 40 C.F.R. §1500.1(a); *see also* 42 U.S.C. §4321 (declaring "a national policy which will encourage productive and enjoyable harmony between man and his environment; to promote efforts which will prevent or eliminate damage to the environment and biosphere and stimulate the health and welfare of man; [and] to enrich the understanding of the ecological systems and natural resources important to the Nation").
6. 40 C.F.R. §1500.1(b).

officials make decisions that are based on understanding of environmental consequences, and take actions that protect, restore, and enhance the environment."[7]

Because NEPA is an action-forcing statute aimed solely at better decisions by public officials,[8] NEPA obligations only attach to proposed decisions of federal agencies, although such decisions frequently include permitting, licensing, funding, or otherwise authorizing activities of private actors. In evaluating any proposed decision, a federal agency must first determine whether NEPA review is required.[9] Although NEPA review is typically required, agencies are exempted in three distinct scenarios: (1) when an agency decision will have no discernable impact on the natural environment; (2) when NEPA compliance would be inconsistent with, or duplicative of, an agency's other statutory mandates;[10] or (3) when a recurring type of agency decision has been determined by the agency to result in such minor and insignificant impacts that the agency has invoked a "categorical exclusion" to exclude such action from NEPA review.[11] Outside of these three narrow exceptions, federal agencies must adhere to several procedures to comply with NEPA, or face potential lawsuits by environmentalists or private parties affected by any alleged failure to satisfy NEPA's procedural requirements.

2. An Environmental Impact Statement or an Environmental Assessment?

Agencies are encouraged to "apply NEPA early in the process" in order to "insure that planning and decisions reflect environmental values, to avoid delays later in the process, and to head off potential conflicts."[12] Early in the process of considering a potential agency action, the first step at which NEPA enters the picture is when the agency determines from the outset that a detailed Environmental Impact Statement (EIS) is necessary, or when the agency prepares a less detailed Environmental Assessment (EA) if it is unsure whether an EIS is required under the circumstances to determine if there is a need for an EIS.[13] In the latter scenario, an agency prepares a brief EA, with public participation allowed "to the extent practicable," for the purpose of analyzing the environmental impacts of the proposed action.[14]

The CEQ describes an EA as "a concise public document for which a Federal agency is responsible," in which an agency "[s]hall include brief discussions of the need for the proposal, of alternatives as required by [NEPA], of the environmental impacts of the proposed action and alternatives, and a listing of agencies and persons consulted."[15] If the EA suggests that the anticipated environmental impacts of the federal decision under review will "significantly affect . . . the quality of the human environment"—a term of art under NEPA described in more detail below—an EIS is required.[16] If, however, the EA suggests that the environmental effects will be insignificant as that term is understood pursuant to NEPA, the agency is authorized to forgo preparing an EIS and must instead issue its EA to the public along with a finding of no significant impact (FONSI) to explain to the public the agency's rationale for not preparing an EIS.[17] Once the EA and FONSI are issued, the agency may proceed with implementing the underlying agency decision it analyzed in the EA.

In the event that an agency determines from the outset that an EIS is necessary, or alternatively in the event that an EA suggests that the effects of a particular decision will be significant thus necessitating preparation of an EIS, the agency must satisfy several regulatory hurdles to meet NEPA's EIS requirement.

---

7. *Id.* §1500.1(c).
8. *Id.* §1500.1(a).
9. *See id.* §1500.3 (explaining that NEPA is "binding on all Federal agencies . . . except where compliance would be inconsistent with other statutory requirements"); 42 U.S.C. §4333 (noting that "[a]ll agencies" must conform with NEPA's mandates). It is important to note that NEPA review is limited to "major Federal actions," 42 U.S.C. §4332(2)(C), which CEQ has defined as those "which are potentially subject to Federal control and responsibility." 40 C.F.R. §1508.18. The limitation to "major" actions has no independent meaning from "significance" under NEPA, which is explained below in detail.
10. *See* 40 C.F.R. §1500.3.
11. *Id.* §1508.4 (defining "categorical exclusion" as "a category of actions which do not individually or cumulatively have a significant effect on the human environment and which have been found to have no such effect . . . and for which, therefore, neither an environmental assessment nor an environmental impact statement is required").
12. *Id.* §1501.2.
13. *See id.* §§1501.3-1501.4.
14. *Id.* §1501.4(b).
15. *Id.* §§1508.9(a)-(b).
16. 42 U.S.C. §4332(2)(C).
17. 40 C.F.R. §1508.13; 40 C.F.R. §1501.4(e).

First, to shed light on *when* an EIS is required, Congress has explained that an EIS is required for any federal action that will "significantly affect" the environment.[18] "Significance" is defined by the CEQ regulations as including considerations of both "context" and "intensity"; therefore, by evaluating the existence of and impact to certain enumerated "significance factors," an agency should be able to determine whether an EIS is necessary.[19] Not surprisingly, the significance factors include key ecological resources that might be impacted by a decision, such as:

1. Impacts that may be both beneficial and adverse. A significant effect may exist even if the Federal agency believes that on balance the effect will be beneficial.

2. The degree to which the proposed action affects public health or safety.

3. Unique characteristics of the geographic area such as proximity to historic or cultural resources, park lands, prime farmlands, wetlands, wild and scenic rivers, or ecologically critical areas.

4. The degree to which the effects on the quality of the human environment are likely to be highly controversial.

5. The degree to which the possible effects on the human environment are highly uncertain or involve unique or unknown risks.

6. The degree to which the action may establish a precedent for future actions with significant effects or represents a decision in principle about a future consideration.

7. Whether the action is related to other actions with individually insignificant but cumulatively significant impacts. Significance exists if it is reasonable to anticipate a cumulatively significant impact on the environment. Significance cannot be avoided by terming an action temporary or by breaking it down into small component parts.

8. The degree to which the action may adversely affect districts, sites, highways, structures, or objects listed in or eligible for listing in the National Register of Historic Places or may cause loss or destruction of significant scientific, cultural, or historical resources.

9. The degree to which the action may adversely affect an endangered or threatened species or its habitat that has been determined to be critical under the Endangered Species Act of 1973.

10. Whether the action threatens a violation of Federal, State, or local law or requirements imposed for the protection of the environment.[20]

Courts have consistently held that the existence of even one of these significance factors is sufficient to trigger an agency's obligation to prepare an EIS.[21] Therefore, where an agency decision will likely implicate any of these ecological constituents, an EIS will be appropriate in most instances.

3. Federal Agency Obligations in an EIS

Once an agency has decided that an EIS is required, the next step is what is referred to as "scoping."[22] During scoping, which is commenced by the agency publishing a formal notice of intent in the *Federal Register*, "[t]here shall be an early and open process for determining the scope of issues to be addressed and for identifying the significant issues related to a proposed action."[23] The scoping process must allow for meaningful public involvement through comment opportunities and scoping meetings. The goal is to "[i]dentify and eliminate from detailed study the issues which are not significant . . . [by] narrowing the discussion of these issues in the statement to a brief presentation of why they will not have a significant effect on the human environment or providing a reference to their coverage elsewhere."[24] In some cases where an EA has already

---

18. 42 U.S.C. §4332(2)(C).
19. 40 C.F.R. §§1508.27(a)-(b).
20. *Id.*
21. Humane Soc'y of the U.S. v. Johanns, 520 F. Supp. 2d 8, 20 (D.D.C. 2007) (explaining that "courts have found that the presence of one or more of [the CEQ significance] factors should result in an agency decision to prepare an EIS") (citations omitted); Fund for Animals v. Norton, 281 F. Supp. 2d 209, 218 (D.D.C. 2003) (same).
22. 40 C.F.R. §1501.4(d).
23. *Id.* §1501.7.
24. *Id.* §1501.7(a)(3).

been completed and has necessitated preparation of an EIS, an agency may forgo the scoping process if the purposes of scoping were fulfilled during preparation of the EA.[25]

After an agency narrows the range of issues through the scoping process, it begins to prepare an EIS. Congress defined the minimum requirements of the EIS as including "a detailed statement on":

(1) the environmental impact of the proposed action,

(2) any adverse environmental effects which cannot be avoided should the proposal be implemented,

(3) alternatives to the proposed action,

(4) the relationship between local short-term uses of man's environment and the maintenance and enhancement of long-term productivity, and

(5) any irreversible and irretrievable commitments of resources which would be involved in the proposed action should it be implemented.[26]

The third component above, the "alternatives to the proposed action," is central to ensuring that an agency decision accurately accounts for the anticipated environmental impacts before the decision is made.[27] As the CEQ has delineated, an EIS "shall provide full and fair discussion of significant environmental impacts and shall inform decisionmakers and the public of the reasonable alternatives which would avoid or minimize adverse impacts or enhance the quality of the human environment."[28] Indeed, the CEQ has singled out the alternatives analysis as "the heart of the [EIS]," because it can "present the environmental impacts of the proposal and the alternatives in comparative form, thus sharply defining the issues and providing a clear basis for choice among options by the decisionmaker and the public."[29]

Agencies are required to first prepare a draft EIS, which must be made available to the public for comment.[30] Because public participation is a crucial element of the NEPA process,[31] the agency is required to consider all comments made on a draft EIS and either modify the final EIS (and the proposed action) in accordance with those comments, or otherwise address those comments by directly responding to each of them in the final EIS and explaining why the comments do not warrant modification of the proposed action.[32]

Assuming that the agency has satisfied all of its NEPA obligations, it may issue its final EIS, selecting the agency's "preferred alternative" at the conclusion of the NEPA process.[33] In addition to the final EIS, an agency must issue a record of decision (ROD), which is a concise document summarizing the agency's decision, the alternatives considered in the final EIS, and any measures taken to minimize or mitigate impacts to the environment.[34] If, at any point after a final EIS is issued, an agency "makes substantial changes to the proposed action" or "[t]here are significant new circumstances or information relevant to environmental concerns," the agency is required to prepare a supplemental EIS (SEIS) to account for the changes that were not analyzed in the original document, and that document is also subject to the requirements for public involvement described above.[35]

4. NEPA Litigation

Legal challenges to agencies' compliance with NEPA are commonplace in the federal courts. Such challenges range from the failure to prepare an EIS, to the failure to prepare an SEIS, to the improper application of a categorical exclusion, to the failure to adequately examine specific components of a proposed action (e.g., alternatives, cumulative impacts), to the failure to prepare site-specific NEPA review pursuant to an earlier programmatic review by an agency. Many of these challenges have reached the Supreme

---

25. *Id.* §1501.7(b)(3).
26. 42 U.S.C. §4332(2)(C).
27. *Id.* §4332(2)(E).
28. 40 C.F.R. §1502.1.
29. *Id.* §1502.14.
30. *Id.* §§1502.9, 1502.19, 1503.1(a)(4).
31. *Id.* §1506.6.
32. *Id.* §1503.4.
33. *Id.* §1502.14(e).
34. *Id.* §1505.2.
35. *Id.* §1502.9(c)(1); *id.* §1502.9(c)(4).

Court, where NEPA plaintiffs have consistently been unsuccessful.[36] However, despite a well-documented and longstanding losing streak at the nation's highest court, plaintiffs have obtained many important NEPA victories elsewhere, maintaining NEPA's status as an important tool for compelling well-informed agency decisionmaking, as Congress intended.

Unlike it did with most other environmental statutes, Congress did not create a private right of action under NEPA by way of a citizen suit provision. Rather, because there exists no express right of action, judicial review of final agency actions under NEPA occurs under the Administrative Procedure Act (APA) because "there is no other adequate remedy in a court."[37] As with any APA lawsuit, an agency's NEPA compliance is adjudicated based on the administrative record that existed with the agency at the time of the decision to determine whether the decision was arbitrary and capricious.[38] In addition, because a NEPA action is essentially an APA action, the appropriate remedy in most circumstances when a NEPA plaintiff prevails is vacatur and remand of the unlawful agency action unless and until NEPA's procedures are fully complied with, as required by the APA.[39] Vacatur and remand simply means to set aside an action and thereby nullify its validity by giving the agency another opportunity to address the legal deficiencies.

While courts have generally held that the presumptive remedy under NEPA (by way of the APA) is vacatur and remand,[40] there are certain scenarios in which more drastic remedies such as preliminary or permanent injunctive relief might be appropriate under NEPA. In 2008, in reviewing a NEPA challenge to the U.S. Navy's failure to prepare an EIS for specific military training exercises off the coast of California, the Supreme Court reinforced that injunctions in the NEPA context (and other contexts) are proper where a plaintiff can establish "that he is likely to succeed on the merits, that he is likely to suffer irreparable harm in the absence of preliminary relief, that the balance of equities tips in his favor, and that an injunction is in the public interest."[41] While the Court ultimately held that the substantial government interest in military preparedness in that case outweighed the asserted injuries to plaintiffs, the Court recognized the legitimacy of plaintiffs' claimed recreational and aesthetic interests that would be harmed, which, in other cases, could lead to an injunction under a different set of countervailing circumstances.[42]

After more than four decades, NEPA still looms large on the environmental landscape for agencies and regulated private parties alike. An agency's NEPA review, which is frequently a highly fact-specific inquiry, can delay or even ultimately derail government proposals ranging from broad, programmatic plans to site-specific decisions about a particular highway, military facility, bridge, airport, or trail. Accordingly, because an agency's NEPA obligations are fact-dependent, the remainder of this chapter examines the past, present, and anticipated future application of NEPA to agriculture and food in the United States.

B. NEPA as Applied to Agriculture and Food

U.S. agricultural policy is heavily shaped by the recurring legislation, enacted by Congress typically on a five-year cycle, known as the farm bill. (For a detailed discussion of the farm bill, see Chapters 1 and 2). In

---

36. David C. Shilton, *Is the Supreme Court Hostile to NEPA? Some Possible Explanations for a 12-0 Record*, 20 ENVTL. L. 551, 565-66 (1990); Jason J. Czarnezki, *Revisiting the Tense Relationship Between the U.S. Supreme Court, Administrative Procedure, and the National Environmental Policy Act*, 25 STAN. ENVTL. L.J. 3 (2006).
37. 5 U.S.C. §704.
38. *E.g.*, Citizens to Preserve Overton Park, Inc. v. Volpe, 401 U.S. 402, 420 (1971) (explaining that judicial review "is to be based on the full administrative record that was before the Secretary at the time he made his decision"); Florida Power & Light Co. v. Lorion, 470 U.S. 729, 743-45 (1985). For a more detailed discussion of judicial review under NEPA, see Susannah T. French, *Judicial Review of the Administrative Record in NEPA Litigation*, 81 CAL. L. REV. 929 (1993); Sam Kalen, *The Devolution of NEPA: How the APA Transformed the Nation's Environmental Policy*, 33 WM. & MARY ENVTL. & POL'Y L. REV. 483 (2009).
39. 5 U.S.C. §706(2) ("The reviewing court shall . . . hold unlawful and set aside agency action, findings, and conclusions found to be arbitrary, capricious, an abuse of discretion, or otherwise not in accordance with law.").
40. *E.g.*, Citizens to Preserve Overton Park, Inc. v. Volpe, 401 U.S. 402, 413-14 (1971) ("In all cases agency action must be set aside if the action was 'arbitrary, capricious, an abuse of discretion, or otherwise not in accordance with law' or if the action failed to meet statutory, procedural, or constitutional requirements."); Humane Soc'y of the U.S. v. Johanns, 520 F. Supp. 2d 8, 37 (D.D.C. 2007) ("Pursuant to the case law in this Circuit, vacating a rule or action promulgated in violation of NEPA is the standard remedy." (citing American Bioscience, Inc. v. Thompson, 269 F.3d 1077, 1084 (D.C. Cir. 2001)); Reed v. Salazar, 744 F. Supp. 2d 98, 119 (D.D.C. 2010) (explaining that, "having found that Defendants violated NEPA in entering into the AFA without properly considering its potential environmental impact, the default remedy is to set aside Defendants' action, thereby rescinding the 2008 AFA").
41. Winter v. Natural Res. Def. Council, 555 U.S. 7, 20 (2008).
42. *Id.* at 25-26; *see generally* Eubanks, *Damage Done*, *supra* note 3; Daniel Mach, *Rules Without Reason: The Diminishing Role of Statutory Policy and Equitable Discretion in the Law of NEPA Remedies*, 35 HARV. ENVTL. L. REV. 205 (2011).

that omnibus legislation, Congress delineates what on-farm and off-farm activities farmers must undertake to receive subsidies under various farm bill payment, insurance, and loan programs, which necessarily mold agriculture activities by the substantial majority of our nation's farms—including practically all of the largest farms—and therefore incentivizes the environmental impacts stemming from those agricultural practices. In addition to the farm bill and its extensive reach into the health of our nation's ecology, there are several independent regulatory mechanisms that contain their own unique federal processes and result in significant impacts to the natural environment. In all of these situations, agencies must decide whether NEPA applies and, if so, must determine what the agency's obligations are under the particularized circumstances. These decisions, and the adequacy of any review undertaken pursuant to them, are subject not only to public scrutiny but also to possible litigation in federal court. The following is a discussion of NEPA's application to the spectrum of programmatic and individualized federal actions related to food and agriculture.

1. Programmatic NEPA Review of the Farm Bill

The starting point for any discussion about the devastating environmental impacts flowing from agriculture is the farm bill. Indeed, commentators have argued that the farm bill is the federal statute with "the most significant environmental impact of any statute enacted by Congress . . . [because it] affects *all* aspects of the natural environment."[43] Although NEPA is not mentioned in any of the four corners of the 2008 farm bill,[44] nor has NEPA been applied consistently to farm bill programs in the past, legal scholarship has nevertheless suggested that NEPA review of the farm bill itself or at least of implementation of certain individual farm bill programs would be lawful and appropriate to address the environmental harms encouraged by the bill.[45] The legality of these theoretical approaches is discussed below.

a. NEPA Review of Farm Bill Legislation

The CEQ regulations encourage not only routine site-specific NEPA review of individual projects but also of "national program[s] or policy statements."[46] Where programmatic NEPA review of an overarching policy or program is undertaken, the CEQ encourages the respective agency to prepare a programmatic EIS or EA, which can then be tiered to when the agency implements the program at the site-specific level and prepares a more detailed NEPA document analyzing the impacts to be expected at the local or regional level that necessarily could not have been considered in the programmatic NEPA document.[47]

As legislation enacted by Congress with assistance from the U.S. Department of Agriculture (USDA), the farm bill would facially appear to be subject to NEPA's mandate since NEPA applies not only to federal agency actions but also to "every recommendation or report on proposals for legislation . . . significantly affecting the quality of the human environment."[48] This type of document is called a legislative EIS (LEIS). The CEQ has further clarified that "legislation" for purposes of NEPA "includes a bill or legislative proposal to Congress developed by or with the significant cooperation and support of a Federal agency, but does not include requests for appropriations."[49] The CEQ has explained that "[t]he test for significant

---

43. William S. Eubanks II, *A Rotten System: Subsidizing Environmental Degradation and Poor Public Health With Our Nation's Tax Dollars*, 28 STAN. ENVTL. L.J. 213, 214 (2009); *see also* William S. Eubanks II, *The Sustainable Farm Bill: A Proposal for Permanent Environmental Change*, 39 ELR 10493 (June 2009); William S. Eubanks II, *Paying the Farm Bill: How One Statute Has Radically Degraded the Natural Environment and How a Newfound Emphasis on Sustainability Is the Key to Reviving the Ecosystem*, 27 ENVTL. F. 56 (2010); Mary Jane Angelo, *Corn, Carbon, and Conservation: Rethinking U.S. Agricultural Policy in a Changing Global Environment*, 17 GEO. MASON L. REV. 593 (2010).
44. *See* Food, Conservation, and Energy Act of 2008, *available at* http://www.govtrack.us/congress/bills/110/hr2419.
45. *E.g.*, Carrie Lowry La Seur & Adam D.K. Abelkop, *Forty Years After NEPA's Enactment, It Is Time for Comprehensive Farm Bill Environmental Impact Statement*, 4 HARV. L. & POL'Y REV. 201 (2010); Jennifer Hoffpauir, *The Environmental Impact of Commodity Subsidies: NEPA and the Farm Bill*, 20 FORDHAM ENVTL. L. REV. 233 (2009).
46. 40 C.F.R. §1508.28.
47. *Id.*; *see also id.* §1502.20; La Seur & Abelkop, *supra* note 45, at 213 ("Agencies may offer a 'tiered' analysis to tackle complex national policies in a single EIS (or series of related EISs) by presenting the impact calculus in a general manner first, and subsequently referencing these overall findings within the context of each specific component of the overall policy analysis.").
48. 42 U.S.C. §4332(2)(C).
49. 40 C.F.R. §1508.17.

cooperation is whether the proposal is in fact predominantly that of the agency rather than another source, [and] [d]rafting does not by itself constitute significant cooperation."[50]

Some commentators have referred to the rarely invoked LEIS as NEPA's "forgotten clause," particularly in recent decades as LEISs have been prepared with diminishing frequency.[51] With that said, in the few instances where courts have had an opportunity to review the adequacy of an EIS for proposed legislation, or alternatively the failure to conduct NEPA review at all for a legislative proposal, courts have held that such claims are judicially reviewable under the APA.[52] The merits of the handful of LEIS cases to date suggest that the legal inquiry turns on the fact-specific circumstances of the legislative proposal (or legislation) at issue and the agency's involvement in the legislation. In a highly publicized case in which environmental organizations sought to compel preparation of an LEIS for the North American Free Trade Agreement (NAFTA) that had yet to be submitted to Congress for a final vote by the Office of the U.S. Trade Representative, the U.S. Court of Appeals for the District of Columbia (D.C.) Circuit reversed the district court's order requiring an LEIS, ruling that NAFTA was not yet a "final agency action" granting judicial review under the APA because it had not yet been submitted to Congress and therefore the agency still had discretion to renegotiate NAFTA's terms.[53] However, the D.C. Circuit explained that the fact-specific ruling in the case would not result in "the death knell of the legislative EIS."[54]

So how would a farm bill LEIS process work? Because of the length and complexity of the farm bill and the variety of sectors affected, many agencies would necessarily be involved, with USDA taking the lead. One scholar recently explained:

> A farm bill EIS will require the cooperation of several agencies, with USDA acting as lead agency. Other participating agencies should include the Department of Energy, which generally enforces biofuel policies; the Treasury Department, and specifically the Internal Revenue Service division, which administers the Volumetric Ethanol Excise Tax Credit; and U.S. Customs and Border Protection, which enforces the import duty for fuel ethanol. Federal agencies concerned with climate, the environment, fisheries, and land use (such as the Environmental Protection Agency, U.S. Geological Survey, and National Oceanic and Atmospheric Administration) may also participate in acquiring relevant data and analyzing cumulative environmental impacts, particularly global warming.[55]

Indeed, USDA's own regulations expressly contemplate NEPA review for legislative proposals, in addition to other farm bill programs discussed in the next section.[56]

Despite NEPA's LEIS mandate, and USDA's regulations reinforcing this requirement, "[n]o comprehensive EA or EIS has attempted to balance the benefits and harms of the environmental and economic impacts of any Farm Bill since 1970" when NEPA was enacted.[57] Since USDA and other federal agencies with responsibility for crafting and implementing the farm bill have long avoided a holistic NEPA review of the legislation, it will likely require litigation to force the relevant agencies to take a programmatic hard look at the farm bill that NEPA requires by considering diverse alternatives and by allowing meaningful public involvement in the NEPA process.[58] Thus, to determine whether litigation to compel a farm bill LEIS would be successful, we must explore the legal hurdles that would face such a challenge.

First, as a threshold matter, a plaintiff would have to determine the proper time in the farm bill's legislative process to bring suit in order to avoid ripeness and mootness concerns related to justiciability. If a chal-

---

50. *Id.*
51. *E.g.*, La Seur & Abelkop, *supra* note 45, at 216; Silvia L. Serpe, Note, *Reviewability of Environmental Impact Statements on Legislative Proposals After* Franklin v. Massachusetts, 80 Cornell L. Rev. 413, 413 (1995); Ian M. Kirschner, Note, *NEPA's Forgotten Clause: Impact Statements for Legislative Proposals*, 58 B.U. L. Rev. 560, 560 (1978).
52. *E.g.*, Atchison, T. & S.F. Ry. Co. v. Callaway, 431 F. Supp. 722, 726 (D.D.C. 1977) (finding that "the §102(2)(C) EIS requirement for legislative proposals is enforceable by a private right of action, and that private right of action includes challenges to the adequacy of, as well as to the absence of, an EIS"); Natural Res. Def. Council v. Lujan, 768 F. Supp. 870, 878 (D.D.C. 1999); Citizens for Mgmt. of Alaska Lands v. Department of Agric., 447 F. Supp. 753 (D. Alaska 1978).
53. Public Citizen v. U.S. Trade Rep., 5 F.3d 549 (D.C. Cir. 1993).
54. *Id.* at 552.
55. La Seur & Abelkop, *supra* note 45, at 216.
56. 7 C.F.R. §799.9(c).
57. La Seur & Abelkop, *supra* note 45, at 220.
58. Hoffpauir, *supra* note 45, at 262 ("For [an LEIS to be prepared], an environmental group would likely have to force the agency to take action via litigation.").

lenge is filed too early in an ongoing farm bill proposal and drafting process, USDA and other defendant agencies are certain to argue that the claim is unripe. As the Supreme Court has explained:

> Ripeness is a justiciability doctrine designed "to prevent the courts, through avoidance of premature adjudication, from entangling themselves in abstract disagreements over administrative policies, and also to protect the agencies from judicial interference until an administrative decision has been formalized and its effects felt in a concrete way by the challenging parties."[59]

Likewise, if a suit is filed once Congress enacts a new farm bill and agencies have already started implementation of farm bill programs, defendant agencies will likely argue that such a claim is not justiciable because the enactment of the legislation rendered moot an EIS of a proposal for legislation due to the finality of the legislative process. Stated differently, the government would likely argue that the court would be unable to issue any effective relief because requiring an LEIS after legislation is enacted would be a futile makework exercise.[60]

Despite the apparent Catch-22, a plaintiff would have persuasive arguments in favor of judicial review of such a claim, although the plaintiff would be in a much stronger position where a programmatic LEIS challenge is brought to a farm bill soon *after* the legislation is enacted by Congress. Indeed, at least one federal district court has held that a pre-enactment LEIS challenge is unripe in any case because the "proposals are in the lap of Congress [and] [t]heir effect on the contents of any subsequent legislation and the date of the ultimate enactment of such legislation are speculative to say the least."[61] That court ruled that the relief requested by the plaintiffs, "if granted, would disrupt the legislative process," and therefore "redress of plaintiff's grievance must await definitive Congressional action."[62]

In contrast, once Congress enacts a new farm bill, a plaintiff need not address the ripeness concerns that might otherwise loom large since Congress and the pertinent agencies would no longer retain discretion to renegotiate the provisions of the farm bill as enacted in final form.[63] With such a post-enactment challenge, the only relevant justiciability consideration would be mootness. A case is moot when "'there is no reasonable expectation . . .' that the alleged violation will recur . . . [and] interim relief or events have completely and irrevocably eradicated the effects of the alleged violation."[64] Therefore, even if USDA and other agencies are arguably required to undertake NEPA review of the farm bill, the agencies would likely argue that this type of claim is moot upon enactment by Congress. Scholars who have contemplated such a challenge have argued that plaintiffs would be able to overcome the mootness hurdle because certain features of the farm bill—which includes staggered program implementation, annual appropriations, and subsidy programs—allow for intermediate NEPA review of the farm bill, despite that fact that final legislation has been enacted, because many concrete implementation steps will still have adverse environmental impacts as a result of the legislation.[65] Therefore, from a temporal standpoint, a plaintiff would have the highest likelihood of success in challenging USDA's failure to conduct NEPA review of the farm bill by bringing a challenge immediately upon enactment of the legislation, in order to encompass review of as much of the farm bill's program implementation as possible before the next farm bill is enacted roughly five years later.[66]

Second, as with any NEPA challenge, a plaintiff must sufficiently demonstrate Article III standing and that the plaintiffs' asserted injuries fall within the zone of interests protected by NEPA.[67] While there

---

59. National Park Hospitality Ass'n v. Department of Interior, 538 U.S. 803, 807-08 (2003) (quotation marks and citation omitted).
60. 42 U.S.C. §4332(2)(C).
61. Chamber of Commerce v. Department of Interior, 439 F. Supp. 762, 766 (D.D.C. 1977).
62. *Id.* at 767; *see also* La Seur & Abelkop, *supra* note 45, at 219; Dana Butler, Note, *The Death Knell of the Legislative Environmental Impact Statement: A Critique of Public Citizen v. U.S. Trade Representative*, 17 LOY. L.A. INT'L & COMP. L. REV. 121, 121 (1994) (arguing that the NAFTA "decision casts doubt upon the role of the EIS in many administrative agency-related proposals for legislation because no agency-prepared materials are truly final until they are implemented or enacted by Congress").
63. *See* La Seur & Abelkop, *supra* note 45, at 216-26 (arguing that a plaintiff should bring a challenge to the 2008 farm bill despite the fact that it has already been enacted by Congress).
64. County of L.A. v. Davis, 440 U.S. 625, 631 (1979) (quoting United States v. W.T. Grant Co., 345 U.S. 629, 633 (1953)); *see also* La Seur & Abelkop, *supra* note 45, at 218.
65. La Seur & Abelkop, *supra* note 45, at 218-19; *see also* RONALD E. BASS ET AL., THE NEPA BOOK: A STEP-BY-STEP GUIDE ON HOW TO COMPLY WITH THE NATIONAL ENVIRONMENTAL POLICY ACT 177 (2d ed. 2001) (citing West v. Secretary of the Dep't of Transp., 206 F.3d 920 (9th Cir. 2000)) ("A NEPA case may not be moot if completed phases of a federal agency action can be operated to reduce environmental effects, or if future phases of an action have not yet begun.").
66. La Seur & Abelkop, *supra* note 45, at 216-26 (advocating for post-enactment litigation to challenge an existing farm bill for failure to comply with NEPA).
67. *See generally* La Seur & Abelkop, *supra* note 45, at 220-25.

exist different standing tests in the various federal judicial circuits with respect to programmatic NEPA challenges,[68] a plaintiff will ultimately have to establish a concrete procedural injury stemming from an agency's failure to prepare an LEIS, that the injury is reasonably traceable to the farm bill, and that the remedy sought—preparation of an LEIS for the farm bill—would redress the plaintiffs' injury.[69] So long as the plaintiff can document specific environmental harms that will flow from the farm bill and how they will affect the plaintiff in a site-specific manner—something never considered by the agencies since they failed to conduct any NEPA review—a court would likely find that an affected plaintiff had established standing for such a challenge.[70]

Assuming a plaintiff can withstand the ripeness, mootness, and standing barriers, the case would finally turn to the merits. Under the APA, a plaintiff must prove that the failure to conduct an LEIS for the farm bill was arbitrary and capricious under NEPA. Because this is an area never before litigated, it is difficult to predict where a court would come down on this issue. On the one hand, NEPA has a categorical requirement to prepare an EIS (or at least an EA) for "every recommendation or report on proposals for legislation . . . significantly affecting the quality of the human environment."[71] On the other hand, the CEQ regulations implementing NEPA limit NEPA review to legislation "developed by or with the significant cooperation and support of a Federal agency"[72]—a clause that would likely prove critical, depending on whether a court would find USDA or other agency cooperation significant enough to trigger NEPA obligations. Because the limited case law on LEISs provides no assistance in discerning the outcome of this crucial inquiry, the question will remain unanswered until such time as a challenge might be brought and decided by the presiding court. While abstaining at this juncture from looking into the hypothetical crystal ball, there is no question that "[s]uccess [if obtained] would likely necessitate a programmatic farm bill EIS, to be conducted as a practical matter before the enactment of the succeeding farm bill."[73]

b. NEPA Review of Individual Farm Bill Programs

Perhaps, more realistic than seeking to compel a programmatic NEPA challenge of the omnibus farm bill as a whole would be NEPA challenges to individual programs that undoubtedly have significant adverse environmental impacts but that have escaped NEPA review. There is a short list of the types of individualized challenges that could be fashioned against farm bill programs to achieve NEPA's core purposes of public participation, environmental impact assessment, and alternatives analysis for the farm bill.

At the outset, it is important to give credit to the federal government where credit is due. For each of the past several farm bills, USDA's Farm Service Agency has prepared an EIS to analyze the impacts of the Conservation Reserve Program before implementing the program.[74] The most recent version, the 2008 Supplemental Programmatic EIS, is a 710-page document considering various alternatives and incorporating extensive public comments related to USDA's Conservation Reserve Program, which is "the Federal Government's largest conservation program for private lands [achieved] [t]hrough voluntary partnerships between individuals and the Federal Government" where farmers are rewarded monetarily for engaging in certain resource-protective measures (as described in Chapters 1 and 2).[75] USDA's preparation of an EIS for this program is especially encouraging, considering that the ultimate goal of the Conservation Reserve Program is to generally create a net *beneficial* effect on the environment, and compels the conclusion that NEPA review is likewise required where USDA implements a program with a net *adverse* effect on the envi-

---

68. La Seur & Abelkop, *supra* note 45, at 221-22.
69. *Id.* at 220-25.
70. *But see id.* at 223 & n.179 (discussing D.C. Circuit NEPA challenge where the court dismissed on causation grounds a case seeking to challenge an increase in tax credits for ethanol because the chain of events was too attenuated compared to the alleged environmental harm).
71. 42 U.S.C. §4332(2)(C).
72. 40 C.F.R. §1508.17.
73. La Seur & Abelkop, *supra* note 45, at 217.
74. *E.g.*, U.S. Dep't. Agric. (USDA), Farm Service Agency, Final Supplemental Environmental Impact Statement for the Conservation Reserve Program Based on Changes Authorized by the 2008 Farm Bill (June 2010), *available at* http://www.fsa.usda.gov/Internet/FSA_File/crpfinalseismaster61010.pdf; USDA, Farm Service Agency, Record of Decision for the Programmatic Environmental Impact Statement on the Conservation Reserve Program Based on Changes Authorized by the 2002 farm bill, 68 Fed. Reg. 24848 (2003).
75. USDA, Farm Service Agency, Record of Decision on the Final Supplemental Environmental Impact Statement for the Conservation Reserve Program Based on Changes Authorized by the 2008 farm bill (July 22, 2010), *available at* http://www.fsa.usda.gov/Internet/FSA_File/rodcrp.pdf.

ronment, in order to achieve the goals of NEPA.[76] Unfortunately, USDA has not applied the rule evenly, meaning that programs with far greater environmental impacts than the Conservation Reserve Program have obtained a virtual free pass to degrade resources without any NEPA analysis or public involvement.[77]

To challenge USDA's apparent failure to consistently apply NEPA from program to program, scholars agree that the most important—and seemingly winnable—individualized NEPA lawsuit would be a challenge to the USDA Farm Service Agency's application of a categorical exclusion under NEPA to the commodity title of the farm bill.[78] The drastic and widespread environmental impacts of the commodity program's billions of dollars spent annually result in irreparable effects to the nation's land, soil, water, air, and wildlife.[79] Moreover, there is no dispute that these commodity subsidies (including direct payments, counter-cyclical payments, nonrecourse marketing loans, and crop insurance payments) individually and cumulatively constitute the definition of a "major federal action" subject to NEPA. As one scholar has explained, "the subsidies involve a great deal of money; . . . the subsidies are national in scope, given to farmers located in all states; . . . [and] the subsidies are given to a large number of farms, covering half of the nation's 938 million acres of farmland."[80] Yet, despite clearly satisfying NEPA's threshold criteria, USDA's Farm Service Agency has long declared via regulation that a categorical exclusion is applicable to "Commodity Income and Support and Disaster Protection Programs" in order to avoid any NEPA review whatsoever.[81]

It is virtually impossible to square USDA's invocation of a categorical exclusion in this instance with the facts on the ground, particularly since a categorical exclusion may only be invoked where an action or "a category of actions . . . do not individually or cumulatively have a significant effect on the human environment and which have been found to have no such effect."[82] Because monoculture cultivation of a handful of commodities (namely corn, soy, wheat, cotton, and rice) is arguably the single most environmentally damaging activity encouraged, funded, and supported by the federal government, there is no conceivable basis for USDA's determination that such subsidies to a large farm, or particularly when such payments to all farms are aggregated on a cumulative level, do not have a significant effect on the human environment.[83] Therefore, challenging USDA's application of a categorical exclusion would be relatively straightforward under NEPA and APA principles, whereby "[a]n agency's determination that a particular action falls within one of its categorical exclusions is reviewed under the arbitrary and capricious standard."[84]

The fact that the commodity program has been in existence for many farm bills is no help to the government in defeating such a challenge because USDA's own regulations expressly contemplate an EIS or at least an EA not only for "[i]nitial program implementation" but also for "[m]ajor changes in ongoing programs" and for "[m]ajor environmental concerns with ongoing programs."[85] Moreover, the fact that part of the commodity title is an appropriation of federal monies should not be a bar to success considering that the commodity program as a whole has a direct and controlling effect on how farmers cultivate their lands, with which crops, and in what quantities, thus overcoming any suggestion that these are merely fiscal appropriations without direct environmental impacts. In addition, USDA would likely find no shelter from the fact that it promulgated the categorical exclusion in 1980,[86] because the current application of the categorical exclusion to thousands of farmers annually itself constitutes a final agency action that can be challenged without statute of limitations concerns. And, in any case, even if a categorical exclusion was

---

76. While it is encouraging that USDA is in fact preparing NEPA documents for this program, there have been concerns with the legality of USDA's NEPA compliance in the past, including one instance in which environmental organizations successfully sued USDA for failing to adequately analyze the effects of a policy change in the farm bill allowing managed haying and grazing of farmland enrolled in the Conservation Reserve Program. *See* National Wildlife Fed'n v. Johanns, Civ. No. 04-2169, 2005 WL 1189583 (W.D. Wash. May 19, 2005).

77. There are other situations in which an agency has undertaken NEPA review for particular farm bill programs, resulting in an EA or EIS. One example is the Environmental Quality Incentives Program, where USDA's Natural Resources Conservation Service usually prepares an EIS (as in 2002), or at least an EA and accompanying FONSI (as in 2008). *See, e.g.*, Interim Rule for Changes to the Environmental Quality Incentives Program, 74 Fed. Reg. 2293 (Jan. 15, 2009).

78. Hoffpauir, *supra* note 45, at 242-65; La Seur & Abelkop, *supra* note 45, at 214-16.

79. Eubanks, *A Rotten System*, *supra* note 43, at 251-73; Hoffpauir, *supra* note 45, at 244-56.

80. Hoffpauir, *supra* note 45, at 244.

81. 7 C.F.R. §799.10(b)(2).

82. 40 C.F.R. §1508.4.

83. *See* Eubanks, *A Rotten System*, *supra* note 43, at 251-73; Hoffpauir, *supra* note 45, at 244-56.

84. Alaska Ctr. for the Env't v. U.S. Forest Serv., 189 F.3d 851, 857 (9th Cir. 1999).

85. 7 C.F.R. §799.9(b).

86. USDA, Environmental Quality and Related Concerns, 45 Fed. Reg. 32312 (May 16, 1980).

appropriate in 1980, USDA has wholly failed to document how a categorical exclusion is legally permissible as implemented under the 2008 farm bill (or subsequent farm bills for which USDA will inevitably invoke the categorical exclusion) as the size of the commodity program continues to increase and in turn so do the adverse environmental impacts.[87] Accordingly, because there is no doubt that programs under the commodity title are causing significant environmental impacts—indeed, impacts that are far greater than those caused by the Conservation Reserve Program for which USDA *does* prepare an EIS each farm bill—the odds are relatively strong that USDA's categorical exclusion would be struck down as arbitrary and capricious under the APA in the event of such a challenge, so long as a plaintiff can prove that USDA retains some level of discretion in providing the commodity payments to farmers or in formulating conditions attached to the payments.

While most of the legal scholarship analyzing the federal government's misapplication of NEPA, or its failure to apply NEPA at all, to specific farm bill programs with significant environmental impacts has rightfully focused on the commodity program, there are other programs that merit similar consideration, and to which the implementing agencies might be vulnerable to challenges. The farm bill directly affects and incentivizes ethanol production,[88] crop insurance, foreign trade of surplus crops, nutrition programs with a variety of environmental justice concerns in communities that lack access to healthy foods, and research related to organic agriculture and local and regional food systems. All of these programs have varying levels of environmental impacts depending on the levels of funding that an agency provides to recipients during program implementation and on the conditions imposed on recipients by the agency, and analysis of these alternatives and their impacts is precisely the review that Congress envisioned when it created NEPA as "our basic national charter for protection of the environment."[89] At this juncture, however, agencies implementing farm bill programs have not taken the initiative to prepare NEPA documentation for these and other programs, but the growing public interest in transparent food systems is likely to lead to a slew of lawsuits, rulemaking petitions, and other protests to bring NEPA to bear on what very well might collectively constitute the most environmentally impactful set of activities authorized by Congress and implemented by the executive branch.[90]

2. NEPA Review of Independent Statutory Processes

Aside from a handful of farm bill programs, there are other instances in which NEPA intersects with federal agency involvement in agricultural and food issues. There are several prominent examples of such NEPA interplay and relevant court decisions interpreting NEPA's mandates in these particular scenarios.

a. Pesticide Registration Under the Federal Insecticide, Fungicide, and Rodenticide Act

Pursuant to the Federal Insecticide, Fungicide, and Rodenticide Act (FIFRA), the Environmental Protection Agency (EPA) reviews applications for new pesticides by undertaking a review of statutory criteria to ensure that certain standards are met before a pesticide is made available for sale to the public.[91] Under FIFRA, EPA can only register a pesticide (i.e., authorize its use, sale, or distribution for commercial, personal, or any other purpose) if it is not anticipated to result in "unreasonable adverse effects on the environment."[92] To rise to that level means "(1) any unreasonable risk to man or the environment, *taking into account the economic, social, and environmental costs and benefits of the use of any pesticide*, or (2) a human dietary risk from residues that result from a use of a pesticide in or on any food."[93] Thus, whether NEPA review applies to FIFRA registration has essentially boiled down in several cases to whether FIFRA's

---

87. *See, e.g.*, 7 C.F.R. §799.9(b) (requiring USDA to prepare an EIS where there exist "[m]ajor changes in ongoing programs" or "[m]ajor environmental concerns with ongoing programs").
88. La Seur & Abelkop, *supra* note 45, at 204-05.
89. 40 C.F.R. §1500.1(a); *see also* 42 U.S.C. §4321 (declaring "a national policy which will encourage productive and enjoyable harmony between man and his environment; to promote efforts which will prevent or eliminate damage to the environment and biosphere and stimulate the health and welfare of man; [and] to enrich the understanding of the ecological systems and natural resources important to the Nation").
90. Eubanks, *A Rotten System*, *supra* note 43, at 214.
91. *See* Chapter 8 for more detail on the FIFRA process.
92. 7 U.S.C. §136a(c)(5).
93. *Id.* §136(bb) (emphasis added).

cost-benefit environmental analysis is the functional equivalent of NEPA review, rendering NEPA review duplicative and unnecessary, or whether Congress displaced NEPA's mandates in the FIFRA registration process when it enacted FIFRA.

In 1973, the first major case involving this question concerned EPA's withdrawal of pesticide registrations for dichlorodiphenyltrichloroethane (DDT). Pesticide companies argued that EPA could not withdraw the registrations without going through a comprehensive NEPA process separate from the FIFRA withdrawal process.[94] The D.C. Circuit ultimately ruled in EPA's favor, finding that "the functional equivalent of a NEPA investigation was provided" during the FIFRA withdrawal process and therefore "[t]he law requires no more."[95] In 1986, the U.S. Court of Appeals for the Ninth Circuit went one step further, concluding that EPA is always exempted from NEPA review when initially registering pesticides under FIFRA because "Congress refrained from incorporating the NEPA standard" into FIFRA when it enacted amendments to the statute in 1972, meaning that "Congress believes that analyses in support of registration currently are an adequate substitute for an EIS in the FIFRA context."[96] Without expressly adopting the term "functional equivalence" as their rationale, the court explained that "we are confident that Congress did not intend NEPA to apply to FIFRA registrations."[97]

In contrast to EPA's registration process under FIFRA, which courts have exempted from NEPA review,[98] actions or programs by federal agencies to spray pesticides on public lands must comply with NEPA, despite the fact that the pesticide has already satisfied FIFRA's standards. The Ninth Circuit resolved this issue in a 1983 case, remanding a programmatic EIS for the spraying of the pesticide carbaryl as inadequate for failure to analyze site-specific impacts, and specifically holding that "[o]ne agency cannot rely on another's examination of environmental effects under NEPA . . . [t]hus, the mere fact that a program involves use of substances registered under FIFRA does not exempt the program from the requirements of NEPA."[99] In a subsequent case, the Ninth Circuit once again held that in implementing pesticide spraying programs, federal agencies must comply with NEPA regardless of FIFRA registration because "[t]he EPA registration process for herbicides under FIFRA is inadequate to address environmental concerns under NEPA," and the court enjoined the U.S. Forest Service and the Bureau of Land Management from implementing its joint spraying program at issue until the agencies achieved full NEPA compliance.[100]

In sum, because of the balance struck by Congress in enacting FIFRA, NEPA plays a small role in the realm of pesticides. While federal agencies are required to conduct legally appropriate NEPA review when implementing a spraying program that will result in environmental impacts, EPA is entirely exempt from NEPA's mandates in registering, cancelling, or withdrawing pesticides as part of the FIFRA process unless it opts to conduct such reviews voluntarily.

b. Discharge and Fill Permits Under the Clean Water Act

The two types of federal authorization for agricultural operations under the Clean Water Act are permits under the National Pollutant Discharge Elimination System (NPDES) and fill permits under §404 of the Act. (See Chapter 9.) While Congress has created many agricultural exemptions for each of these programs (again, as detailed in Chapter 9), when agricultural activities are subject to either of these permitting schemes, the lead agency must determine what level of NEPA review applies since federal authorization of the activity at issue likely has environmental impacts.[101]

Under the NPDES program, which most frequently arises with respect to concentrated animal feeding operations (CAFO) and aquaculture operations, "EPA must prepare environmental assessments or envi-

---

94. Environmental Def. Fund v. EPA, 489 F.2d 1247 (D.C. Cir. 1973).
95. *Id.* at 1256.
96. Merrell v. Thomas, 807 F.2d 776, 780 (9th Cir. 1986); *see also* Wyoming v. Hathaway, 525 F.2d 66, 71-72 (10th Cir. 1975) (EPA need not prepare an EIS before cancelling or suspending registrations of three coyote poisons); Environmental Def. Fund, Inc. v. Blum, 458 F. Supp. 650, 661-62 (D.D.C. 1978) (EPA need not prepare an EIS before granting an emergency exemption to a state to use an unregistered pesticide.).
97. *Id.* at 781.
98. It is important to note that while EPA is not *required* to comply with NEPA in the FIFRA registration process, the agency often voluntarily prepares a NEPA document before approving a new pesticide. *See* Notice of Policy and Procedures for Voluntary Preparation of National Policy Act (NEPA) Documents, 63 Fed. Reg. 58045, 58046 (Oct. 29, 1998).
99. Oregon Envtl. Council v. Kunzman, 714 F.2d 901, 905 (9th Cir. 1983) (citations and quotation marks omitted).
100. Save Our Ecosystems v. Clark, 747 F.2d 1240, 1248 (9th Cir. 1984).
101. *See, e.g.*, EPA, Animal Feeding Operations—Best Management Practices, http://www.epa.gov/oecaagct/anafobmp.html.

ronmental impact statements, as required by [NEPA] to assess the potential for impacts."[102] The NPDES requirement and the accompanying NEPA review process must be initiated for *all* new or expanding CAFOs or aquaculture operations.[103] EPA has determined that, regardless of the size of, or pollution caused by, an older CAFO or aquaculture operation not seeking to expand, it need not undertake NEPA review in such scenarios, promulgating a categorical exclusion in 2007 for "[a]ctions involving re-issuance of a NPDES permit for a new source providing the conclusions of the original NEPA document are still valid (including the appropriate mitigation), there will be no degradation of the receiving waters, and the permit conditions do not change or are more environmentally protective."[104] While the logic behind this categorical exclusion makes superficial sense because it is based on prior NEPA review, it fails to account for long-term changes to the local environment and also circumvents NEPA's requirement to analyze cumulative impacts, which of course change—often in an adverse manner—over time.[105] Likewise, in the relatively rare instances in which the provision applies to agriculture (e.g., deep ripping and certain types of wetlands conversion to agricultural land, as described in Chapter 9), permits under the Clean Water Act's §404 fill permit process are subject to NEPA's requirements by EPA and the U.S. Army Corps of Engineers.[106] Therefore, although the Clean Water Act does not arise as frequently in the agricultural context as some other federal environmental statutes, there is no "functional equivalence" performed in the permit application review process, nor has Congress expressly or implicitly exempted NPDES or fill permits from the NEPA process, meaning that NEPA applies with full force regardless of which Clean Water Act provision is triggered.

c. GE Seed Deregulation Under the Plant Protection Act

Another federal process where NEPA applies is in the USDA Animal and Plant Health Inspection Service's (APHIS) deregulation process under the Plant Protection Act.[107] In deciding whether to deregulate a genetically engineered (GE) seed crop—which means to authorize its use commercially and otherwise—courts have held that USDA has an unequivocal obligation under NEPA to prepare at least an EA, and in many cases an EIS, because of the diverse and significant ecological impacts that result from the introduction and extensive planting of a new GE crop. Cases addressing USDA's NEPA compliance as part of its deregulation process have made it as far as the Supreme Court.

The first case to raise these issues involved a challenge to USDA's authorization of field testing of GE crops under the Plant Protection Act before USDA made a decision on deregulation of such crops.[108] In that case, USDA undertook no NEPA review before approving the applicant's request to conduct limited field testing as contemplated by the Plant Protection Act to observe the crop's traits in the field, in order to lead to a more well-informed deregulation decision by USDA.[109] The court, relying on basic APA principles, ultimately sided with environmental organizations that brought suit, finding that "[t]here is nothing in the administrative record to indicate that, contemporaneously with the issuance of the four permits, APHIS considered the applicability of NEPA, categorical exclusions, or the exceptions to those exclusions, . . . [thus fail[ing] to provide a reasoned explanation for its apparent determinations."[110] Because of the temporal posture of the case when the court finally decided the merits (the field testing permits had already expired), the court provided only declaratory relief since an injunction would make little sense under the circumstances.[111] Another case with a similar factual basis (field testing without NEPA review) was also resolved in favor of environmental groups challenging USDA's failure to prepare even an EA for GE creeping bentgrass because "[t]he record contains substantial evidence that the field tests may have had the potential to affect significantly the quality of the human environment, and that the tests may have

---

102. *Id.*
103. *Id.*
104. 40 C.F.R. §6.204(a)(1)(iv).
105. *See* 40 C.F.R. §1508.7.
106. 33 C.F.R. §325, app. B, §7(b)(1).
107. 7 U.S.C. §§7701-7786.
108. Center for Food Safety v. Johanns, 451 F. Supp. 2d 1165 (D. Haw. 2006).
109. *Id.*
110. *Id.* at 1183.
111. *Id.* at 1195-96.

involved, at the least, novel modifications . . . that raised new environmental issues."[112] These cases firmly establish that USDA cannot authorize field testing of any kind for GE crops under the Plant Protection Act without engaging in some level of appropriate NEPA review or thorough documentation of its rationale for opting not to engage in NEPA review.

Another case dealt with the application of NEPA to the final deregulation decision of GE alfalfa. In that case, USDA opted to prepare only an EA, determining that the level of impact of nationwide planting of GE alfalfa would not result in significant environmental impacts triggering the need for an EIS.[113] A federal district court disagreed with USDA's conclusion, finding that USDA "failed to adequately analyze the risk of gene flow and to what extent, if any, certain measures could be implemented to effectively prevent such contamination," and further that an EIS was required because the anticipated impacts would undoubtedly be significant.[114] Notably, the court admonished USDA for attempting to circumvent the EIS requirement by instead imposing nondiscretionary conditions on anyone planting the crop, explaining that USDA cannot "skip the EIS process and decide without any public comment that deregulation with certain conditions is appropriate."[115] As a remedy, the judge not only vacated and remanded the decision to enable USDA to comply with NEPA, but also enjoined USDA from making any interim decisions on more limited planting of GE alfalfa prior to completion of an EIS.[116] The latter part of the district court's remedy was challenged all the way to the Supreme Court, which held that the injunction against the agency went too far by precluding USDA from exercising its expert discretion in fashioning interim planting rules that do not run afoul of NEPA. Importantly, the district court's remedy of vacatur and remand of the deregulation decision was never challenged by the government nor questioned by the Supreme Court.[117] In January 2011, USDA released its final EIS for GE alfalfa, which a district court upheld in 2012 as adequate under NEPA and the Endangered Species Act, but which, at the time of writing, is on appeal to the Ninth Circuit by many of the same organizations that prevailed in overturning USDA's failure to prepare an EIS in the first instance.[118]

Most recently, there has been an ongoing litigation battle over GE sugar beets, and it too has focused heavily on the appropriate remedy for USDA's failure to comply with NEPA in deregulating a GE crop. In that case, USDA once again issued an EA with conclusory statements on key environmental impacts, but opted not to prepare an EIS, which a federal district court determined to be a NEPA violation.[119] As a remedy, the district court vacated and remanded USDA's deregulation decision to give USDA an opportunity to prepare an EIS in compliance with NEPA.[120] Once the full deregulation decision was vacated and while USDA was in the process of preparing an EIS, USDA issued permits to certain companies for limited planting of GE sugar beets without preparing any NEPA review for the permits, instead determining that a categorical exclusion applied to the permits.[121] This too was challenged; the court agreed with the plaintiffs that USDA's authorization of the GE sugar beet planting without any NEPA review whatsoever was unlawful, enjoined all planting under the permits, and went so far as to order that all GE sugar beet stecklings (immature beets) planted under the permits be removed from the ground to avoid the environmental impacts of their planting.[122] Similar to the broad remedy with respect to GE alfalfa, the Ninth Circuit struck down the district court's preliminary injunction and order to remove stecklings from the ground as too overreaching a remedy for the NEPA violation at issue, but did not disturb the lower court's vacatur

---

112. International Ctr. for Tech. Assessment v. Johanns, 473 F. Supp. 2d 9, 30 (D.D.C. 2007).
113. Geertson Seed Farms v. Johanns, Civ. No. 06-1075, 2007 WL 1302981, at *1 (N.D. Cal. May 3, 2007).
114. *Id.* at *4.
115. *Id.* at *5.
116. *Id.* at **6-8.
117. Monsanto Co. v. Geertson Seed Farms, 130 S. Ct. 2743 (2010).
118. USDA, Final EIS for Roundup Ready Alfalfa (Dec. 2010), *available at* http://www.aphis.usda.gov/biotechnology/downloads/alfalfa/gt_al-falfa%20_feis.pdf; *see also* Center for Food Safety v. Vilsack, Civ. No. 11-1310, 2012 WL 27787 (N.D. Cal. Jan. 5, 2012) (granting summary judgment to government on adequacy of EIS).
119. Center for Food Safety v. Vilsack, Civ. No. 08-484, 2009 WL 3047227, at **8-9 (N.D. Cal. Sept. 21, 2009).
120. Center for Food Safety v. Vilsack, 734 F. Supp. 2d 948 (N.D. Cal. 2010).
121. Center for Food Safety v. Vilsack, Civ. No. 10-4038, 2010 WL 3835699, at **7-8 (N.D. Cal. Sept. 28, 2010).
122. Center for Food Safety v. Vilsack, 753 F. Supp. 2d 1051 (N.D. Cal. 2010).

and remand of USDA's earlier deregulation decision.[123] The NEPA litigation challenging GE sugar beets continues at the time of writing.[124]

As USDA, GE seed companies, and farmers have learned the hard way, USDA is required to comply with NEPA in all facets of deregulating GE crops. Not only is appropriate NEPA review required for any field testing of GE crops during the deregulation process, but USDA is also obligated to conduct a compliant NEPA review—in most cases an EIS—to accompany any ultimate deregulation decision under the Plant Protection Act. The conclusions to the two ongoing cases of GE alfalfa and GE sugar beets will likely continue to inform the question of what, if any, remedies in addition to vacatur and remand are permissible in the GE crop context when NEPA is violated.

d. Authorization of GE Animals for Human Consumption

In contrast to GE seed crops where there are longstanding judicial directives requiring USDA to comply with NEPA early and often throughout the Plant Protection Act deregulation process, to date there is no precedent for the emerging field of GE animals for human consumption. However, based on the similarities between GE animals and GE seed crops—in particular the potential environmental concerns of introducing species with modified genetic structures into the ecosystem—it is safe to assume that NEPA applies in full force where an agency, likely the U.S. Food and Drug Administration (FDA), approves or otherwise authorizes a private company to proceed with commercial production of a GE animal.

The guinea pig, so to speak, on this issue is GE salmon, which is nearing final approval from FDA, and authorization would ultimately fall into the category of a "new animal drug" under FDA's statutory authority in the Federal Food, Drug, and Cosmetic Act. The well-documented environmental impacts of GE salmon are discussed in Chapter 6, but questions remain as to whether FDA's NEPA review, which currently appears to encompass only preparation of an EA,[125] will meet NEPA's requirement to account for and consider all of the relevant environmental impacts of what would be a precedent-setting decision. In 2010, the applicant prepared its version of an EA and submitted it to FDA for agency review.[126] Several environmental organizations submitted a citizen petition in 2011 to FDA requesting that the agency withhold action on the application until it completes an EIS.[127] However, in December 2012, FDA issued a 145-page draft EA and preliminary FONSI.[128]

Based on general NEPA jurisprudence, as well as rulings in the agricultural realm,[129] it seems likely that FDA's decision will be vacated and remanded if the agency continues to refuse to prepare an EIS for this landmark decision, which will inevitably create a precedent for the production of GE animals for human consumption. In any event, the writing is on the wall—GE foods (and their inevitable environmental impacts) are here to stay, and NEPA is yet another tool in the legal toolbox to ensure that their impacts on our natural ecology, food supply, and public health are considered and analyzed with meaningful public involvement before decisions are made that will forever change the way in which we consume food.

e. Miscellaneous Federal Permits

There are several other federal approval processes where courts have held that NEPA applies, although they are less prominent than the examples cited above. At least one federal court has held that when an agency

---

123. Center for Food Safety v. Vilsack, 636 F.3d 1166 (9th Cir. 2011).
124. Center for Food Safety v. Vilsack, Civ. No. 11-831, 2011 WL 996343 (N.D. Cal. Mar. 17, 2011) (transferring the case to federal district court in the District of Columbia).
125. AquaBounty, Environmental Assessment for AquAdvantage® Salmon (Aug. 25, 2010), *available at* http://www.fda.gov/downloads/AdvisoryCommittees/CommitteesMeetingMaterials/VeterinaryMedicineAdvisoryCommittee/UCM224760.pdf.
126. *Id.*
127. Center for Food Safety, Environmental Groups Petition Federal Government to Complete Full and Transparent Review of Environmental Impacts Associated With "Frankenfish" (May 25, 2011), *available at* http://www.centerforfoodsafety.org/wp-content/uploads/2011/05/Final-GE-Salmon-Citizen-Petition-5.25.11.pdf.
128. U.S. Food & Drug Administration (FDA), Draft Environmental Assessment for AquAdvantage Salmon, *available at* http://www.fda.gov/downloads/AnimalVeterinary/DevelopmentApprovalProcess/GeneticEngineering/GeneticallyEngineeredAnimals/UCM333102.pdf; FDA, Preliminary Finding of No Significant Impact, *available at* http://www.fda.gov/downloads/AnimalVeterinary/DevelopmentApprovalProcess/GeneticEngineering/GeneticallyEngineeredAnimals/UCM333105.pdf.
129. *See, e.g.*, Foundation on Econ. Trends v. Heckler, 587 F. Supp. 753 (D.D.C. 1984) (requiring EIS for experiments introducing recombinant DNA (i.e., genetic modification) into potato and other food crops).

permits any planting of GE crops on federal lands, the agency must prepare an EA or EIS, as appropriate under the circumstances.[130] In that case, the court held that the failure of the U.S. Fish and Wildlife Service to engage in any NEPA review when it authorized private farmers to plant GE corn and soybeans at the Prime Hook Wildlife Refuge was arbitrary and capricious in violation of NEPA.[131]

Second, with respect to FDA approval of growth hormones used to artificially fatten livestock to reduce time before slaughter, one court has held that an EIS is not necessary as long as FDA's statutory analysis under the Federal Food, Drug, and Cosmetic Act appropriately considers and analyzes all of the factors required by NEPA.[132] Importantly, however, in that case FDA had not only conducted its parallel statutory analysis under the Act, but it had also prepared an EA, meaning that the agency recognized the need for NEPA review even if it would result in some level of duplication.[133]

Third, in regard to federally funded research with future (albeit attenuated) implications for our nation's food system, courts have been more reluctant to require an agency to prepare an EIS. In two cases dealing with such research—one on animal productivity and one on aquaculture—the courts respectively held in the first case that a court is unable to require an EIS for animal productivity research objectives that are not presently having any environmental impacts,[134] and in the second case that where the U.S. Army Corps of Engineers had in fact completed an EA for an aquaculture research project, there was nothing more required of NEPA since no significance factors were triggered.[135]

Fourth, a court has held that any major changes in the federal inspection process for horses slaughtered at foreign-owned facilities in the United States for human consumption abroad must fully comply with NEPA.[136]

Conclusion

While the application of NEPA is a fact-specific inquiry that does not result in an EIS or even an EA in all cases, agencies that disregard its statutory mandates do so at their own peril because the failure to fully comply with NEPA can delay an agency's decisionmaking by months or even years. Particularly since the Supreme Court wrote that "NEPA itself does not impose substantive duties mandating particular results, but simply prescribes the necessary process . . . [for preventing] uninformed—rather than unwise—agency action,"[137] it behooves agencies to engage in NEPA review where there is any doubt that environmental impacts will result, including in the agricultural and food arenas where almost all decisions will likely have widespread and significant ecological effects.

While it might not mean the avoidance of "unwise" agency actions,[138] more regular and comprehensive NEPA review by USDA, FDA, and other pertinent agencies will certainly lead to a more transparent, effective, and meaningful process whereby public involvement and informed decisionmaking in our nation's agricultural and food system are paramount. Such practices would ensure Congress' collective goals in NEPA that the statute "encourage productive and enjoyable harmony between man and his environment . . . [and] promote efforts which will prevent or eliminate damage to the environment and biosphere and stimulate the health and welfare of man."[139]

---

130. Delaware Audubon Soc'y v. Sec. of Dept. of Interior, 612 F. Supp. 2d 442 (D. Del. 2009).
131. *Id.* at 450-51.
132. Stauber v. Shalala, 895 F. Supp. 1178, 1193-96 (W.D. Wis. 1995).
133. *Id.*
134. Foundation on Econ. Trends v. Lyng, 817 F.2d 882 (D.C. Cir. 1987).
135. Food & Water Watch v. U.S. Army Corps of Eng'rs, 570 F. Supp. 2d 177 (D. Mass. 2008).
136. Humane Soc'y v. Johanns, 520 F. Supp. 2d 8 (D.D.C. 2007).
137. Robertson v. Methow Valley Citizens Council, 490 U.S. 332, 350-51 (1989).
138. *Id.*
139. 42 U.S.C. §4321.

Chapter 13
The Food Statutes
Jason J. Czarnezki and Elena M. Mihaly

Food and agriculture policy does not exist in a vacuum, but is instead inextricably connected to a host of other environmental legislation. While the fields of food and drug law and agriculture law have existed for some time, food law is a field that only recently grew out of both of those disciplines, among others. This chapter discusses a set of statutes that are specifically geared toward food, organized into three categories, (1) public health and safety statutes, (2) organic and origin labeling statutes, and (3) legislation related to the National School Lunch Program.

The chapter focuses on statutes that deal with some of the most contentious issues surrounding the relationship between food and the environment, including pesticide residue in foods and other health and safety issues associated with the proliferation of commodity-driven industrial agriculture, and the future of our school lunch programs, which rely on government-subsidized agriculture. The first section of the chapter discusses the Federal Food, Drug, and Cosmetic Act, the Food Quality Protection Act, and the Food Safety Modernization Act of 2010. The next section discusses the Organic Foods Production Act, and country of origin labeling requirements that were first enacted in the 2002 farm bill. The final section addresses the National School Lunch Program as it has developed under the Richard B. Russell National School Lunch Act, the Child Nutrition Act of 1966, and the Healthy, Hunger-Free Kids Act of 2010.

It is important to understand the environmental and social repercussions of these major federal food statutes. An expanding food market has stimulated increased government regulation of food, as well as increased the awareness of citizens and scholars alike. The attention that both the government and the public are now giving to the safety of our food, how it is grown, where it originates, and its nutritional content is a sign that food and agriculture policy has secured a firm place in current public discourse.

A. Public Health and Safety

At its most basic level, federal law and regulation seek to ensure that our food is safe. In doing so, federal law not only provides for inspection of food facilities and recall of tainted food products, but also implicates some of the social and safety concerns that have become commonplace in the industrial food system such as the consequences of pesticide usage and the impact of federal regulation on small farmers. The key laws in these areas are the Federal Food Drug, and Cosmetic Act, the Food Quality Protection Act, and the recent passage of the Food Safety Modernization Act.

The authors wish to thank Allison Marshall and Andrew Homan for their research assistance, and Mary Jane Angelo for allowing them to take advantage of her expertise on the subject of pesticides, and to rely on so much of her earlier work, especially Mary Jane Angelo, *Regulating Evolution for Sale: An Evolutionary Biology Model for Regulating the Unnatural Selection of Genetically Modified Organisms*, 42 WAKE FOREST L. REV. 93 (2007). The chapter also relies significantly on JASON J. CZARNEZKI, EVERYDAY ENVIRONMENTALISM: LAW, NATURE & INDIVIDUAL BEHAVIOR (ELI Press 2011); Jason J. Czarnezki, *Food, Law & the Environment: Informational and Structural Changes for a Sustainable Food System*, 31 UTAH ENVTL. L. REV. 263 (2011); and Jason J. Czarnezki, *The Future of Food Eco-Labeling: Organic, Carbon Footprint, and Environmental Life-Cycle Analysis*, 30 STAN. ENVTL. L.J. 3 (2011).

1. The Federal Food, Drug, and Cosmetic Act and the Food Quality Protection Act

The Federal Food, Drug, and Cosmetic Act (FFDCA) is "the primary food law in the United States."[1] The FFDCA is primarily administered by the Food and Drug Administration (FDA), an agency charged with ensuring that most domestic and imported foods products are "safe, nutritious, wholesome, and accurately labeled."[2] The FFDCA,[3] in conjunction with its amendment, the Food Quality Protection Act (FQPA)[4] (and other statutes), seeks to regulate pesticides and defects in food.

a. Pesticides

The U.S. Environmental Protection Agency (EPA) registers pesticides, assesses their risk, and establishes maximum residue limits (or "tolerances") for quantities in food. The Food Safety and Inspection Service of the U.S. Department of Agriculture (USDA) monitors and enforces pesticide tolerance levels in meat and poultry. FDA enforces tolerance levels in imported and domestic food, primarily fruits and vegetables.

The Federal Insecticide, Fungicide, and Rodenticide Act (FIFRA), first known as the Insecticide Act of 1910, began as a labeling and consumer protection statute attempting to abate false claims about pesticide effectiveness. (See Chapter 8 for a full discussion of FIFRA.) Its modern formulation, passed in 1947 and significantly amended in the 1970s, mandates pesticide registration and the setting of tolerance levels for pesticide in food. If a pesticide is not registered, it is prohibited from being sold or distributed.[5] FIFRA directs EPA to register pesticides only to the extent that their use will not cause "(1) any unreasonable risks to man or the environment, taking into account the economic, social, and environmental costs and benefits of the use of any pesticide, or (2) a human dietary risk from residues that result from a use of a pesticide in or on any food inconsistent with the standard [under the FFDCA]."[6] The Act's definition of "environment" includes "water, air, land, and all plants and man and other animals living therein, and interrelationships which exist among these."[7] Registration encompasses providing proper labeling and directions for use.

When EPA reviews a pesticide for registration under FIFRA, risk determination is only one consideration among others. EPA will not register a pesticide under FIFRA unless the applicant receives tolerance level approval or an exemption from establishing tolerance under the FFDCA. In addition to regulating pesticides under FIFRA, EPA is also responsible for regulating pesticide residues in human food or animal feed under FFDCA.[8] Pursuant to §408(a) of FFDCA, a pesticide chemical residue in or on food is not considered to be safe unless EPA has issued a tolerance for such residue and the residue is within the tolerance limits.[9] EPA may issue an exemption from the requirements of a tolerance if it determines that "there is a reasonable certainty that no harm will result from aggregate exposure to the pesticide chemical residue, including all anticipated dietary exposures and all other exposures for which there is reliable information."[10]

---

1. Patricia A. Curtis, Guide to Food Laws and Regulations 57 (2005); *see also* 21 U.S.C. §§301 et seq.
2. Susan A. Schneider, Food, Farming, and Sustainability: Readings in Agricultural Law 619 (2010). FDA is not responsible for meat and poultry safety. *Id.*
3. Federal Food, Drug, and Cosmetic Act, 21 U.S.C. §§301 et seq.
4. Food Quality Protection Act, Pub. L. No. 104-170 (1996).
5. *Id.* at 1.
6. Section 136(bb) defines the term "unreasonable adverse effects on the environment" as any "unreasonable risk to man or the environment, taking into account the economic, social, and environmental costs and benefits of the use of any pesticide. . . ." *Id.* §136(bb).
7. Federal Insecticide, Fungicide, and Rodenticide Act, §§2(j), 3; 7 U.S.C. §§136(a), 136(j).
8. Mary Jane Angelo, *Regulating Evolution for Sale: An Evolutionary Biology Model for Regulating the Unnatural Selection of Genetically Modified Organisms*, 42 Wake Forest L. Rev. 93, 128 (2007), citing 21 U.S.C. §346 (2000). The Reorganization Plan No. 3 of 1970, which created EPA, granted EPA authority to establish tolerances for residues of pesticide chemicals in foods and animal feeds. Reorganization Plan No. 3 of 1970, 3 C.F.R. §199 (1970 Comp.), *reprinted in* 5 U.S.C. app. 184, and in 84 Stat. 2086 (1970-1971). Regulatory authority over other nonpesticidal substances in foods and animal feeds was left within the jurisdiction of FDA.
9. Angelo, *supra* note 8, at 128, citing 21 U.S.C. §346a(a)(1).
10. Angelo, *supra* note 8, at 128, citing 21 U.S.C. §346a(c)(2)(A). In 2001, EPA adopted an exemption under this standard for pesticidal residue in plant-incorporated protectants. 40 C.F.R. §174.508.

b. Cosmetic Standards

The existence of "cosmetic standards" can result in excess pesticide use on agricultural products for largely, and sometimes purely, aesthetic purposes. These cosmetic standards take the form of FDA defect action levels (DALs) and USDA grading standards. DALs put a limitation on the number of somewhat odd food defects—such as how many rodent hairs can be present in a certain amount of flour, how many insects can be present in a certain amount of fruit, or how much mold can be on fruit and vegetables.[11] USDA grading standards are voluntary guidelines that provide the produce industry with a "uniform language for describing the quality and condition of commodities in the marketplace."[12] DALs[13] represent "the limits at which FDA will regard a food product as "adulterated" and subject to enforcement action under the FFDCA."[14] The FFDCA regulates filth, decomposition, and extraneous matter, and states that a food is adulterated if it consists in whole or in part of a filthy, putrid, or decamped substance.[15] Yet, while DALs are the maximum levels of natural or unavoidable defects in foods, such defects do not necessarily present any significant health hazards.[16] Even so, DALs have lowered over time, fostered by new pesticide technologies and continued efforts by FDA and USDA to limit insect and mite levels in foods and vegetables.[17] In addition, wholesalers, processors, and retailers are increasing their industries' cosmetic standards,[18] which can put pressure on some small growers who might lack the means to meet the higher standards.[19] In order to meet these cosmetic standards, producers use additional insecticide and miticides on fruits and vegetables.[20]

The need for rigorous DALs is suspect. "[M]ost insects found on produce are probably not any more of a health hazard than beef or chicken."[21] Despite the lack of health hazards, DALs do exist and lead to increased pesticide use,[22] and FDA is clearly aware of this concern. In the *Defects Levels Handbook*, FDA states:

> It is FDA's position that pesticides are not the alternative to preventing food defects. The use of chemical substances to control insects, rodents and other natural contaminants has little, if any impact on natural and unavoidable defects in foods. The primary use of pesticides in the field is to protect food plants from being ravaged by destructive plant pests (leaf feeders, stem borers, etc.). A secondary use of pesticides is for cosmetic purposes—to prevent some food products from becoming so severely damaged by pests that it becomes unfit to eat.[23]

But commercial buyers strongly prefer blemish-free fruit even though blemished foods are still nutritious. This can lead to absurd results, such as juice makers refusing to take blemished fruit even though the fruit will be ground up for juice.[24] The achievement of producing "perfect" produce, and the resulting

---

11. *See* FDA, DEFECT LEVELS HANDBOOK (2011), *available at* http://www.fda.gov/food/guidancecomplianceregulatoryinformation/guidancedocuments/sanitation/ucm056174.htm.
12. USDA Agricultural Marketing Service, *Grading, Certification and Verification*, http://www.ams.usda.gov/AMSv1.0/ams.fetchTemplateData.do?template=TemplateN&navID=U.S.GradeStandards&rightNav1=U.S.GradeStandards&topNav=&leftNav=&page=FreshGradeStandardsIndex&resultType=&acct=freshgrdcert.
13. Adopted pursuant to 21 C.F.R. §110.110. The levels themselves are found in the *Defect Levels Handbook*, not within regulations. FDA, DEFECT LEVELS HANDBOOK, *available at* http://www.fda.gov/Food/GuidanceComplianceRegulatoryInformation/GuidanceDocuments/Sanitation/ucm056174.htm. 21 C.F.R. §110.110 allows FDA "to establish maximum levels of natural or unavoidable defects in foods for human use that present no health hazard. Defect action levels represent levels of natural or unavoidable defects or contaminants that can be found in a food product produced from acceptable quality raw materials using current processing technologies and sanitation practices. The levels represent limits at which FDA regards food products to be adulterated and subject to regulatory action under Section 402(a)(3) of the FD&C Act." John S. Gecan, *Food Defect Action Levels*, *in* FOOD PLANT SANITATION 79 (Yiu H. Hui et al. eds., 2003).
14. FFDCA §402(a), 21 U.S.C. §342(a); FDA, DEFECT LEVELS HANDBOOK (2011), at http://www.fda.gov/Food/GuidanceComplianceRegulatoryInformation/GuidanceDocuments/Sanitation/ucm056174.htm#intro.
15. Gecan, *supra* note 13, at 77, citing FFDCA §402(a)(3).
16. *See id.* "The FDA set these action levels because it is economically impractical to grow, harvest, or process raw products that are totally free of non-hazardous, naturally occurring, unavoidable defects. Products harmful to consumers are subject to regulatory action whether or not they exceed the action levels." *Id.*
17. David Pimentel et al., *The Relationship Between "Cosmetic Standards," for Foods and Pesticide Use*, *in* THE PESTICIDE QUESTION: ENVIRONMENT, ECONOMICS, AND ETHICS 85, 87 (David Pimentel & Hugh Lehman eds., 1993). "During the past 40 years, as the FDA has been lowering these tolerance levels, more pesticides have been used to insure that crop produce meet the more stringent defect levels." *Id.* at 85.
18. *Id.* at 85.
19. *Id.* at 90.
20. *Id.* at 100.
21. *Id.* at 93.
22. *Id.* at 88.
23. FDA, DEFECT LEVELS HANDBOOK, *supra* note 14.
24. Pimentel et al., *supra* note 17, at 91.

demand for such produce, was made possible because of the increased availability and use of insecticides and miticides.[25] Thus, most fruits and vegetables in American supermarkets have little or no insect damage on their surfaces,[26] and "growers must apply more pesticide to achieve these marketplace demands."[27] Prof. David Pimentel of Cornell summarizes the problem in this way:

> From the 1930's to 1976, the FDA and USDA gradually reduced the defect action levels (DALs) for insects and mites found in foods, even though there was no proven health hazard associated with the presence of small herbivorous anthropods in food. Since 1976, both the FDA and USDA have maintained the established DALs. This is encouraging. However, food processors, wholesalers, and retailers seem to be placing even greater emphasis on blemish-free, perfect produce. Not only has this pressure caused substantial crop losses because large portions of food crops are now being classified as unsuitable for commercial sale, but it has also contributed to heavy pesticide usage by farmers who feel compelled to spray to reduce the incidence of insects and mites in foods to meet these "cosmetic appearance" standards.[28]

Like DALs, USDA grading standards are designed to limit insects, blemishes, and other defects in produce. But, "[t]he establishment of voluntary grading standards by federal marketing orders has resulted in grading of produce (e.g., USDA Extra Fancy) that, over time, has evolved into mandatory industry requirements."[29] And, again like DALs, USDA grading standards could result in increased pesticide usage. However, consumers may not be aware of the connection between cosmetic appearance and increased pesticide use, though increased awareness may encourage consumers to purchase produce that is not cosmetically perfect because it has less or no pesticide residues.[30]

2. The Food Safety Modernization Act of 2010

The Food Safety and Modernization Act (FSMA), passed by the U.S. Congress and signed into law on January 4, 2011, is a major overhaul and expansion of FDA's food safety responsibilities.[31] FDA itself explains that the law "aims to ensure the U.S. food supply is safe by shifting the focus of federal regulators from responding to contamination to preventing it."[32]

The FSMA has three major provisions[33] that (1) attempt to prevent food safety hazards through procedural safeguards, recordkeeping, and the promulgation of regulations on sanitary food transportation practices[34]; (2) detect and respond to food safety problems by identifying high-risk facilities, developing science-based methods of testing, disposing of contaminated foods, and improving training of food safety officials[35]; and (3) improve safety of imported food.[36]

A great deal of controversy has arisen concerning the impact that the FSMA will have on small farms and food processors, some of whom feel that the provisions are too onerous and will threaten small farms and producers, both from a regulatory and financial standpoint.[37] The FSMA mandates inspection frequencies,

---

25. *Id.* at 89.
26. *Id.*
27. *Id.* at 90.
28. *Id.* at 99-100.
29. *Id.* at 90.
30. *Id.* at 89.
31. Food Safety and Modernization Act, Pub. L. No. 111-353, 124 Stat. 3885 (2011). Full text available at http://www.fda.gov/Food/FoodSafety/FSMA/ucm247548.htm. *See also* Gardiner Harris & William Neuman, *Senate Passes Overhaul of Food Safety Regulations*, N.Y. TIMES, Nov. 30, 2010, at http://www.nytimes.com/2010/12/01/health/policy/01food.html.
32. FDA, *About FSMA*, http://www.fda.gov/Food/FoodSafety/FSMA/ucm247546.htm. It is important to note that the law only covers foods under FDA's jurisdiction (including produce and processed foods). It does not cover meat, poultry, or certain egg products that are regulated by USDA.
33. There is also a miscellaneous fourth provision. Title IV, Miscellaneous Provisions, provides protections to whistleblowers and addresses compliance with international agreements and determination of budgetary effects. FSMA §§401-405, codified as amended at 21 U.S.C. §§399(d), 2251, and 2252. Summary at http://www.ift.org/public-policy-and-regulations/~/media/Public%20Policy/PolicyDevelopment/FDAFoodSafetyModernizationAct_summary.pdf.
34. FSMA §§101-116, codified as amended by 21 U.S.C. §§350(g)-(i), 379(j-31), 2201-2206.
35. Title II, Improving Capacity to Detect and Respond to Food Safety Problems. FSMA §§201-211, codified as amended by 7 U.S.C. §7625, 21 U.S.C. §§350(j)-(l-1), 399(b), 2221-2225, 42 U.S.C. §280(g-16).
36. Title III, Improving the Safety of Imported Food, requires importers, with certain exemptions, to verify compliance of foreign suppliers with food safety regulations. It also gives FDA authority to require import certifications for food and to inspect foreign facilities. FSMA §§301-309, codified as amended by 21 U.S.C. §§384(a)-(d), 2241-2243.
37. *Will the Food Safety Modernization Act Harm Small Farms or Producers?* (Nov. 16, 2010), *available at* http://www.grist.org/article/food-2010-11-15-food-fight-safety-modernization-act-harm-small-farms.

based on risk, for food facilities and requires the frequency of inspection to increase immediately.[38] Food facilities that require registration, and thus inspection, include factories, warehouses, or establishments that manufacture, process, pack, or hold food.[39] Under the current definition, farms, restaurants, and retail facilities are exempt from the registration requirement.[40] However, the exemption may be amended by FDA regulations. FDA is required to issue regulations within 18 months of the passage of the FSMA specifying what activities constitute on-farm packing or holding of food, and on-farm manufacturing or processing of food.[41] FDA is required to conduct a science-based risk analysis of the types of manufacturing, processing, packing, and holding activities that occur on farms and, based on that risk analysis, exempt (or modify the requirements for) facilities that are defined as "small businesses" or "very small businesses" engaged in specific activities FDA determines to be low risk.[42]

FDA regulations are required to provide sufficient flexibility to be practicable for all sizes and types of facilities[43]; however, without concrete parameters, small farmers may struggle in making certain management decisions. This would be especially true if the FSMA's hazard analysis and risk-based preventive controls (HARPC) requirement is applied to small farm processing activities. A study done by a small farmer in North Carolina to assess the costs and time required for a small farm to establish a HARPC plan showed that a typical small farm doing on-farm processing would need 150 hours to create, implement, and monitor the plan, and spend $9,500 per year on consulting and testing costs.[44] If the farm hired a consultant to create the plan, the first year costs soared to $20,000.[45]

FDA published guidelines for fruit and vegetable production, called "good agricultural practices" (GAPs), but the FSMA now mandates FDA to apply those rules on fruit and vegetable growers, no matter the size of the farm.[46] Small-scale and organic farms report that the application of these guidelines often compromises their ability to operate sustainably.[47] For instance, FDA's existing GAP guidelines for growing produce treat manure and compost as a major threat to food safety.[48] The prescriptive one-size-fits all production standards that GAPs impose on growers threaten the economic viability of small farms that practice diversified crop management, because replicating GAP paperwork for dozens of different crops is a financial burden.[49]

In response to concerns over standards and inspections, Congress passed the Tester-Hagan Amendment to create exemptions for small-scale producers and processors.[50] To be eligible for this exemption, a farm or business must gross less than $500,000 per year and sell a majority of its food directly to consumers, restaurants, and grocery stores within a 275-mile radius from its place of business or within the same state.[51] It must also demonstrate that it is in compliance with state and local food safety laws.[52] FDA can remove the exemption if any of its products are implicated in a food-borne illness outbreak, or if it is determined that removal is "necessary to protect the public health and prevent or mitigate a foodborne illness outbreak based on conduct or conditions associated with a qualified facility that are material to the safety of the food manufactured, processed, [or] packed."[53]

---

38. FDA, *Background on the FDA Food Safety and Modernization Act (FSMA)*, http://www.fda.gov/Food/FoodSafety/FSMA/ucm239907.htm.
39. Amending §415 of the FFDCA, 21 U.S.C. §350d(b). *See also* Institute of Food Technologists, *Summary of the FDA Food Safety Modernization Act*, http://www.ift.org/public-policy-and-regulations/recent-news/2011/january/summary-of-the-fda-food-safety-modernization-act.aspx.
40. *Id.*
41. FSMA §103, amending 21 U.S.C. §§341 et seq.
42. 21 C.F.R. §§1.226(b), 1.227(b)(3).
43. *Id.* §§1.226(b), 1.227(b)(3).
44. Roland McReynolds, Carolina Farm Stewardship Association, Hurting NC's Local Food Harvest: The Unintended Consequences of Federal Food Safety Legislation on North Carolina's Small Agricultural Enterprises (2010), *available at* http://www.carolinafarmstewards.org/wp-content/uploads/2012/04/Hurting_NCs_Local_Food_Harvest042010.pdf.
45. *Id.*
46. FSMA §105(e), amending 21 U.S.C. §350(h).
47. *See* Elanor Starmer, Food & Water Watch, and Marie Kulick, Institute for Agriculture and Trade Policy, Bridging the GAPs: Strategies to Improve Produce Safety, Preserve Farm Diversity and Strengthen Local Food Systems (2009), *available at* http://www.iatp.org/files/258_2_106746.pdf.
48. *Id.* at 11.
49. *Id.* at 13. *See also* Food and Agriculture Organization of the United Nations, *Good Agricultural Practices*, http://www.fao.org/prods/gap/ (2008).
50. Pub L. No. 111-353; Tester Amendment Agreement, http://www.jontester.com/news/press-releases/2012/key-legislation-by-jon-tester-now-public-law/.
51. *Id.*
52. *Id.*
53. *Id.*

B. Organic Food and Labeling

The past decade has seen both an increased interest in food labeling and a growing market for organic food.[54] Under the Organic Foods Production Act (OFPA) of 1990 and the National Organic Program (NOP), the U.S. government establishes production, handling, and labeling standards for organic agricultural products.[55] The United States has also enacted country of origin labeling legislation that requires retailers to inform consumers about the source of certain foods. This section addresses organic food labeling in the United States, the dominant environmental label that has earned great cache with consumers, and the origin label which, while created for consumer protection reasons, can serve as a proxy for information about "food miles."

1. The OFPA

Because people want chemical-free foods for personal health and environmental reasons, the organic food market is flourishing.[56] Organic food has almost quadrupled its market share in the last decade, and sales of organic food have grown from $1 billion in 1990 to over $20 billion today. In light of the profits available from organic production—organic products sell at a higher retail price than conventional ones—the modern organic production and distribution system is now dominated by large-scale industrial producers. While large-scale production, even if organic, causes increased greenhouse gas emissions, it also yields food produced and processed without synthetic pesticides or fertilizers. All of this may not have happened without a regulatory model creating a value-added food label like "organic."

a. Standards and Recordkeeping

Under the OFPA and the NOP, the U.S. government creates production, handling, and labeling standards for organic agricultural products.[57] A goal is to assure that "[o]rganic farming emphasizes the use of renewable sources, land management that maintains natural soil fertility, water conservation, rich biodiversity, and long term sustainability."[58]

The OFPA establishes a national organic certification program in which agricultural products may be labeled as organic if produced and handled without the use of certain synthetic substances. The program prohibits using synthetic pesticides, fertilizers, growth hormones and antibiotics in livestock,[59] or adding synthetic ingredients during processing.[60] However, exceptions exist, and some nonagricultural products and synthetics can be used on organic produce if they are on the National List of Allowed and Prohibited Substances.[61] Such products include waxes (carnauba and wood rosin) on organic fruit and fruit products; ethylene for postharvest ripening of tropical fruit and citrus degreening; and citric acid and ascorbic acid for fresh-cut fruits.[62] Chlorine is permitted and is the most commonly used synthetic for sanitation of organic fruit and vegetable surfaces.[63]

---

54. *See* Jason J. Czarnezki, *Food, Law & the Environment: Informational and Structural Changes for a Sustainable Food System*, 31 Utah Envtl. L. Rev. 263 (2011); Jason J. Czarnezki, *The Future of Food Eco-Labeling: Organic, Carbon Footprint, and Environmental Life-Cycle Analysis*, 30 Stan. Envtl. L.J. 3 (2011).
55. Organic Foods Production Act of 1990, 7 U.S.C. §§6501 et seq.
56. Organic products are available in more than 20,000 natural food stores and over three-quarters of conventional grocery stores. Catherine Greene & Carolyn Dimitri, USDA, Economic Research Service, *Organic Agriculture: Gaining Ground*, Amber Waves (2003), *available at* http://www.ers.usda.gov/amberwaves/feb03/findings/organicagriculture.htm. The total number of certified organic growers increased from 3,587 in 1992 to more than 8,000 in 2006. USDA, Organic Production Overview, Data Sets, Table 2, U.S. Certified Organic Farmland Acreage, Livestock Numbers, and Farm Operations, http://www.ers.usda.gov/data-products/organic-production.aspx. The number of farmland acres devoted to organic production, including both cropland and rangeland, exploded from 935,450 acres in 1992 to more than four million acres in 2005. *Id.*.
57. Organic Foods Production Act of 1990, 7 U.S.C. §§6501 et seq.
58. Anne Plotto & Jan A. Narciso, *Guidelines and Acceptable Postharvest Practices for Organically Grown Produce*, 41 HortSci. 287 (2006) (citing 7 C.F.R. §205.2).
59. National Organic Program; Access to Pasture (Livestock), 75 Fed. Reg. 7154 (Feb. 17, 2010), to be codified at 7 C.F.R. pt. 205.
60. 7 U.S.C. §§6508(b)(1), 6509(c)(3), 6510. *See also* Plotto & Narciso, *supra* note 58, at 287: "Food must be produced without synthetic chemicals, except those specifically allowed by regulations, and without substances (nonsynthetic and nonagricultural) prohibited by regulations, including no sewage sludge, ionizing radiation or bioengineering (7 C.F.R. 205.105)."
61. Catherine Greene et al., Emerging Issues in the U.S. Organic Industry, Economic Information Bulletin No. 55, at 2 (June 2009), *available at* http://www.ers.usda.gov/Publications/EIB55/.
62. Plotto & Narciso, *supra* note 58, at 288. *See* National List of Allowed and Prohibited Substances, 7 C.F.R. §§205.605, 205.606.
63. Plotto & Narciso, *supra* note 58, at 290, citing 7 C.F.R. §205.605(b).

Agricultural practices must follow an organic plan approved by an accredited certifying agent and the producer and handler of the product.[64] The OFPA establishes standards like "certified organic" labeling that informs consumers about the food production process, but does not require direct testing of all food or indicate a lack of land degradation, though organic food still is likely to have fewer chemicals than conventional counterparts.[65]

The NOP also has detailed recordkeeping requirements to meet the production process-based standards of the OFPA.[66] Due to the recordkeeping requirements and detailed standards, third-party certifiers work with state and federal governments to oversee organic certification:

> Despite meager funding and a small staff, the National Organic Program operates by accrediting nearly 100 third-party certifying agencies, who in turn provide the oversight required to certify farms and businesses as organic. Since standards cover the materials and processes used for both growing and processing food, not only does NOP regulate which pesticides a farmer may use, it also specifies which cleaning solution a processor may use to clean his equipment.[67]

Small farmers who gross less than $5,000 annually and only sell directly to consumers (e.g., via farmers markets and family farm stands) can avoid the certification process by simply signing a declaration stating that they comply with organic standards.[68] If these small farmers sell any of their products through conventional distribution channels, they may use the term "organic" but may not use the term "certified organic" or the USDA organic label on products without also obtaining official certification, a process that can be expensive and time-consuming.[69]

There is a concern that small farmers may have trouble coming up with the funds to receive organic certification, and may also lack the resources to fully promote and market their chemical-free and sustainably grown products.[70] In recognition of the costs of organic certification for small farmers, sliding scales for payment are the norm, and the cost of payment may be subsidized.[71] Organic certification fees, based on total sales, usually are below $1,000, except for large processors with far greater sales.[72] And costs can be as much as 75% less after government reimbursement if a state participates in the federal cost-share assistance program discussed below. But perhaps due to sliding scale differences (and thus fee differences), it has been claimed that organic certifiers largely ignore issues pertaining to small-scale farmers, placing a greater emphasis on enlisting larger producers who pay larger certification fees.[73] Cost-sharing resources help small farmers with the organic certification process. The Agricultural Management Assistance Organic Certification Cost Share Program, established in 2001, authorizes cost-share assistance to producers of organic agricultural products in most states.[74] In fiscal year 2008, Congress allocated on a one-time basis $22 million for this program, to be distributed to states until the funds are exhausted.[75] In 2010, the program was funded at $1.45 million. The program authorizes cost-share assistance to producers and handlers of organic agricultural products in each participating state. The states reimburse each eligible producer or handler up

---

64. 7 U.S.C. §§6504, 6505.
65. Michelle T. Friedland, *You Call That Organic?—The USDA's Misleading Food Regulations*, 13 N.Y.U. ENVTL. L.J. 379, 398-99 (2005). However, "because food produced in accordance with the NOP regulations will not be intentionally sprayed with pesticides or intentionally grown or raised using genetically engineered seed or other inputs, the likelihood of the presence of pesticide residue or genetically engineered content will clearly be lower than in foods intentionally produced with pesticides and genetic engineering techniques. But organic food will not be free of such contamination. Evidence clearly indicates that both pesticides and genetically engineered plant materials often drift beyond their intended applications, and organic food, like any food, may be accidentally contaminated." *Id.* at 399-400.
66. Plotto & Narciso, *supra* note 58, at 287-88, citing 7 C.F.R. §§205.103, 205.201.
67. JILL RICHARDSON, RECIPE FOR AMERICA: WHY OUR FOOD SYSTEM IS BROKEN AND WHAT WE CAN DO TO FIX IT 63 (2009).
68. Kate L. Harrison, *Organic Plus: Regulating Beyond the Current Organic Standards*, 25 PACE ENVTL. L. REV. 211 (2008), at 219, citing Andrew J. Nicholas, *As the Organic Industry Gets Its House in Order, the Time Has Come for National Standards on Genetically Modified Foods*, 15 LOY. CONSUMER L. REV. 277, 285 (2003).
69. *Id.*
70. RICHARDSON, *supra* note 67, at 63-64.
71. Ariana R. Levinson, *Lawyers as Problem-Solvers, One Meal at a Time: A Review of Barbara Kingsolver's Animal, Vegetable, Miracle*, 15 WIDENER L. REV. 289 (2009).
72. *See, e.g.,* Vermont Organic Farmers, LLC—Timeline and Fees for Certification, http://nofavt.org/programs/organic-certification/application-deadline-and-fees.
73. Denis A. O'Connell, *Shade-Grown Coffee Plantations in Northern Latin America: A Refuge for More Than Just Birds & Biodiversity*, 22 UCLA J. ENVTL. L. & POL'Y 131 (2003/2004), citing RUSSELL GREENBERG, CRITERIA WORKING GROUP THOUGHT PAPER 4 (2001).
74. USDA, AGRICULTURAL MARKETING SERVICE, NATIONAL ORGANIC PROGRAM COST SHARE PROGRAMS 2010 REPORT TO CONGRESS (2010), *available at* http://www.ams.usda.gov/AMSv1.0/getfile?dDocName=STELPRDC5084541&acct=nopgeninfo.
75. *Id.* "To prevent duplicate assistance payments, producers participating in the AMA program are not eligible to participate in the producer portion of the National program."

to 75% of its organic certification costs, not to exceed $750.[76] Thus, significant subsidies are available, at least in the short term.

While the costs of organic certification are expensive, they are not prohibitive; the real barriers to entry may be the costs of monitoring and recordkeeping. For example, applicants for certification must keep accurate post-certification records for five years concerning the production, harvesting, and handling of agricultural products that are to be sold as organic.[77] In addition to making organic certification more affordable for small farmers, some states are also providing property tax rebates for farmers who convert from conventional to organic farming practices, and attempting to lower the tax burden on small farmers.[78]

There has been a concern that organic standards would somehow limit imports and adversely affect the global food market. However, as described in a USDA report:

> The U.S. National Organic Program (NOP) streamlined the certification process for international as well as domestic trade when it was implemented in 2002. Organic farmers and handlers anywhere in the world are permitted to export organic products to the United States if they meet NOP standards, along with other regulatory standards, and are certified by a public or private organic certification body with USDA accreditation. In 2007, USDA accredited groups certified 27,000 producers and handlers worldwide to the U.S. organic standard, with approximately 16,000 in the United States and 11,000 in over 100 foreign countries. Farmers and handlers certified to NOP standards are most numerous in Canada, Italy, Turkey, China, and Mexico, which together accounted for half the total foreign organic farmers/handlers in 2007."[79]

b. Labeling

The OFPA monopolizes the use of the term "organic," requiring all products labeled as organic to be certified through the government-approved certifiers who comply with all OFPA regulations under the NOP.[80] Under one view, it is effective to have a single government label bringing singular meaning to a word developing significant cache in the food market:

> The OFPA, from the point of view of regulatory design and administrative law, was strikingly innovative. At the same time that alternatives to traditional command-and-control regulation such as risk-based decision-making and market-like incentives were drawing so much attention, the OFPA created a system that could tie public environmental and ethical values into existing, real markets; that informed the development of governmental organic standards with input from a National Organic Standards Board composed of nongovernmental representatives from different facets of the organic industry; and that centered regulatory compliance on a system of approved private-sector certification rather than a large federal bureaucracy. The OFPA is a marketing-oriented statute designed to regularize what was at the time a potentially confusing Babel of competing standards with an official federal "organic" label. Not only was a federal label thought useful in promoting consumer confidence in the growing organic industry within the United States, but it was also viewed as helpful in facilitating trade in "a potentially lucrative international organic market.[81]

> Despite this approach, there is still consumer confusion about the meaning of the term "organic."

What counts as organic? For many, the organic label means healthy, environmentally friendly, safe, and pesticide-free. While in some cases these characteristics are true, they are not elements of the legal definitions of organic. And legal definitions matter. The NOP created a four-tiered labeling system for organic foods.[82] As seen in Table 13.1, all organics are not created equally.

---

76. Press Release, USDA, USDA Amends National Organic Certification Cost Assistance Program (Nov. 7, 2008), *available at* http://www.ams.usda.gov/AMSv1.0/ams.fetchTemplateData.do?template=TemplateU&navID=Newsroom&page=Newsroom&resultType=Details&dDocName=STELPRDC5073574&dID=103098&wf=false&description=USDA+Amends+National+Organic+Certification+Cost+Assistance+Program&to;.
77. USDA, Agricultural Marketing Service, *National Organic Program: Certification*, http://www.ams.usda.gov/AMSv1.0/getfile?dDocName=STELDEV3004346&acct=nopgeninfo.
78. *See, e.g.,* Woodbury County Organics Conversion Policy (June 28, 2005), *available at* http://www.woodburyiowa.com/attachments/2636_Organics%20Conversion%20Policy.pdf; 1997 Georgia H.B. 1350 (1998).
79. GREENE ET AL., *supra* note 61, at 8.
80. 7 U.S.C. §6505(a)(1)(A).
81. Donald T. Hornstein, *The Road Also Taken: Lessons From Organic Agriculture for Market- and Risk-Based Regulation*, 56 DUKE L.J. 1541, 1549-50 (2007) (citing JEAN M. RAWSON, CONG. RESEARCH SERV., ORGANIC AGRICULTURE IN THE UNITED STATES: PROGRAM AND POLICY ISSUES 3 (2006)).
82. 7 C.F.R. §205.301. In addition to looking for "organic" labeled foods, consumers can look at five-digit PLU codes. Organic foods all start with 9.

Table 13.1
Categories of USDA Organic Foods

Content of organic ingredients (%)	Organic seal?	Permitted label phrases
100	Yes	100% organic
95-99	Yes	Organic
70-94	No	Made with organic Ingredients
Less than 70	No	Can only list organic ingredients

First, a product can be labeled "100% organic" and carry the USDA and certifying agent seals if its entire ingredients are organically produced as defined by the OFPA (i.e., without any synthetic substances).[83] Second, a product must contain at least 95% organic ingredients to be labeled simply "organic" and use the USDA and private certifying agent seals.[84] Third, a product with at least 70% organically produced ingredients can be labeled "made with organic ingredients" and carry the seal of a private certifying agent.[85] For products containing less than 70% organic ingredients, organic ingredients may be listed on the label, but neither the word "organic" nor any seal can be used. Although individual U.S. states have the right to seek approval of stricter standards,[86] to date, none have exercised this right (as discussed in Chapter 17).

Two key questions arise in determining the effectiveness of organic labeling. First, when an average consumer sees one of the labels in Table 13.1, are the different meanings clear? Second, does "certified organic" mean what consumers think it means?

Potentially adding to the confusion, agribusiness has sought a watered-down definition of "organic" so it can benefit from the growing popularity of organic products. The secretary of agriculture, after lobbying by industry officials to loosen the standard for organic, created rules allowing nonorganic feed to be used in dairy cattle herds that were transitioning to an organic diet and permitted the use of synthetic substances in the handling of products labeled as organic. The U.S. Court of Appeals for the First Circuit declared, in *Harvey v. Veneman*, that these rules contravened the plain language of the OFPA.[87] Despite this, producers can use chemicals in the production and handling stages if the synthetics are not harmful and are necessary because no natural substitute exists.[88] For example, carbon dioxide can be used in post-harvest activities like ripening.[89] That said, and despite attempts to the contrary, current NOP rules ban genetically modified organisms, sewage sludge, and irradiation in certified organic foods.[90] The organic brand also excludes poultry, eggs, or milk from animals raised with antibiotics or growth hormones.

2. Country of Origin Labeling Provisions of the 2002 Farm Bill

Country of origin labeling (COOL) requires that a food label notify consumers of its source location.[91] While COOL increases the information available to consumers, it is unclear what changes in consumer behavior manifest. COOL information does allow consumers to choose food products that did not travel so far to market and thus may have a lower carbon footprint (i.e., lower food miles). Also, COOL may provide implicit information to buyers because educated consumers may know, for example, whether produce was grown out of season in a greenhouse or came from an unsustainable or depleted fishery. COOL

---

83. 7 C.F.R. §205.301(a); 7 C.F.R. §205.303. The OFPA defines "synthetic" as "a substance that is formulated or manufactured by a chemical process or by a process that chemically changes a substance extracted from naturally occurring plant, animal, or mineral sources, except that such term shall not apply to substances created by natural occurring biological processes." 7 U.S.C. §6502(12); 7 C.F.R. §205.2.
84. 7 C.F.R. §205.301(b); 7 C.F.R. §205.303.
85. 7 C.F.R. §205.301(c).
86. Under 7 U.S.C. §6507(a) of the OFPA, individual U.S. states have the right to seek approval for their own organic certification program. The statute reads: "A State organic certification program . . . may contain more restrictive requirements governing the organic certification of farms and handling operations and the production and handling of agricultural products that are to be sold or labeled as organically produced under this chapter than are contained in the program established by the Secretary."
87. Harvey v. Veneman, 396 F.3d 28 (1st Cir. 2005).
88. 7 U.S.C. §6517.
89. 7 C.F.R. §205.605(b).
90. Friedland, *supra* note 65, at 384, 388. The regulations also prohibit most uses of ionizing radiation, the application of sewage sludge as fertilizer, and the use of drugs or hormones to promote growth in livestock. 7 C.F.R. §§205.105(f), 205.105(g), 205.237(b)(1).
91. For a discussion of COOL, see Peter Chang, *Country of Origin Labeling: History and Public Choice Theory*, 64 Food Drug L.J. 693, 702 (2009); Anastasia Lewandoski, *Legislative Update: Country-of-Origin Labeling*, 9 Sustainable Dev. L. & Pol'y 62 (2008).

requirements were enacted in American law under the 2002 farm bill[92] and its implementing regulations,[93] and reauthorized by the 2008 farm bill.[94]

Despite objections to COOL by powerful producers and retailers, the idea had much support from consumer and product-safety organizations.[95] COOL retailers such as grocery stores, supermarkets, and club warehouse stores must provide customers with information regarding the source of certain foods.[96] Food products subject to the legislation currently include "covered commodities," such as cut and ground meats (beef, veal, pork, lamb, goat, and chicken), wild and farm-raised fish and shellfish, fresh and frozen fruits and vegetables, nuts (peanuts, pecans, and macadamia nuts), and ginseng.[97]

Under the COOL program four labeling categories exist to indicate a product's source, depending on its country of origin and whether U.S. products are mixed with foreign products. Difficulties arise in designating the country of origin because many food products today are produced in multiple countries, particularly meats. For example, beef might come from a cow that was born and fed in Canada, but slaughtered and processed in the United States. Similarly, products from several countries often are mixed, such as for ground beef. For covered red meats and chicken, the COOL law:

- Permits the U.S. origin label to be used only on items from animals that were exclusively born, raised, and slaughtered in the United States;

- Permits meat or chicken with multiple countries of origin to be labeled as being from all of the countries in which the animal may have been born, raised, or slaughtered;

- Requires meat or chicken from animals imported for immediate U.S. slaughter to be labeled as from both the country the animal came from and the United States;

- Requires products from animals not born, raised, or slaughtered in the United States to be labeled with their correct country(ies) of origin; and

- Requires, for ground meat and chicken products, that the label list all countries of origin, or all "reasonably possible" countries of origin.[98]

These meat-labeling requirements have proven to be quite controversial because of the steps that U.S. feeding operations and packing plants have had to adopt to segregate, hold, and slaughter foreign-origin livestock from U.S. livestock.[99] The catch-all label (see second bullet, above) was a favorite to many meat processors and retailers, even on products that would qualify for the U.S.-only label, because it was both allowed and the easiest requirement to meet.[100] After objections from COOL supporters that the label would be overused and thus undermine COOL's intent—to distinguish between U.S. and non-U.S. meats—a final rule, promulgated in August 2008, clarified the "multiple countries of origin" language.[101] The rule stated that meats derived from both U.S. and non-U.S. origin animals may carry a mixed-origin claim (e.g., "Product of the United States, Canada, and Mexico"), but that the mixed-origin label cannot be used if only U.S.-origin meat was produced on a production day.[102]

To pacify continued concerns that the COOL label's purpose was being evaded, Secretary of Agriculture Tom Vilsack asked industry representatives in a February 2009 letter to voluntarily provide additional information. He stated that

---

92. Farm Security and Rural Investment Act of 2002, Pub. L. No. 107-171, §10816, 116 Stat. 134, 533 (codified at 7 U.S.C. §§1638 et seq.).
93. *See* 7 C.F.R. §§60, 65.
94. Food, Conservation, and Energy Act of 2008 (Farm Bill, 2008), Pub. L. No. 110-234, §11002, 122 Stat. 923, 1352-1354 (codified at 7 U.S.C. §§1638 et seq.).
95. Chang, *supra* note 91, at 702.
96. USDA, Agricultural Marketing Service, *Country of Origin Labeling*, http://www.ams.usda.gov/AMSv1.0/Cool.
97. *Id. See also* 7 C.F.R. §§60, 65.
98. 7 C.F.R. §65.3000(e)-(h). *See also* Remy Jurenas, Congressional Research Service, Country-of-Origin Labeling for Foods (2010), available at http://assets.opencrs.com/rpts/RS22955_20100715.pdf.
99. Jurenas, *supra* note 98.
100. *Id.*
101. Mandatory Country of Origin Labeling of Beef, Pork, Lamb, Chicken, Goat Meat, Perishable Agricultural Commodities, Peanuts, Pecans, Ginseng, and Macadamia Nuts, 73 Fed. Reg 45106-01 (Aug. 1, 2008).
102. Jurenas, *supra* note 98, at 6, citing USDA statement, *Country of Origin Labeling (COOL) Frequently Asked Questions* (Sept. 26, 2008).

processors should voluntarily include information about what production step occurred in each country when multiple countries appear on the label. For example, animals born and raised in Country X and slaughtered in Country Y might be labeled as "Born and Raised in Country X and Slaughtered in Country Y." Animals born in Country X but raised and slaughtered in Country Y might be labeled as "Born in Country X and Raised and Slaughtered in Country Y."[103]

For perishable agricultural commodities, ginseng, peanuts, pecans, and macadamia nuts, retailers may only claim U.S. origin if they were exclusively produced in the United States.[104] For farm-raised fish and shellfish, a U.S.-labeled product must be derived exclusively from fish or shellfish hatched, raised, harvested, and processed in the United States; wild fish and shellfish must be derived exclusively from those either harvested in U.S. waters or by a U.S.-flagged vessel, and processed in the United States or on a U.S. vessel.[105] Also, labels must differentiate between wild and farm-raised seafood.[106]

Future COOL legislation may seek to add this labeling scheme to essentially require all labels to identify the country in which the final processing occurs, and mandating that manufacturer websites identify the country (or countries) of origin for each ingredient.[107] However, despite the covered commodities, there is a substantial list of products that do not require the COOL label, including processed food items.[108] USDA has broadly defined processing to include any item "undergoing a specific processing to change the character of the commodity or combining it with at least one other covered commodity or substantive food component"[109]—peanut butter, thus, is exempt from COOL requirements. Items are also exempt if they enter hotels and restaurants, small retail outlets, butcher shops, and fish markets.[110] Thus, large retailers selling covered commodities to grocery shoppers are likely to be the only place COOL will be required.[111]

C. The National School Lunch Program

Increasing concerns over the health of our nation's children have caused many to scrutinize what children are eating, especially in school. The National School Lunch Program (NSLP) is the most comprehensive food program aimed at school children, serving approximately 31 million children in about 100,000 schools each day.[112] The NSLP has endured longer than any other 20th century federal welfare initiative,[113] and school food purchases occupy a large slice of the consumer market, costing American taxpayers more than $10 billion a year.[114] Thus, the laws that govern school food have the potential to shape the agricultural supply market and food policy in many ways. For instance, because schools often purchase surplus commodity crops like corn or soy, or processed foods made primarily from them, school food policy could greatly impact the corn or soy market.

This section discusses the law governing school lunches by tracing the NSLP initiatives through the Richard B. Russell National School Lunch Act (NSLA) of 1946, the Child Nutrition Act of 1966, and the recent enactment of the Healthy, Hunger-Free Kids Act in 2010. The section then discusses the current role of USDA in administering the NSLP, and addresses a number of challenges the NSLP poses for schools.

---

103. Letter from Thomas Vilsack, Secretary of Agriculture, to Industry Representatives (Feb. 20, 2009), *available at* http://www.usda.gov/documents/0220_IndustryLetterCOOL.pdf.
104. 7 C.F.R. §65.3000(g).
105. *Id.* §60.101.
106. *Id.*
107. Chang, *supra* note 91, at 713, citing Draft Bill, Food and Drug Globalization Act of 2009, H.R. 759, 111th Cong. (2009).
108. Matt Mullins, *Not Cool: The Consequences of Mandatory Country of Origin Labeling*, 6 J. Food L. & Pol'y 89, 89-90 (2010) (citing Mandatory COOL, 74 Fed. Reg. 2682 (Jan. 15, 2009) (to be codified at 7 C.F.R. pts.60 and 65)).
109. Mullins, *supra* note 108, at 90 (citing Produce Marketing Association, *PMA Analysis: USDA Final Rule for Mandatory Country of Origin Labeling*, http://www.pma.com/resources/issues-monitoring/country-origin-labeling/analysis-usda-final-rule-mandatory-country-origi (last visited Nov. 28, 2012)).
110. C. Parr Rosson III & Flynn J. Adcock, *The Potential Impacts of Mandatory Country-of-Origin Labeling on U.S. Agriculture*, in International Agricultural Trade Disputes: Case Studies in North America 38 (Andrew Schmitz et al., eds. 2005).
111. *Id.*
112. U.S. Dep't. Agric, Economic Research Service Briefing Room, *Child Nutrition Programs*, *available at* http://www.ers.usda.gov/Briefing/Child-Nutrition/lunch.htm (last modified June 14, 2010).
113. Susan Levine, School Lunch Politics: The Surprising History of America's Favorite Welfare Program 2 (2008).
114. *Child Nutrition Programs*, *supra* note 112.

I. Origins of the National School Lunch Program

Responding to heightened public concern about adequate nutrition after World War II, Congress recognized the need for permanent legislation on the issue.[115] In June 1946, Congress passed the NSLA, which authorized the NSLP.[116] The program's stated purpose was to

> safeguard the health and well-being of the nation's children and to encourage the domestic consumption of nutritious agricultural commodities and other food, by assisting the States, through grants-in-aid and other means, in providing adequate supply of food and other facilities for the establishment, maintenance, operation and expansion of nonprofit school lunch programs.[117]

The original NSLP created a three-tiered system of free and reduced-price lunches.[118] It required that the program operate as a nonprofit, serve meals meeting minimum nutritional standards, and use surplus commodities "to the extent practical."[119] In 1966, Congress reauthorized the NSLP by enacting the Child Nutrition Act (CNA).[120]

From the outset, the NSLP has been subject to the competing interests of farmers, agribusiness, school administrators, nutritionists, and parents. Chronic underfunding of school meals by states and the federal government put pressure on local school officials to seek alternative means of funding, one of which was to permit the sale of "competitive," non-NSLP food items on school premises.[121] Competitive, or "a la carte" food, refers to any food that competes with the federally funded meals.[122] Virtually all schools offer competitive foods, and the types of food available are often of low nutritional quality.[123] One study observed that the most prevalent options available are soft drinks, fruit drinks containing less than 100% juice, candy, chips, cookies, and snack cakes.[124] The presence of these foods on school campuses causes several economic, health, and environmental effects.[125] First, sales of these foods may infuse the school environment with commercialism and marketing that affect food choices into adulthood.[126] Second, the increased consumption of highly processed food directly and indirectly strains the nation's health care system.[127] Obesity is the most directly related health crisis, but a wide variety of hidden health crises indirectly result from the current industrial agricultural practices used to produce these products (see Chapters 1 and 2). These impacts include "water and air pollution, decline of socioeconomic health, pesticide-related health impacts, methicillin-resistant Staphylococcus aureus (MRSA), and antibiotic resistance."[128]

---

115. Prior to World War II, the school lunch program was a temporary government tactic to find a market for surplus agricultural products. Kathryn L. Plemmons, *The National School Lunch Program and USDA Dietary Guidelines: Is There Room for Reconciliation?*, 33 J.L. & EDUC. 181, 185 (2004).
116. National School Lunch Act, Pub. L. No. 79-396, 60 Stat. 231 (1946), 42 U.S.C. §§1751 et seq.
117. *Id.* §2.
118. Currently, the eligibility criteria for each tier are based on the federal income poverty guidelines and are stated by household size. The most recent Income Eligibility Guidelines for July 1, 2011, through July 30, 2012, are available at http://www.fns.usda.gov/cnd/governance/notices/iegs/IEGs11-12.pdf.
119. National School Lunch Act, Pub. L. No. 79-396, §9, 60 Stat. 231 (1946), 42 U.S.C. §1758.
120. Child Nutrition Act, Pub. L. No. 89-642, 80 Stat. 885-890 (1966), 42 U.S.C. §§1771 et seq.
121. KEVIN MORGAN & ROBERTA SONNINO, THE SCHOOL FOOD REVOLUTION: PUBLIC FOOD AND THE CHALLENGE OF SUSTAINABLE DEVELOPMENT 47 (2008).
122. Under the National School Lunch Act, "competitive food" means any food sold in competition with the NSLP to children in food service areas during lunch periods. 7 C.F.R. §210.11(a)(1) (2011).
123. Nicole Larson & Mary Stone, School Foods Served Outside of Meals, Research Brief conducted by the Healthy Eating Program of the Robert Wood Johnson Foundation (May 2007), *available at* http://www.healthyeatingresearch.org/images/stories/her_research_briefs/hercompetfoodsresearchbrief.pdf.
124. *Id.* (citing CENTER FOR SCIENCE IN THE PUBLIC INTEREST, DISPENSING JUNK: HOW SCHOOL VENDING UNDERMINES EFFORTS TO FEED CHILDREN WELL (May 2004), *available at* http://www.cspinet.org/new/pdf/dispensing_junk.pdf).
125. Ellen Fried & Michele Simon, *The Competitive Food Conundrum: Can Government Regulations Improve School Food?*, 56 DUKE L.J. 1491, 1497 (2007).
126. "One longitudinal study among 594 fourth and fifth-grade students showed that, as fourth-grade students transitioned from elementary school to middle school and gained access to school snack bars at lunch, they decreased their consumption of fruits by 33 percent, regular (not fried) vegetables by 42 percent and milk by 35 percent. The study also found that students gaining access to snack bars increased their consumption of sweetened beverages (e.g., soft drinks) and high-fat vegetables (e.g., french fries and tater tots)." LARSON & STONE, *supra* note 123 (citing K.W. Cullen & I. Zakeri, *Fruits, Vegetables, Milk, and Sweetened Beverages Consumption and Access to a la Carte/Snack Bar Meals at School*, 94 AM. J. PUB. HEALTH 463 (2004)).
127. *See* Jamie Harvie et al., *A New Health Care Prevention Agenda: Sustainable Food Procurement and Agricultural Policy*, 4 J. HUNGER & ENVTL. NUTRITION 409 (2009).
128. LARSON & STONE, *supra* note 123, citing Harvie et al., *supra* note 127; *see also* Jason J. Czarnezki, *Food, Law & the Environment*, *supra* note 54.

While the federal government had under the NLSA the authority to regulate the nutritional content of the NSLP meals it funded, it had no such control over the privately funded competitive foods. Concern about unhealthy competitive food options at schools led Congress to grant authority in 1977 to USDA to regulate competitive foods in or near cafeterias during mealtimes.[129] When the secretary of agriculture eventually promulgated a regulation that eliminated the sale of foods of minimal nutritional value on school premises until after the last lunch period,[130] the National Soft Drink Association (NSDA, later called the American Beverage Association) and others sued.[131]

In *National Soft Drink Association (NSDA) v. Block*, the NSDA filed suit to set aside the secretary's "time and place" regulation on the ground that he exceeded the authority granted to him by the NSLA.[132] The U.S. Court of Appeals for the District of Columbia (D.C.) Circuit sided with the NSDA, finding that the secretary had statutory authority only to promulgate "time and place" restrictions on sales of minimal nutritional value in food service areas during meal times, not all throughout the school day or throughout the school.[133]

Later amendments to the school meal laws improved the nutritional content of school meals. The Healthy Meals for Healthy Americans Act of 1994 requires schools to provide meals that meet the Institute of Medicine's Dietary Guidelines for Americans.[134] The nutrition title of the 2002 farm bill provided $6 million in funding for a fruit and vegetable pilot program to provide free fresh and dried fruits to designated schools.[135] By 2007, however, survey results revealed that less than one third of U.S. schools were able to serve lunches that met the new guidelines.[136]

2. 2010 Reauthorization of the Child Nutrition Act

In December 2010, President Obama signed into law the Healthy, Hunger-Free Kids Act (HHFKA) of 2010,[137] setting new standards for the NSLP, improving the nutritional quality of the meals, and increasing the number of children served by child-nutrition programs. The HHFKA reauthorized the Child Nutrition Act, which comes up for reauthorization every five years. Since the NSLP alone feeds over 31 million school children a day,[138] the new HHFKA has the potential to impact the lives of many families in at least three key ways.

First, one of the provisions hailed by nutritionists grants the secretary of agriculture clearer authority to regulate competitive foods.[139] The nutrition standards will apply to all foods sold on the school campus at any time during the school day.[140] If accepted as written, the standards will radically change the NSLP and essentially render the decision in *Block* moot.[141]

Second, in an effort to increase the transparency of food nutrient content, the HHFKA also requires the secretary to undertake a study to analyze the quantity and quality of nutritional information available to school food authorities (SFAs) about food service products and commodities.[142]

---

129. National School Lunch Act and Child Nutrition Amendments of 1977, Pub. L. No. 95-166, 91 Stat. 1325 (codified as amended at 42 U.S.C. §§1751, 1753-1759a, 1760-1762a, 1763, 1766, 1769, 1771-1776, 1779, 1784, 1786, 1788 (2000)).
130. 7 C.F.R. §§210-220, and app. B thereto (2011). The NSLA defines foods of minimal nutritional value as those that provide less than 5% of the RDI for each of eight specified nutrients per serving. The specified nutrients are protein, vitamin A, vitamin C, niacin, riboflavin, thiamine, calcium, and iron. App. B of 7 C.F.R. §210 (2011).
131. Nat'l Soft Drink Ass'n v. Block, 721 F.2d 1348 (D.C. Cir. 1983).
132. 7 C.F.R. §§210.15h and 220.12 (2011).
133. Nat'l Soft Drink Ass'n v. Block, 721 F.2d 1348, 1353 (D.C. Cir. 1983).
134. The Healthy Meals for Healthy Americans Act of 1994, Pub. L. No.103-448 (1999) (codified as amended at 42 U.S.C. §1751). The latest *Guidelines* released were in 2005. MANUEL P. BORGES, NATIONAL SCHOOL LUNCH PROGRAM ASSESSMENT: CHILDREN'S ISSUES, LAWS AND PROGRAMS SERIES 15 (Nova Science Publishers 2009).
135. The Child Nutrition and WIC Reauthorization Act of 2004 made this a permanent program. BORGES, *supra* note 134, at 14.
136. BORGES, *supra* note 134, at 14.
137. Pub. L. No. 111-296.
138. USDA, Economic Research Service Briefing Room, *Child Nutrition Programs*, http://www.ers.usda.gov/Briefing/ChildNutrition/lunch.htm (last modified June 6, 2012).
139. Specifically, the bill gives the secretary the authority to establish "science-based nutrition standards for foods sold in schools *other than* provided under this act" (emphasis added). Healthy, Hunger-Free Kids Act of 2010, Pub. L. No. 111-296, §208(A), 124 Stat. 3183 (codified as amended at 42 U.S.C. §1779 (2011)).
140. Healthy, Hunger-Free Kids Act of 2010, Pub. L. No. 111-296, §208; 42 U.S.C. §1779 (2011).
141. Nat'l Soft Drink Ass'n v. Block, 721 F.2d 1348, 1353 (D.C. Cir. 1983).
142. Healthy, Hunger-Free Kids Act of 2010, Pub. L. No. 111-296, §242(C), 42 U.S.C. §1758(a)(4)(C) (2011).

Third, the HHFKA amends the NSLA to foster the growth of the burgeoning organic food and "farm to school" movements. Under the Organic Food Pilot Program, the secretary provides grants on a competitive basis to SFAs to establish programs to increase the quantity of organic foods provided to schoolchildren under the NSLP.[143] Congress authorized $10 million to carry out this program.[144] The rise of the local foods movement has spurred interest in finding ways to integrate local foods into institutional settings such as hospitals, prisons, and schools. Farm-to-school programs, whereby farms deliver produce directly to schools, have become a popular way to institutionalize local food systems since the early 2000s. According to recent estimates, over 2,000 farm-to-school projects now exist.[145] To help promote this growth, Congress amended the NSLA to provide $5 million annually in mandatory funding for farm-to-school programs, starting October 1, 2012.[146]

3. Role of USDA in the NSLP

To assist states in providing low cost or free meals, USDA gives states money for each school lunch served to students under the NSLP. In addition, USDA provides states with USDA-procured foods (USDA food) for use in preparing these lunches.[147] Each state receives a dollar value of product, referred to as the state's "entitlement," based on meals served in the previous year.[148] Each state orders USDA food against its entitlement until the dollar balance is depleted. Typical entitlement foods include cheeses, grains, frozen hamburger patties, canned fruits, and pasta.[149] Milk is the only entitlement food that school lunch legislation mandates be a part of every reimbursable lunch.[150] When there is an oversupply in the market, each state's entitlement for USDA foods is supplemented by donations of "bonus" products purchased by USDA.[151] An average school lunch is composed of 15-20% USDA food.[152] The remaining amount is purchased from commercial markets using the cash assistance provided by USDA, funds provided by state and local governments, childrens' payments for reduced price and paid lunches, proceeds from vending machines, and catering activities.[153]

The Food and Nutrition Service of USDA determines how government funds should be spent on food purchases.[154] To save money, the foods incorporated into this purchase plan are often those commodity crops that the government subsidizes. At the school district level, school food programs are administered by SFAs who are responsible for procuring food, approving student eligibility, and keeping financial records.[155] While smaller school districts do not have much sway over what type of food they purchase or where it comes from, larger authorities, like New York City, do. In fact, with 860,000 meals served every day and an annual budget of $450 million, the New York City Board of Education is the largest institutional food buyer in the United States.[156]

Schools can opt to have commodity items sent directly to the school, or have them sent to a processor first,[157] which will turn them into more readily servable products, such as potato puffs or beef patties.[158] Because SFAs fear losing reimbursements if meals are found wanting in nutrients, they often prefer the option of buying processed products that are affixed with the federal child nutrition label guaranteeing that the producer has met the standard for reimbursable meals.[159] As food manufacturers have caught on to the implications of these

---

143. Healthy, Hunger-Free Kids Act of 2010, Pub. L. No. 111-296, §210, codified as amended at 42 U.S.C. §1769(k) (2011).
144. *Id.*
145. National Farm-to-School Network, http://www.farmtoschool.org/Nat/media.htm.
146. Healthy, Hunger-Free Kids Act of 2010, Pub. L. No. 111-296. §243, 42 U.S.C. §1769(g) (2011).
147. USDA, Food and Nutrition Service White Paper, USDA Foods in the National School Lunch Program 2-3 (May 2010), *available at* http://www.fns.usda.gov/fdd/foods/healthy/WhitePaper.pdf.
148. *Id.* at 4.
149. Borges, *supra* note 134, at 29.
150. 42 U.S.C. §1758(a)(2) (2011).
151. Borges, *supra* note 134, at 29.
152. USDA Foods in the National School Lunch Program, *supra* note 147, at 3.
153. *Id.*
154. *Id.*
155. Morgan & Sonnino, *supra* note 121, at 51.
156. *Id.* at 52.
157. Janet Poppendieck, Free for All: Fixing School Food in America 79-80 (2010).
158. *Id.* at 29.
159. *Id.* at 5.

current regulations, they have begun to tweak and fortify products, creating items such as mineral-fortified donuts and muffins that make it easier for menu planners to meet the guidelines.[160]

A key question is whether the government, through its purchase of food, its subsidy programs, and its commodity donations for the NSLP, is affecting supply and demand in the agricultural market. And if the government is influencing the agricultural market, what are the public health and environmental effects? One way to answer these questions is to analyze the purchase and sale of commodity items produced by modern industrial, carbon-heavy food systems. One study has shown that "[i]n 2005, the USDA spent $975.1 million on commodity purchases for the NSLP,"[161] and regulations require participating schools to "accept and use" these commodities in "as large of quantities as may be efficiently utilized . . . ."[162] However, such massive USDA purchases represent only 17% or less of the total food budgets of SFAs. Nevertheless, USDA is often able to purchase the commodities at lower prices than those available on the open market,[163] and as a result, those dollars purchase a larger volume of commodities than the schools would be able to purchase with the same amount of money.[164] This suggests that as government spending on school food procurement increases, so does the percentage of commodity foods found in school lunches. This increase in commodity food donations may influence whether schools decide to plan their menus using commodity items or by using more locally sourced and less-processed products.

In addition, as farm bill subsidies have lowered prices of commodity crops over the past 30 years (though current prices are high), the food industry has invested heavily in an infrastructure that turns these commodities into profitable value-added products with long shelf lives.[165] SFAs are interested in products with a long shelf life because the products can be recycled if not consumed by students.[166] While there is only a slight fiscal impact on the agricultural market observed as a result of the NSLP and related government procurement practices,[167] thousands of childrens' food preferences and thousands of acres of cropland are being shaped by what is served in school meals.

4. Challenges to the NSLP

The NSLP faces many challenges. Childhood obesity is on the rise,[168] as is Type 2 diabetes.[169] Food safety scares are more frequent than ever; in 2008, for example, USDA issued the nation's biggest beef recall of 143 million pounds, much of which went to school lunch programs.[170] What barriers do SFAs face in their efforts to change the status quo? One obstacle to revamping school lunch programs is attempting to navigate through and abide by USDA's complex procurement policies. The 2004 NSLA requires the Secretary of Agriculture to encourage institutions participating in the NSLP to purchase "locally produced foods for school meal programs to the maximum extent practicable and appropriate."[171] Despite this encouragement, there was confusion over whether SFAs could lawfully specify local food in their school meal contracts.[172]

---

160. Mary Kay Crepinsek et al., *Meals Offered and Served in Public Schools: Do They Meet Nutritional Standards?*, 109 J. Am. Dietetic Ass'n S31 (2009).
161. Borges, *supra* note 134, at 19.
162. 7 C.F.R. §210.9.
163. USDA Foods in the National School Lunch Program, *supra* note 147, at 3.
164. *Id.*
165. Poppendieck, *supra* note 157, at 99.
166. *Id.* at 100.
167. A study done by the Economic Research Service of USDA in 2003 reveals that the NSLP and the School Breakfast Program contributed about $870 million in additional farm production in 2001, or about 0.3% of U.S. farm cash receipts. Kenneth Hanson, USDA, Economic Research Service, Importance of Child Nutrition Programs to Agriculture, Food Assistance Research Brief, Food Assistance and Nutrition Research Report Number 34-12, at 2 (2003), *available at* http://www.ers.usda.gov/publications/fanrr34/fanrr34-12/fanrr34-12.pdf.
168. Obesity now affects 17% of all children and adolescents in the United States—triple the rate from just one generation ago. Centers for Disease Control and Prevention, *Childhood Overweight and Obesity*, http://www.cdc.gov/obesity/childhood/index.html (last accessed Nov. 4, 2011).
169. In the last two decades, type 2 diabetes (formerly known as adult-onset diabetes) has been reported among U.S. children and adolescents with increasing frequency. Centers for Disease Control and Prevention, *Diabetes Public Health Resource*, http://www.cdc.gov/diabetes/projects/cda2.htm (last accessed Nov. 28, 2012).
170. *Meat Inspectors Suspended*, N.Y. Times, Mar. 1, 2008, at http://query.nytimes.com/gst/fullpage.html?res=9E00E5D7153BF932A35750C0A96E9C8B63&ref=westlandhallmarkmeatcompany.
171. Child Nutrition and WIC Reauthorization Act of 2004, Pub. L. No. 108-265 §112 (2004) (codified as amended at 42 U.S.C. §1758(j)(2)(A) (2011)).
172. Forum on Democracy & Trade, *Can States and Cities Give a Preference for Local Food?*, http://www.forumdemocracy.net/section.php?id=238 (last visited Nov. 4, 2011).

In addition, because the statute did not specify that the phrase "locally produced" meant locally *grown*, foods that were locally produced but contained processed products grown hundreds of miles away still satisfied the statute.[173]

Seeking to clarify this ambiguity, Congress amended the NSLA to require the secretary to encourage participating institutions to purchase "unprocessed agricultural products, both locally grown and locally raised, to the maximum extent practicable and appropriate."[174] While the NSLA made clear that geographic preference could only be applied to the procurement of locally grown and unprocessed agricultural products, some uncertainty remained over the definition of "unprocessed." As a result, USDA issued a final rule in April 2011, clarifying that unprocessed products meant only those agricultural products whose "inherent character" had not been altered.[175] The rule specified food handling and preservation techniques that are permitted, such as cooling, refrigerating, freezing, size adjustment, packaging, or the addition of ascorbic acid.[176] In addition, Congress amended the NSLA to allow participating institutions to "use a geographic preference for the procurement of unprocessed agricultural products, both locally grown and locally raised."[177] This provision allows SFAs to give preference to locally grown foods in their contracts with producers and suppliers, but not to make it a requirement.

SFAs can prescribe the geographic preference in the form of preference points or as a percentage in their solicitation for bids.[178] Federal regulations do not prescribe the number of points or maximum percentage an SFA can assign to geographic preference.[179] The only requirement is that the solicitation document (usually an invitation for bid from a producer) clearly outlines the "significant evaluation factors" and "mechanisms for technical evaluation" of the evaluated factors.[180] Furthermore, it is imperative that SFAs solicit bids in a manner that will "provide maximum open and free competition."[181] As a result, SFAs' application of the geographic preference option must leave an appropriate number of "qualified sources to permit reasonable competition consistent with the nature and requirements of the procurement."[182] The exact number of sources is not specified by regulation.[183] The solicitation must be publicized, and "reasonable requests by other sources to compete must be honored to the maximum extent practicable."[184] Unfortunately, SFAs cannot use commodity entitlement dollars to purchase locally grown products; instead, USDA uses those funds to purchase commodity foods.[185]

Under the principles of federalism, a state has the right to create regulations requiring state governmental entities to give geographic preference to farmers within a state.[186] However, the application of

---

173. Child Nutrition and WIC Reauthorization Act of 2004, Pub. L. No. 108-265 §112 (2004) (codified as amended at 42 U.S.C. §1758(j)(2)(A) (2011)).
174. The geographic preference option was authorized by §4302 of Public Law No. 110-246, the Food, Conservation, and Energy Act of 2008, which amended §9(j) of the Richard B. Russell National School Lunch Act allowing institutions receiving funds through the Child Nutrition Programs to apply an optional geographic preference in the procurement of unprocessed locally grown or locally raised agricultural products.
175. Geographic Preference Option for the Procurement of Unprocessed Agricultural Products in Child Nutrition Programs, 7 C.F.R. Parts 210, 215, 220, 225, and 226, *available at* http://www.gpo.gov/fdsys/pkg/FR-2011-04-22/pdf/2011-9843.pdf.
176. *Id.*, 7 C.F.R §210.21.
177. Pub. L. No, 110-246, §15422 (codified as amended at 42 U.S.C.A. §1758 (2011)).
178. USDA, Food and Nutrition Service, *Farm to School FAQs*, http://www.fns.usda.gov/cnd/f2s/faqs_procurement.htm (last modified May 25, 2011). The USDA geographic preference guidance sheet provides the following example as one approach to incorporate geographic preference points: After the SFA puts out an invitation for bid (IFB) and determines the three lowest cost bidders, each bidder that meets the SFA's predetermined and communicated geographic preference criteria is given 10 geographic preference points (in the scheme, each point equals one cent). So if a bidder offers $2.00 but meets the geographic preference criteria, that bidder's price is reduced to $1.90, and may outcompete another bidder with an original price of $1.95 who does not meet the geographic preference criteria. However, this price deduction only applies to determining the winning bidder and does not affect the actual price paid to a bidder. The actual price remains the original bid price. USDA, Memo Regarding Procurement Geographic Preference Q&As (2011), *available at* http://www.fns.usda.gov/cnd/governance/Policy-Memos/2011/SP18-2011_os.pdf.
179. *Farm to School FAQs*, *supra* note 178.
180. 7 C.F.R. §226.22(i)(3)(ii-iii).
181. *Id.* §226.22(g).
182. *Id.* §226.22(i)(3)(i).
183. While the regulations do not specify how many sources an SFA must solicit in order to meet the "maximum open and free competition" burden, a USDA memorandum released in February 2011 gives some instruction to SFAs making "small purchase threshold" procurements, which are any procurements under $100,000 in value. The memorandum says "[w]hen using the small purchase threshold, we recommend that at least three sources be contacted who are eligible, able and willing to provide the unprocessed locally grown or locally raised agricultural product. Contacting a minimum of three sources ensures that an adequate number of potential offerors will be afforded the opportunity to respond to the solicitation." *Farm to School FAQs*, *supra* note 178.
184. 7 C.F.R. §226.22(i)(3)(i).
185. *Farm to School FAQs*, *supra* note 178.
186. For example, Oregon HB 2763, passed in 2009, allows contracting agencies using public funds to procure goods for public use to give preference to procure an agricultural product that is produced and transported entirely within the state if the product costs not more than 10% more than

such state law to the Federal Child Nutrition Programs (including the NSLP) is an entirely different matter. Because the NSLA grants the authority of whether or not to apply a geographic preference when conducting procurements for school food directly to the purchasing institution (i.e., SFA), states cannot mandate through law or policy that SFAs apply the state's adopted geographic preference regulation.[187] Thus, no state geographic preference regulation can ever be applied to procurements made under the NSLA.

While the amendment to NSLA's geographic preference provision certainly gives SFAs freedom to purchase local unprocessed products, SFAs face another barrier in attempting to integrate those fresh ingredients into school lunch menus. Decades of prepackaged and prepared menu items have rendered full facility kitchens obsolete, and today it is hard to find a school that has more than bulk freezers and massive ovens.[188] Remodeling those kitchens would prove costly.[189]

5. Successful Reform Programs

Factory food and prepackaged meals have become dominant in school meals, but several players—SFAs, parents, teachers, and health and legal professionals—are working to reverse this trend. The New York City Office of School Food and Nutrition Services (SchoolFood) has implemented a controversial strategy to improve NSLP enrollment: rather than impose a healthy eating regime that history shows attracts few students, SchoolFood has chosen to serve healthy meals disguised as fast food.[190] For instance, healthy burritos are wrapped in aluminum foil to resemble the burritos children are used to eating at a fast food restaurant.[191] SchoolFood brings in people from the Institute of Culinary Education to train cooks to view the cafeteria as a franchise and the students as customers. As part of its farm-to-school efforts, the New York State Department of Agriculture convinced SchoolFood to promote New York farm products; through the farm bill's geographic preference procurement procedures, SchoolFood wrote a specification for a variety of apples grown only in New York State.[192] During the first year, 5.5 million pounds of local apples were purchased, providing an economic benefit to the state's agricultural economy of almost $1.5 million."[193]

An elementary school in Portland, Oregon, where 43% of the children were enrolled in the NSLP decided in 2005 to embark on its own school food revolution.[194] Beginning with a kitchen renovation and a new chef, the school partnered with Ecotrust, a Portland-based conservation and economic development nonprofit organization, to implement a farm to school program that included a school garden, classroom lessons involving the garden, and lunches made with predominantly local foods. The meals were all cooked on-site from raw ingredients, and the cost of the average meal actually decreased by five cents from the frozen lunch of the past.[195] But, as predicted, the cost of labor went up by an average of $1.85 per meal, partly because the kitchen needed two chefs instead of one, and the start-up costs had to be accounted for.[196] The school continues its "School Kitchen Garden" program.

---

a similar product grown out of the state. H.B. 2763, 75TH LEGIS. ASSEM. REG. SESS. (Or. 2009). *See also* Brannon P. Denning et al., *Laws to Require Purchase of Locally Grown Food and Constitutional Limits on State and Local Government: Suggestions for Policymakers and Advocates*, 1 J. AGRIC., FOOD SYS., & COMMUNITY DEV. 139 (2010), *available at* http://www.agdevjournal.com/attachments/115_JAFSCD_Laws_on_Locally_Grown_Food_Corrected_10-10.pdf.

187. *Farm to School FAQs*, supra note 178.
188. AMY KALAFA, LUNCH WARS: HOW TO START A SCHOOL FOOD REVOLUTION AND WIN THE BATTLE FOR OUR CHILDREN'S HEALTH 15 (2011).
189. The Marshall Associates, Inc., a food service consulting and design firm in Oakland, California, estimates that a minimal "fresh food" production kitchen to serve 200 to 1,000 meals would require 1,000 square feet of space. The firm further estimates that building costs would be about $300 per square foot for remodeling, while building a new kitchen will cost an additional $50 per square foot for the outer shell. Center for Ecoliteracy, *Answers to Basic Facilities Questions*, http://www.ecoliteracy.org/essays/answers-basic-facilities-questions.
190. MORGAN & SONNINO, *supra* note 121, at 59.
191. *Id.*
192. *Id.* at 61.
193. *Id.*, citing MARKET VENTURES, INC., KARP RESOURCES, AND NEW YORK UNIVERSITY CENTER FOR HEALTH AND PUBLIC SERVICE RESEARCH, SCHOOLFOOD PLUS EVALUATION: INTERIM REPORT: PHASE 3, SCHOOL YEAR 2005-2006, 86 (2005).
194. ABERNETHY ELEMENTARY, PORTLAND PUBLIC SCHOOLS NUTRITION SERVICES, INJURY FREE COALITION FOR KIDS, AND ECOTRUST, NEW ON THE MENU: DISTRICTWIDE CHANGES TO SCHOOL FOOD START IN THE KITCHEN AT PORTLAND'S ABERNETHY ELEMENTARY (Oct. 2006), *available at* http://www.ecotrust.org/farmtoschool/downloads/Abernethy_report.pdf.
195. *Id.* at 8.
196. *Id.*

A school in Greeley, Colorado, where 60% of the 19,500 students qualify for free or reduced-price meals, launched an initiative to revitalize the use of fresh ingredients in the school cafeteria kitchen. In a story about the initiative, the *New York Times* reported that "[g]etting ready for the counterrevolution in Greely involved a weeklong boot camp to relearn forgotten arts like kitchen math—projecting ingredients to scale when making, say, 300 pans of lasagna, which cooks were doing this week—and to brush up on safe cooking temperatures for meat."[197]

These programs show that, although the NSLP regulations are convoluted, schools can navigate through them in order to serve healthy meals, and that states are passing their own school lunch legislation.[198] While many schools around the nation would like to adopt similar programs to improve their meals, the reality of tight budgets, structural inadequacies, and legal technicalities still stand in the way.

Conclusion

Today's increasing population, migration towards urban centers, and the effects of climate change have made food policy, and by extension agricultural policy, a leading societal concern. To engage in a robust dialogue on the topic of agricultural policy, it is imperative to have a general understanding of how the federal government regulates food safety, food labeling, and the food served in schools through the major statutes that control each of these food-related categories. The substantive and procedural requirements of the statutes are important to know; but it is equally important to understand the environmental and social repercussions of these statutes.

While agriculture has historically received preferential treatment under many of the nation's environmental statutes, an expanding food market and an increasing number of mouths to feed have led to increased government regulation of food, as well as an increased awareness of citizens and scholars alike. The attention that both the government and the public are now giving to the safety of our food, how it is grown, where it originates, and what its nutritional content may be shows that food and agriculture policy is no longer a niche field, but a growing and important facet of our nation's regulatory structure.

---

197. Kirk Johnson, *Schools Restore Fresh Cooking to the Cafeteria*, N.Y. Times, Aug. 16, 2011, at http://www.nytimes.com/2011/08/17/education/17lunch.html.

198. As of November 2010, at least 33 states have farm-to-school legislation supporting diverse strategies, from budget appropriations and grant programs and promotional events. Deborah Kane et al., Ecotrust, The Impact of Seven Cents 6 (2011), *available at* http://www.farmtoschool.org/files/publications_386.pdf.

Chapter 14
Agriculture and Ecosystem Services: Paying Farmers to Do the *New* Right Thing
J.B. Ruhl

American agriculture policy has been built for decades around the core principle of paying farmers to do the right thing. The policy question, therefore, has not been whether to pay, but what is the right thing—what do we want to pay farmers to do? For the most part, the answer has been to produce food and fiber commodities. But a new "right thing" vision has taken hold since the mid-1990s. This new vision sees farms as holding tremendous untapped value as providers of ecosystem services to local, regional, and national communities. The goal in this new policy movement is to unlock the multi-functional capacity of farms to contribute to the environmental and economic well-being of the landscape while continuing to serve as our primary source of food and fiber, and it is playing out with promise at federal, state, and local levels.[1]

One might think implementing this win-win for agriculture and the environment is a policy "no-brainer," but agriculture has long been the Rubik's Cube of environmental policy. Although agriculture is a leading cause of pollution and other environmental harms in the United States,[2] it has been resistant to regulation and, for the most part, remarkably successful at getting the public to pay it to do the right thing.[3] While other industries have advanced to flexible, market-based "second generation" environmental policies and beyond, agriculture somehow keeps dodging the regulatory bullet.[4] Federal and state agencies have tried to overlay small pieces of conventional regulation on farms, which farm interests have resisted at every turn,[5] and the U.S. Congress opens debate on farm bills every few years with promises of innovative policy reform, only to drift back into business as usual.[6] Seldom has so much time, money, and energy been

---

1. *See* Scott M. Swinton, *Reimagining Farms as Managed Ecosystems*, 23(2) Choices 28-31 (2008), *available at* http://www.choicesmagazine.org/magazine/pdf/article_17.pdf. The development of farm multifunctionality policy began in earnest with the European Union's Agenda 2000 reforms of the Common Agricultural Policy. *See* Thomas L. Dobbs & Jules N. Pretty, *Agri-Environmental Stewardship Schemes and "Multifunctionality,"* 26 Rev. Agric. Econ. 220 (2004). Extensive background and evaluation of the topic can be found in Org. for Economic Co-Operation & Development, Multifunctionality: Towards an Analytical Framework (2001) [hereinafter OECD]. The discussion of the farm multifunctionality theme in this chapter draws from earlier work by the author published in connection with the NYU Law School Symposium on Breaking the Logjam: Environmental Reform for the New Congress and Administration. *See* J.B. Ruhl, *Agriculture and Ecosystem Services: Strategies for State and Local Governments*, 17 NYU Envtl. L.J. 424 (2008).
2. For an inventory of environmental harms agriculture has caused and is continuing to cause in the United States, see J.B. Ruhl, *Farms, Their Environmental Harms, and Environmental Law*, 27 Ecology L.Q. 263, 272-92 (2000). The trend is not abating as "recent scientific assessments have alerted the world to the increasing size of agriculture's footprint, including its contribution to climate change and degradation of natural resources." E. Toby Kiers et al., *Agriculture at a Crossroads*, 320 Sci. 320, 320 (2008).
3. For a survey of this policy failure, describing the "safe harbor" agriculture enjoys from environmental regulation and the subsidy programs that pay farms to meet minimal baseline standards other industries are mandated to achieve, see Ruhl, *Farms*, *supra* note 2, at 293-316, 325-27. *See also* Mary Jane Angelo, *Corn, Carbon, and Conservation: Rethinking U.S. Agricultural Policy in a Changing Global Environment*, 17 Geo. Mason L. Rev. 593 (2010) (evaluating environmental impacts of historical and current agricultural policy); J.B. Ruhl, *Three Questions for Agriculture About the Environment*, 17 J. Land Use & Envtl. L. 395, 404-05 (2002); J.B. Ruhl, *Farmland Stewardship: Can Ecosystems Stand Any More of It?*, 9 Wash. U. J.L. & Pol'y 1 (2002).
4. Agriculture "never had coherent first-generation environmental protection programs" and "no significant environmental controls have been placed on farm practices even where agricultural activities are a primary cause of pollution problems." C. Ford Runge, *Environmental Protection From Farm to Market*, *in* Thinking Ecologically: The Next Generation of Environmental Policy 200, 200-01 (Marian R. Chertow & Daniel C. Esty eds., 1997); *see also* Ruhl, *Farms*, *supra* note 2, at 268 n.6.
5. For several examples of regulatory controls on agriculture, including regulation of concentrated animal feeding operations under the Clean Water Act and regulation of habitat disturbance under the Endangered Species Act, see Ruhl, *Farms*, *supra* note 2, at 316-27.
6. For a thorough examination of farm bill politics, see Woods Inst. for the Env't., U.S. Agricultural Policy and the 2007 Farm Bill (Kaush Arha et al. eds., 2007), *available at* http://woods.stanford.edu/docs/farmbill/farmbill_book.pdf [hereinafter 2007 Farm Bill].

expended year after year, decade upon decade, to keep policy of any other kind exactly where it started out. Agricultural economist David Freshwater sums up this history well:

> With each farm bill cycle there are calls for a major rethinking of U.S. farm policy to make it better suit current farm conditions and the expectations of the broader American public about the roles of agriculture. These calls for reform have been for the most part unsuccessful because there has been no argument compelling enough to overcome advocates of the status quo. But as time passes the wisdom of maintaining a set of policies that have their basis in the 1930s and were designed to support a structure of agriculture that no longer exists becomes more questionable.[7]

Paying farmers to do the "right thing" environmentally has been a theme of federal farm policy for decades, embodied in programs such as the Conservation Reserve Program (CRP), which pays farmers to take land out of production for defined periods to enhance its conservation values, and the Conservation Stewardship Program (CSP) and Environmental Quality Incentives Program (EQIP), which pay farmers to employ better practices on working lands.[8] And either paying or forcing farmers to preserve agricultural land uses at the urban fringe has become a standard component of state and local land use policy.[9] In this sense, farms have long been understood as land units that have the capacity to contribute to environmental and cultural values.

In recent years, however, ecologists and economists focusing on agriculture have forged a more complete vision of the capacity of agricultural lands. They see farms as housing the natural capital capable of providing a stream of diverse good and services, including ecosystem services such as increased biodiversity, carbon sequestration, pollination, groundwater recharge, and improvement of water quality.[10] To be sure, farms taking this model to heart would look and behave differently from conventional operations based on intensive monoculture and concentrated livestock, but they unmistakably would be active and potentially prosperous agricultural operations. It is not overly optimistic to think that "the scientific and political planets are aligning to create both the demand for policy-relevant research into the [ecosystem services] available from agriculture and the means to create incentives for farmers to provide those services."[11]

Unfortunately, federal policy has been slow to move in this direction. While it has become a rite of passage to begin each five-year cycle of farm bill work with great fanfare over the prospect of stepping up the green subsidy and farm preservation programs, the rhetoric and content each time are steadily watered down until the programs look about as they started. The long-prevailing system of farm income supports, including green subsidies, simply does not tap into or promote a sense that there is more to agriculture than supplying food, fiber, and energy commodities, with a dose of cultural nostalgia.[12]

It is unlikely, therefore, that federal farm policy alone will align these interests. It will be important for farm multifunctionality to respond to demand-driven signals, whereas even the green subsidy component of federal farm policy is supply-driven and tailored to what is possible and convenient for conventional agriculture. As Katherine Smith explains:

---

7. David Freshwater, *Applying Multifunctionality to U.S. Farm Policy* 1 (Univ. of Ky., Econ. Staff Paper No. 437, 2002) (unpublished manuscript on file with author), *available at* http://www.uky.edu/Ag/AgEcon/pubs/staff/staff437.pdf. *See also* J.P. Reganold et al., *Transforming U.S. Agriculture*, 332 Sci. 670 (2011) (discussing the history of farm bill environmental policy and the need for reform).
8. For thorough reviews of agricultural land retirement and working land conservation subsidy programs, see Craig Cox, *U.S. Agriculture Conservation Policy & Programs: History, Trends, and Implications*, *in* 2007 Farm Bill, *supra* note 6, at 113; Neil Hamilton, *Feeding Our Green Future: Legal Responsibilities and Sustainable Agricultural Land Tenure*, 13 Drake J. Agric. L. 377 (2008).
9. For a comprehensive overview of this state and local land use regulation trend, see Julian Conrad Juergensmeyer & Thomas E. Roberts, Land Use Planning and Development Regulation Law 815-71 (2d ed. 2007).
10. Ecosystem services are economically valuable benefits humans derive from ecological resources directly, such as storm surge mitigation provided by coastal dunes and marshes, and indirectly, such as nutrient cycling that supports crop production. Natural capital consists of the ecological resources that produce these service values, such as forests, riparian habitat, and wetlands. For descriptions of natural capital and ecosystem services, see Millennium Ecosystem Assessment, Ecosystems and Human Well-Being: Synthesis (2005), *available at* http://www.wri.org/publication/millennium-ecosystem-assessment; Nature's Services: Societal Dependence on Natural Ecosystems (Gretchen C. Daily ed. 1997); Robert Costanza et al., *The Value of the World's Ecosystem Services and Natural Capital*, 387 Nature 253 (1997). For coverage of the emergence of the ecosystem services concept in law and policy, see J.B. Ruhl, Steven E. Kraft & Christopher L. Lant, The Law and Policy of Ecosystem Services (2007); James Salzman, *A Field of Green? The Past and Future of Ecosystem Services*, 21 J. Land Use & Envtl. L. 133 (2006); J.B. Ruhl & James Salzman, *The Law and Policy Beginnings of Ecosystem Services*, 22 J. Land Use & Envtl. L. 157 (2007).
11. Scott M. Swinton et al., *Ecosystem Services From Agriculture: Looking Beyond the Usual Suspects*, 88 Amer. J. Agric. Econ. 1160, 1164 (2006). An excellent survey of the literature supporting this movement is found at G. Philip Robertson & Scott M. Swinton, *Reconciling Agricultural Productivity and Environmental Integrity: A Grand Challenge for Agriculture*, 3 Frontiers Ecology & Env't 38 (2005).
12. *See* David Abler, *Multifunctionality, Agricultural Policy, and Environmental Policy*, 33 Agric. & Resource Econ. Rev. 8 (2004); Katherine R. Smith, *Public Payments for Environmental Services From Agriculture: Precedents and Possibilities*, 88 Am. J. Agric. Econ. 1167, 1167-68 (2006).

These land retirement, working lands, and land use preservation programs' payment priorities are agriculture-centric. They are based on what the producers of the agri-environmental benefits can supply, rather than what is necessarily demanded by the population that would benefit from ensuing environmental service enhancement.... [T]he choice of how benefits are targeted derives from a universe of acreage-based attributes; in other words, what existing, independent farm production facilities can supply.... There is no good evidence that any existing public agri-environmental payment program purchases a given "environmental service."[13]

This chapter focuses on how to move farm policy past this history and into an era of multifunctionality. It explores the emerging theme of farms as multifunctional land uses and suggests that federal, state, and local governments can best help ground it through flexible, efficient, incentive-based policy instruments. At the federal level, Congress can realign federal farm policy to facilitate the delivery of a more sustainable profile of farm goods and services through state and local programs. Although federal farm subsidies surely could be repositioned to better promote farm multifunctionality directly,[14] the benefits of multifunctional agricultural production, compared to the conventional commodity production orientation, are primarily local. Federal policy should therefore support state and local innovations rather than dominate the field as has been the case historically.

The first section of the chapter examines the theme of farms as multifunctional production units as it is developing in ecological literature, juxtaposing the ecological theory with the economic reality of agriculture. This section then examines the potential future ecological/economic scenarios of agricultural land uses and the tools that could be used to help break the logjam of agriculture-environment policy by promoting the multifunctionality of farms.

The second section of the chapter surveys conventional agricultural policy instruments used in the context of environmental conservation goals. These range from highly regulatory approaches such as zoning, to incentive-based approaches such as payments for development rights. By and large, none of these programs has been put to use with a focus on the provision of ecosystem services from farms.

The final section focuses on two incentive-based tools that seem particularly well-suited to responding to the emerging goal of promoting farm multifunctionality—payments for ecosystem services (PES) and transferable development rights (TDR). The PES approach is predicated on the opportunity to reduce infrastructure spending associated with residential and commercial development, such as the need for increased water supply and maintaining water quality, by paying agricultural operations directly to deliver equivalents at lower cost in the form of ecosystem services.[15] Although PES programs defined broadly include conventional green subsidy programs such as the CRP, as well as payments for environmental amenities such as conservation of endangered species' habitat, both of which can generate incidental ecosystem service benefits, agricultural PES payments ideally would be based on demand-driven, market-priced transactions for ecosystem services.[16]

---

13. Smith, *supra* note 12, at 1167-68.
14. The 2008 Farm Bill, popularly known as the Food Conservation, and Energy Act of 2008, did not alter the structural features of the green subsidy programs. Farm, Nutrition and Bioenergy Act of 2007, H.R. 2419, 110th Cong. (2007). For a discussion of how federal "green subsidy" farm payments could be reconfigured to promote farms' multifunctionality, see Kaush Arha et al., *Conserving Ecosystem Services Across Agrarian Landscapes*, in 2007 FARM BILL, *supra* note 6, at 207; William J. Even, *Green Payments: The Next Generation of U.S. Farm Programs?*, 10 DRAKE J. AGRIC. L. 173 (2005).
15. For general background on PES programs, see B. Kelsey Jack et al., *Designing Payments for Ecosystem Services: Lessons From Previous Experience With Incentive-Based Mechanisms*, 105 PROC. NAT'L ACAD. SCI. 9465 (2008). For discussion of specific PES initiatives, mainly from other nations, see James Salzman, *Creating Markets for Ecosystem Services: Notes From the Field*, 80 N.Y.U. L. REV. 870 (2005); Brian C. Steed, *Government Payments for Ecosystem Services—Lessons From Costa Rica*, 23 J. LAND USE & ENVTL. L. 177 (2007).
16. Indeed, some advocates of PES programs in agricultural settings emphasize this feature to differentiate PES programs from subsidy programs. *See, e.g.*, JOHN M. ANTLE, PAYMENTS FOR ECOSYSTEM SERVICES AND U.S. FARM POLICY 17 (Am. Enter. Inst. Forum on the 2007 Farm Bill and Beyond, Working Paper, 2006), *available at* http://aic.ucdavis.edu/research/farmbill07/aeibriefs/20070515_antlefinal.pdf. Ecoagriculture. org is a Washington, D.C.-based organization providing information about agricultural PES programs generally, including in other nations, through a periodic newsletter, *Ecoagriculture PES Newsletter*, available at http://www.ecoagriculture.org/page.php?id=50&name=Newsletter. This chapter focuses on the promise of agricultural PES programs in the United States given the significant differences in economic incentives and trade offs between agricultural land uses in developed versus developing nations.

The TDR is a technique well-known in land use law[17] and gaining traction in agriculture policy as a means of preventing farmland from being devoured by the suburban amoeba.[18] TDRs, which have a long history in local historic preservation[19] and environmental protection[20] programs, are a way of rewarding a landowner for foregoing development (either voluntarily or by regulatory force) in one area (the "sending area") by providing a building density or other development "credit" that can be applied to exceed the default development limits in another area (the "receiving area").

Although distinct in several ways, including fiscal impact, the role of regulation, and the medium through which provision of ecosystem services is rewarded, these two approaches share design issues being worked in a number of different applications. The discussion compares and contrasts PES and TDR programs in the context of farm multifunctionality policy.

PES and TDR programs would comprise only a small component of a shift toward an overall agriculture policy of promoting farm multifunctionality. The closing section of the chapter, therefore, explores some of the more overarching policy design questions that will arise as farm multifunctionality becomes a more pressing objective.

A. Promoting Farm Multifunctionality

The cultural image of agriculture has vacillated in the public eye over time. In the mid-1800s, George Perkins Marsh revealed the opportunity costs of converting natural habitat to agriculture in his epic book *Man and Nature*[21]; yet, Aldo Leopold's equally influential *Sand County Almanac*[22] later offered a "poetic evocation of agriculture as part of a larger ecosystem community."[23] In 1962, Rachel Carson's *Silent Spring*[24] then returned "scientific and public attention to the negative externalities of farming."[25] Yet, "if the harbinger of the last intellectual wave to wash over agriculture was *Silent Spring*, the bellwether of the next wave may be *Nature's Services*, edited by Gretchen Daily."[26] Published in 1997, *Nature's Services*[27] was the first comprehensive treatment of the ecosystem services concept grounded in practical ecological foundations. It quickly became the impetus for a broad movement toward integrating ecological economics and ecology across a spectrum of policy fronts. One emerging focal point of the new intellectual wave is the concept of farm multifunctionality.

1. The Ecology and Economics of Farm Multifunctionality

Following the lead of *Nature's Services*, the growing science and policy literature divides ecosystem services into five types: *provisioning services* that underlie the production of commodities; *regulating services* that moderate dynamic natural phenomena; *cultural services* that provide human psychic satisfaction; *preserving services* that maintain ecological diversity and resilience; and *supporting services* that promote the capacity of ecosystems to produce the other service types.[28] The story of conventional crop and livestock agriculture has been largely one of managing provisioning (food and fiber) and cultural (farmland character) services and their associated supporting services, primarily because these are essential for farms to produce marketable commodities and retain their charmed status in the public eye. Only recently has the focus turned to

---

17. *See* JUERGENSMEYER & ROBERTS, *supra* note 9, §9.10, at 546 (TDR "programs are frequently incorporated into growth management programs"); Julian Conrad Juergensmeyer et al., *Transferable Development Rights and Alternatives After* Suitum, 30 URB. LAW. 441 (1998) (surveying several prominent programs).
18. For surveys of the use of TDRs in agricultural land policy, see Elisa Paster, *Preservation of Agricultural Lands Through Land Use Planning Tools and Techniques*, 44 NAT. RESOURCES J. 283, 306-08 (2004); Edward Thompson Jr., *"Hybrid" Farmland Protection Programs: A New Paradigm for Growth Management?*, 23 WM. & MARY ENVTL. L. & POL'Y REV. 831 (1999).
19. *See, e.g.*, Penn Cent. Transp. Co. v. City of New York, 438 U.S. 104 (1978).
20. *See, e.g.*, Suitum v. Tahoe Reg'l Planning Agency, 520 U.S. 725, 728-32 (1997).
21. *See generally* GEORGE PERKINS MARSH, MAN AND NATURE (David Lowenthal ed., Univ. Wash. Press 2003) (1864).
22. *See generally* ALDO LEOPOLD, A SAND COUNTY ALMANAC (Oxford Univ. Press 2001) (1949).
23. Swinton et al., *supra* note 11, at 1161.
24. RACHEL CARSON, SILENT SPRING (Houghton Mifflin 1994) (1962).
25. Swinton et al., *supra* note 11, at 1161.
26. *Id.*
27. *See* NATURE'S SERVICES, *supra* note 10.
28. *See* MILLENNIUM ECOSYSTEM ASSESSMENT, ECOSYSTEMS AND HUMAN WELL-BEING: A FRAMEWORK FOR ASSESSMENT 57 (2003), *available at* http://www.wri.org/publication/millennium-ecosystem-assessment.

expanding agriculture's position as a source of regulating services valuable to surrounding local, regional, and national communities, the problem being how to provide farmers the incentive to manage for such services when no market yet exists for them. As Scott Swinton et al. explain:

> Agriculture (including planted forests) conventionally supplies food, fiber, and fuel—"provisioning services" in [ecosystem services] parlance. Farmers also help maintain the natural "supporting" [ecosystem services] that make agriculture productive, such as pollination, biological pest regulation, and soil nutrient renewal. In theory, the same managed ecosystems that provide these marketed products could produce other types of [ecosystem services] if suitable incentives existed. The broad class of "regulation [ecosystem services]" covers climate regulation, water purity, surface water flows, groundwater levels, and waste absorption and breakdown. All of these offer benefits that are poorly captured by current markets, yet which managed agricultural and forest ecosystems could potentially provide.[29]

The problem, however, goes well beyond how services could be captured in markets. Nicholas Jordan et al. explain that agricultural "research and development . . . and policy have focused on maximizing biomass production and optimizing its use, with far less emphasis on evaluation of environmental, social, and economic performance."[30] Similarly, "current federal programs and policy on environmental quality in agricultural landscapes mainly subsidize retirement of land from active production."[31]

By contrast, agricultural multifunctionality emphasizes "the joint production of standard commodities (e.g., food or fiber) and 'ecological services'" on the premise that "major additional gains may result from a 'working landscape' approach that improves environmental performance of active farmland by rewarding farmers for delivering environmental benefits, as well as food and biomass."[32] The methods a multifunctional farm would use to achieve this more balanced production profile would include precision farming, no-till farming, organic farming, rotational cropping, crop residue usage, bio-pest controls, riparian cover, filter strips, contour farming, incorporated pollinator habitat, and water retention and recharge ponds.[33] Table 14.1 illustrates the different profiles of ecosystem service production employed by conventional and multifunctional agricultural land uses.

Table 14.1
Ecosystem Services of Conventional Farming Versus Multifunctional Farming

Ecosystem Service	Conventional Farming	Multifunctional Farming
Provisioning—food, fiber, energy sources, pharmaceuticals, and other consumed commodities supplied by nature.	Land and resources are managed primarily to produce food and fiber commodities and, increasingly, biomass fuels.	Food, fiber, and fuel production remain a primary purpose of land and resource management.
Regulating—services that modulate ecosystem processes with economic relevance to humans, such as gas composition, air and water temperature, nutrient flows, and waste decomposition.	Land unsuitable for cultivation or grazing and land taken out of production through CRP and other subsidy programs will provide incidental regulating service benefits; land in cultivation and active grazing has diminished capacity to provide regulating services.	Riparian habitat is actively managed to promote nutrient and sediment capture, provide flood control, and provide thermal regulation of stream flows; interior wetland areas are managed to promote groundwater recharge and suppress dry freeze effects; woody and grassy biomass is managed for carbon sequestration.

---

29. Swinton et al., *supra* note 11, at 1160 (citation omitted).
30. Nicholas Jordan et al., *Sustainable Development of the Agricultural Bio-Economy*, 316 SCI. 1570, 1570 (2007) (footnotes omitted).
31. *Id.*
32. *Id.* (footnote omitted).
33. *See* Rebecca L. Goldman et al., *Managing for Ecosystem Services on U.S. Agricultural Lands*, in 2007 FARM BILL, *supra* note 6, at 97, 106.

Ecosystem Service	Conventional Farming	Multifunctional Farming
Cultural—services that enhance human use and appreciation of natural resources and the built environment, including recreation, aesthetic appreciation, scientific research, and cultural, spiritual, and intellectual inspiration.	Active farmlands are devoted primarily to food and fiber production and not generally open to public; existence of farming lands in community provides some background cultural significance.	Active farmlands could be opened to public cultural activities such as stay-and-work, school visits, or bed-and-breakfast; areas managed for regulating and supporting services could provide eco-tourism, recreational, and scientific opportunities.
Preserving—services that maintain ecological resilience and the diversity of ecological futures.	None of significance.	Areas are actively managed as seed banks, wildlife habitat, and to restore native grasses and other vegetation.
Supporting—services that sustain other forms of service flows.	Land unsuitable for cultivation or grazing and land taken out of production through CRP and other subsidy programs will provide incidental supporting service benefits such as pollination, seed dispersal, and biological pest control; land in cultivation and active grazing has diminished capacity.	Areas are actively managed with the specific purpose of enhancing pollination, pest control, seed dispersal, and other supporting services.

The problem, of course, is that farmers have no inherent incentive to move from the conventional model to the multifunctional model. Farms are highly-managed ecosystems. They alter and then manage ecological processes and functions on small and large scales. In so doing, farms reconfigure ecological attributes to maximize *provisioning services*—the food, fiber, energy, and other commodities supplied by nature.[34] Farms manage these provisioning services to optimize on-site farm production, often at the expense of off-site environmental conditions. Farms can cause, for example, with soil erosion, nutrient and pesticide runoff, groundwater depletion, and a host of other off-site environmental degradations. In short, almost nothing takes place on a farm without ecological impacts somewhere else.

One off-site impact of farming heretofore little noticed, however, is the depletion of *regulating services*. Unlike provisioning services, the market value of which is embedded in commodity prices and thus easily measured and monitored, regulating services tend to behave more like nonmarket public goods. Farms thus have all the incentive to optimize provisioning services available to them, but little incentive to provide regulating services that benefit other lands.

In this respect, a farm is like any other ecological unit—changes in one ecosystem usually affect other ecosystems, however we draw the boundaries. But as highly managed ecological units, farms significantly tilt the production frontier for ecosystem services toward provisioning services and away from regulating services. Ecological practices at a cornfield are designed to produce corn efficiently within the relevant regulatory and economic environment. Putting aside the question whether regulation of farms has established appropriate environmental performance baselines, unless paid to provide regulating services such as carbon sequestration, farmers cannot be expected to provide significant flows of off-site regulating services except as incidental to management of provisioning services.

We need to know more about the geographic and economic contexts before we can assess the prospects of realigning the ecological profile. Agriculture presents a difficult geographic scenario for developing generalized strategies for ecosystem services. Farms are numerous, dispersed, come in all sizes, and produce many different commodities under many different climate and landscape conditions. Farms also manage ecological resources for relatively small spatial scales (the farm) and short temporal scales (the next harvest). The focus on optimizing on-site provisioning services also tends to sever farms and larger agricultural

---

34. For a thorough discussion of these economic incentive structures in the context of agricultural land use, see Marc Ribaudo et al., *Ecosystem Services From Agriculture: Steps for Expanding Markets*, 69 ECOLOGICAL ECON. 2085 (2010).

landscapes from surrounding ecological resources. Managing ecosystem services sustainably, by contrast, requires multiscalar approaches that integrate connected ecological units across space and time.

These geographic disconnects strongly influence the economics of farming and the bias toward provisioning services. The payoff for providing regulating services, assuming some mechanism for compensation, is likely to be marginal compared to commodity production or, worse, selling to urban development interests. In the absence of any compensation, in other words, economically rational farmers will not provide free regulating services to off-site lands unless doing so is incidental to optimization of on-site commodity production or is forced by regulation. Promoting farm multifunctionality, therefore, is a balancing exercise between providing farms the flexibility to continue benefitting from their skill at managing provisioning services on the one hand, and providing the impetus to produce more regulating services for society on the other. Moreover, market distortions from subsidies, which have promoted intensive production on marginal and environmentally sensitive lands, have made it much more difficult to integrate ecosystem service values into agricultural production decisions. Society cannot assume that the flow of regulating services off farms (or any land for that matter) will continue to be provided for free, lest they not be provided at all, nor can we expect farmers to forego the incentives that the collection of production and insurance subsidies deliver. Ideally, the economics of farming, including market-distorting subsidy policies, can be altered to change the flow of services, rather than forcing the issue through command-and-control regulation.

To put the policy problem in standard economic terms, farmers view the provision of regulating services to outside communities as a positive externality—doing so benefits others—but nobody has been willing to pay for the benefits.[35] The market primarily rewards farmers for producing commodities, and federal farm subsidies force a trade off between commodity production and ecological conservation. Instead of asking how to balance that trade off, we should be asking a key question: Why is it that farms cannot be rewarded for producing commodities *and* ecosystem services?

2. Conceiving Alternative Futures for Agricultural Lands

Even if farming as usual is a superior land use option for a community as compared to, say, cookie-cutter sprawl,[36] those do not exhaust the alternatives. Rather, a spectrum of potential future scenarios presenting different trade offs must be considered before land use policy can make sensible comparisons. Those additional scenarios can be simplified to the following four, the advantages and disadvantages of which are explored in the next section:

1. *Agricultural Use With Increased Environmental Performance Baseline.* Under this scenario farms are regulated more heavily than is the current practice, primarily to enhance environmental performance. For example, riparian buffers would be mandated, onsite water recharge features would be required, and tillage practices would be specified. Of course, this is the scenario agriculture has steadfastly and thus far successfully resisted, but it is nonetheless an option.

2. *Conversion to Multifunctional Working Landscape.* A baseline performance level of agricultural practices would first be specified, either at conventional levels or through regulation at more demanding levels (as above), then incentive programs would be designed to compensate farmers for enhancing the flow of regulating ecosystem services above the baseline to identified off-farm populations and areas. For example, if riparian buffers and onsite recharge features were not required under the baseline, providing them would entitle a farmer to some compensation in return.

3. *Conversion to Open Space.* Public or private interests would simply buy out all, or substantially all, of the land use rights associated with agricultural lands, either through conservation easements or

---

35. *See* OECD, *supra* note 1, at 13; Jules Pretty et al., *Policy Challenges and Priorities for Internalizing the Externalities of Modern Agriculture*, 44 J. Envtl. Plan. & Mgmt. 263 (2001). The literature summarizing the economic incentives associated with ecosystem services generally, particularly regulating services that flow from land where natural capital is located and benefits users of other land parcels, is reviewed in Ruhl, Kraft & Lant, *supra* note 10, at 57-83.
36. *See, e.g.*, Paster, *supra* note 18, at 283. ("[P]roductive agricultural lands are an irreplaceable natural resource being lost to sprawling subdivisions throughout the country.").

fee title. From there, the land management regime might include management for ecosystem service flows, perhaps even selling them where markets or other compensatory incentives can be identified.

4. *Conversion to Planned Mixed-Use, Mixed-Density Development.* Under this scenario agricultural lands are converted to development, but not as uniform low-density sprawl. Rather, either through land use regulation or in response to market demand, the buildout is comprehensively planned and includes clustered high-density development; mixed commercial, office, and residential uses; and substantial recreational and conservation open space. Some working agricultural uses might be retained, and the planning of land use could take into account the location of natural capital and its associated ecosystem services flows.

Which of the alternative futures is "better" is by no means obvious because opening up "the multifunctional set of services provided by farmland complicates the task of identifying which farmland should be preserved,"[37] and expanding the alternatives to farming beyond sprawl suggests a spectrum of public trade offs, any one of which might, in context, be preferable to agriculture. For present purposes, it is not necessary to decide which of these scenarios best fits; rather, the question is what instruments state and local governments have at their disposal to pursue a multifunctional agricultural land policy.

3. Policy Instruments

With the expanded slate of scenarios in hand, regulatory authorities wishing to favor one or another must review the policy instruments at their disposal and the advantages and disadvantages of using particular instruments to achieve the desired scenario. This section provides a brief inventory of methods that state and local jurisdictions can use toward that objective, followed by an integrated assessment of scenarios and tools.

Many of the tools are designed to preserve existing farmland "as is"—a "save farming" premise that permeates federal and state policy. The American Farmland Trust (AFT) in particular has been a vocal advocate on behalf of farmland conservation.[38] AFT has been quite successful, helping to bring about the Farmland Protection Policy Act[39] in the 1981 Farm Bill and a host of farmland protection measures in the Farm and Ranch Lands Protection Program[40] renewed in the 2002 Farm Bill. But many state and local programs are designed to serve the same objective.

For example, one way farmland can be "saved" in this sense is to configure local zoning regulations to prohibit it from being converted from agricultural uses, or to impose insurmountable barriers to converting it to suburban development, a method some states and localities have used over vociferous objections of the very landowners ostensibly being protected.[41] Whether this status quo lock-in approach saves farms or saves existing suburbanites from yet more suburban development is, of course, a matter for debate and is largely

---

37. B. James Deaton et al., *Setting the Standards for Farmland Preservation: Do Preservation Criteria Motivate Citizen Support for Farmland Preservation?*, 32 AGRIC. & RESOURCE ECON. REV. 272, 272 (2003).
38. *See* AMERICAN FARMLAND TRUST, http://www.farmland.org (last visited Nov. 28, 2012). Based on the National Resources Inventory, the Department of Agriculture reports that "46 percent of the land converted to urban and built-up uses comes from cropland and pasture, while 38 and 14 percent comes from forest land and range land, respectively. Much of the land being lost is prime, unique, or important farmland located near cities." 7 C.F.R. §1491. Some critics have portrayed the farmland preservation movement as an alliance between agricultural landowners seeking to be paid to keep farming and local antidevelopment, pro-open space interests seeking to thwart urban growth. *See, e.g.*, William A. Fischel, *The Urbanization of Agricultural Land: A Review of the National Agricultural Lands Study*, 58 LAND ECON. 236 (1982); Jesse Richardson, *Farmland Protection*, AGRIC. L. UPDATE, Oct. 2006, at 4. Nevertheless, many different federal, state, and local programs have been implemented to respond to AFT's call, and the trend is on the rise. *See* David C. Levy & Rachel P. Melliar Smith, *The Race for the Future: Farmland Preservation Tools*, 18 NAT. RESOURCES & ENV'T 15, 15 (2003).
39. *See* 7 U.S.C. §§4201-4209; 7 C.F.R. §658.1. The legislation requires federal agencies to ensure their respective programs avoid unnecessarily contributing to the loss of farmlands and to ensure that they act compatibly with state and local policies designed to protect farmland; no funding for or regulation of farmland preservation is provided. For a critique of the legislation as largely ineffective, see Robert M. Ward, *The U.S. Farmland Protection Policy Act: Another Case of Benign Neglect*, 8 LAND USE POL'Y 63 (1991).
40. *See* Farm Security and Rural Investment Act of 2002, Pub. L. No. 107-171 §2503; 7 C.F.R. §1491 (2003). The program provides matching funding to states and local governments to purchase conservation easements from farmers and ranchers to limit conversion to nonagricultural land uses. *See generally* Renee Johnson, CRS Report for Congress, Farm Protection Program: Status and Current Issues, RS22565 (Jan. 5, 2007); Michael R. Eitel, *The Farm and Ranch Lands Protection Program: An Analysis of the Federal Policy on United States Farmland Loss*, 8 DRAKE J. AGRIC. L. 591 (2003).
41. *See* JUERGENSMEYER & ROBERTS, *supra* note 9, §13.8, at 852-55. For a survey of techniques, including exclusive use zoning, large lot zoning, and cluster zoning, see Peggy Kirk Hall, *Approaches to Zoning That Support and Protect Agriculture*, AGRIC. L. UPDATE, May 2007, at 6.

in the eyes of the beholder.[42] In any event, courts have generally rejected the argument that these exclusive agricultural use zoning restrictions constitute regulatory takings.[43]

Another technique is to use tax policy to favor continuation of agricultural land uses.[44] For example, many state and local governments adopt differential property tax assessment provisions that provide lower assessment rates for agricultural land uses and thereby, in theory, deter conversion to higher-rate land uses. But the evidence is that these measures do not deter conversion to development at the urban fringe, where returns on development frequently more than offset the higher tax rates.[45]

By contrast to the state and local exclusive agricultural use zoning and tax relief programs, the early thrust of state and local efforts, later supported by the federal farmland protection initiative, was funding of programs for purchase of development rights (PDR) and purchase of agricultural conservation easements (PACE), the effect being to preclude conversion to more intense development. For perhaps obvious reasons, AFT has strongly advocated PDR/PACE programs, with over half the states and 50 local governments adopting such programs and the 2002 Farm Bill providing $600 million in federal matching dollars for PDR/PACE acquisitions, as implemented by USDA's Commodity Credit Corporation.[46]

Zoning, tax breaks, and PDR/PACE programs involve either regulation or public financing. Another alternative for farmland preservation, one that neither regulates farms directly nor demands public revenue, is the local use of TDRs to reward an agricultural landowner who withdraws land from potential conversion to development with credits for developers that can be used in other areas to go above and beyond the baseline of allowable development parameters, such as density of units.[47] The obvious attraction to TDRs for purposes of farmland preservation is that they impose no fiscal burden on the public; on the other hand, the potential weakness of TDRs is that they depend on developer demand for the credits.

The techniques mentioned thus far may be useful in maintaining agricultural land uses in status quo, but they do not inherently promote better farming practices to reduce environmental harms or enhance regulating and supporting services. The chief method of improving the baseline environmental performance of farms has been through the promulgation of best management practices (BMPs), such as tillage methods, integrated pest management, and retention of riparian habitat.[48] To be comprehensively effective, these would have to be regulatory mandates, whereas they have been employed mostly as voluntary guidelines[49] or as the cross-compliance condition to receive subsidies or other incentives.[50]

The underlying assumption of this collection of instruments is that farming remains on the landscape in some substantial form, whereas some of the alternative scenarios involve removing agricultural uses altogether. At one extreme, the conversion to open space can be accomplished through purchase of permanent conservation easements restricting all but passive uses, or by acquisition of title with similar deed restrictions. Some state and local governments, as well as private land trusts, have been aggressive at accomplishing these conversions, though often some level of agricultural use is contemplated.[51] Agricultural interests have not always been keen about programs designed to convert agriculture into open space, concerned that

---

42. For a series of articles presenting contrasting perspectives, see Mark W. Cordes, *Takings, Fairness and Farmland Preservation*, 60 OHIO ST. L.J. 1033 (1999); Jesse J. Richardson Jr., *Downzoning, Fairness and Farmland Protection*, 19 J. LAND USE & ENVTL. L. 59 (2003); Mark W. Cordes, *Fairness and Farmland Preservation: A Response to Professor Richardson*, 20 J. LAND USE & ENVTL. L. 371 (2005).
43. *See, e.g.*, Gardner v. N.J. Pinelands Comm'n, 593 A.2d 251 (N.J. 1991).
44. *See* JUERGENSMEYER & ROBERTS, *supra* note 9, §13.14, at 866-69.
45. *See* Sandra A. Hoffman, Note, *Farmland and Open Space Preservation in Michigan: An Empirical Analysis*, 19 U. MICH. J.L. REFORM 1107 (1986).
46. 7 C.F.R. §1491.
47. *See* JUERGENSMEYER & ROBERTS, *supra* note 9, §13.11, at 860-62.
48. *See* David Zaring, *Best Practices*, 81 N.Y.U. L. REV. 294, 331-34 (2006).
49. *See* Even, *supra* note 14, at 180-83; Zaring, *supra* note 48, at 326-39.
50. *See* Sandra S. Batie & Alyson G. Sappington, *Cross-Compliance as a Soil Conservation Strategy: A Case Study*, 68 AM. J. AGRIC. ECON. 880 (1986); Ramu Govindasamy & Mark J. Cochran, *The Conservation Compliance Program and Best Management Practices: An Integrated Approach for Economic Analysis*, 17 REV. AGRIC. ECON. 369 (1995); Drew Kershen, *Sustainable Intensive Agriculture: High Technology and Environmental Benefits*, 16 KAN. J. L. & PUB. POL'Y 424 (2007). For a comprehensive case study of the potential environmental benefits of agricultural BMP policy in practice, see Alfred Light, *Reducing Nutrient Pollution in the Everglades Agricultural Area Through Best Management Practices*, 25 NAT. RESOURCES & ENV'T 26 (Fall 2010).
51. *See* JUERGENSMEYER & ROBERTS, *supra* note 9, §§13.12-13.13, at 862-66. One of the largest such programs in the world is the Florida Forever land acquisition program, which has put into conservation status over 535,000 acres of land at a cost of $1.8 billion through December 2006. *See* Fla. Dep't Envtl. Prot., *Florida Forever*, http://www.dep.state.fl.us/lands/fl_forever.htm.

the agricultural land base in an area may fall below the critical mass necessary to support a cohesive agricultural economy, including seed and equipment suppliers and produce distributors.[52]

At the other conversion extreme, the image AFT and other "save farming" advocates portray as the inevitable alternative to farming is conversion to the uniform low-density residential buildout characteristic of conventional zoning—the scenario most associated with sprawl—even though mixed-use, mixed-density development scenarios are viable options in many agricultural localities.[53] At the core of either kind of buildout scenario is the local zoning power, in these cases exercised not to restrict agricultural landowners to farming but to liberate them from it.

Nowhere in the list thus far have ecosystem services been the central focus. To be sure, agricultural BMPs, though directed primarily at environmental quality, will in many instances incidentally enhance regulating services, and as discussed below could even be designed more purposefully for that effect. Indeed, ecosystem service delivery can be integrated into any of the described programs as an output goal.

Two instruments in particular have become most closely associated with proposals for promoting farm multifunctionality. One is obvious: pay for enhanced environmental services directly through PES programs tied to the costs local jurisdictions avoid by substituting regulating ecosystem services for technological service infrastructure. Used this way, PES programs offer neither a subsidy nor a payment for intrinsic or ecological benefits such as endangered species habitat; rather, they are what the name implies—a demand-driven payment for a valuable service rendered.

In areas where the development market has put extreme pressure on agricultural lands, however, PES payment rates may not be adequate to compete with alternative land uses to preserve agricultural uses. In that scenario, TDRs, because they tap into development market values, may provide sufficient incentive to retain some agricultural integrity in land use. Here, the TDR credit calculus is not limited to preservation of farmland or cultural amenities, but includes also the level of ecosystem service delivery expected from the natural capital that is secured through altered agricultural practices. Either instrument, therefore, can promote ecosystem service delivery to an important, if not driving, component of the valuation calculus on which the PES transfer or TDR credit is based.

The different scenarios and policy instruments, with their associated advantages and disadvantages, are matched up in Table 14.2.

Table 14.2
Future Scenarios and Policy Tolls: Advantages and Disadvantages

Future Scenario	Policy Tools	Advantages	Disadvantages
Compete in Land Market to Maintain Status Quo Agricultural Use	Tax incentives; subsidies	Maintains agricultural land uses	May not compete successfully against high value suburban development; potentially expensive to maintain competitive edge; does not alter ecosystem service profile
Lock-in of Status Quo Agricultural Use	Exclusive agricultural use zoning districts; purchase of development rights (e.g., PDR/PACE programs)	Maintains agricultural land uses	Politically controversial if zoning used; does not alter ecosystem service profile; restricts land market

---

52. See Juergensmeyer & Roberts, supra note 9, §13.5, at 838.
53. See id. §13.10, at 857-60.

Future Scenario	Policy Tools	Advantages	Disadvantages
Agricultural Use with Increased Environmental Performance Baseline	Command-and-control regulation mandate of best management practices; incentives such as subsidies and tax relief; possibly also zoning	Maintains agricultural land uses; reduces environmental harms; possible shift of ecosystem service profile toward regulating and supporting services	Potentially undermines financial stability of agricultural uses by increasing compliance costs and reducing production potential; requires new managerial skills; politically controversial; requires more regulatory infrastructure; expensive if incentives used; could prompt conversions to development scenarios if exclusive agricultural use zoning not also used
Transformation to Multifunctional Working Landscape	Payment for environmental services; transferable development rights; pollutant trading programs; certification programs; planned unit development zoning for any areas being developed	Maintains some agricultural land uses; likely to increase open space and associated ecosystem services; likely to significantly shift ecosystem service profile toward regulating and supporting services; requires only moderate use of regulation	Possible reduction in food and fiber production; requires public expenditures for PES (potentially offset by cost savings); requires new managerial skills; possible increased density of development within community if transferable development rights are used
Conversion to Open Space	Purchase of conservation easement with enforceable terms or fee simple title with deed restrictions	Eliminates environmental harms; nonregulatory; responds to land market; likely to significantly shift ecosystem service profile toward regulating and supporting services	Expensive; loss of agricultural land; reduction in food and fiber production; restricts future land market if terms or restrictions are comprehensive and permanent
Conversion to New Urbanism Mixed-Use, Mixed-Density Development	Planned unit development zoning; transferable development rights	Responds to land market; likely to increase open space and associated ecosystem services; promotes affordability of housing stock	Loss of rural and agricultural land; reduction in food and fiber production; loss of opportunity to enhance ecosystem service flows; increased fiscal and infrastructure demands on local community
Conversion to Uniform Low-Density Development	Conventional uniform, low-density residential district zoning	Responds to land market; promotes affordability of housing stock	Loss of rural and agricultural land; reduction in food and fiber production; loss of opportunity to enhance ecosystem service flows; increased fiscal and infrastructure demands on local community

B. Designing PES and TDR Programs for Agricultural Ecosystem Services

Because of their potentially prominent role in encouraging the conversion of conventional farming to multifunctional agricultural land uses, this section focuses on the use of PES and TDR programs built around ecosystem service values. While distinct in many respects, the two approaches share several general design features. If appropriately designed and managed, PES and TDR programs can contribute significantly to state and local policies designed to enhance farm multifunctionality.

1. General Design Features

Promoting the shift from conventional farming to multifunctional farming, particularly when incentives are used to enhance delivery of regulating services to surrounding communities as a primary goal, presents a number of threshold design issues for the managing jurisdiction regardless of the incentive mechanism. First, the baseline expectations of agricultural land uses must be defined so that the managing jurisdiction can identify appropriate incentives and when to apply them.[54] As noted previously, the regulatory baseline for agriculture has been set quite low, meaning farmers have relatively high expectations for when they deserve incentives to push them toward improved performance. An incentive program could be designed, however, to leave a performance gap between the regulatory baseline and the performance levels that trigger eligibility for incentives, providing a pull toward a more realistic baseline before the push of incentives encourages even more improvement. In either approach, the managing jurisdiction must form a clear understanding of existing agricultural practices, the desired practices (e.g., riparian buffers, wetland recharge features, native vegetation open space), and performance levels that trigger incentives.

Next, the goal of enhancing regulating ecosystem services should be based on a known present or expected future demand that can be assigned a value with reasonable geographic and economic specificity. Where are the expected ecosystem services likely to produce benefits, in what form, and how valuable are they in present and expected future land use scenarios within the jurisdiction? These are macro-level questions that require assessment of the potential capacity of existing and restored natural capital on the target agricultural lands, the present and future configuration of land uses in the managing jurisdiction, and the geographic match between the two. Valuation of the ecosystem service flows then can be based on expected avoided costs—for example, the cost savings of avoided flood control or recharge capital expenditures and the avoided costs of flood and drought damages.[55] Comprehensive jurisdiction-wide inventories of natural and built capital will be needed, as will well-conceived future land use planning projections, both of which will necessarily rely heavily on geographic information system (GIS) modeling.

With the demand side analysis in place, a supply-side assessment also is necessary to identify the most effective and efficient incentive distribution. In all likelihood, the agricultural lands identified as having the capacity to enhance ecosystem service values will be held by numerous owners. Yet, the delivery of ecosystem services off the landscape is unlikely to be linear and proportionate, such that securing 10% of the targeted lands will yield 10% of the ecosystem services. Natural capital might provide services that substitute for the services technological capital provides, but natural capital is an ecological resource that behaves according to complex ecosystem properties.

For example, the connections between ecological resources and the delivery of ecosystem service benefits to human populations is known in many contexts to operate at landscape scales, to involve an array of ecological attributes, and to behave in nonlinear relationships over space and time.[56] This raises the difficult question of how precisely to define the proxy for natural capital the program is designed to maintain and for which the incentive is paid or doled out.[57] For example, is it a riparian habitat, or does the type and density of vegetation matter?[58] Moreover, securing 10% of the natural capital capacity of the targeted agricultural lands might produce zero improvement in ecological service flows—it might take half of the

---

54. *See* Even, *supra* note 14, at 197.
55. Assigning these values is a major research effort today among agricultural economists. *See* John M. Antle & Jetse J. Stoorvogel, *Predicting the Supply of Ecosystem Services From Agriculture*, 88 Am. J. Agric. Econ. 1174, 1174 (2006).
56. *See, e.g.*, Edward B. Barbier et al., *Coastal Ecosystem-Based Management With Nonlinear Ecological Functions and Values*, 319 Sci. 321, 321 (2008) (demonstrating that wave attenuation benefits of coastal wetlands do not respond linearly to surface area of the wetlands). The complex relationship between ecological resources and ecosystem service values is explored in Ruhl, Kraft & Lant, *supra* note 10, at 15-35.
57. Jack et al. note that "when marginal benefits from service provision are not constant, more complex incentive schemes are needed to achieve environmental effectiveness." Jack et al., *supra* note 15, at 9466.
58. *See* Francisco Alpizar et al., *Payments for Ecosystem Services: Why Precision and Targeting Matter*, Resources, Spring 2007, at 20, 20-21.

targeted natural capital enhancements before the jurisdiction realizes any measurable ecosystem service benefits. In that event, providing incentives to cover anything less than half the resources, while it would secure ecological resources, would not have the desired payoff in ecosystem service value. Incentive programs thus must be carefully designed to correspond with the ecological properties of the targeted natural capital resources on a landscape level first, from which incentives can be provided to specific parcel owners in a coordinated manner.

Finally, any incentive system designed to enhance targeted ecosystem services must account for the larger physical and political systems in which it is operating. Aligning farmer incentives to provide, say, increased wetland recharge resources or riparian habitat necessarily imposes some trade offs both ecologically and economically. Those trade offs must be recognized and considered. Also, other land use and farm policies must be considered. Will the retention of agricultural land and open space have impacts on the stock of affordable housing in the jurisdiction? Are federal farm policies competing with state and local incentive programs, making it difficult for one or the other set of incentives to gain traction?[59] The point is simply that the program, even assuming it has been thought through with respect to baseline performance expectations and the demand for and supply of services, must be integrated into the larger picture, and its consequences and conflicts fully considered.

2. *Designing Agricultural PES Programs*

Assuming a jurisdiction has a firm handle on the general design issues outlined above, the choice of a PES program presents additional considerations. For example, the Florida Ranchlands Environmental Services Project (FRESP), launched in 2005 by the World Wildlife Fund (WWF) and private and public partners, is a PES program to pay ranchers in an 850,000-acre area of central Florida to enhance delivery of three regulating ecosystem services—water retention, phosphorous load reduction, and wetlands habitat expansion.[60] The target area is located north of Lake Okeechobee, with cattle operations the dominant agricultural land use. A 2004 study that WWF conducted for state agencies[61] concluded that changing water management practices in the ranchlands could be a cost-effective alternative to constructing regional water treatment facilities in moderating water flows and phosphorous loads to Lake Okeechobee.[62] Most significantly, the study demonstrated that "the agencies could buy these services from cattle ranchers at a lower cost than producing the services by building new public works projects."[63] And the ranchers could be better off as well:

> Under the program, ranchers will sell environmental services to agencies of the state and other willing buyers. The public will benefit when services are provided at a lower cost than can be secured from public investment in regional water storage and water treatment facilities. And ranchers, who face low profit margins and fluctuations in the price of beef, will be provided with another source of income, creating a financial incentive for land to remain in ranching rather than be converted to more intensive agriculture and urban development—land uses that will further aggravate water flow, pollution, and habitat problems.[64]

Whereas WWF might normally have targeted payments from its limited funds for wildlife habitat conservation, the idea behind the program is to identify cost savings to local jurisdictions and state agencies that make paying for ecosystem services an efficient expenditure of public resources, with the incidental benefit of increased wildlife habitat conservation. Nevertheless, design issues identified in

---

59. *See* Jack et al., *supra* note 15, at 9467 (discussing the potential for different incentive and subsidy programs to work at cross purposes).
60. *See* Florida Ranchlands Environmental Services Project, http://www.fresp.org. For a thorough review of the FRESP program, see Jacob T. Cremer, *Tractors Versus Bulldozers: Integrating Growth Management and Ecosystem Services to Conserve Agriculture*, 39 ELR 10541 (June 2009). *See also* Sarah Lynch & Leonard Shabman, *The Florida Ranchlands Environmental Services Project: Field Testing a Pay-for-Environmental-Services Program*, Resources, Spring 2007, at 17. Funding for the three-year FRESP pilot study is $2.3 million from USDA, Florida Department of Agriculture, and South Florida Water Management District, plus an additional $2 million from the State of Florida. *See* Memorandum from Deena Reppen, Director, Office of Government and Public Affairs, to South Florida Water Management District Governing Board Members, re FRESP (Aug. 6, 2007) (on file with author).
61. *See* Sarah Lynch et al., Assessing On-Ranch Provision of Water Management Environmental Services (June 2005) (on file with author).
62. *See* Lynch & Shabman, *supra* note 60, at 17.
63. *Id.*
64. *Id.* at 18.

the report led WWF and its partners to test the concept through the FRESP pilot program, involving eight ranches. Chief among these issues was how to document that the payment has produced the benefit, which requires finding the right "trade off between the cost of documentation and the accuracy of measurements that is acceptable to buyers and sellers."[65] For example, the payments for water retention service will compensate ranchers who rehydrate drained wetlands and raise the height of the water table in the ranch soil profile and drainage network. When that is done, remote instruments will monitor data on rainfall, water stages, and flow, allowing a before-and-after comparison.[66] Once this relationship between changed ranching practices and enhanced service flows is identified, measuring on-site changes in ranching practices can provide the pricing proxy for ecosystem service enhancement and the documentation to support buyer confidence. In short, any PES program must devise a way for the buyer and seller to know that payment X yields service value Y, and that this is a rational economic move for both parties.[67] Building on this experience, the FRESP program ultimately led to solicitations in 2011 by the South Florida Water Management District for rancher payment requests based on different water management proposals.[68]

3. Designing Agricultural TDRs for Ecosystem Service Enhancement

Relatively new to agricultural settings[69] and showing only limited success thus far,[70] farmland TDR programs are for the most part constructed around an "old agriculture" model—one based largely on preserving farm "character"[71] and which neither recognizes nor promotes farmland multifunctionality.[72] Jesse Richardson's analysis of agricultural TDR programs[73] identifies several design challenges that have grown out of this experience. First, and most obviously, the program depends on supply of and demand for the TDRs. Neither is as easy to make happen as it seems. Agricultural property owners need to view the TDR as more attractive than either their conventional agriculture or "last harvest" option of selling their land for development, and there must be "communities willing to accept designation as a receiving area for higher-density development."[74] Even when the supply and demand communities are identified as willing to engage in the transaction, the balance between the two is delicate. By contrast to PES, the "market" for TDRs is a regulatory construct, not a true market, and thus depends on some finely tuned government intervention to make demand in the receiving area strong and the supply of the TDRs just right to keep them valuable in that market:

> If too many development rights are created or if the incentives in the receiving areas are insufficient, the price of the development rights will be too low.... If not enough development rights are distributed or if the incentives in the receiving areas are too great, the price of development rights will be very high.... The number of development rights and the incentives for both sides must be "just right."[75]

But getting that balance right is just the beginning. Housing markets fluctuate and cross local political boundaries. How will a local TDR program coordinate with other local governments in the region to keep

---

65. *Id.*
66. *See id.* at 18-19. The FRESP pilot study was initiated in 2006. However, severe drought in 2007 limited the collection of data and no conclusions have been drawn yet as to these critical relationships. Interview with Sarah Lynch, Program Director, WWF FRESP (Feb. 5, 2008). For a general discussion of the challenges of performance monitoring, quality control, and information costs in the context of agricultural ecosystem service markets, see Ribaudo et al., *supra* note 34, at 2088-90.
67. This design need, as well as other economic aspects of PES design, is covered in more detail in Antle, *supra* note 16, at 13-17.
68. *See* Sarah Lynch & Leonard Shabman, *Designing a Payment for Environmental Services Program for the Northern Everglades*, NAT'L WETLANDS NEWSL., July-Aug. 2011, at 12.
69. *See* JUERGENSMEYER & ROBERTS, *supra* note 9, §13.11, at 860 ("[T]he application of the transferable development approach to agricultural land use preservation is of relatively recent origin").
70. *See* LINCOLN INST. OF LAND POLICY, TRANSFER OF DEVELOPMENT RIGHTS FOR BALANCED DEVELOPMENT 3-5 (May 1998) (suggesting that, outside of a few success stories, "the overall picture is ambiguous"), *available at* http://pb.state.ny.us/pbc/conference_9805_transferdevelopment.pdf; Jesse J. Richardson Jr., *Goldilocks, the Three Bears and Transfer of Development Rights*, AGRIC. L. UPDATE, Dec. 2006, at 4 (noting that only eight of 23 farmland preservation TDR programs active in 2004 protected more than 1,000 acres).
71. *See* Rick Pruetz & Erica Pruetz, *Transfer of Development Rights Turns 40*, 59 PLAN. & DEV. L. 3, 5 (2007).
72. *See* Richardson, *supra* note 70, at 5-6.
73. *See id.*
74. *Id.* at 4-5.
75. *Id.* at 6.

a handle on those trends?[76] Farmers who sell TDRs cannot sell their land to developers, but might not necessarily stay in active farming. "No one has investigated whether these programs actually promote and aid farm production."[77] And if the TDR program is aimed at rewarding farmers for conserving environmental values, how are those values calculated in sending areas and then converted into density development rights in receiving areas? How many apples get you so many oranges? Based on these challenges, Richardson concludes that "the theoretical beauty of TDR programs lures many to the tool. However, the complexity makes implementation difficult."[78]

Although Richardson's assessment of agricultural TDR programs is a sobering reminder that the simple elegance of TDR theory ultimately gives way to the utter complexity of their implementation, some states have forged ahead with what could be promising structural advances, particularly with respect to enhancing the delivery of ecosystem services. For example, in 2001, the Florida Legislature enacted the Rural Land Stewardship Act (RLSA),[79] which allows a county to designate all or portions of agricultural and rural lands within its jurisdiction as a rural land stewardship (RLS) area.[80] Within an RLS area, the local government applies planning and economic incentives to encourage the implementation of innovative and flexible planning and development strategies and creative land use planning techniques, with TDRs as the primary policy mechanism.

Like any TDR-based program, RLS areas contain "stewardship sending areas" within which natural resources and rural land values are conserved, and "receiving areas" within which development is authorized to occur, with the TDR linking the two areas.[81] A landowner who conserves rural and natural resource values in the sending area accrues "stewardship credits" entitling the landowner to TDRs, known in RLSA parlance as transferable rural land use credits, allowing greater development densities in receiving areas than would apply under the otherwise applicable zoning rules.[82] These credits

> may be assigned at different ratios of credits per acre according to the natural resource or other beneficial use characteristics of the land and according to the land use remaining following the transfer of credits, with the highest number of credits per acre assigned to the most environmentally valuable lands or, in locations where the retention of open space and agricultural land is a priority, to such lands.[83]

RLSA stands apart from most TDR programs in two respects. First, it is entirely voluntary on the credit generating side. Most TDR programs, particularly those focused on historic and environmental preservation, regulate activities in the sending area and provide TDRs as the purported quid pro quo.[84] Understandably, this leads to resentment among the landowners regulated in the sending area who receive what they may believe is inadequate value in the TDR to compensate for the lost development potential, even so far as to frequently lead to takings claims.[85] By contrast, RLSA uses TDRs purely as an incentive to alter land use practices and deter conversion to suburban development in the sending area.

Second, RLSA strikes a chord very close to the farm multifunctionality theme. The statute specifies six goals that must be served by creation and operation of a RLS area: (1) restoration and maintenance of the economic value of rural land; (2) control of urban sprawl; (3) identification and protection of ecosystems, habitats, and natural resources; (4) promotion of rural economic activity; (5) maintenance of the viability of the state's agricultural economy; and (6) protection of property rights in rural areas of the state.[86] These goals evidence an advance in thinking beyond prior practice in agricultural TDRs. On its face at least, RLSA thus is more than a farmland status quo or cultural amenity preservation program—it focuses on

---
76. *See id.* at 4.
77. *Id.* at 5.
78. *Id.* at 6.
79. Fla. Stat. §163.3248. RLSA originally appeared in another provision of the Florida statutes, but amendments in 2011 moved it to its current location. *See* Fla. H.B. 7207 (2011). For background on the legislative purpose and amendment history of the original version of RLSA, see Dep't of Urban & Reg'l Planning, Fla. State Univ., Rural Land Stewardship 2007 Annual Report 1-4 (2007).
80. For a comprehensive review of the RLSA program, see Daniel S. Stringer, *Rural Land Stewardship: Reinventing Development From the Grassroots With a Localized, Long-Term, Incentive-Based Program*, 17 Penn St. Envtl. L. Rev. 225 (2009).
81. Fla. Stat. §163.3248(8)(b).
82. *Id.* §163.3248(7).
83. *Id.* §163.3248(8)(k).
84. *See* Juergensmeyer et al., *supra* note 17, at 448-55 (surveying several prominent programs).
85. *See* Andrew J. Miller, *Transferable Development Rights in the Constitutional Landscape: Has* Penn Central *Failed to Weather the Storm?*, 39 Nat. Resources J. 459, 459 (1999).
86. Fla. Stat. §163.3177(11)(d)(2).

providing incentives tied to the economic value of rural land and natural resources integrated within working landscapes.[87]

Nevertheless, although the credit-generating side of RLSA is nonregulatory and innovatively ties in the concept of economic value of rural lands, the credit consumption side has the look and feel of conventional TDR programs in that it relies on a default rule for development density and units that can be exceeded through purchase of credits. Hence, in addition to the agriculture and ecosystem-service design issues mentioned already generally and for PES programs, RLSA leaves many land use policy and implementation questions unanswered. For example: What is the methodology for identifying and designing development in receiving areas? How is demand for RLSA land use credits maintained in the receiving area? What is it that land use credits are "buying" in the way of number of units, density of development, mixed uses, and so on?[88]

Although RLSA makes no specific mention of farm multifunctionality or farm provision of ecosystem services, the statute's multifactored set of goals clearly opened the door to organizing RLS areas and TDRs around those principles. For example, integrating the capacity for ecosystem service production into the calculus for stewardship credit would support and reward the "restoration and maintenance of the economic value of rural land" and contribute to the "promotion of rural economic activity." Providing farms a means of capitalizing on their production of regulating services would contribute to the "maintenance of the viability of Florida's agricultural economy." Providing incentives to conserve the agricultural land capital producing those services would "support the identification and protection of ecosystems, habitats, and natural resources," and the consequence of doing all of the foregoing could only contribute to the "control of urban sprawl" and the "protection of property rights in rural areas of the state." The potential fit between RLSA and the farm multifunctionality movement thus seems as tight as a glove.

Moreover, by linking the value of the TDR to ecosystem service production, RLSA could test the farm stewardship claim—the better the stewarding for the greater community, the more value in the TDR. This approach would thus make the trade off between provisioning and regulating services explicit and transparent. To the extent the TDR contains an increment of value clearly attributable to provision of regulating services, farmers in RLS areas can evaluate the consequences of emphasizing continued commodity production over conservation of agricultural land capital capable of supplying regulating services to surrounding communities. The scale of the RLSA program, if so configured, would operate from local to national, as TDR values could reflect services such as local groundwater recharge to global carbon sequestration.

RLSA potentially was headed in this direction. The original version of the statute required the Florida Department of Community Affairs, a now-defunct state agency, to promulgate rules local governments had to follow in their RLSA initiatives. As part of its RLSA rulemaking process, the DCA commissioned the Florida Planning and Development Laboratory at Florida State University (FSU Laboratory) to prepare a RLSA program evaluation study.[89] The final report from that study recognized the importance of integrating ecosystem services in the RLSA TDR calculus. In particular, two of RLSA's goals—restoration and maintenance of the economic value of rural land and the identification and protection of ecosystems, habitats, and natural resources—invite attention to ecosystem service values that farms can provide. The FSU Laboratory's final report[90] thus identified "captur[ing] the value of environmental services" as one of the "core principles" of successful agriculture TDR programs:

> Successful programs are those that account not only for the aesthetic aspects of agricultural land but also for environmental services agricultural lands provide. These would include the provisioning of non-land resources, like water, and the land's participation in environmental regulation processes (like water purification) that would have to be otherwise acquired in the marketplace.[91]

---

87. For general policy discussion of valuing rural land ecosystem service values, see Adam Davis, *Ecosystem Services and the Value of Land*, 20 DUKE ENVTL. L. & POL'Y F. 339 (2010).
88. A DCA staff brainstorming session the author attended in the spring of 2007 developed a long list of such issues, which focus primarily on land use in the receiving area and thus are outside the scope of this work. *See also* NATHANIEL REED, CHAIRMAN EMERITUS, 1000 FRIENDS OF FLORIDA, WORKING TO SUSTAIN FLORIDA'S RURAL AND NATURAL LANDS: A CALL TO ACTION 10-11 (2007); Letter from Charles Pattison et al. to the Hon. Thomas Pelham, Secretary, Florida Department of Community Affairs 2-5 (June 6, 2007).
89. The author served on the panel of experts the FSU Laboratory formed to provide input and to critique early drafts of the report.
90. *See* TIM CHAPIN & HARRISON HIGGINS, RURAL LAND STEWARDSHIP AREAS (RLSA) PROGRAM EVALUATION FRAMEWORK (2007) (on file with author).
91. *Id.* at 13.

To ensure that these values are "captured" in a way that properly aligns incentives toward farm multifunctionality, the FSU Laboratory suggested several program evaluation indicators and metrics for RLS areas that focus on ecosystem service values. One such indicator appears in connection with the goal of restoration and maintenance of the economic value of rural land:

Indicator 1.3. Environmental service values delivered by rural lands in sending areas are reflected in the RLSA system.

Metric 1.3.1. Stewardship credits reflect the value of conserved environmental services that are bought and sold outside of the RLS program.

Metric 1.3.2. The stewardship credit system provides market incentives to maintain and enhance capacity of rural lands in sending areas to provide environmental services and to monitor the provision of those services.

Metric 1.3.3. Economic values of rural lands in sending areas are enhanced by the use of environmental service values in RLSA system.[92]

The other indicator appears under the goal of identifying and protecting ecosystems, habitats, and natural resources:

Indicator 3.5. The capacity for rural lands in stewardship sending areas to provide, maintain, and enhance environmental services is enhanced, as measured by

Metric 3.5.1. Delivery and value of environmental services within the potential and approved stewardship sending areas.[93]

The FSU Laboratory report did not go further in outlining how to design RLSA implementation to accomplish these goals. In particular, unlike the FRESP PES program, the RLSA TDR program involves two related pricing decisions, the relationship between which is not a market-based outcome. As structured, RLSA can be thought of as an accounting mechanism that correlates the public benefits of enhanced ecosystem services in sending areas with the public impacts of increased density in receiving areas. On the one hand, therefore, like a PES program, the RLSA program must calibrate the award of credits to the value of the ecosystem services being delivered through altered land uses. In addition, however, RLSA implementation requires a method for controlling the value of the TDR credits in the receiving areas, because that provides the financial basis for the incentive in sending areas to change land use. But it is the value of enhanced land development opportunities, not the enhanced ecosystem services, that drives TDR values in the receiving areas. Balancing these two markets when there is no market-based way of equating a development opportunity in one market with an ecosystem service value in the other market presents the difficult "apples for oranges" conversion calibration for RLSA.

It may be necessary for local jurisdictions to develop proxies for keeping the exchange between the two markets in synch. For example, based on the macro-analysis of natural capital potential in the jurisdiction and the present and expected jurisdictional land uses, it may be possible to define TDR premiums assigned to different sets of agricultural land practices and conservation measures that enhance ecosystem service flows above a defined baseline. A conservation easement might define the standard credit, and restoration of riparian habitat might earn a set premium. For ongoing agricultural land uses, preservation of the status quo might define the standard credit, and sets of management and restoration practices—the silver, gold, and platinum levels, so to speak—might be used to define levels of premiums. For lands moved into conservation status, premiums above the reward for simple open space could be based on the qualitatively described connection between measurable geographic (e.g., acreage of wetlands), service (e.g., flood protection), demographic (e.g., benefited population), and economic (e.g., replacement value to the benefitted built environment) factors, even if precise quantification is not possible. While these or other proxies might not precisely calibrate ecosystem service benefits with development density impacts, RLSA provides an accounting mechanism that is more transparent and planned than a trade off negotiated between landowners and local governments as part of a zoning decision.

---

92. *Id.* at 15.
93. *Id.* at 19.

Leaving ecosystem services out of a program like RLSA will render it an "old agriculture" program that continues to drive farming toward the production of commodities and rural character, with the provision of regulating services to surrounding communities an accidental and incidental benefit that farmers will view, if at all, as a positive externality for which they receive nothing in return. On the other hand, the design issues identified above suggest it will be difficult to fashion general formulae for integrating ecosystem services into specific local settings. The recommendations of the FSU Laboratory provided useful guidelines for doing so. Alas, the DCA rules did not incorporate the recommendations in this regard, and in the 2011 RLSA amendments the legislature nullified the rules altogether, leaving it to local governments to design the TDR programs. On the other hand, the amendments integrated a new ecosystem services provision specifically allowing that

> Owners of land within rural land stewardship sending areas should be provided other incentives, in addition to the use or conveyance of stewardship credits, to enter into rural land stewardship agreements, pursuant to existing law and rules adopted thereto, with state agencies, water management districts, the Fish and Wildlife Conservation Commission, and local governments to achieve mutually agreed upon objectives. Such incentives may include, but are not limited to, the following:
>
> . . .
>
> (d) Compensation for the achievement of specified land management activities of public benefit, including, but not limited to, facility siting and corridors, recreational leases, water conservation and storage, water reuse, wastewater recycling, water supply and water resource development, nutrient reduction, environmental restoration and mitigation, public recreation, listed species protection and recovery, and wildlife corridor management and enhancement.[94]

RLSA thus uses a PES approach as an added incentive for entering a TDR program, a marriage which may prove to deliver the best of both approaches.

4. Matching PES and TDR Programs With Context

One unmistakable theme from the preceding sections is that, by comparison, PES programs are simple and TDR programs are complex. A PES program in essence is simply a market exchange bringing willing buyers and sellers together, providing the information and market monitoring and enforcement both parties need to enter confidently into transactions.

Of course, that is precisely the trick with ecosystem services—finding buyers and sellers who can exchange in a market. The FRESP had that good fortune. Motivated by the strong national and state desire to improve water quality in the Everglades and to manage water resources better in central Florida, FRESP seized on a golden opportunity to match demand for and supply of ecosystem services. In rural agricultural areas distant from populated service demand markets, however, PES programs will often consist of sellers without buyers. Even where urbanization is expected over time, a PES program may lack public and private buyers to secure ecosystem services today for the urban populations of the future.

By contrast, TDR programs require no public expenditure to generate credits, though by all means they depend on demand for urban development to make the credits valuable. By appropriately placing receiving areas closer to the urbanizing fringe, however, TDR programs such as RLSA may be able to leverage demand for development into demand for credits, thereby promoting enhancement of ecosystem service flows sooner than would be the case under a PES program. Moreover, because a RLSA program is tied closely to a local jurisdiction's future land use plan, the jurisdiction is more likely to appreciate the long-term need for ecosystem service green infrastructure and view the TDR trade off between development and conservation as a worthy investment. In other words, allowing more development to secure enhanced ecosystem service flows may be a better option for a local jurisdiction than less development with degraded ecosystem services flows.

The point is that PES and TDR programs are different, and as such may be suited to different contexts. Table 14.3 summarizes some of those differences.

---

94. FLA. STAT. §163.3248(9).

Table 14.3
Comparison of PES and TDR Programs

	PES	TDRs
Advantages	Provides opportunity for public capital infrastructure cost savings; based purely on market incentives; can be applied at large scales where regional ecosystem services are valued; simple by comparison.	No expenditure of public resources required for creation of credits; provides opportunity for public capital infrastructure cost savings; receiving area can be positioned near market demand for development while sending area can be rural; value of credits can be managed through regulation of receiving areas and kept sufficiently high to deter conversion to development.
Disadvantages	Requires expenditure from public (or private) resources; less likely to be viable in rural areas where no immediate market exists for the services; payments may not be sufficient to deter conversion to development scenarios; requires new managerial skills.	Requires regulation in receiving area; results in increased development and costs associated with resulting demand on public infrastructure; depends on active development market demand in receiving area; increased development intensity in receiving areas could diminish ecosystem services from those areas; requires new managerial skills.
Major Design Issues	Deciding the baseline expected performance levels of agricultural land uses; identifying the economic values of the enhanced ecosystem services and their pathways of delivery; downscaling macro-level to parcel level; calibrating altered land use practices with enhanced ecosystem service flows; documenting the altered management practices.	All of the PES design issues plus: managing supply and demand equilibrium between the sending and receiving markets; setting conversion rates between enhanced ecosystem service values and development rights in receiving areas; meeting fiscal and infrastructure demands imposed by the increased development rights in receiving areas.

Conclusion

Robert Wolcott has aptly described the threshold at which American farm policy finds itself:

> Agriculture occupies the high ground of comparative advantage in supplying socially demanded, low-cost ecosystem services. Agriculture is accustomed to publicly funded incentives, and private markets to signal supply value sought by society whether corn, soybeans, or wildlife habitat. The level and composition of the demand side is increasingly evident, though in flux as well.... The prospect of broad-scale compensation of agricultural producers for supplying ecosystem services is real.[95]

The question is whether federal, state, and local farm policy will seize this opportunity. Can the farm bill ever break out of its commodity support/land retirement mold? Will federal, state, and local policy converge on the vision of farms as multifunctional production units? Federal policy can set the stage by moving the green subsidy program toward a demand-driven multifunctionality model. And state and local policy, through PES, TDR, and similar techniques, can have a significant role to play in moving toward that vision.

---

95. Robert M. Wolcott, *Prospects for Ecosystem Services in the Future Agricultural Economy: Reflections of a Policy Hand*, 88 AM. J. AGRIC. ECON. 1181, 1182-83 (2006).

Federal farm policy should encourage and support such state and local initiatives, as it is in the national interest to maintain and enhance the natural capital that agricultural lands contain and can deliver locally across the landscape. Congress could take several measures:

- Fund research to determine how to calibrate farm practices with ecosystem service delivery at local scales, as USDA has done with FRESP.
- Develop national standards for quantifying ecosystem service values associated with agricultural lands, including the development of proxies that can inexpensively be measured to estimate service delivery potential.
- Give preference in federal green subsidy payments programs for farms that would actually deliver ecosystem service values to identifiable local and regional populations.
- Fund pilot and permanent demand-based state and local farm multifunctionality programs such as FRESP.

The 2008 Farm Bill took a modest step in this direction. Section 2709 of the Food, Conservation, and Energy Act requires the Department of Agriculture to "establish technical guidelines that outline science-based methods to measure the environmental services benefits from conservation and land management activities in order to facilitate the participation of farmers, ranchers, and forest landowners in emerging environmental services markets," and to establish guidelines to develop a procedure to measure environmental services benefits, a protocol to report environmental services benefits, and a registry to collect, record, and maintain the benefits measured.[96] To implement this innovative provision, the Forest Service established the Office of Ecosystem Services and Markets, and the multiagency Conservation and Land Management Environmental Services Board was established in December 2008 to assist the secretary of agriculture in adopting the technical guidelines to assess ecosystem services provided by conservation and land management activities.[97] The board's guidelines were intended to focus on scientifically rigorous and economically sound methods for quantifying carbon, air and water quality, wetlands, and endangered species benefits in an effort to facilitate the participation of farmers, ranchers, and forest landowners in emerging ecosystem markets.[98] Unfortunately, as of this writing the agency's initiative, now known as the Office of Environmental Markets, has stalled, producing no concrete policies or guidance for implementing the program.[99]

Conventional agriculture thus remains at a crossroads, facing pressure to improve its environmental performance profile at the same time it is facing pressure to produce more food, fiber, and fuel commodities on the one hand or to give way to urban development on the other. In the best of all worlds, markets would fully recognize the value of ecosystem service flows and farms could make appropriate balances between providing services, commodities, or land development opportunities. But hoping for this seems quixotic, as markets have proven time and again to be poor at valuing the multifunctional capacity of ecological landscapes.

Understanding the multifunctional capacity of agricultural lands, however, provides insight into how to promote alternatives that blend enhanced environmental performance with better development planning. PES programs such as FRESP, and TDR programs such as RLSA, could become model farm policy programs in this respect, or they could recede into the ways of "old agriculture." Whatever their future, however, it is promising to find regulatory authorities beginning to act strategically to influence the future scenarios of existing agricultural land uses, despite the substantial design challenges these techniques face.

Of course, it is important to stay focused on what the goal of agricultural multifunctionality is. We *do* want farms effectively to manage provisioning services to provide food, fiber, and other commodities. And we should *not* force farms unfairly to bear the cost of supplying regulating services to society. We pay

---

96. Food, Conservation & Energy Act of 2008, H.R. 2419, 110th Cong. §2709.
97. USDA, Office of Environmental Markets, *Ecosystems and the Farm Bill*, http://www.fs.fed.us/ecosystemservices/OEM/index.shtml/index.shtml (last visited Nov. 28, 2012).
98. *See* Conservation and Land Management Environmental Services Board Charter 1-2 (Dec. 5, 2008), *available at* http://www.fs.fed.us/ecosystemservices/pdf/farmbill/ESB_Charter.pdf.
99. For developments in this program, see U.S. Dep't of Agric., Office of Environmental Markets, *supra* note 97.

farmers for corn; how much should we also pay them for supplying carbon sequestration and groundwater recharge? The answer must start with identifying the drivers at the interface between agriculture and ecosystem services and developing a model of how these drivers operate. How do farm subsidy programs influence farm behavior toward ecosystem services? How do the upstream and downstream food and fiber industries affect farm behavior toward ecosystem services? If we were to change these or other conditions, how would farmers respond with respect to ecosystem services? And which regulating ecosystem services do we wish to promote?

As we understand more about how and why farms manage ecosystem services in particular ways, we must then widen the lens to consider the trade offs associated with different policy approaches. How would encouraging farms to shift toward greater production of regulating services, however that might be accomplished, affect farm income, food prices, and land costs? Who would benefit, and by how much, where, and when? Would moving a significant portion of existing agricultural lands into, say, carbon sequestration, simply prompt conversion of undisturbed lands into farming to replace lost food supply? Would promoting a particular regulating service, such as carbon sequestration, have a trade off effect with other regulating services, such as groundwater recharge? How will other services that farms might provide, such as providing cultural and historical context for surrounding communities, be enhanced or degraded by moving to greater farm multifunctionality?

Once these trade offs are better understood, the difficulties of transitioning to new policy regimes can be identified. The costs and benefits of new policies are almost never evenly distributed. For example, are global, national, regional, or local regulating services to be favored, and which interests are affected positively and negatively by that choice? What new skill sets will farmers need to acquire to take advantage of the new policies, and how much will gaining them cost? Will agricultural communities prosper with increased farm multifunctionality? Those who stand to lose under new policy regimes are likely to oppose them unless their interests are appropriately accounted for in the transition. After decades of habituating farms (and farm communities) to subsidies designed around provisioning services, it may be unfair and unwise to shift to new policies without addressing the impact to those interests most affected. Should those farms be exempt from new programs, or compensated for losses suffered, or simply forced to play under the new rules?

Ultimately, if promoting greater production of regulating services is the goal for agricultural policy over the next decade, we must choose the instruments and institutions to make it happen. As with almost all else in agricultural policy, political expediency will point toward incentive programs administered through federal agencies. Indeed, putting aside the politically charged question of what baseline of performance to demand from farms, a strong case can be made for incentive-based approaches, as it is appropriate for farms to receive at least some compensation for satisfying public demand for economically valuable regulating services.

Although federal policy can support such programs broadly, federal agencies may be poorly equipped to administer the incentives for all relevant services. Ecosystem services are, after all, benefits to human populations, meaning they satisfy demand at different scales. Some services relevant at national and global scales, such as carbon sequestration, seem well-suited for incorporation into federal incentive programs designed to influence land retirement or crop selection. By contrast, ecosystem services such as groundwater recharge, water quality control, and sediment capture are most valuable to local populations. Farmers should be paid in such cases to provide local services, but only based on local demand, meaning that locally administered PES and TDR programs are more likely to calibrate compensation for local services efficiently. The point is to ensure that incentives for ecosystem services such as PES are driven by demand, not by supply. In this sense, policies designed to promote farm production of regulating services may give multifunctionality a renewed purpose and goal at local scales, connecting farms to their urban and suburban surroundings in ways that enable all interests to recognize the advantages of maintaining working agricultural landscapes.

Chapter 15
Achieving a Sustainable Farm Bill
William S. Eubanks II

When advocating a path forward to modernize domestic farm and food policy, an organization or individual must decide whether to press for a systematic overhaul or incremental change. Indeed, there is no question that a fundamental shift in the structure of U.S. agricultural policy could enable our nation to achieve a sustainable, environmentally sound, and nutritious food system.[1] Accordingly, there would be substantial value in advocating for such an immediate, fundamental farm bill reform because the process of making that argument provides a glimpse of that objective, and it is important that the ultimate goal not be lost.

However, the enormous political and financial power of agribusiness might be better challenged through incremental reforms targeting specific farm bill programs on issues that have the support of large segments of the American public. In effect, this approach allows the millions of interested Americans, and the organizations that advocate on their behalf, to slowly and strategically chip away at the outdated, sometimes illogical, components of U.S. agricultural policy. These small but critical reforms would breathe new life into the farm bill. Although reform would take longer under such an incremental approach, the goal is still the same; indeed, these vitally significant and targeted reforms along the way will allow for improvement at many levels even if fundamental farm bill reform never comes to pass.

This chapter first presents the argument for a major and urgent shift in U.S. agriculture and food policies to achieve sustainability. It then provides a counterview, arguing instead for several narrower and more gradual reforms to achieve many of the same goals, highlighting examples of targeted challenges and ways to strengthen support for existing programs that sorely need the public's backing to achieve a healthier food system.

A. Seeking a Truly "Green" Revolution: Large-Scale Reform for Widespread Problems

While arguably somewhat overstated, the author James H. Kunstler has starkly drawn this picture of American agriculture:

> We have to produce food differently. The [Archer Daniels Midland]/Cargill model of industrial agribusiness is heading toward its Waterloo. As oil and gas deplete, we will be left with sterile soils and farming organized at an unworkable scale. Many lives will depend on our ability to fix this. Farming will soon return much closer to the center of American economic life. It will necessarily have to be done more locally, at a smaller and finer scale, and will require more human labor.[2]

This chapter is, with permission, an updated and adapted version of an article previously published as William S. Eubanks II, *A Rotten System: Subsidizing Environmental Degradation and Poor Public Health With Our Nation's Tax Dollars*, 28 Stan. Envtl. L.J. 213 (2009). Copyright © 2009 by the Board of Trustees of the Leland Stanford Junior University.

1. Additional arguments in favor of fundamental changes to farm bill policy are set forth in William S. Eubanks II, *A Rotten System: Subsidizing Environmental Degradation and Poor Public Health With Our Nation's Tax Dollars*, 28 Stan. Envtl. L.J. 213 (2009); William S. Eubanks II, *The Sustainable Farm Bill: A Proposal for Permanent Environmental Change*, 39 ELR 10493 (June 2009); William S. Eubanks II, *Paying the Farm Bill: How One Statute Has Radically Degraded the Natural Environment and How a Newfound Emphasis on Sustainability Is the Key to Reviving the Ecosystem*, 27 Envtl. F. 56 (2010).
2. James H. Kunstler, *Ten Ways to Prepare for a Post-Oil Society*, Canadian Nat'l Newspaper, Jan. 12, 2008, at http://www.agoracosmopolitan.com/home/Frontpage/2008/01/12/02127.html.

There is at least some truth to this depiction. Despite what many scientists, farmers, and ranchers think to be the best available agricultural practices for environmental protection and a nutritious food supply,[3] U.S. agriculture and food policies under the farm bill have generally strayed from these practices to placate the agribusiness and food-processing industries. The average commodity crop farmer now produces enough corn and soybeans to feed hundreds, or even thousands, of Americans each year from food items processed from his crops.[4] However, that same commodity farmer sends no healthy fruits and vegetables to the market and amazingly can no longer feed his own family from his massive fields because of inflexible planting rules and encouragement of monocrop production through various types of farm bill subsidies.[5] Heavy corn-producing states such as Iowa now import, on average, *more than 80% of the food consumed by the residents of those states.*[6] The U.S. food production system under the farm bill, which should ideally encourage production of healthy food, is instead creating a plethora of "food deserts"—even in rural areas where the local economy is dependent on farming—composed of locations where food is difficult to come by and much of the food that is available consists of processed commodities, saturated fats, and little to no nutrition.[7] Moreover, as Part II of this book highlights, the environmental impacts of the current industrial model are significant. And that model is founded on the farm bill—the programs authorized, and the agriculture and food system encouraged by the U.S. Congress in that omnibus legislation every five years.

The successive farm bills have promoted larger and larger farms and the inherent adverse consequences of monocrop production and market consolidation. Scholars have noted how the stability of the Soviet Union "foundered precisely on the issue of food" as it tried to force a transition to industrial agriculture.[8] That policy contributed to the Soviet collapse because the program "sacrificed millions of small farms and farmers," but the system of industrial agriculture "never managed to do what a food system has to do: feed the nation."[9] Indeed, with each passing farm bill, it can be argued that the domestic farming and food system is gradually moving toward its own failure to accomplish the fundamental objective of feeding the nation, at least in terms of providing nutritious food grown in an ecologically resilient manner that seeks to preserve our natural resource for the long term.[10]

One promising change that could mitigate the primary problems of industrial commodity crop agriculture in the United States would be incentivizing sustainable agriculture to assist in normalizing the market so that the gap in supermarket price could close somewhat between the handful of heavily subsidized commodities and all other foods that receive little or no financial incentives and thus appear more expensive than would otherwise be the case in a free market. Although a truly free market without subsidies would be ideal,[11] such as the system currently operating in New Zealand,[12] the vast subsidy infrastructure currently embedded in the farm bill would be difficult to pull out from under the feet of farmers that depend on those subsidies to survive, and upon which farmers benefiting from that system have made long-term

---

3. As described later in this chapter, some of the many farming practices that best preserve and enhance soil, water, and other natural resources are no-till farming, cover cropping, crop rotation, residue mulching, elimination of most or all agrochemical fertilizers, significant reduction per acre of water usage, nitrogen fixing through on-farm manure use, measurable energy reduction per acre farmed, greater use of integrated pest management, contour farming, and increased direct sales from farm to consumer or intermediated sources to reduce transportation. Indeed, these and many other practices are incentivized through an existing farm bill program, discussed later in the chapter, called the Conservation Stewardship Program, which lays out a more detailed list of practices that have been determined through consensus as key farming practices that maximize environmental benefits. *See* Natural Resources Conservation Service, CSP (Dec. 9, 2011), *available at* http://www.nrcs.usda.gov/Internet/FSE_DOCUMENTS/stelprdb1046211.pdf.
4. Michael Pollan, The Omnivore's Dilemma: A Natural History of Four Meals 34-35 (2006).
5. *Id.*
6. *Id.*
7. *See generally id.*
8. *Id.* at 256; Robert W. Campbell, The Soviet-Type Economics: Performance and Evolution 65 (1974) (calling the inefficient Soviet industrial agricultural system "unreliable, irrational, wasteful, unprogressive—almost any pejorative adjective one can call to mind would be appropriate"); Campbell R. McConnell, Economics: Principles, Problems and Policies 900 (1975) (arguing that the Soviet industrial agricultural system was "something of a monument to inefficiency").
9. Pollan, The Omnivore's Dilemma, *supra* note 4, at 256.
10. *See id.*, at 256-57 (discussing the benefits of a more local agricultural system as compared to the current industrial agricultural system).
11. *See, e.g.,* Eliot Coleman, *Four Season Farm*, http://www.fourseasonfarm.com/ (last visited Nov. 9, 2012). Many scholars such as Eliot Coleman believe that any nationalized system of agriculture—conventional or organic—is inefficient. Thus, these critics advocate for a localized agricultural system with no national standards, subsidies, or framework for regulating agriculture. *Id.*
12. Daniel Imhoff, Food Fight: The Citizen's Guide to a Food and Farm Bill 80-83 (2007). New Zealand is one of the few nations that has eliminated agricultural subsidies altogether. In 1984, New Zealand eliminated all subsidies for farming and the results have been very positive. In fact, New Zealand has seen "an energizing transformation of the food and farming sectors . . . [and] [p]rofitability, innovation, and agricultural diversity have returned to farming." *Id.* Both farm output and farm income are on the rise in New Zealand. *Id.*

machinery and other capitalized purchases based on the assumption that such subsidies would continue to exist.

Therefore, instead of immediately eliminating the farm bill subsidies on which many farms now rely for survival, Congress should instead shift a substantial portion of these subsidies, in phases, to farmers implementing sustainable agricultural methods. As detailed in Part I, past and current conservation programs often had a major flaw: they targeted only large commodity crop growers. A more workable policy would be to offer a predetermined share of subsidy incentives to *all* farmers based on their farming practices, irrespective of crops cultivated or farm size. This would create a more just system than the current subsidy framework that excludes 60% of American farmers from any subsidies whatsoever.[13]

Farmers who never see farm bill subsidies in our current system are typically those who grow most of the nation's fruits, vegetables, and nuts, which are called "specialty crops" in the farm bill, but yet are critical for good health, and the farmers who grow crops using environmentally sustainable agricultural methods—although it should be noted that the two sets of farmers are not necessarily the same. Growers in California provide a vivid example of the current failures of the farm bill's subsidy program to reward farmers for growing healthy food for our nation. With nearly 30,000 farms, and nearly $38 billion in annual on-farm revenues, California is the leading state in annual agricultural sales.[14] Despite this, more than 90% of California's farmers receive no agricultural subsidies.[15] (Of the few Californian farmers that do receive farm bill subsidies, most are cotton and rice farmers.[16]) Yet, these subsidy-neglected California farmers are invaluable to our nation's agricultural system because the state contributes more than 16% of the total U.S. agricultural market value and nearly half of all fruits, nuts, and vegetables.[17] By ignoring these farmers and precluding them from receiving farm bill subsidies, Congress is prioritizing monocultures of corn, soybean, wheat, cotton, and rice at the expense of sound agricultural, nutritional, and environmental practices.[18]

Sustainable agriculture, however, can serve as a first step in changing these policies for the better. What is "sustainable agriculture"? According to the scholar James Horne, sustainable agriculture "encompasses a variety of philosophies and farm techniques . . . [that] are low chemical, resource and energy conserving, and resource efficient."[19] Ironically (because it did little to encourage such agriculture), the 1990 farm bill defined sustainable agriculture as

> an integrated system of plant and animal production practices having a site-specific application that will, over the long term, satisfy human food and fiber needs; enhance environmental quality and the natural resource base upon which the agricultural economy depends; make the most efficient use of nonrenewable resources and on-farm/ranch resources; integrate, where appropriate, natural biological cycles and controls; sustain the economic viability of farm/ranch operations; and enhance the quality of life for farmers/ranchers and society as a whole.[20]

As most agricultural experts note, it is important to understand that "[s]ustainable agriculture does not mandate a specific set of farming practices."[21] Rather, sustainable practices vary from place to place depending on the ecosystem, climate, and other factors, but "[t]here are myriad approaches to farming that may be sustainable."[22] The more important overarching goal of sustainable agriculture is the "stewardship

---

13. The Conservation Stewardship Program that was implemented by the 2008 farm bill does in fact reward farmers—including vegetable and fruit growers, as well as organic producers—with incentives for using ecologically sustainable practices. That program is a small-scale version of what would be necessary on a much larger subsidy scale to drive the change necessary to transform the food system in a measurable and even more meaningful way.
14. California Dept. of Food and Agriculture, *California Agricultural Production Statistics* [hereinafter CA Statistics], http://www.cdfa.ca.gov/statistics/.
15. Kari Hamerschlag, Environmental Working Group, Farm Subsidies in California: Skewed Priorities and Gross Inequities (2010), *available at* http://farm.ewg.org/pdf/california-farm.pdf.
16. *See generally* Mark Arax & Rick Wartzman, The King of California: J.G. Boswell and the Making of a Secret American Empire (2005) (describing J.G. Boswell's cotton production methods that led to nearly $18 million in farm bill subsidies between 1994 and 2004).
17. CA Statistics, *supra* note 14.
18. *Id.*
19. James E. Horne & Maura McDermott, The Next Green Revolution: Essential Steps to a Healthy, Sustainable Agriculture 55 (2001).
20. President's Council on Sustainable Dev., Sustainable Agriculture: Task Force Report 3 (1997), *available at* http://clinton2.nara.gov/PCSD/Publications/TF_Reports/ag-top.html.
21. Horne & McDermott, *supra* note 19; Imhoff, *supra* note 12, at 59.
22. Horne & McDermott, *supra* note 19.

of both natural and human resources . . . includ[ing] concern over the living and working conditions of farm laborers, consumer health and safety, and the needs of rural communities."[23]

Despite the promise of sustainable agriculture to solve the multifaceted ecological problems discussed in Part II, the farm bill has been surprisingly silent as to *how* to encourage farmers to engage in such practices. As early as 1994, the President's Council on Sustainable Development chartered the Sustainable Agricultural Task Force, composed of agricultural experts, to present strategies to alleviate the problems identified in Part II.[24] In the mid-1990s, the task force outlined goals and made policy recommendations that were intended to serve as updates to the farm bill the next time the legislation came up for reauthorization.[25] In particular, the task force reached consensus on nine key policy recommendations: (1) integrate pollution prevention and natural resource conservation into agricultural production, (2) increase the flexibility for participants in commodity programs to respond to market signals and adopt environmentally sound production practices and systems, thereby increasing profitability and enhancing environmental quality, (3) expand agricultural markets, (4) revise the pricing of public natural resources, (5) keep prime farmlands in agricultural production, (6) invest in rural communities' infrastructure, (7) continue improvements in food safety and quality, (8) promote the research needed to support a sustainable U.S. agriculture, and (9) pursue international harmonization of intellectual property rights.[26] Since that time, Congress has reauthorized three farm bills (1996, 2002, and 2008), and is currently in the process of reauthorizing a fourth. Yet, these recommendations have been given little, if any, consideration by Congress. After ignoring such experts for nearly two decades, it is now time for Congress to listen to the proponents of sustainable agriculture in order to address the environmental and health problems triggered by the farm bill.

I. Why a Fundamental Shift Will Work: Sustainable Agriculture Already Exists on a Small Scale

Of the nearly $20 billion in annual farm bill subsidies, 84% currently go to the five primary commodity crops of corn, rice, wheat, cotton, and soybeans.[27] Shifting a sizeable portion of these subsidies (billions, not mere millions, of dollars) to farmers who implement sustainable farming practices would greatly impact the market by bringing down the supermarket prices of sustainably farmed goods—which are almost invariably more labor-intensive—while nudging up supermarket prices of foods based on industrial-farmed corn and soybeans to a level that would more closely reflect the market prices that would appear in the absence of the heavy subsidies that artificially deflate market prices of corn and other commodities. A critical step would involve tapping into the knowledge of scientists, experts from the U.S. Department of Agriculture (USDA), nonprofit advocates, farmers, and other key stakeholders in order to set specific standards of what constitutes a sustainable agricultural practice for purposes of receiving these incentives.[28] Although this approach would require time to reach consensus among those varied interests, it is clear that such incentives would better protect the natural environment and the public's health than continuing to maintain the status quo. Indeed, this expert panel could use the Conservation Stewardship Program's grading system as a starting point for discussion.[29]

---

23. *Id.*
24. President's Council on Sustainable Dev., *supra* note 20, at 1.
25. *Id.* at 3-8.
26. *Id.*
27. Imhoff, *supra* note 12, at 60.
28. This is a very important step that would have to be developed thoroughly prior to implementation. In addition to setting concrete standards for sustainable agricultural practices, experts and regulators would also have to create a defined spectrum on which the environmental and public health benefits of a farmer's sustainable practices can be measured in order to receive one's fair share of subsidies. For example, a large corn farm in Iowa might allege that it uses a single practice deemed "sustainable" by the regulatory scheme such as crop rotation, which benefits both the soil and local water sources as runoff is reduced. Although this farm would likely receive subsidies for undertaking this practice because it is "sustainable" and benefits the environment, the farm would likely receive considerably less in subsidies than a similarly-situated large corn farm that instead decides to implement crop rotation *and* to diversify its crops, reduce pesticide use, utilize integrated pest management, and begin selling to local markets to reduce transportation and fossil fuel use. Despite the fact that both are benefiting the environment and public health, the second farm clearly has undertaken sustainable practices that are not only greater in number but, more important, greater in positive impact to the natural environment and public health. Due to this difference in magnitude, the second farm would receive greater rewards for its efforts.
29. *See, e.g.*, Michael Pollan, *The Food Issue: Farmer in Chief*, N.Y. Times Mag., Oct. 9, 2008, at http://www.nytimes.com/2008/10/12/magazine/12policy-t.html. In fact, the 2008 farm bill might have taken the first step toward such a sustainable subsidy system with the creation of the Conservation Stewardship Program, which rewards farmers for making wise agricultural decisions that provide off-farm benefits. Despite the program's promise, however, Pollan notes that legislators "need to move this approach from the periphery of our farm policy to the very

Agricultural methods that could fall into the category of sustainable agriculture for subsidy purposes are no-till farming, cover cropping, crop rotation, residue mulching, elimination of most or all agrochemical fertilizers, significant reduction per acre of water usage, nitrogen fixing through on-farm manure use, measurable energy reduction per acre farmed, greater use of integrated pest management, contour farming, and increased direct sales from farm to consumer to reduce transportation.[30] Each of these farming practices promotes sustainability by eliminating harmful inputs in the soil, reducing pollution in our ecosystem, or preventing some other harmful result. Not only would these practices create a healthier environment in which to live, but they would also almost certainly produce a healthier food product for the consumer, therefore allowing us to address public health concerns such as obesity.[31]

Because many Americans associate the sustainable practices listed above with "organic" agriculture, it is necessary to tackle the controversial "organic" certification label under USDA's National Organic Program (NOP), the existence of which might or might not be included as one of the many factors entitling a farmer to subsidies under any new incentive system. Since it is uncertain how an expert panel would define the conditions for eligibility to a new farm subsidy system, a producer that is USDA-certified organic might be automatically eligible for such a program on the grounds that organic certification denotes certain of the practices listed in the previous paragraph, consistent with the NOP's implementing regulations. On the other hand, however, an expert panel might decide—in part due to the fact that inputs for certified organic foods can change over time pursuant to regulation—that, instead of granting eligibility based solely on organic certification, all farms, including organic producers, must demonstrate the on-farm practices and techniques carried out to achieve ecological protection in order to satisfy the eligibility requirements for the program.

Historically, organic products have been generally grown and raised using sustainable agricultural methods, and are then certified by an entity that has been authorized by USDA to ensure that the regulatory labeling standards are satisfied.[32] As discussed in Chapter 13, there is a very important and distinct difference, however, between sustainable agriculture and organic agriculture: sustainable agricultural practices *always* have the goal of protecting public health and preserving the environment because sustainability is its very foundation.[33] In contrast, since what constitutes "organic" produce is a construct of federal regulation, the standards imposed *may* be ecologically protective, but also may not reflect sound agricultural, environmental, or health-based decisionmaking because of the influence of agribusiness or other interested parties that lobby the agency and its National Organic Standards Board to modify standards.[34]

In any event, from the beginning of the NOP, USDA has been careful not to endorse organic products as superior to their nonorganic counterparts. For example, in 2000, USDA Secretary Dan Glickman "went out of his way to say that organic food is no better than [industrial-farmed] conventional food."[35] Secretary Glickman made clear that in his opinion "[t]he organic label is a marketing tool . . . [and] is not a statement about food safety . . . nutrition or quality."[36]

---

center." *Id.* Until such a system becomes the foundation of the farm bill, the United States will not maximize its agricultural potential to "grow crops and graze animals in systems that will support biodiversity, soil health, clean water and carbon sequestration." *Id.* For a discussion of CSP's grading system, see U.S. Dept. Agric., Natural Res. Conservation Serv., *CSP 2011 Ranking Period One Enhancement Activity Job Sheets*, http://www.nrcs.usda.gov/wps/portal/nrcs/detailfull/national/programs/financial/csp/?&cid=stelprdb1045117.

30. David Pimentel et al., *Environmental, Energetic, and Economic Comparisons of Organic and Conventional Farming Systems*, 55 BIOSCIENCE 573, 573 (2005); *see generally* HORNE & MCDERMOTT, *supra* note 19. "Cover cropping" means planting certain plants on lands not in production during a given season with plants that are known to replenish critical nutrients to the soil; "crop rotation" means growing a series of crops from different families in consecutive seasons to ensure that nutrients are balanced in the soil; "residue mulching" means returning crop residues or unused portions of previous crops as mulch onto current crops to recycle the vital nutrients already absorbed by the plant without having to use chemical fertilizers; and "contour farming" means plowing a slope in contour lines that prevent soil erosion and efficiently store water for crop use.

31. *See, e.g.*, Alyson E. Mitchell et al., *Ten-Year Comparison of the Influence of Organic and Conventional Crop Management Practices on the Content of Flavonoids in Tomatoes*, 55 J. AGRIC. FOOD CHEM. 6154 (2007), *available at* http://pubs.acs.org/cgi-bin/sample.cgi/jafcau/2007/55/i15/pdf/jf070344+.pdf?isMac=706237 (concluding that sustainable organic farming practices with tomatoes resulted in much higher levels of healthy flavonoids as compared to nitrogen-fertilized, conventionally produced tomatoes).

32. U.S. Dept. Agric., *National Organic Program*, http://www.ams.usda.gov/AMSv1.0/ams.fetchTemplateData.do?template=TemplateA&navID=NationalOrganicProgram&leftNav=NationalOrganicProgram&page=NOPNationalOrganicProgramHome&acct=nop (last visited Nov. 9, 2011).

33. HORNE & MCDERMOTT, *supra* note 19.

34. *See* POLLAN, THE OMNIVORE'S DILEMMA, *supra* note 4, at 178-79.

35. *Id.* at 178.

36. *Id.* at 179.

In spite of these public pronouncements from USDA, mounting evidence gathered from recent studies is increasingly illustrating that organic produce, when farmed using sustainable methods, appears to have more key nutrients, vitamins, minerals, and health benefits than its industrial-farmed conventional counterparts.[37] While more research is needed to comprehensively verify these results, there is no question that, because of the constraints imposed on organic farmers as a result of their certification, organic foods are produced with far less pesticides, and in some cases none at all, meaning that both the risk of environmental harm as a result of pesticide application and the health risk of consumption of pesticide residues are considerably lower with organic crops.[38]

In recent years, many consumers have become aware of the purported benefits of buying organic: in 2011, 78% of U.S. families acknowledged that they purchased some organic foods (the highest percentage ever), with 48% of those families buying organic indicating that their "strongest motivator . . . is their belief that organic products 'are healthier for me and my children.'"[39] In 2009, nearly $24.8 billion in U.S. sales were attributed to organics, which was a 5.1% increase from 2008.[40]

Despite these accomplishments, less than 4% of food sales in 2009 were for organic products,[41] due in part to the price distortion caused by farm bill subsidies that prioritize nonorganic commodity crops and make them appear cheaper at the market than their organic counterparts. For years, Congress, USDA, and agribusiness have used subsidies as a way to keep commodity crops cheap compared to organic alternatives,[42] but a new trend has taken hold that might be just as troubling—large agribusiness companies such as Monsanto, Wal-Mart, and Cargill are recognizing the growing success of organic agriculture and are not only joining the market but consolidating it the way those companies consolidated conventional markets in the post-World War II era (as described in Chapter 1).[43] Although this may be a positive development because it should lead to greater overall production of organic foods farmed with sustainable methods and in turn make such foods more affordable to consumers, it also provides a potential avenue for agribusiness to commandeer organic standards in order to water them down (e.g., advocate for inclusion of pesticides and other chemicals not currently on the list of approved organic inputs), much in the way that it has used the farm bill's commodity provisions for several decades to encourage farm and market consolidation.[44] Moreover, organic farming on an industrial scale, as many of the largest companies do, presents many of the same ecological problems (e.g., monoculture, soil erosion, and overtilling) that plague conventional farming on an industrial scale, therefore potentially undermining the value and purpose of having an organic label in the first place.[45]

---

37. *Id.*; Mitchell et al., *supra* note 31; *Study Hails Organic Food Benefits: Organic Food Has a Higher Nutritional Value Than Ordinary Produce, a Study by Newcastle University Has Found*, BBC News, Oct. 29, 2007, at http://news.bbc.co.uk/2/hi/uk_news/england/tyne/7067226.stm (reporting that one of the largest studies of sustainably-farmed organic agriculture ever conducted has found up to 40% more of healthful antioxidants in organic fruit and vegetables as compared to nonorganic competitors farmed alongside their organic counterparts); *see also* Organic Trade Ass'n, Nutritional Consideration, *available at* http://www.ota.com/organic/benefits/nutrition.html (summarizing 20 recent scientific studies that have all found measurable increases in nutrients in organically produced foods compared to conventional counterparts).
38. *See, e.g.*, Crystal Smith-Spangler et al., *Are Organic Foods Safer or Healthier Than Conventional Alternatives? A Systematic Review*, 157 Annals Internal Med. No. 5 (Sept. 4, 2012); Dr. Charles Benbrook, *Initial Reflections on New Organic Study* (Sept. 4, 2012), *available at* http://www.tfrec.wsu.edu/pdfs/P2566.pdf.
39. Organic Trade Association, *Seventy-Eight Percent of U.S. Families Say They Purchase Organic Foods*, http://www.organicnewsroom.com/2011/11/seventyeight_percent_of_us_fam.html.
40. Organic Trade Association, *Organic Food Facts*, http://www.ota.com/organic/mt/food.html.
41. *Id.*
42. Pollan, The Omnivore's Dilemma, *supra* note 4, at 178-79.
43. *See id.* at 145-84; *see* Samuel Fromartz, Organic Inc.: Natural Foods and How They Grow 188 (2006) (discussing the multibillion-dollar organic food business, in which more than half of all organic sales in 2006 came from only the largest 2% of organic farms owned or controlled by Kraft, General Mills, Monsanto, and other corporations); Philip H. Howard, Organic Processing Industry Structure, *available at* https://www.msu.edu/~howardp/organicindustry.html (demonstrating that very few independent organic companies exist because many have been purchased or consolidated by large food processors or retailers).
44. *See* Pollan, The Omnivore's Dilemma, *supra* note 4, at 145-84. *See also* Howard, Organic Processing Industry Structure, *supra* note 43.
45. *See, e.g.*, Eliot Coleman, *Four Season Farm*, http://www.fourseasonfarm.com/ (last visited Nov. 9, 2011) ("Now that the food-buying public has become enthusiastic about organically grown foods, the food industry wants to take over. Toward that end the USDA-controlled national definition of 'organic' is tailored to meet the marketing needs of organizations that have no connection to the agricultural integrity 'organic' once represented. We now need to ask whether we want to be content with an 'organic' food option that places the marketing concerns of corporate America ahead of nutrition, flavor and social benefits to consumers."); *see generally* Fromartz, *supra* note 43 (highlighting the controversies surrounding organic certification that have been caused in large part due to the emergence of big corporations in the organic market and the stark contrasts between these corporations and the small growers that initially sparked the organic movement); Joel Salatin, Holy Cows and Hog Heaven: The Food Buyer's Guide to Farm Friendly Food (2004) (encouraging consumers to purchase foods from local buyers as opposed to purchasing organic foods from large corporations because of the politicization of the organic label).

Therefore, the public must stay vigilant in protecting the integrity of organic standards as part of a renewed push to subsidize sustainable agriculture, in order to ensure that foods labeled as organic protect the key values for consumers that the label was created to safeguard. And, in any event, if Congress does endeavor to incentivize sustainable practices, agricultural experts, as described above, will have to determine how the preexisting organic program fits, if at all, within the parameters of the new subsidy framework and its eligibility requirements.

2. Scaling Up Sustainable Agriculture With Significant Reform of Farm Bill Commodity Subsidies

As seen with our nation's massive corn production tied tightly to subsidies, farmers will farm wherever the money is. If subsidies were available for sustainable agriculture, regardless of crop produced, data suggests that farmers would undertake sustainable agricultural practices in order to survive financially. Further, all available data indicates that many farmers genuinely want to grow healthier foods, maintain their communities, and conserve their natural ecosystems, but they have been pressured to farm corn and other commodity crops at the expense of those values because that is where profits could be garnered under the existing subsidy framework.[46] Although most farmers in the United States do not want farm bill subsidies eliminated or phased out,[47] farmers "show[ ] strong support for programs focused on conservation" and seem very concerned about the status of the natural environment.[48] This is not surprising considering the interdependent relationship between healthy farms and a healthy environment: long-term farm health requires a functioning local ecosystem that can sufficiently supply all of a farm's needs. To prevent degradation of this important ecosystem, which suffers from a classic "tragedy of the commons"[49] problem under the current farm bill subsidy regime, the proposed sustainable agriculture subsidy system would pay farmers to protect this common pool resource for society and for the farmers themselves for future crop years to avoid passing on environmental externalities as has typically been the case under federal farm policies.

A related question that is often asked is whether farmers are willing to make the transition from solely growing corn or other commodity crops to planting a diversity of fruits and vegetables under a sustainable agriculture subsidy program. Based on available research, it seems that farmers would be willing to do so for financial reasons as well as for the viability of their farms and families. Financially speaking, a farmer receives only four cents out of every consumer dollar spent on a corn-based product in the supermarket because of the large number of middlemen such as Cargill, ADM, Coca-Cola, and PepsiCo.[50] The return is starkly different for whole foods such as green vegetables, fruits, and eggs, where the respective farmer receives 40 cents for every supermarket dollar spent, or 10 times the amount of return on investment.[51] Thus, it makes financial sense for farmers to indulge in the cultivation of healthier produce and unprocessed whole foods once sustainable agriculture subsidies are put into place, not to mention the ability to feed one's family with the farm's diverse crops rather than purchasing food at the supermarket that was produced and processed hundreds, if not thousands, of miles away.

With respect to anticipated environmental impacts, sustainable agriculture will greatly help to repair local ecosystems, boost farmers' yields as the ecology and soil improve, and mitigate the degradation caused by decades of mechanized agriculture under the farm bill, as described in Part II. As farmers well know, sustainable agriculture includes polycultures and crop rotations that are essential to protect soils from ero-

---

46. NAT'L PUB. POL'Y EDUC. COMMITTEE, THE 2007 FARM BILL: U.S. PRODUCER PREFERENCES FOR AGRICULTURAL, FOOD, AND PUBLIC POLICY v-viii (2007), *available at* http://digitalcommons.unl.edu/cgi/viewcontent.cgi?article=1057&context=ageconfacpub(illustrating that many farmers support the current commodity subsidy program despite the fact that such a program undermines other values highly supported by the same farmers such as environmental protection, financial payments for small farms, compliance with WTO rules, and better food safety); *see also* Timothy A. Wise, *Identifying the Real Winners From U.S. Agricultural Policies* 9 (Global Dev. and Env't Inst. Working Paper No. 05-07, 2005), *available at* http://ase.tufts.edu/gdae/Pubs/wp/05-07RealWinnersUSAg.pdf (concluding that, despite revenues garnered through subsidized corn and soybean production in the past, "diversified family farms [would be much] more competitive relative to [food processors and] industrial livestock operations" if agricultural subsidies were altered so that the price of crops "more accurately reflected costs [paid by the farmer]").
47. NAT'L PUB. POL'Y EDUC. COMMITTEE, *supra* note 46, at vi.
48. *Id.* at vi-vii.
49. Garrett Hardin, *The Tragedy of the Commons*, 162 SCI. 1243 (1968) (explaining that a "tragedy of the commons" occurs when a common resource (e.g., an ecosystem, air, or water) is degraded by individual users of that resource (e.g., farmers) as each user maximizes his personal benefit while sharing the burden of his resource use (e.g., pollution) among all of users of the commons).
50. POLLAN, THE OMNIVORE'S DILEMMA, *supra* note 4, at 95.
51. *Id.*

sion and streambeds from sedimentation.[52] Farmers have long recognized the need for better farming practices to enhance environmental protection.[53] When USDA has given farmers flexibility to diversify their crops into polycultures and yet retain their full direct payment of commodity subsidies, many farmers have taken advantage of this flexibility and planted noncommodity crops on nearly one-half of the land available for diversification, indicating a desire to move towards a more ecologically protective cultivation scheme within the parameters of the farm bill's commodity title.[54] Additionally, sustainable agricultural systems do not rely on harmful chemical inputs of synthetic fertilizers or pesticides that pose serious threats to humans and wildlife.[55] And studies indicate that sustainable farming systems "use 30% to 70% less energy per unit of land than conventional systems, a critical factor in terms of climate change and eventual fossil fuel shortages."[56] Since subsidizing sustainable agriculture will result in more polycultures and thus more robust and diverse local food supplies, less transportation will be needed, and the result will be "reduced energy consumption, less processing and packaging, and higher nutritional values" lost during storage and transportation.[57]

Additionally, as money is drawn away from subsidizing corn and spent instead on subsidizing sustainable farming practices, the decreasing amount of corn grown will gradually force a reconfiguration of the CAFO industry, which (as described in Chapter 5) is built almost entirely on large volumes of highly subsidized corn. Fewer livestock animals would be bred for meat production as corn prices return closer to nonsubsidized market rates, which will likely result in an increased proportion of cattle being transitioned back to their native grass-fed diets because of the prohibitive cost of raising grain-fed cattle in the face of decreasing corn subsidies. To encourage the transition, a certain proportion of farm bill subsidies could be allocated to farmers transitioning from concentrated livestock production to more traditional grazing patterns. Such an incentive approach would vastly improve not only water and air quality, but also the health of Americans consuming meat and meat-based products, according to studies that have compared grass-fed animals with their corn-fed counterparts.[58] Of course, the supermarket price of meat and meat-based products will rise as the agricultural market normalizes under this new policy, but the health benefits, and thus the reduction in medical costs, that will be gained from shifting from corn-fed meat to grass-fed meat would be expected to mitigate, if not outweigh, the expected supermarket price increase that is likely to occur.[59]

Finally, and importantly, rural farming communities will be able to sustain some semblance of their past strength, which author and agriculturist Wendell Berry argued could only be regained with a "revolt of local small producers and local consumers against the global industrialism of the corporation."[60] Thus, assuming large-scale reform of the farm bill is more than an idealistic pipe dream, the time is now for a revolution—a truly "green" revolution—against current agricultural policies, which can only end when the farm bill once again protects our nation's farmers, the natural environment, and ultimately, the American public by substantially reworking the commodity program to infuse a level of sustainability that powerful interests have attempted to shut out for far too long.

B. Breathing New Life Into the Farm Bill: Life by a Thousand Cuts

While seeking a major overhaul to the farm bill to achieve a sustainable nationwide agricultural system is a laudable objective, the tense partisan climate in Congress and the political and financial power of the

---

52. *See generally* Pimentel et al., *supra* note 30.
53. PRESIDENT'S COUNCIL ON SUSTAINABLE DEV., *supra* note 20, at 5 ("In 1990, Congress passed legislation that allowed farmers who had signed up for a particular commodity program—for example, the wheat program—to plant some of their land in a crop other than that specified by the program. In response, farmers reduced the number of acres under monoculture and diversified their crops. By 1994, approximately 42 percent of the land on which farmers were allowed to grow whatever they chose was planted in crops other than those specified by the commodity program in which the farmers were enrolled.").
54. *Id.*
55. IMHOFF, *supra* note 12, at 143.
56. *Id.*
57. *Id.*
58. E.N. Ponnampalam et al., *Effect of Feeding Systems on Omega-3 Fatty Acids, Conjugated Linoleic Acid and Trans Fatty Acids in Australian Beef Cuts: Potential Impact on Human Health*, 15 ASIA PAC. J. CLINICAL NUTRITION 21, 21 (2006) (concluding that grass-fed cattle have much higher levels of healthy fats and other compounds while grain-fed cattle have much higher levels of unhealthy fats and compounds).
59. *Id.*
60. POLLAN, THE OMNIVORE'S DILEMMA, *supra* note 4, at 254.

agribusiness industry suggest that a more pragmatic course of action would be to seek targeted reforms and to enhance existing initiatives that can incrementally, but steadily, enhance sustainable agriculture and improve consumer choices. By engaging in this more cautious approach, success would, at the least, result over time in a dual system in which sustainable agriculture can thrive alongside an industrial system, and could push the entire system towards a more ecologically balanced and healthful equilibrium. Below are brief examples of programs that, if shaped properly and funded sufficiently, would move the needle much farther towards sustainability in the U.S. farming and food system in the short term.

1. Eliminating or Limiting Commodity Payments and Crop Insurance Payments

An issue that has aroused vigorous debate is the extent to which Congress should limit commodity and crop insurance payments (or other subsidy vehicles under the farm bill), especially given the tenuous financial situation of the U.S. government and the level of federal debt. Everyone from President Barack Obama to members of Congress have taken aim at the most-known subsidy payment scheme—direct payments under the commodity program. The 2008 farm bill limited direct payments, which are provided based on a fixed payment rate, to a maximum of $40,000 per farm, processor, or other eligible entity each year.[61] In 2011, facing an uncertain financial future and an increasing debt load, President Obama proposed, as part of his budget reduction plan, to eliminate direct payments entirely, which do not vary with "prices, yields, or producers' farm incomes" but rather "provide[ ] producers fixed annual income support payments for having historically planted crops that were supported by Government programs, regardless of whether the farmer is currently producing those crops—or producing any crop, for that matter."[62] As the president's plan explained, "[e]conomists have shown that direct payments have priced young Americans out of renting or owning the land needed to enter into farming."[63]

On the heels of this budget plan and in the midst of the debate over the 2012/2013 farm bill, elimination of direct payments received bipartisan support from some members of the U.S. House of Representatives, who urged that, if Congress "takes even a single action to reduce federal farm subsidies, it should eliminate the direct payment program and apply the savings to reducing the deficit."[64] While a bill introduced in the U.S. Senate with bipartisan support stopped short of recommending elimination of direct payments entirely, it nonetheless urged a hard per-farm cap on direct payments of $20,000, a 50% reduction from the 2008 farm bill, regardless of farm size or income.[65] Myriad environmental benefits would result from a system with fewer and smaller direct payments (or, ideally, none), since such incentives severely restrict planting flexibility (as described in more detail below) and thus encourage large-scale and ecologically devastating monocultures. But any attempt to eliminate or further cap direct payments will inevitably be met with fierce resistance from large farms and processors that benefit from the current system. At the time of writing, it remains to be seen whether this important step will be achieved in the 2012/2013 farm bill, behind the weight of the White House and congressional members on both sides of the aisle.

In addition to elimination of direct payments, hard caps are also necessary to limit other forms of subsidy payments that deplete public funds and foster environmental degradation, and to ensure that these payments are allocated only to provide a safety net to farmers in need, instead of helping large farms get even wealthier on the taxpayer's dime. A few areas where caps have been proposed include counter-cyclical subsidy payments that are paid to eligible farmers in years where the actual price paid for a commodity is

---

61. Food, Conservation, and Energy Act of 2008, Pub. L. No. 110-246, §1603 (2008), *available at* http://www.gpo.gov/fdsys/pkg/PLAW-110publ246/pdf/PLAW-110publ246.pdf.
62. Office of Mgmt. & Budget, Living Within Our Means and Investing in Our Future: The President's Plan for Economic Growth and Deficit Reduction 17 (Sept. 2011), *available at* http://www.whitehouse.gov/sites/default/files/omb/budget/fy2012/assets/jointcommitteereport.pdf.
63. *Id.*
64. Jeff Flake & Earl Blumenauer, Letter to the Joint Select Committee on Deficit Reduction (Oct. 18, 2011), *available at* http://www.taxpayer.net/library/article/congressmen-support-common-sense-cuts-to-wasteful-ag-spending.
65. Charles Grassley & Tim Johnson, Letter to the Joint Select Committee on Deficit Reduction Concerning the Rural America Preservation Act of 2011 (Oct. 14, 2011), *available at* http://www.grassley.senate.gov/judiciary/upload/Agriculture-10-14-11-Grassley-Johnson-Letter-to-Jt-Cmte-re-Payment-Limits.pdf; *see also* Rural America Preservation Act of 2011, S. 1161, 112th Cong., 1st Sess., *available at* http://www.gpo.gov/fdsys/pkg/BILLS-112s1161is/pdf/BILLS-112s1161is.pdf.

less than a target set by USDA,[66] and nonrecourse commodity marketing loans and loan deficiency payments that "provide[ ] an influx of cash when market prices are typically at harvest-time lows, which allows the producer to delay the sale of the commodity until more favorable market conditions emerge."[67] As with direct payments, Congress authorized in the 2008 farm bill sizeable disbursements for both counter-cyclical payments, which were capped at $65,000 annually per farm, and marketing loans/loan deficiency payments, which had no limits on their allocation.[68] There are two different ways that such payments can be reduced to more reasonable levels—on the front end by capping them on a per-farm basis like the above proposals for direct payments, or on the back end by lowering eligibility requirements far below the current limit for most commodity subsidies set at an Adjusted Gross Income (AGI) of $1.25 million by the 2008 farm bill.[69] Recent efforts have seen members of Congress advocate for both approaches; one proposed bill calls for a hard per-farm cap of $30,000 for counter-cyclical payments (down 54% from the 2008 farm bill) and $75,000 for loan deficiency payments and marketing loans (down from no limit at all).[70] Another proposal offered with bipartisan support advocated that "farm subsidy payments be limited to those with an [AGI] of less than $250,000," which is an 80% reduction in the maximum eligibility for subsidies.[71] Despite their different approaches, both proposals agreed that "the aggregate of agriculture subsidies any one . . . entity [or married couple] can receive be capped at . . . $250,000 annually."[72]

The well-documented environmental devastation encouraged by commodity payment incentives compels the conclusion that the time and effort of conservation and sustainable agriculture advocates would be well spent by continuing to press for even lower caps and stricter eligibility requirements so that subsidy payments as well as crop insurance payments provided by the federal government, if any at all, are used only for their original purpose of buttressing small family farmers in need of supplemental income or disaster relief.[73]

2. Putting the Flexible Back in Planting Flexibility

Another area of intense debate in the agriculture community is over what, if any, conditions should be placed on planting flexibility as part of the farm bill's commodity and crop insurance programs. In the 2008 farm bill, Congress restricted all payments under the commodity title of the farm bill (direct payments, counter-cyclical payments, and average crop revenue election payments) to only those farms that "comply with the planting flexibility requirements of §1107" of the legislation.[74] In effect, §1107 provides that farmers enrolled in any of the three types of commodity payment programs may not plant fruits, vegetables, or wild rice on any portion of their base acres.[75] As a result, farmers have all of their eligible base acres available for commodity cultivation that in turn translates to a per-bushel or per-weight payment to the farmer each year on October 1 as specified in §§1103, 1104 and 1202 of the legislation. If a farmer whose farm has not historically produced fruits or vegetables opts to grow fruits or vegetables on his base acres, the farmer is ineligible for commodity payments that year.[76]

---

66. *See* U.S. Dept. Agric., Economic Research Service, *Farm and Commodity Policy: Counter-Cyclical Payments*, http://www.ers.usda.gov/briefing/farmpolicy/countercyclicalpay.htm (last visited Nov. 9, 2011).
67. U.S. Dept. Agric, Farm Serv. Agency, *Non-Recourse Marketing Assistance Loan*, http://www.fsa.usda.gov/FSA/webapp?area=home&subject=prsu&topic=col-nl (last visited Nov. 9, 2011).
68. Food, Conservation, and Energy Act of 2008, Pub. L. No. 110-246, §1603 (2008), *available at* http://frwebgate.access.gpo.gov/cgi-bin/getdoc.cgi?dbname=110_cong_public_laws&docid=f:publ246.110.pdf.
69. *See* Flake & Blumenauer, *supra* note 64.
70. Grassley & Johnson, *supra* note 65; *see also* Rural America Preservation Act of 2011, S. 1161, 112th Cong., 1st Sess., *available at* http://www.gpo.gov/fdsys/pkg/BILLS-112s1161is/pdf/BILLS-112s1161is.pdf.
71. Flake & Blumenauer, *supra* note 64.
72. *Id.*; Grassley & Johnson, *supra* note 65.
73. For a detailed discussion of the environmental impacts incentivized by commodity subsidies, see Eubanks, *A Rotten System*, *supra* note 1.
74. Food, Conservation, and Energy Act of 2008, Pub. L. No. 110-246, §1106(a)(1) (2008), *available at* http://frwebgate.access.gpo.gov/cgi-bin/getdoc.cgi?dbname=110_cong_public_laws&docid=f:publ246.110.pdf.
75. *Id.* §1107.
76. *See* Nat'l Sustainable Agric. Coal., *Planting Flexibility for Fruits and Vegetables*, http://sustainableagriculture.net/publications/grassrootsguide/competitive-markets-commodity-program-reform/planting-flexibility-for-fruits-vegetables/.

The idea of planting flexibility—or the ability to diversify one's crops for environmental or other sustainability purposes—was first adopted in the Integrated Farm Management Program in the 1990 farm bill.[77] As the National Sustainable Agriculture Coalition notes:

> The adoption of planting flexibility [in the 1990 farm bill] was important to farmers utilizing sustainable farming methods. Producers who for environmental, health or economic reasons were adopting diversified resource-conserving crop rotations or were adding grass-based livestock production with continuing grain production activities found themselves enormously disadvantaged by the traditional commodity program structure. As these farmers added forages and soil-building crops to their rotations or converted marginal or hilly crop acres to grass-based production systems—all very positive practices for the environment—they lost government payments. The advent of planting flexibility rules . . . at least provided for a prospective elimination of a significant barrier to the adoption of more sustainable and diversified systems.[78]

Unfortunately, recognizing that an increase of noncommodity crop production on commodity crop lands formerly ineligible for fruit and vegetable production would harm their interests, the lobby for fruit and vegetable growers sought to ensure that the new planting flexibility would not jeopardize their profit and market share.[79] Thus, before the new planting flexibility rules could be enacted, several restrictive limits were incorporated in the 1990 farm bill to prohibit commodity-eligible farms from "growing fruits and vegetables as an alternative crop on base acres."[80] As noted, the original intent behind the restrictions, which still persist today, was "to protect fruit and vegetable growers who do not receive government payments."[81] Indeed, fruit and vegetable farmers—who are almost categorically excluded from subsidy programs under the farm bill—"generally oppose [measures to allow more planting flexibility] . . . because the change would weaken protections that have existed since planting flexibility was created in the 1990 farm bill," meaning that fruit and vegetable farmers could "face unfair competition if producers of [commodity] program crops were allowed to plant [fruits and vegetables] on program base acres and still receive government payments" by way of commodity subsidies.[82]

Despite the concerns voiced by fruit and vegetable growers, sustainable agriculture and environmental advocates have strongly urged Congress for more than two decades to lift these growing restrictions on the tens of thousands of farms enrolled in the commodity program because of the enormous conservation and health benefits that would result from the ability to diversify crops, enhance stewardship efforts, and bolster local food systems and farm-to-consumer sales. Indeed, some farmers who have attempted to expand their sustainable farming operations by leasing land formerly enrolled in the commodity program for organic fruit and vegetable production have found themselves subject to stifling financial penalties simply "[b]ecause national fruit and vegetable growers based in California, Florida and Texas fear competition from regional producers . . . [and] they have been able to virtually monopolize the country's fresh produce markets" regardless of the ecological impacts of that monopolization.[83]

To close this loophole, there have been renewed calls during the 2012/2013 farm bill cycle to reexamine and greatly expand planting flexibility for farmers enrolled in commodity (or crop insurance) programs.[84] Research has suggested the need for more flexible planting requirements: a 2010 study indicated that, in order to meet USDA's daily recommended nutritional guidelines for each American, the United States needs an additional 13 million acres of farmland growing fruits and vegetables.[85] However, at the time of writing, Congress has elected to ignore those requests in favor of supporting the status quo at the expense of a more sustainable, transparent, and balanced food system.

---

77. Id.
78. Id.
79. Union of Concerned Scientists, Ensuring the Harvest: Crop Insurance and Credit for a Healthy Farm and Food Future 1, 19 (2012), available at http://www.ucsusa.org/assets/documents/food_and_agriculture/ensuring-the-harvest-full-report.pdf.
80. Jim Monke, Congressional Research Serv., Farm Commodity Programs: Base Acreage and Planting Flexibility 3, available at http://www.nationalaglawcenter.org/assets/crs/RS21615.pdf.
81. Id.
82. Id. at CRS-4.
83. Jack Hedin, My Forbidden Fruits and Vegetables, N.Y. Times, Mar. 1, 2008, at http://www.nytimes.com/2008/03/01/opinion/01hedin.html.
84. See, e.g., Local Farms, Food, and Jobs Act of 2011, S. 1773, 112th Cong., 1st Sess., §1001 (2011), available at http://www.gpo.gov/fdsys/pkg/BILLS-112s1773is/pdf/BILLS-112s1773is.pdf.
85. American Farmland Trust, The United States Needs 13 Million More Acres of Fruits and Vegetables to Meet RDA, http://www.farmland.org/news/pressreleases/13-Million-More-Acres.asp (last visited Nov. 9, 2011).

3. Reestablishing Conservation Compliance Conditions on Federal Crop and Revenue Insurance Payments

Conservation compliance is not a stand-alone conservation program in which a farm may choose to enroll but rather imposes certain environmental requirements on all farms that participate in most other farm bill programs and, thus, receive federal incentives through those programs.[86] The two key conservation compliance provisions "encourage greater soil conservation and wetland protection," and those respective provisions are commonly referred to as Sodbuster and Swampbuster.[87]

To maintain eligibility for the various programs to which conservation compliance attaches (e.g., direct payments and countercyclical payments through commodity programs, payments through disaster programs, and incentives through working lands conservation programs), a farm must: (1) implement a soil conservation plan on "highly erodible" land that has been approved by USDA's Natural Resources Conservation Service (NRCS); (2) refrain from planting and harvesting on highly erodible land without implementing a conservation plan approved by NRCS; and (3) refrain from draining any wetlands for crop production purposes.[88] If a farm violates any of these conditions, it "could lose some or all of [its] commodity, conservation, and disaster payments; access to USDA farm loan and loan guarantee programs; and other agriculture-related benefits," which is why conservation compliance is such "a potent incentive for soil and wetland conservation."[89] Conservation compliance is of critical importance in terms of agriculture-related conservation efforts because approximately 100 million acres of U.S. cropland, or roughly 25% of cropland in production, is highly erodible and thus subject to these conditions, and because more than 1.5 million acres of wetlands have been saved from crop production through these provisions.[90]

The one glaring omission in major farm bill programs subject to conservation compliance is the crop insurance program, which provides annual crop and revenue insurance payments to approximately 80% of eligible acres for the four major commodity crops (corn, wheat, soybeans, and cotton), thereby significantly reducing their insurance premiums.[91] When Congress enacted the conservation compliance provisions in 1985, crop insurance payments were subject to them. However, in 1996, in an effort to appease large commodity growers, Congress decoupled conservation compliance from crop insurance payments, meaning that for the past 16 years farms have collected crop insurance subsidies without complying with the conditions set forth above concerning highly erodible land and wetlands.[92]

Because there is currently a strong push toward reallocating the farm bill's incentives from direct payments to risk management incentives heavily dependent on crop and revenue insurance payments, it is vital that Congress reestablish the link between conservation compliance and crop insurance.[93] Indeed, crop insurance payments "are now the largest farm program public benefits" and as a result must "be part of the same social contract as commodity and conservation support."[94] A recent study commissioned by USDA found that if Congress moves to a more risk management based system (i.e., abandoning direct payments in lieu of crop insurance) without relinking conservation compliance to crop insurance payments, at least 181,000 farms consisting of 141 million acres, or 36% of U.S. cropland, would no longer be subject to any of the conservation compliance provisions that have had invaluable ecological benefits since their inception in 1985.[95] Accordingly, it is crucial that Congress reestablish the link between these two programs in

---

86. *See, e.g.*, Roger Claassen, USDA, Econ. Research Serv., The Future of Environmental Compliance Incentives in U.S. Agriculture: The Role of Commodity, Conservation, and Crop Insurance Programs at iii (Mar. 2012), *available at* http://www.ers.usda.gov/publications/eib-economic-information-bulletin/eib94.aspx.
87. *Id.* at 1.
88. *Id.*
89. *Id.*
90. *Id.*
91. *Id.*
92. *Id.* at 4 n.3.
93. *See* Nat'l Sustainable Agric. Coal., Farming for the Future: A Sustainable Agriculture Agenda for the 2012 Food and Farm Bill 41 (Mar. 2012), *available at* http://sustainableagriculture.net/wp-content/uploads/2008/08/2012_3_21NSACFarmBillPlatform.pdf ("The 2012 Farm Bill should re-establish compliance requirements for federal crop and revenue insurance benefits so that all existing or new crop and revenue insurance or other risk management programs are subject to conservation compliance provisions.").
94. *Id.*
95. Claassen, *supra* note 86, at 5.

order to avoid potentially devastating environmental effects to our nation's soils and wetlands when that reallocation occurs.

4. Ensuring Adequate Funding for the Conservation Stewardship Program and Eliminating Barriers to Enrollment in the Program

In the 2008 farm bill, Congress created the Conservation Stewardship Program (CSP) (successor to the Conservation Security Program), a small environmental protection program to pay farmers for "operation-level environmental benefits they produce; . . . the higher the operational performance, the higher their payment."[96] The program is small (a total of 12.8 million new acres each year, added to existing CSP-enrolled acres from previous years). Enrolled farmers participate under five-year contracts, and CSP payments are "capped at $200,000 over the life of a five-year contract, which is equivalent to $40,000 per year."[97] Program payments are made based on a ranking system of the most critical conservation needs, with allotments for such activities as converting cropland to grass-based forage, employing continuous cover cropping, extending riparian buffers, and establishing windbreaks or shelterbelts.[98] CSP incorporates many of the ideas that would be included in any fundamental and immediate reform of the farm bill commodity provisions (such as described above), although the current CSP operates on a much smaller scale and without any correlative dramatic changes to the commodity payment structure that would be part of a fundamental shift from incentivizing commodities to encouraging sustainable practices. At the time of writing, the CSP was not funded at levels that would allow all (or even most) interested farmers to enroll in the program.[99] It is important that Congress continue to support this ecologically protective approach in future farm bills by funding it at no less than current levels and, in better financial times, at significantly enhanced levels to open enrollment to more of our nation's farms to encourage widespread conservation practices.

A glaring omission with CSP as designed in the 2008 farm bill is that the program does not have a minimum annual per-contract payment.[100] Therefore, while "CSP is size-neutral," meaning that smaller farms are eligible to apply for enrollment in the program, very small farms "even if producing very high value conservation per acre . . . still can only earn a certain amount of environmental benefit payment points when multiplying value times acres."[101] The end result of this omission is that a small farm "may only be able to earn a few hundred dollars per year from CSP . . . [which] may not be worth the paperwork" involved in enrolling in the program.[102] Congress has already eliminated this barrier for socially disadvantaged,[103] beginning, and limited-resource farmers by ensuring a minimum annual payment of $1,000 for CSP-enrolled farms meeting any of those three criteria.[104] In the future, efforts to expand this minimum annual payment for these farmers, as well as measures to extend the predetermined minimum annual payment to *all* farmers to ensure that CSP does not inadvertently discriminate against small farms engaged in sustainable practices, would support a more ecologically balanced and transparent food system.

Likewise, it would bolster the program's goals if Congress amended CSP contract terms to allow for more than a single contract renewal. At present, CSP contracts are for five years, and a farm that has increased its environmental benefit score during the contract term may apply for and receive a single five-year renewal contract.[105] But once the contract renewal period ends, farms are precluded from any addi-

---

96. U.S. Dept. Agric., Natural Res. Conservation Serv., *Conservation Stewardship Program*, http://www.nrcs.usda.gov/wps/portal/nrcs/main/?ss=1 6&navid=100120300000000&pnavid=100120000000000&position=SUBNAVIGATION&ttype=main&navtype=SUBNAVIGATION& pname=Conservation%20Stewardship%20Program%20|%20NRCS (last visited Nov. 9, 2011).
97. Ctr. for Rural Affairs, *New Conservation Stewardship Program*, http://www.cfra.org/csp-new-improved (last visited Nov. 28, 2012).
98. U.S. Dept. Agric., Natural Res. Conservation Serv., *CSP 2011 Ranking Period One Enhancement Activity Job Sheets*, http://www.nrcs.usda.gov/ wps/portal/nrcs/detailfull/national/programs/financial/csp/?&cid=stelprdb1045117 (last visited Nov. 28, 2012).
99. *See* Nat'l Sustainable Agric. Coal., Farmer's Guide to the Conservation Stewardship Program: Rewarding Farmers for How They Grow What They Grow 12 (2011), *available at* http://sustainableagriculture.net/wp-content/uploads/2011/09/NSAC-Farmers-Guide-to-CSP-2011.pdf (explaining that only 64% of "beginning, social disadvantaged and limited resource farmers and ranchers" that applied to CSP in 2009 and 2010 received grants, and that the overall grant ratio for all farmers was significantly lower than 64%).
100. *Id.* at 17.
101. *Id.*
102. *Id.*
103. 7 U.S.C. §2003 (defining "socially disadvantaged farmer or rancher" as an individual belonging to "a group whose members have been subjected to racial, ethnic, or gender prejudice because of their identity as members of a group without regard to their individual qualities").
104. *Id.*
105. *See* Nat'l Sustainable Agric. Coal., Farming for the Future, *supra* note 93, at 54.

tional renewals, even if they have satisfied all contractual obligations and increased their environmental benefit scores.[106] Because the CSP is oriented toward encouraging *long-term* programs that enhance the natural environment, the one-renewal limit "is counter-productive to the program's goal to advance ongoing, iterative land stewardship to improve and maintain environmental performance."[107] Therefore, it is imperative that Congress revisit CSP contract renewal terms in order to harmonize them with the underlying purpose of the program.

5. Prioritizing Organic Agriculture Through Funding, Research, and Targeted Set-Asides

The conservation and even arguably the nutrient benefits of organic agriculture are now well understood in the scientific community.[108] In the face of strong political pressure from conventional growers who want to maintain the status quo, the primary challenge is redirecting precious commodity dollars from supporting large farms and processors to programs that benefit organic growers committed to earth-friendly practices, and to research ways to make organic production even more competitive in the market. In addition, efforts to create set-asides for organic growers in existing programs (i.e., a predetermined amount of money only available to a subset of organic producers meeting certain eligibility requirements) would serve to expand the proportion of organic agriculture within our nation's food system.

The National Organic Certification Cost-Share Program (NOCCSP) partially reimburses farms and ranches for the cost of USDA organic certification, making it more likely that those farms can afford certification and thus to have a better chance for financial security.[109] The 2008 farm bill funded NOCCSP at $22 million, or approximately $4.4 million per year, during the life of the legislation, which was a substantial increase from the $5 million ($1 million annually) authorized for the same program during the life of the 2002 farm bill.[110] Sustainable agriculture advocates have called for funding at $30 million in the 2012/2013 farm bill, or an increase of 36%.[111]

Another key requirement to provide stability for organic producers is an equitable organic insurance scheme. In the 2008 farm bill, the insurance plans and premiums offered to organic farmers differed little from those available to conventional farmers, which wholly failed to account for the unique risks and challenges facing organic producers. As the Organic Farming Research Foundation explains:

> USDA currently does not provide appropriate risk management tools for organic producers. The agency charges an unjustified surcharge to organic farmers, and does not pay organic farmers at the organic price after a loss for most commodities. The agency does not provide appropriate tools for diversified farmers.[112]

It is critical to remedy this inequity so that organic producers can compete on the open market with conventional farmers who are also backed by government insurance, and that they have organic insurance plans that sufficiently protect their crop investments in the event of disaster or crop loss. In the end, a policy change on this front would result in organic producers being paid fair market *organic* prices for their crops when insurance claims are paid out in the event of covered crop losses or disasters, as opposed to lower prices that reflect the conventional values (i.e., price of nonorganic counterparts) as is currently the case.

Another centerpiece of reform that would demonstrate Congress' commitment to organic agriculture would be setting aside certain mandatory funding in existing competitive programs for organic producers. CSP exemplifies this need. Indeed, because CSP rewards farmers undertaking substantial conservation efforts—generally a key tenet of organic production in any event—"organic producers are very likely

---

106. *Id.*
107. *Id.*
108. *See, e.g.*, Organic Trade Ass'n, *Nutritional Considerations*, http://www.ota.com/organic/benefits/nutrition.html (summarizing 20 recent scientific studies that have all found measurable increases in nutrients in organically produced foods compared to conventional counterparts) (last visited Nov. 9, 2011).
109. Nat'l Sustainable Agric. Coal., *Organic Certification Cost Share*, http://sustainableagriculture.net/publications/grassrootsguide/organic-production/organic-certification-cost-share/.
110. *Id.*; Food, Conservation, and Energy Act of 2008, Pub. L. No. 110-246, §10301 (2008), *available at* http://frwebgate.access.gpo.gov/cgi-bin/getdoc.cgi?dbname=110_cong_public_laws&docid=f:publ246.110.pdf.
111. Organic Farming Research Found., *Opportunities to Invest in the Growing Organic Sector Through the 2012 Farm Bill*, http://ofrf.org/sites/ofrf.org/files/docs/pdf/2012FarmBillOpportunities2page.pdf.
112. *Id.*

to have extensive conservation systems in place and . . . [t]hus organic farmers may rank high and earn good payments" under CSP.[113] However, despite the clear match between CSP goals and organic producers, "CSP does not have a separate pool of funds for organic producers," which means that many eligible organic producers are excluded from the program because "CSP has proven to be very popular, and thus entry is quite competitive."[114] This could be cured, to some extent, by authorizing a set-aside for organic farmers, just as Congress has already done for beginning farmers (5% set-aside within CSP) and for socially disadvantaged or resource-limited farmers (5% set-aside within CSP).[115] Such a set-aside would ensure that organic producers are being compensated in some way for the choices they make in production methods to better the planet.

One last crucial piece of the organic puzzle is increasing funding for research. An organic research initiative that has produced invaluable information is the Organic Agriculture Research and Extension Initiation (OREI), which is "[u]nique in its scope and function," which "funds research and extension projects to help meet the production, marketing, and policy needs of the growing organic industry," and which "helps farmers be successful and improve and increase production."[116] OREI is a competitive grant program, and only funds a small percentage of eligible proposals each year.[117] The 2008 farm bill authorized $18 million in 2009, and $20 million annually in 2010-2012.[118] To address the number of innovative organic research projects turned away at that funding level, leading scientific and advocacy organizations called for $30 million of mandatory annual funding for OREI in the 2012/2013 farm bill.[119] Whether Congress will agree to fund the program at that level remains to be seen, but the intense competition for OREI grants in the past indicates that organic research is sorely needed to protect our ecosystems and organic producers, and taxpayer funding is necessary to ensure that these research vehicles are prioritized.

6. Bolstering Local and Regional Food Systems

Finally, there are many laudable farm bill programs in their relative infancy that, if sustained and funded adequately, have the potential to support a drastically different food system—one that is based, in large part, on local and regional production and distribution rather than the industrial model promoted by the farm bill for the past several decades. Congress allocated funding for the Farmers' Market Promotion Program (FMPP) in the 2008 farm bill, authorizing $3 million in 2008, $5 million annually in 2009 and 2010, and $10 million annually in 2011 and 2012.[120] FMPP is a competitive grant program "targeted to help improve and expand domestic farmers' markets, roadside stands, community-supported agriculture programs, agri-tourism activities, and other direct producer-to-consumer market opportunities," which inevitably results in less environmental damage due to reduced transportation and a fresher and more nutritious end product for the consumer, as described in Chapter 7.[121] In particular, "[s]pecific grant uses include developing relevant financial and marketing information, business planning, improving market access and education for consumers, organizing markets and direct marketing networks, and supporting innovative approaches to market management and operations."[122] Because FMPP is essential to ensuring that local and regional food systems can persist against the competition of cheap processed supermarket foods, this is a program that is worthy of congressional support through future funding increases. During consideration of the 2012/2013 farm bill, several members of Congress and policy advocates requested that

---

113. Nat'l Sustainable Agric. Coal., Farmer's Guide to the Conservation Stewardship Program: Rewarding Farmers for How They Grow What They Grow 17 (2011), *available at* http://sustainableagriculture.net/wp-content/uploads/2011/09/NSAC-Farmers-Guide-to-CSP-2011.pdf.
114. *Id.* at 17, 20.
115. *Id.* at 12.
116. Organic Farming Research Found., *supra* note 111.
117. *Id.*
118. Food, Conservation, and Energy Act of 2008, Pub. L. No. 110-246, §7206(f) (2008), *available at* http://www.gpo.gov/fdsys/pkg/PLAW-110publ246/pdf/PLAW-110publ246.pdf.
119. Organic Farming Research Found., *supra* note 111.
120. Food, Conservation, and Energy Act of 2008, Pub. L. No. 110-246, §10106 (2008), *available at* http://www.gpo.gov/fdsys/pkg/PLAW-110publ246/pdf/PLAW-110publ246.pdf.
121. U.S. Dept. Agric., Agricultural Marketing Service, *Farmers Market Promotion Program*, http://www.ams.usda.gov/AMS v1.0 /FMPP (last visited Nov. 9, 2011).
122. Nat'l Sustainable Agric. Coal., *Farmers' Market Promotion Program*, http://sustainableagriculture.net/publications/grassrootsguide/local-food-systems-rural-development/farmers-market-promotion-program/.

FMPP be "refashioned" to "do everything FMPP does, but also [to] provide grants to scale up local and regional food enterprises, including processing, distribution, aggregation, storage, and marketing," with the money equally allocated between traditional direct-market FMPP activities and scaled-up activities for local and regional food systems, such as retail and institutional markets.[123]

Another competitive grant program that bolsters local (as well as nonlocal) food systems is the Value-Added Producer Grant program (VAPG), which provides grants to farmers to produce "value-added" products.[124] What constitutes a value-added product is quite expansive; examples that fall within the definition and are thus eligible for grants include wine, flour, cheese, jam, organic grass-fed beef, GE-free foods, non-rBGH dairy products, or business entities selling directly from farm to institution (for example, to schools, prisons, and hospitals).[125] Congress authorized $40 million annually under the 2008 farm bill, with grants of up to $50,000 per grantee, although it ultimately appropriated only $20.4 million in 2010 and just under $19 million in 2011, meaning that legislative funding fell short of what the farm bill promised.[126] This program has been instrumental in helping various organic and other sustainable farms and businesses add value to their products and maintain thriving operations without compromising their environmental ethics, and many of these products (although not all) are sold in local and regional food markets.[127] Accordingly, securing additional funding from Congress for this program—since farmers selling locally can submit grants to take advantage of these incentives—is critically important.

Yet another creative policy solution to enhance local and regional food systems is community food projects (CFPs), which are funded through competitive grants as "proactive approaches to making communities more self reliant at maintaining their food systems while addressing food, nutrition, and farm issues."[128] These projects are "designed to increase food security in communities by bringing the whole food system together to assess strengths, establish linkages, and create systems that improve the self-reliance of community members over their food needs."[129] Congress authorized approximately $5 million annually in the 2008 farm bill, and a CFP grant can last up to three years.[130] To make CFPs even more effective, several senators and congressmen introduced a bill in 2011 calling for a doubling of mandatory annual funding to $10 million in the 2012/2013 farm bill, as well as an increase in the term of CFP grants from three years to five years to allow time for proper implementation of ideas developed under the grants.[131]

One legislative step for which support is needed is an amendment to the National School Lunch Program (NSLP) to allow schools participating in the program to utilize a certain amount of their allocated entitlement dollars—which are normally used to purchase highly processed foods made from surplus commodities such as corn and soybeans[132]—to instead purchase locally produced fruits and vegetables, as well as local value-added products, in lieu of highly processed commodities that travel thousands of miles. While this type of local and value-added set-aside does not currently exist in the NSLP, there is growing pressure on Congress from sustainable agriculture organizations and progressive elected officials for precisely this type of "local food credit program," which, if adopted, would enable schools to spend up to

---

123. Local Farms, Food, and Jobs Act of 2011, S. 1773, 112th Cong., 1st Sess., §7004 (2011), *available at* http://www.gpo.gov/fdsys/pkg/BILLS-112s1773is/pdf/BILLS-112s1773is.pdf.
124. Nat'l Sustainable Agric. Coal., *Value-Added Producer Grants*, http://sustainableagriculture.net/publications/grassrootsguide/local-food-systems-rural-development/value-added-producer-grants/.
125. *Id.*
126. Food, Conservation, and Energy Act of 2008, Pub. L. No. 110-246, §6202 (2008), *available at* http://www.gpo.gov/fdsys/pkg/PLAW-110publ246/pdf/PLAW-110publ246.pdf; Nat'l Sustainable Agric. Coal., *Value-Added Producer Grants*, http://sustainableagriculture.net/publications/grassrootsguide/local-food-systems-rural-development/value-added-producer-grants/.
127. Nat'l Sustainable Agric. Coal., *Value-Added Producer Grants*, *supra* note 124.
128. U.S. Dept. Agric., National Institute of Food and Agriculture, *Community Food Projects*, http://www.csrees.usda.gov/nea/food/sri/hunger_sri_awards.html (last visited Nov. 9, 2011).
129. U.S. Dept. Agric., National Institute of Food and Agriculture, *Community Food Projects Competitive Grants*, http://www.csrees.usda.gov/nea/food/in_focus/hunger_if_competitive.html (last visited Nov. 9, 2011).
130. Food, Conservation, and Energy Act of 2008, Pub. L. No. 110-246, §4402 (2008), *available at* http://www.gpo.gov/fdsys/pkg/PLAW-110publ246/pdf/PLAW-110publ246.pdf.
131. Local Farms, Food, and Jobs Act of 2011, S. 1773, 112th Cong., 1st Sess., §3008 (2011), *available at* http://www.gpo.gov/fdsys/pkg/BILLS-112s1773is/pdf/BILLS-112s1773is.pdf.
132. *See generally* U.S. Dept. Agric., Food and Nutrition Service, *White Paper: USDA Foods in the National School Lunch Program* (May 2010), *available at* http://www.fns.usda.gov/fdd/foods/healthy/WhitePaper.pdf.

15% (or some other defined percentage) of their entitlement dollars on locally produced foods.[133] If such a program were created, our nation's school foods would take a significant step towards better nutritional, environmental, and economic sustainability.

Conclusion

The alarming environmental, health, and economic toll of our nation's industrialized food system is reason enough to wish for a radical change in the status quo whereby our elected representatives appear to favor agribusiness interests at the expense of the needs of the American public and our shared natural resources. While a truly green revolution could be achieved if the will of the American people were to fully endorse it (which at this time seems unlikely), the more likely pathway to solving many problems that directly or indirectly result from our agricultural policies is to embrace the long-standing policy structure and to gradually mold and shape that omnibus policy until our food system mirrors the system that reflects a true "green." There are myriad existing programs and creative ideas for new programs that have extraordinary potential to reward farmers for implementing sound ecological practices and cultivating nutritious products for consumers, to fund research on key scientific and economic objectives to ensure a stable and fair farm economy, and to establish the critical connection between consumers, their farmers, and the lands upon which our daily meals are grown. If we succeed in breathing renewed life into the farm bill in such a manner after having endured decades of legislative bias towards a certain form of agriculture and of our communal unlearning of our nation's agrarian roots, we will shed the outdated and antiquated mid-20th century farming and food model and instead build a more equitable, just, and sustainable 21st century food system unlike any the modern world has ever seen.

---

133. Local Farms, Food, and Jobs Act of 2011, S. 1773, 112th Cong., 1st Sess., §3004 (2011), *available at* http://www.gpo.gov/fdsys/pkg/BILLS-112s1773is/pdf/BILLS-112s1773is.pdf.

Chapter 16
Regulating Transgenic Crops Pursuant to the Plant Protection Act
George A. Kimbrell

The U.S. government's oversight of genetically engineered (GE) crops can be charitably described as limited. Over the past 15 years, commercial approval has opened the door to the planting of millions of transgenic acres, yet the environmental and health impacts of this widespread change in our agricultural landscape are not being adequately studied or regulated. The U.S. Department of Agriculture (USDA), entrusted with the chief responsibility for regulation of these transgenic crops, has not contained them, and as a result they have caused significant economic harm and transgenic pollution of both conventional and wild plant species. Government investigations, courts, and even the U.S. Congress have all found USDA's practices inadequate.

Similarly, agricultural biotechnology itself has not increased yields, reduced world hunger, or mitigated global warming. Instead, there is growing evidence that these crops carry with them significant adverse environmental and intertwined socioeconomic impacts: transgenic contamination (gene flow from GE crops to related conventional or organic cultivars or wild species); significant increases in herbicides used in engineered herbicide-resistant (HR) cropping systems; and the creation of weeds resistant to these herbicides. The vast majority of transgenic crops are genetically engineered to be resistant to the direct application of herbicides, mainly Monsanto's now-ubiquitous Roundup, leading to increased and indiscriminate usage. Through the widespread adoption of GE "Roundup Ready" cropping systems, Roundup has now become the most heavily used herbicide in the history of agriculture. In addition, USDA has declined to act as crops under its watch continue to exacerbate one of the greatest threats to agriculture in a generation: an epidemic of HR "superweeds" requiring ever more numerous applications of increasingly toxic herbicide cocktails. Instead, USDA has thus far self-limited its own review, providing no post-market monitoring or restrictions on planting.

Revising the 1986 Coordinated Framework for the Regulation of Biotechnology,[1] which laid out a mosaic of federal agencies to oversee the development of genetically engineered organisms, remains an important focus of the U.S. biotech policy dialogue. While expert analysts have pointed out the Framework's shortcomings, numerous congressional bills intended to remedy them have failed to pass.[2]

---

1. Coordinated Framework for Regulation of Biotechnology, 51 Fed. Reg. 23302 (June 26, 1986).
2. *See, e.g.*, Genetically Engineered Foods Act, S. 2546, 108th Cong. (2004); S. 3095, 107th Cong. (2002); S. 3184, 106th Cong. (2000); H.R. 713, 107th Cong. (2002); Genetically Engineered Food Safety Act, S. 2315, 112th Cong. (2011); H.R. 3883, 112th Cong. (2011); H.R. 5268, 109th Cong. (2006); H.R. 2917, 108th Cong. (2003); H.R. 4813, 107th Cong. (2002); Genetically Engineered Food Right to Know Act, H.R. 5577, 111th Cong. (2010); H.R. 6636, 110th Cong. (2008); H.R. 5269, 109th Cong. (2006); H.R. 5269, 109th Cong. (2006); H.R. 2916, 108th Cong. (2003); H.R. 4814, 107th Cong. (2002); Genetically Engineered Pharmaceutical and Industrial Crop Safety Act of 2003, H.R. 2921, 108th Cong. (2003); Real Solutions to World Hunger Act of 2003, H.R. 2920, 108th Cong. (2003); The Genetically Engineered Crop and Animal Farmer Protection Act, H.R. 5266, 109th Cong. (2006); The Genetically Engineered Pharmaceutical and Industrial Crop Safety Act of 2005, H.R. 5267, 109th Cong. (2006); Real Solutions to World Hunger Act of 2005, H.R. 5270, 109th Cong. (2006); The Genetically Engineered Organism Liability Act, H.R. 5271, 109th Cong. (2006); Genetically Engineered Technology Farmer Protection Act, H.R. 6637, 110th Cong. (2008); Genetically Engineered Safety Act, H.R. 5578, 111th Cong. (2010); H.R. 6635, 110th Cong. (2008); Genetically Engineered Technology Farmer Protection Act, H.R.5579, 111th Cong. (2010); A Bill to Amend the Federal Food, Drug, and Cosmetic Act to Prevent the Approval of Genetically Engineered Fish, S.230, 112th Cong. (2011); H.R. 521, 112th Cong. (2011); A Bill to Amend the Federal Food, Drug, and Cosmetic Act to Require Labeling of Genetically Engineered Fish, S. 229, 112th Cong. (2011); H.R.520, 112th Cong. (2011).

An analysis of the Framework, on which there is already a rich academic literature,[3] is beyond the scope of this chapter, which focuses only on USDA's oversight. That said, imagine that an obscure law already exists that could better address many of the above issues, should it only be fully implemented. Imagine further that this law provides USDA broad authority to protect the environment, agriculture, and health from direct and indirect harms. This law exists and is known as the Plant Protection Act of 2000 (PPA).[4] Yet thus far, USDA has continued to regulate GE crops under regulations promulgated under earlier, narrower laws, and has not applied the full mandate delegated to it in the PPA.

This chapter argues that USDA should revise its regulation of transgenic crops to apply its full PPA authority. It first provides an overview of agricultural biotechnology's development and current agronomic reality, then focuses on some of the main impacts of transgenic crops. The chapter outlines USDA's current oversight structure and provides a blueprint for needed regulatory improvement. The chapter concludes that USDA's oversight could be greatly improved, and as a consequence the impacts of transgenic crops better analyzed, regulated, and prevented, under a robust application of the PPA that takes into consideration the direct and indirect harms of transgenic crops in order to protect all interests of agriculture, public health, and the environment.

A. Agricultural Biotechnology

As discussed in Chapter 6, agricultural biotechnology generally refers to the use of recombinant DNA techniques and related tools of biotechnology to genetically engineer crops used in the production of food, feed, and fiber. The resulting products are interchangeably referred to as "transgenic" or "genetically engineered" (GE). Genetic engineering is not the same as traditional plant breeding. The latter process involves identifying similar, related plants with useful traits and crossing these plants to produce offspring with the desired characteristics of both parents. Genetic engineering is a powerful technology that allows scientists to combine genetic material from widely dissimilar and unrelated organisms—for example, bacterial genes with alfalfa genes or chicken genes with maize genes. In other words, scientists can produce combinations of genetic material that have never before occurred in nature.[5]

In the 1980 landmark case *Diamond v. Chakrabarty*, the U.S. Supreme Court ruled by a 5-4 margin that living organisms could be patented.[6] Because the patentee had introduced new genetic material within the bacterium cell, the Court held that he had produced something that was not a product of nature and was thus patentable subject matter.[7] That decision paved the way for the U.S. Patent and Trademark Office to decide in the 1985 case *Ex parte Hibberd* that sexually reproducing plants are patentable under the Patent Act, providing stronger protection and greater profit potential for seed companies.[8] Previously, such plants were only protected under the 1970 Plant Variety Protection Act (PVPA), which provided temporary exclusivity, but exempted farmers, who could then save seed and replant, and plant researchers, who could use protected varieties to breed improved plants.[9] In 2001, another 5-4 Supreme Court decision in *J.E.M Ag Supply v. Pioneer Hi-Bred International* upheld the granting of utility patents, which do not have these exemptions, for plants.[10] These decisions opened the door to expansive intellectual property rights in genetically engineered organisms and crops.

As a consequence, firms raced to patent genetic resources and plant breeding technologies and to purchase existing seed companies; the agricultural biotechnology industry emerged through the rapid acquisi-

---

3. *See, e.g.*, Mary Jane Angelo, *Regulating Evolution for Sale: An Evolutionary Biology Model for Regulating the Unnatural Selection of Genetically Modified Organisms*, 42 WAKE FOREST L. REV. 93, 112 (2007); Douglas A. Kysar, *Preferences for Processes: The Process/Product Distinction and the Regulation of Consumer Choice*, 118 HARV. L. REV. 525 (2004); Gregory N. Mandel, *Gaps, Inexperience, Inconsistencies, and Overlaps: Crisis in the Regulation of Genetically Modified Plants and Animals*, 45 WM. & MARY L. REV. 2216 (2004); John Charles Kunich, *Mother Frankenstein, Doctor Nature, and the Environmental Law of Genetic Engineering*, 74 S. CAL. L. REV. 807 (2001); Rebecca Bratspies, *Some Thoughts on the American Approach to Regulating Genetically Modified Organisms*, 16 KAN. J.L. & PUB. POL'Y 393 (2007).
4. 7 U.S.C. §§7701 et seq. (2011).
5. *See, e.g.*, Stanley Cohen et al., *Construction of Biologically Functional Bacterial Plasmids in Vitro*, 70 PROC. NAT'L ACAD. SCI. 3240-44 (1973).
6. 447 U.S. 303 (1980).
7. *Id.* at 309-10.
8. 227 U.S.P.Q. 443 (Bd. Pat. App. & Interferences 1985).
9. J.E.M. Ag Supply, Inc. v. Pioneer Hi-Bred Intern., Inc., 534 U.S. 124, 140, 122 S.Ct. 593, 603 (2001); *see also* Asgrow Seed Co. v. Winterboer, 513 U.S. 179, 115 S.Ct. 788, 130 L. Ed. 2d 682 (1995); *see also* 7 U.S.C. §2544.
10. *See* J.E.M. Ag Supply, 534 U.S. at 127.

tion of existing seed firms by chemical and pesticide companies such as Monsanto, DuPont, Syngenta, and Dow.[11] Dozens of mergers and acquisitions followed; at least 200 independent seed companies were bought out and consolidated from 1996 to 2009.[12] The four dominant firms in the agricultural chemical market now account for 43% of the global commercial seed market.[13] As smaller and independent companies disappeared, farmers encountered fewer non-GE seed options.[14] Based on the seed patents, the companies require farmers to sign contracts called "technology use agreements," prohibiting them from saving and replanting the seed in the age-old farming tradition and instead requiring them to repurchase it annually from them. Farmers suspected of violating contract terms are prosecuted by seed companies.[15]

U.S. adoption of transgenic crops has been relatively rapid, but limited in the main to the major commodity crops, of which transgenic varieties now make up the vast majority: soybean (93% transgenic in 2010), cotton (88%), corn (86%), and canola (64%).[16] So far, engineered food has been an American-dominated experiment: total acreage in the United States was 158 million acres in 2009, dwarfing that of the next closest countries, Brazil (52 million), Argentina (52 million), Canada (20 million), India (20 million), and China (9 million).[17] These six countries make up 95% of the world's transgenic cultivation.[18]

By contrast, the total global acreage of transgenic crops is less than 3% of all agricultural land.[19] In Europe, where only two GE crops have been permitted, already limited cultivation declined between 2008 and 2010, to a total of approximately 82,000 hectares (200,000 acres),[20] as six European Union (EU) countries (Austria, France, Germany, Greece, Hungary, and Luxembourg) banned the main EU-authorized crop, a Monsanto transgenic corn, and another country (Bulgaria) banned all potential transgenic varieties.[21] Also in 2010, India, where farmers have grown transgenic cotton since 2002, announced a moratorium on what would have been its first transgenic food crop, a transgenic eggplant.[22]

Despite one-quarter century of promises and over 15 years of commercialization, agricultural biotechnology has yet to provide concrete advancements towards reducing world hunger, ameliorating global malnutrition, combating global warming, or creating miracle drugs through GE plant and animal "biofactories." Instead, biotechnology firms have delivered GE commodity crop types that produce pesticides and/or withstand direct application of herbicides. This herbicide resistance lends crops the ability to survive indiscriminate spraying of a broad-spectrum herbicide to kill nearby weeds. Over five of every six acres of transgenic crops worldwide (84%) are engineered for herbicide resistance.[23] Despite claims that these crops increase yields, the only independent study of their results, conducted by the Union of Con-

---

11. Philip H. Howard, *Visualizing Consolidation in the Global Seed Industry: 1996-2008*, 1 SUSTAINABILITY 1266-87 (2009).
12. KRISTINA HUBBARD, FARMER TO FARM CAMPAIGN ON GENETIC ENG'G, NAT'L FAMILY FARM COALITION, OUT OF HAND: FARMERS FACE THE CONSEQUENCES OF A CONSOLIDATED SEED INDUSTRY, at 4 (2009), *available at* http://farmertofarmercampaign.com/Out%20of%20Hand.FullReport.pdf.
13. *Id.* at 8.
14. *Id.* at 25-38.
15. *See generally* CENTER FOR FOOD SAFETY, REPORT, MONSANTO V. U.S. FARMERS 13, 23-48, app. A (2005), *available at* http://www.centerforfoodsafety.org/pubs/CFSMOnsantovsFarmerReport1.13.05.pdf; *see also* Monsanto v. Parr, 545 F. Supp. 2d 836 (N.D. Ind. 2008); Monsanto v. Scruggs, 459 F.3d 1328 (Fed. Cir. 2006); Monsanto v. McFarling, 363 F.3d 1336 (Fed. Cir. 2004); Monsanto v. Trantham, 156 F. Supp. 2d 855 (W.D. Tenn. 2001).
16. Wayne Peng, *GM Crop Cultivation Surges, but Novel Traits Languish*, 29 NATURE BIOTECH. 302 (Apr. 2011), *available at* http://www.nature.com/nbt/journal/v29/n4/box/nbt.1842_BX5.html.
17. *See, e.g.*, *A Recession Ends, the Fall Campaign Revs Up*, WALL ST. J., Sept. 20-24, 2010, at Graph/Image 1, *High Tech Harvest*, *available at* http://online.wsj.com/article/SB10001424052748703384204575510561292390660.html?KEYWORDS=high-tech+harvest (citing International Service for the Acquisition of Agri-Biotech Applications and USDA); *see also* Wayne Peng, *GM Crop Cultivation Surges, but Novel Traits Languish*, 29 NATURE BIOTECH. 302 (Apr. 2011), *available at* http://www.nature.com/nbt/journal/v29/n4/box/nbt.1842_BX5.html.
18. FRIENDS OF THE EARTH, WHO BENEFITS FROM GE CROPS? 7 (Feb. 2011), *available at* http://www.foei.org/en/what-we-do/food-sovereignty/latest-news/who-benefits-from-gm-crops.
19. *Id.* at 7.
20. Friends of the Earth Europe, *Fact Sheet—22nd February 2011 GM Crops Continue to Fail in Europe* (Feb. 2011), http://www.biosafety-info.net/file_dir/1480815384d64d9194a538.pdf.
21. BBC News, *GM Crops: EU Parliament Backs National Bans*, July 6, 2011, http://www.bbc.co.uk/news/world-europe-14045365; Irian Ivanova, *Bulgaria Parliament Bans GMO Crops to Soothe Fears*, REUTERS, Mar. 18, 2010, at http://www.reuters.com/article/2010/03/18/us-bulgaria-gmo-idUSTRE62H3EJ20100318.
22. BBC News, *India Puts on Hold First GM Food Crop on Safety Grounds*, Feb. 9, 2010, http://news.bbc.co.uk/2/hi/8506047.stm; Chetan Chauhan, *Ramesh Faults Regulator, Says Standards Fail Global Norms*, HINDUSTAN TIMES, Feb. 10, 2010, at http://www.hindustantimes.com/News-Feed/newdelhi/Ramesh-faults-regulator-says-standards-fail-global-norms/Article1-507198.aspx.
23. Center for Food Safety, Revised Comments delivered at the Aug. 1, 2007 Meeting of the USDA's Advisory Comm. on Biotechnology and 21st Century Agric., *Genetically Modified (GM) Crops and Pesticide Use* (Feb. 2009), *available at* http://www.co.lake.ca.us/Assets/BOS/GE+Crops+Committee/6.+GM+Crops+and+Pesticide+Use.pdf.

cerned Scientists, concluded that they have not, while at the same time successes in traditional breeding increased yields.[24]

Monsanto, now the world's largest seed company,[25] uses genetic engineering primarily to create patented "Roundup Ready" crops for use in tandem with its Roundup herbicide. American soybeans, corn, cotton, canola, and sugar beets are now largely Roundup Ready.[26] This has made glyphosate (Roundup's active ingredient) the most heavily used chemical pesticide in history, with 180-185 million pounds applied in U.S. agriculture in 2007 alone.[27]

B. The Impacts of Transgenic Crops

At the most basic level, transgenic crops reinforce an industrial agriculture paradigm of questionable sustainability at the expense of more environmentally sound methods of farming. The privatization and concentration of the world's seed supply is an escalating problem, compounded with each new transgenic crop approval. Market concentration has resulted in 10 multinational corporations holding approximately two-thirds (65%) of commercial seed for major crops, reducing choice and increasing prices for the American farmer.[28] In 2009, the U.S. Department of Justice began an investigation into anticompetitive practices resulting in sharply rising GE seed prices and a dwindling supply of non-GE seed due to Monsanto's seed pricing systems and market control.[29] Other controversial issues include the privatization of the millennia-old practice of seed-saving through genetic engineering and patents and Monsanto's subsequent use of lawsuits to sue farmers attempting to continue to save their seed[30]; crop breeding programs that are guided by corporate profit rather than the public interest and the needs of farmers[31]; and the inability of scientists to undertake independent research on the potential adverse health and environmental impacts of transgenic crops due to a lack of access to transgenic seeds and corporate control over their findings.[32]

One major adverse impact stemming from the cultivation of GE crops is transgenic contamination: the unintended, undesired presence of GE crop material in organic or conventional (non-GE) crops, as well as in wild species.[33] "Gene flow" causes contamination when a crop disperses its pollen or seeds to propagate itself over time and space. A GE crop can cross-pollinate a crop or wild plant of the same species (via wind or insect pollinator) and thereby transfer its transgene and associated trait to that plant. GE crop seed can contaminate non-GE crops in numerous ways, via wind (for light seed), flooding, improper seed cleaning of machinery, spillage during transport, and a variety of human errors that may occur at each stage of the crop production process.[34]

Harm from transgenic contamination manifests itself in several ways and includes both an environmental and socioeconomic component.[35] It causes significant widespread economic harm to the agricultural

---

24. UNION OF CONCERNED SCIENTISTS, FAILURE TO YIELD: EVALUATING THE PERFORMANCE OF GENETICALLY ENGINEERED CROPS 1-5 (Apr. 2009), *available at* http://www.ucsusa.org/food_and_agriculture/science_and_impacts/science/failure-to-yield.html.
25. Chittur Subramanian Srinivasan, *Concentration in Ownership of Plant Variety Rights: Some Implications for Developing Countries*, 28 FOOD POL'Y 5-6 , at 519-46 (2003).
26. USDA figures show that 93% of all soybeans, 78% of all cotton, and 70% of all corn grown in the United States in 2010 were genetically engineered, HR varieties—nearly all Roundup Ready. U.S. Dep't Agric., Economic Research Serv., *Adoption of Genetically Engineered Crops in the U.S.*, http://www.ers.usda.gov/Data/BiotechCrops/. *See also* William Neuman & Andrew Pollack, *Farmers Cope With Roundup-Resistant Weeds*, N.Y. TIMES, May 3, 2010, at http://www.nytimes.com/2010/05/04/business/energy-environment/04weed.html?_r=1&pagewanted=all.
27. U.S. EPA, *Pesticide Industry Sales and Usage: 2006 and 2007 Market Estimates*, tbl. 3-6 (Feb. 2011), *available at* http://www.epa.gov/opp00001/pestsales/07pestsales/market_estimates06-07.pdf.
28. *See, e.g.*, HUBBARD, *supra* note 12, at 12-13.
29. *See, e.g.*, William Neuman, *Rapid Rise in Seed Prices Draws U.S. Scrutiny*, N.Y. TIMES, at B1 (Mar. 12, 2010), *available at* http://www.nytimes.com/2010/03/12/business/12seed.html.
30. CENTER FOR FOOD SAFETY, REPORT, *MONSANTO V. U.S. FARMERS* (2005), *available at* http://www.centerforfoodsafety.org/pubs/CFSMOnsantovsFarmerReport1.13.05.pdf.
31. HUBBARD, *supra* note 12, at 12-13.
32. Doug Gurian-Sherman, Op-Ed, *No Seeds, No Independent Research*, L.A. TIMES, Feb. 13, 2011, at A19, *available at* http://articles.latimes.com/2011/feb/13/opinion/la-oe-guriansherman-seeds-20110213; Bruce Stutz, *Companies Put Restrictions on Research Into GM Crops*, YALE ENV'T 360, May 13, 2010, *available at* http://e360.yale.edu/content/feature.msp?id=2273; Andrew Pollack, *Crop Scientists Say Biotechnology Seed Companies Are Thwarting Research*, N.Y. TIMES, Feb. 19, 2009, at http://www.nytimes.com/2009/02/20/business/20crop.html.
33. *See* Geertson Seed Farms v. Johanns, 2007 WL 518624, at *5 (N.D. Cal. Feb. 17, 2007) ("Biological contamination can occur through pollination of non-genetically engineered plants by genetically engineered plants or by the mixing of genetically engineered seed with natural or non-genetically engineered seed.").
34. *See, e.g.*, Michelle Marvier and Rene C. Van Acker, *Can Crop Transgenes Be Kept on a Leash?*, 3 FRONTIERS ECOL. ENV'T 95-100 (2005).
35. Monsanto Co. v. Geertson Seed Farms, 130 S. Ct. 2743, 2756 (2010) (holding that the "injury has an environmental as well as an economic component"); Geertson Seed Farms, 2007 WL 518624, at *8 ("Here, the economic effects on the organic and conventional farmers of the

economy, both domestically and abroad[36]; the fundamental loss of choice for farmers and consumers[37]; and irreparable contamination of the wild.[38] Unlike standard chemical pollution, transgenic contamination is a living pollutant that can propagate itself via gene flow.[39] As one federal court found: "Once the gene transmission occurs and a farmer's seed crop is contaminated with the Roundup Ready gene, there is no way for the farmer to remove the gene from the crop or control its further spread."[40]

A second major adverse impact of transgenic crops is the growing epidemic of HR "superweeds." HR GE crops withstand direct, "over the top" application of a herbicide that is toxic to conventional crops, facilitating season-long application of a herbicide that otherwise is used primarily prior to planting or sprouting of a conventional crop seed in order to remove early season weeds. As a consequence, Roundup Ready crops have fostered an ongoing epidemic of glyphosate-resistant weeds, now regarded by agronomists as one of the most serious challenges facing American agriculture.[41] The weeds evolve most quickly when Roundup Ready crops are grown year after year, without break, on the same fields. Like bacteria exposed to antibiotics, some weeds naturally resistant to glyphosate survive exposure, and then reproduce and flourish.

Glyphosate-resistant weeds were unknown in the two decades from the introduction of glyphosate in 1974 to the introduction of Roundup Ready crops in 1996. Since the year 2000, glyphosate-resistant weeds evolved in an epidemic manner,[42] infesting over 10 million acres of cropland in 26 states.[43] Glyphosate-resistant weed-infested acreage in the United States quadrupled from November of 2007 to summer of 2009,[44] and is projected to nearly quadruple again to 38 million acres by 2013.[45] These superweeds lead to increased use of glyphosate and more toxic herbicide cocktails, greater use of soil-eroding tillage operations to physically remove weeds, and massive deployment of weeding crews to manually remove weeds, all of which increase farmers' weed control costs, often dramatically.[46]

The rapid evolution of glyphosate-resistant weeds also set the stage for rapid adoption of the next generation of transgenic crops, which are engineered for resistance to older, more toxic herbicides like 2,4-D, dicamba, and imidazolinones, often in combination.[47] These multiple HR, "stacked" crops—presented by the pesticide/biotech industry as the "solution" to glyphosate-resistant weeds—will in turn foster multiple HR weeds and a toxic spiral of increased herbicide use in response.[48] Further, 2,4-D, an active ingredient in the Agent Orange defoliant used in the Vietnam War, is a probable human carcinogen, endocrine dis-

---

government's deregulation decision are interrelated with, and, indeed, a direct result of, the effect on the physical environment; namely, the alteration of a plant specie's DNA through the transmission of the genetically engineered gene to organic and conventional alfalfa.").

36. *See, e.g.*, Carey Gillam, U.S. Organic Food Industry Fears GMO Contamination, REUTERS NEWS SERV., Mar. 12, 2008, *available at* http://www.reuters.com/article/idUSN1216250820080312; Andrew Harris & David Beasley, *Bayer Agrees to Pay $750 Million to End Lawsuits Over Gene-Modified Rice*, BLOOMBERG NEWS, July 1, 2011, at http://www.bloomberg.com/news/2011-07-01/bayer-to-pay-750-million-to-end-lawsuits-over-genetically-modified-rice.html; K.L. Hewett, *The Economic Impacts of GM Contamination Incidents on the Organic Sector*, 16th IFOAM Organic World Congress, Modena, Italy (June 16-20, 2008); Stuart Smyth et al., *Liabilities and Economics of Transgenic Crops*, 20 NATURE BIOTECH. 6 (June 2002).
37. *See, e.g.*, Geertson Seed Farms, 2007 WL 518624 at *8; *id.* at *9 ("For those farmers who choose to grow non-genetically engineered alfalfa, the possibility that their crops will be infected with the engineered gene is tantamount to the elimination of all alfalfa; they cannot grow their chosen crop."); *see also* Center for Food Safety v. Vilsack, 2009 WL 3047227, at *9 (N.D. Cal. Sept. 21, 2009).
38. *See generally* CENTER FOR FOOD SAFETY, CONTAMINATING THE WILD? GENE FLOW FROM EXPERIMENTAL FIELD TRIALS OF GENETICALLY ENGINEERED CROPS TO RELATED WILD PLANTS 1 (2006), *available at* http://www.centerforfoodsafety.org/pubs/Contaminating_the_Wild_Report.pdf; *see, e.g.*, Jay R. Reichman et al., *Establishment of Transgenic Herbicide-Resistant Creeping Bentgrass (Agrostis solonifera L.) in Nonagronomic Habitats*, 15 MOLECULAR ECOLOGY 4243-4255, *available at* http://onlinelibrary.wiley.com/doi/10.1111/j.1365-294X.2006.03072.x/abstract.
39. *See, e.g.*, Rachel Bernstein, *Study Details Wild Crop of Genetically Modified Canola*, L.A. TIMES, Aug. 14, 2010, at http://www.post-gazette.com/pg/10226/1079933-115.stm; *New Study Finds GM Genes in Wild Mexican Maize*, NEW SCIENTIST, Feb. 21, 2009; Mitch Lies, *Bentgrass Eradication Plan Unveiled*, CAPITAL PRESS, June 16, 2011, at http://www.capitalpress.com/newest/ml-scotts-061711.
40. Geertson Seed Farms, 2007 WL 518624, at *5.
41. Stephen B. Powles, *Gene Amplification Delivers GR Weed Evolution*, PNAS 107, 955-56 (2010).
42. Robert Service, *A Growing Threat Down on the Farm*, SCI., May 25, 2007, at 1114-17.
43. Jerry Adler, *The Growing Menace From Superweeds*, SCI. AM., May 2011, *available at* http://www.scientificamerican.com/article.cfm?id=the-growing-menace-from-superweeds.
44. Congressional testimony of Penn State weed scientist David A. Mortensen, *available at* http://live.psu.edu/story/48259.
45. U.S. DEP'T AGRIC., APHIS, GLYPHOSATE-TOLERANT ALFALFA EVENTS J101 AND J163: REQUEST FOR NONREGULATED STATUS: FINAL ENVIRONMENTAL IMPACT STATEMENT (Dec. 2010) [hereinafter *Roundup Ready Alfalfa FEIS*], *available at* http://blogs.desmoinesregister.com/dmr/wp-content/uploads/2010/12/AlfalfaEIS.pdf.
46. *See, e.g.*, Georgina Gustin, *Resistant Weeds Leave Farmers Desperate*, ST. LOUIS POST-DISPATCH, July 17, 2011, at http://www.stltoday.com/business/local/article_f01139be-ace0-502b-944a-0c534b70511c.html; Charles Benbrook, The Organic Center, *Impacts of Genetically Engineered Crops on Pesticide Use: The First Thirteen Years* (Nov. 2009), *available at* http://www.organic-center.org/science.pest.php?action=view&report_id=159.
47. *See, e.g.*, S. Kilman, *Superweed Outbreak Triggers Arms Race*, WALL ST. J., June 4, 2010.
48. *See* Bill Freese, Science Policy Analyst, Center for Food Safety, *Response to Questions From Congressional Committee Investigating Herbicide-Resistant Weeds*, *available at* http://www.centerforfoodsafety.org/wp-content/uploads/2011/03/Oversight-hearing-Freese-Response-to-Questions-corrected.pdf.

ruptor, and a possible neurotoxin.[49] Studies show dicamba to be a potential carcinogen and developmental toxin.[50] Both herbicides are also highly volatile, prone to drift and off-target impacts, which can significantly harm nonresistant, conventional crops in the vicinity as well as destroy wildlife habitat.[51] Scientists estimate that use of these older, more toxic herbicides will increase by 55 million pounds a year if soybeans resistant to 2,4-D and dicamba are approved without restrictions and widely adopted.[52]

The unrestricted adoption of transgenic crops also caused significant increases in overall use of herbicides in American agriculture, by 383 million pounds from 1996 to 2008.[53] Much of this increase is attributable to greater use of glyphosate. Roundup Ready crop systems made glyphosate the most heavily used pesticide in the history of agriculture, with 180-185 million pounds applied by American farmers in 2007.[54] Overall glyphosate use in American agriculture jumped tenfold from just 1995 to 2007.[55] While Roundup Ready crops led to glyphosate displacing certain other herbicides, the use of still other toxic herbicides did not diminish; for instance, atrazine use remained relatively constant at 70-82 million pounds per year over the past two decades despite widespread adoption of Roundup Ready crops.[56]

In some cases, adoption of Roundup Ready cropping systems will increase herbicide use because current conventional production involves little herbicide application. Alfalfa, the fourth most widely grown crop in the United States, at over 20 million acres covering all 50 states, is the most prominent example.[57] Because the majority of conventional alfalfa hay growers currently use little or no herbicides,[58] USDA estimated that glyphosate use could increase from under one-half million pounds to as much as 24.5 million pounds with unrestricted approval, assuming 51% adoption.[59]

C. USDA Oversight of Transgenic Crops

In the United States, no single overarching law or federal agency oversees biotechnology. Instead, as discussed in Chapter 6, the U.S. government oversees its products using a mosaic of preexisting laws, implemented by several agencies, known as the Coordinated Framework for the Regulation of Biotechnology (Framework).[60] The U.S. Environmental Protection Agency (EPA), USDA, and the Food and Drug Administration (FDA) share responsibility for regulating products of biotechnology in the United States.[61] FDA is charged with overseeing food safety issues related to nonpesticidal GE foods and genetically engineered animals. EPA oversees the impacts of some genetically engineered organisms, including transgenic microbes, as well as GE plants that have been modified to produce pesticidal substances. USDA regulates all transgenic plants, overseeing their field trials and commercialization.

The Framework called for these agencies to stretch the boundaries of their various existing statutes using existing definitions and authorities to promulgate agency regulations and oversee transgenic products. For example, nonpesticidal transgenic ingredients would be classified as "food additives" by FDA under the Federal Food, Drug, and Cosmetic Act. Transgenic plants resistant to herbicides were to be regulated by

---

49. *See* Natural Res. Defense Council, Petition to Revoke All Tolerances and Cancel All Registrations for the Pesticide 2,4-D, EPA-HQ-OPP-2008-0877-0002 (filed Nov. 6, 2008), *available at* http://www.regulations.gov/#!documentDetail;D=EPA-HQ-OPP-2008-0877-0002.
50. Claudine Samanic et al., *Cancer Incidence Among Pesticide Applicators Exposed to Dicamba in Agricultural Health Study*, 114 Envtl. Health Persp. 1521-26 (2006); Kenneth P. Cantor, *Pesticides and Other Agricultural Risk Factors for Non-Hodgkin's Lymphoma Among Men in Iowa and Minnesota*, 52 Cancer Res. 2447-55 (1992).
51. David Mercer, *Roundup Resistant Weeds Pose Environmental Threat*, Associated Press, June 21, 2010, at http://www.usatoday.com/tech/science/environment/2010-06-21-roundup-weeds_N.htm.
52. *Id.*
53. Benbrook, *supra* note 46.
54. U.S. EPA, Biological and Economic Analysis Div., Office of Pesticide Programs, Pesticide Industry Sales and Usage: 2006 and 2007 Market Estimates, tbl. 3.6 (2011). Total 2007 glyphosate usage in the United States of 198-208 million lbs. is more than twice as high as the second-leading pesticide, and exceeds even the peak U.S. production of DDT. Nat'l Pesticide Info. Ctr., Oregon State Univ., *DDT Technical Fact Sheet*, http://npic.orst.edu/factsheets/ddttech.pdf. Peak DDT production in the United States was 188 million lbs. in 1963. *Id.*
55. Service, *supra* note 42, at 1114-1117.
56. U.S. EPA, Biological and Economic Analysis Div., *supra* note 54, tbl. 3.6; U.S. EPA, Biological and Economic Analysis Div., Office of Pesticide Programs, Pesticide Industry Sales and Usage: 2000 and 2001 Market Estimates, tbl. 3.6 (2004).
57. *Roundup Ready Alfalfa FEIS*, *supra* note 45, at 22-23.
58. U.S. Dep't Agric., National Agricultural Statistics Service, 1999. Agricultural Chemical Usage: 1998 Field Crops Summary, p. 3. (finding that only 7% of alfalfa hay acres are treated with herbicides); *see also Roundup Ready Alfalfa FEIS*, *supra* note 45, at 81, 146, app. N.-12.
59. *Roundup Ready Alfalfa FEIS*, *supra* note 45, at app. J., tbl. 19.
60. *See* Coordinated Framework for Regulation of Biotechnology, 51 Fed. Reg. at 23302.
61. *Id.* at 23302-08, 23309, 23313-14.

USDA as "plant pests" under the former Plant Pest Act.[62] Transgenic plants engineered to express pesticidal properties were to be regulated under Federal Insecticide, Fungicide, and Rodenticide Act (FIFRA) as "pesticides" by EPA.[63] Any transgenic microorganisms would be regulated by EPA as "toxic chemicals" under the Toxic Substances Control Act (TSCA).[64] Transgenic animals are classified by FDA as "new animal drugs."[65]

USDA has regulatory authority over all transgenic plants regardless of whether they are food or feed crops or whether they have been modified to produce pesticidal substances.[66] Until the passage of the PPA in 2000, USDA derived its authority over transgenic plants from the former Federal Plant Pest Act (FPPA) of 1957 and the former Federal Plant Quarantine Act (PQA) of 1912.[67] These laws were passed before the advent of biotechnology. They were intended to prevent the introduction of damaging pests and plant disease agents from abroad, and to mitigate the adverse effects of such pests and pathogens. USDA's regulations on transgenic plants were promulgated in 1987 pursuant to those authorities and amended in 1993.[68] When Congress enacted the PPA in 2000, USDA simply continued to use its preexisting regulations.[69]

Under USDA's transgenic crop regulations, certain genetically engineered plants are presumed to be "plant pests"—and thus "regulated articles"—until USDA determines otherwise.[70] Anyone seeking to introduce (i.e., import, transport interstate, or release into the environment)[71] a regulated article must receive authorization from USDA.[72] The agency retains control over these "regulated article[s]," prescribing how they may be "introduce[d]" into the environment and forbidding their "release" or "move[ment in] interstate [commerce]" absent explicit approval.[73]

A developer may obtain such agency approval in several ways. First, if an applicant complies with performance standards intended to limit the risk of unintentional release, it may secure streamlined permission to conduct experimental field trials of the regulated article.[74] These are known as notifications and the vast majority of field trials are conducted pursuant to them.[75] Second, permits are available for field testing transgenic plants that do not qualify for the notification procedure. Permits require more detailed information on the field trial and the procedures intended to ensure containment during and after.[76] Finally, based on field trial experiment data, developers who want to commercialize a transgenic plant must petition USDA for nonregulated status or "deregulation."[77] If USDA determines that the transgenic crop does not pose a "plant pest risk," it then grants the petition and grants the crop deregulated status, after which

---

62. This is because they often contained engineered sequences from viruses and bacteria that cause plant disease and can be considered plant pests themselves. Introduction of Organisms and Products Altered or Produced Through Genetic Engineering Which Are Plant Pests or Which There Is Reason to Believe Are Plant Pests, 7 C.F.R. Part 340; 58 Fed. Reg. 17044-59 (USDA Mar. 31, 1993).
63. Plant-Incorporated Protectants (Formerly Plant-Pesticides), Supplemental Proposal, Part IV, 66 Fed. Reg. 37855-69 (EPA July 19, 2001).
64. 40 C.F.R. Part 725 (2011).
65. Statement of Policy: Foods Derived From New Plant Varieties, 57 Fed. Reg. 22984-23005 (FDA May 29, 1992).
66. Bd. on Agric. and Natural Res., Div. on Earth and Life Studies, Nat'l Research Council, Environmental Impacts of Transgenic Plants: The Scope and Adequacy of Regulation 101 (2002) [hereinafter *2002 NRC Report*].
67. *See* Coordinated Framework for Regulation of Biotechnology, 51 Fed. Reg. at 23342-43; 7 U.S.C. §§151-164, 166-167; 7 U.S.C. §150aa-jj. The secretary delegated that authority to the Animal Plant Health Inspection Service (APHIS), a division of USDA. 7 C.F.R. §§2.22(a), 2.80(a)(36).
68. 7 C.F.R. §§340-340.9 (2011); Introduction of Organisms and Products Altered or Produced Through Genetic Engineering Which Are Plant Pests or Which There Is Reason to Believe Are Plant Pests, 52 Fed. Reg. 22908 (APHIS June 16, 1987); *Genetically Engineered Organisms and Products; Notification Procedures for the Introduction of Certain Regulated Articles; and Petition for Nonregulated Status*, 58 Fed. Reg. 17044 (APHIS Mar. 31, 1993); *Genetically Engineered Organisms and Products; Simplification of Requirements and Procedures for Genetically Engineered Organisms*, 62 Fed. Reg. 23945 (APHIS May 2, 1997).
69. *See* Plant Protection Act, Revisions to Authority Citations, 66 Fed. Reg. 21049 (APHIS Apr. 27, 2001) (revising the genetically modified plant regulations to change authority citations to the PPA without revising the regulations); *see also* Plant Pest Regulations; Update of Current Provisions, 66 Fed. Reg. 51340 (APHIS Oct. 1, 2001) (noting that "the provisions of this proposed rule do not differ significantly from what we would have proposed under the authority of those applicable provisions of law that were repealed by the Plant Protection Act"). The PPA included a provision stating that the regulations issued under superseded laws would remain in effect until USDA issued new regulations. 7 U.S.C. §7758.
70. 7 C.F.R. §§340.1, 340.2, 340.6.
71. 7 C.F.R. §340.1.
72. *Id.* §340.
73. *Id.* §340.1.
74. *Id.* §340.3(e).
75. 7 C.F.R. §340.3; Info. Syst. For Biotech., Virginia Tech, http://www.isb.vt.edu/search-release-data.aspx.
76. 7 C.F.R. §§340.3(e)(5), 340.4.
77. *Id.* §340.6.

it can be commercialized and USDA does not continue to regulate or monitor it. USDA can grant such a deregulation petition in whole or in part.[78]

A number of government investigations and reports have found USDA's oversight lacking under this regulatory structure.[79] For example, a National Academy of Sciences, National Research Council (NRC) 2002 review of USDA's performance found numerous shortcomings, including: the need for greater transparency[80]; insufficient external scientific and public review[81]; inadequate number, training, and allocation of personnel[82]; deficiencies in environmental analyses and decision documents[83]; inherent weaknesses of pre-commercialization testing[84]; a need for post-commercialization monitoring[85]; and excessive claims of confidentiality by companies.[86] The NRC noted in particular with regard to post-commercialization impacts that "[t]here has been no environmental monitoring of these transgenic crops, so any effects that might have occurred could not have been detected."[87]

In 2005, USDA's Inspector General (IG) audited USDA's oversight of both the field trial notification and permitting processes, finding that "weaknesses in APHIS[88] regulations and internal management controls increase the risk that regulated genetically engineered organisms (GEO) will inadvertently persist in the environment before they are deemed safe to grow without regulation."[89] The IG audit revealed frequent cases where the agency did not know the planting locations of field trials,[90] did not require submission of written protocols prior to approvals,[91] did not maintain a list of planted field trials and did not undertake necessary review of applications,[92] and did not undertake adequate inspections or progress reports.[93] As a result of these failures, the IG issued a series of recommendations aimed at increasing USDA's management and oversight.[94]

Similarly, a 2008 U.S. Government Accountability Office (GAO) study analyzed several major transgenic contamination incidents from the past decade, noting the billions of dollars in economic damages associated with them.[95] The GAO concluded that "the ease with which genetic material from crops can be spread makes future releases likely."[96] It recommended that USDA address issues of escape and contamination, as well as develop and coordinate strategies for post-commercialization monitoring.[97] The report

---

78. *Id.* §340.6(d)(3).
79. *See* U.S. Gov't Accountability Office, Food Safety and Quality: Innovative Strategies May Be Needed to Regulate New Food Technologies 14 (1993), GAO/RCED-93-132; U.S. Gov't Accountability Office, Genetically Engineered Crops: Agencies Are Proposing Changes to Improve Oversight, but Could Take Additional Steps to Enhance Coordination and Monitoring 1, 64, 67 (Nov. 2008) [hereinafter *2008 GAO Report*], *available at* http://www.gao.gov/new.items/d0960.pdf; U.S. Gov't Accountability Office, GAO/RCED-86-59, Biotechnology: Agriculture's Regulatory System Needs Clarification 31, 36, 41, 56 (1986); U.S. Gov't Accountability Office, GAO/RCED-88-27 Biotechnology: Managing the Risks of Field Testing Genetically Engineered Organisms 25 (1988).
80. *2002 NRC Report*, *supra* note 66, at 10-11, 16, 148, 177, 254.
81. *Id.* at 10, 12, 106, 173-75, 211.
82. *Id.* at 12, 182, 187.
83. *Id.* at 132, 134, 179, 189.
84. *Id.* at 193-94.
85. *Id.* at 13-14, 195-96.
86. *Id.* at 11-12, 177.
87. *Id.* at 79.
88. APHIS is the Animal and Plant Health Inspection Service, the subagency of USDA charged with oversight of transgenic crops. In this chapter, the agency acronyms will be used synonymously.
89. U.S. Dep't Agric., Office of the Inspector General, Audit Report: Animal and Plant Health Inspection Service Controls Over Issuance of Genetically Engineered Organism Release Permits i (2005), *available at* http://www.usda.gov/oig/webdocs/50601-08-TE.pdf.
90. *Id.* at 14.
91. *Id.* at 20.
92. *Id.* at 24.
93. *Id.* at 28-35.
94. *Id.*
95. *2008 GAO Report*, *supra* note 79, at 44. The GAO report documented six major events of GE crops contaminating the food and feed supply, including the 2000 StarLink Corn incident, causing between $26 to $288 million in economic damages; the 2002 Prodigene Corn contamination incident, where a variety of GE corn designed to create a pig vaccine protein contaminated non-GE corn; the 2004 Syngenta Bt Corn incident, where a pesticidal Bt corn determined not to be suitable for commercialization was illegally released onto 37,000 acres; the 2006 Event 32 Corn incident, where 72,000 acres were planted to three lines of corn contaminated with regulated GE pesticidal corn; and the 2006 Liberty Link Rice incident, where GE rice contaminated export rice stocks causing economic damages of over $1 billion. *Id.* at 3.
96. *Id.* at 3.
97. *Id.* at 45-47.

also called on USDA "to monitor for other unintended consequences, such as economic impacts on other agricultural sectors, such as organic crops, which may become contaminated by GE crops."[98]

Numerous other nongovernmental reviews also found shortcomings. Reports have noted USDA's failure to assert post-market authority.[99] Other reports analyze how USDA's oversight system impacts the quality of the seed supply and exposes our food supply to novel contaminants such as engineered biopharmaceuticals.[100] Still others have pointed out serious shortcomings in USDA's process for assessing the impacts of gene flow leading to transgenic contamination of wild plants.[101]

With the adoption of the 2008 farm bill,[102] Congress mandated that USDA "improve the management and oversight" of GE crop field trials by implementing more rigorous measures to mitigate the risk of transgenic contamination following the 2006 "Liberty Link" rice contamination. In that case, testing discovered widespread contamination of U.S. southern long-grain rice from a GE variety field tested by Bayer in Louisiana several years before, causing both Japan and the European Union to shut down U.S. rice imports within days and costing U.S. farmers nearly a billion dollars in losses.[103] Although Congress mandated that USDA take action to implement these new directives "[n]ot later than 18 months after the date of enactment of this Act [June, 18, 2008],"[104] USDA has yet to implement this congressional directive. Also, in 2008 and again in 2010, Congress held a series of investigative hearings, on the harm to farmers from USDA policy regarding transgenic contamination[105] and HR weeds, respectively.[106]

Independent governmental reviews were not alone in their censure—courts also found USDA lacking in its oversight. Regulation of GE crops has, as a result, been defined in part by lawsuits brought on behalf of farmers, consumers, and environmental groups. Most of these cases involve the National Environmental Policy Act (NEPA), which, as described in Chapter 12, requires agencies to consider the environmental consequences of their proposed actions.[107] Among other requirements, NEPA instructs agencies to prepare an "environmental impact statement" (EIS) if the agency's action may "significantly affect the quality of the human environment."[108] Two courts have held that USDA violated environmental laws in categorically excluding field trials from any environmental review.[109] Two other courts have held deregulations unlawful

---

98. *Id.* at 48.
99. *See* Pew Initiative on Food and Biotech., Post-Market Oversight of Biotech Foods: Is the System Prepared (2003), *available at* http://www.pewtrusts.org/our_work_report_detail.aspx?id=33352; Pew Initiative on Food and Biotech., Issues in the Regulation of Genetically Engineered Plants and Animals (2004), *available at* http://www.pewtrusts.org/our_work_report_detail.aspx?id=17976.
100. Union of Concerned Scientists, Gone to Seed: Transgenic Contaminants in the Traditional Seed Supply (2004), *available at* http://www.ucsusa.org/food_and_agriculture/our-failing-food-system/genetic-engineering/gone-to-seed.html (finding that about 50% or more of the certified non-GE corn, canola, and soybean seed have been contaminated with transgenes); Union of Concern Scientists, A Growing Concern: Protecting the Food Supply in an Era of Pharmaceutical and Industrial Crops (2004), *available at* http://www.ucsusa.org/food_and_agriculture/our-failing-food-system/genetic-engineering/a-growing-concern-protecting.html.
101. Center for Food Safety, *supra* note 38.
102. Food, Conservation, and Energy Act of 2008 (2008 Farm Bill), Pub. L. No. 110-246, tit. X, §10204, 122 Stat. 1651, 2105 (2008); *see* 7 U.S.C. §7701. USDA's "Lessons Learned" recommended new measures that APHIS must require to avoid the pitfalls it discovered during the rice contamination investigation. USDA, Lessons Learned and Revisions Under Consideration for APHIS' Biotechnology Framework (2007) [hereinafter Lessons Learned], *available at* http://www.aphis.usda.gov/newsroom/content/2007/10/content/printable/LessonsLearned10-2007.pdf. Section 10204(b) of the 2008 Farm Bill requires the secretary to take nine actions to make the improvements suggested in "Lessons Learned." Section 10204(c) of the 2008 Farm Bill requires the secretary to consider ten additional improvements. Pub. L. No. 110-234, tit. X, §10204, 122 Stat. 1343 (May 22, 2008), and Pub. L. No. 110-246, tit. X, §10204, 122 Stat. 2105 (June 18, 2008); *see* 7 U.S.C. §7701 (listed under statutory notes).
103. Harris & Beasley, *supra* note 38; In re Genetically Modified Rice Litigation, 666 F. Supp. 2d 1004 (E.D. Mo. 2009); In re Genetically Modified Rice Litigation, 2009 WL 4801399 (E.D. Mo. Dec. 9, 2009); *Japan Bans "Contaminated" US Rice*, BBC News, Aug. 21, 2006, http://news.bbc.co.uk/2/hi/science/nature/5271384.stm; Press Release, Europa, *Commission Requires Certification of US Rice Exports to Stop Unauthorised GMO Entering the EU*, Aug. 23, 2006, http://europa.eu/rapid/pressReleasesAction.do?reference=IP/06/1120.
104. 2008 Farm Bill, Pub. L. No. 110-246, tit. X, §10204, 122 Stat. 1651, 2105 (2008).
105. *Is USDA Accounting for Costs to Farmers Caused by Contamination From Genetically Engineered Plants?: Hearing Before the Subcomm. on Domestic Policy of the H. Comm. on Oversight and Gov't Reform*, 110th Cong. 110-165 (Mar. 13, 2008), *available at* https://house.resource.org/110/gov.house.ogr.20080313b_hrs03RFM2154.1.raw.txt.
106. *07-28-2009—Domestic Policy—Are "Superweeds" an Outgrowth of USDA Biotech Policy? (Part I): Hearing Before the H. Comm. on Oversight and Government Reform*, 111th Cong. (2009), *available at* http://oversight.house.gov/hearing/are-superweeds-an-outgrowth-of-usda-biotech-policy-part-i/; *07-28-2009—Domestic Policy—Are "Superweeds" an Outgrowth of USDA Biotech Policy? (Part II): Hearing Before the H. Comm. on Oversight and Government Reform*, 111th Cong. (2009), *available at* http://oversight.house.gov/hearing/are-superweeds-an-outgrowth-of-usda-biotech-policy-part-ii/.
107. 40 C.F.R. §1500.1(a); *see, e.g.*, Sierra Club v. U.S. Army Corp of Engineers, 295 F.3d 1209, 1214 (11th Cir. 2002).
108. 42 U.S.C. §4332(2)(C); *see, e.g.*, Blue Mountains Biodiversity Project v. Blackwood, 161 F.3d 1208, 1216 (9th Cir. 1998); Klamath Siskiyou Wildlands Ctr. v. Boody, 468 F.3d 549, 562 (9th Cir. 2006).
109. Center for Food Safety v. Johanns, 451 F. Supp. 2d at 1182, 1183, 1184-85 (D. Haw. 2006); Int'l Ctr. for Tech. Assessment (ICTA) v. Johanns, 473 F. Supp. 2d 9, 29-31 (D.D.C. 2007).

for failing to take a "hard look"[110] at potential impacts, and ordered USDA to complete EISs pursuant to NEPA on new decisions, vacating deregulations in the interim.[111] (Remarkably, in over 15 years of approving dozens of transgenic crops planted on millions of acres, these two court-ordered EISs are the first two USDA has ever completed for any transgenic crop.)

The judicial language of these decisions is at times notably harsh. In 2006, a federal court in Hawaii found that USDA's approval of field tests of transgenic, pharmaceutical-producing plants violated environmental laws, describing USDA's arguments as "utterly without merit," its actions as evincing "utter disregard," and constituting an "unequivocal violation of a clear congressional mandate," and "abdication" of its responsibilities.[112] In 2007, another federal court in Washington, D.C., held that USDA violated environmental laws in approving field trials of genetically engineered grasses, finding the record "devoid of any evidence" that USDA had analyzed environmental risks.[113] Similarly in 2007, a federal court in California described USDA's attitude toward environmental risk assessment as "cavalier"[114] and concluded that USDA "simply ignore[d]"[115] the risks in question or "refused" to analyze them.[116] In 2009, yet another court wrote of USDA's position that it showed an "apparent perception that conducting the requisite comprehensive review is a mere formality, caus[ing] some concern that Defendants are not taking this process seriously."[117]

D. Applying the Plant Protection Act to Transgenic Crops

The above-enumerated impacts and agency track record raise the question of whether USDA needs to request from Congress further statutory authority in order to improve its oversight. It does not. On June 22, 2000, Congress repealed the former Plant Quarantine Act, the Federal Plant Pest Act, and the Federal Noxious Weed Act and replaced them with the Plant Protection Act (PPA), as part of the Agricultural Risk Protection Act.[118] The PPA consolidated these previous statutes and enhanced USDA's authority to regulate plants and related plant items in order to prevent the introduction or spread of plant pests or noxious weeds within the United States.[119] Its overarching purpose is to prevent and halt the spread of these items as necessary for "the protection of the agriculture, environment, and economy of the United States."[120] The new PPA adopted the old FPPA's broad definition of plant pests,[121] which enabled USDA to continue regulation of transgenic plants based on the fact that many are genetically engineered with the aid of a plant pest, or incorporate DNA from a plant pest, and as such are designated as "regulated articles."[122]

The PPA also created an affirmative obligation for USDA to prevent the spread of noxious weeds.[123] Inclusion of the PPA's noxious weed authority significantly expanded USDA's powers; the definition of "noxious weed" is quite broad, including plants or plant products that can directly or indirectly injure a wide array of subjects, including the interests of agriculture, natural resources, public health, and the environment.[124]

The PPA provided USDA multiple new tools with which it can carry out its mandate, including the power to hold, seize, quarantine, or apply other remedial measures to destroy or dispose of any plant, plant pest, noxious weed, plant product, article, or means of conveyance if necessary to prevent the spread of new plant pests or noxious weeds[125]; issue subpoenas[126] and conduct warrantless inspections[127]; provide

---

110. *See* Kleppe v. Sierra Club, 427 U.S. 390, 410 (1976) (for a discussion of what has come to be known as the "hard look doctrine").
111. Geertson Seed Farms, 2007 WL 518624 (N.D. Cal. 2007); Center for Food Safety, 2009 WL 3047227 (N.D. Cal. Sept. 21, 2009).
112. Center for Food Safety, 451 F. Supp. 2d at 1182, 1183, 1184-85.
113. *ICTA*, 479 F. Supp. 2d at 29.
114. Geertson Seed Farms, 2007 WL 518624, at *10 (N.D. Cal. Feb. 17, 2007).
115. *Id.* at *7.
116. Geertson Seed Farms, 2007 WL 1302981, at *1 (N.D. Cal. May 3, 2007).
117. Center for Food Safety, 734 F. Supp. 2d at 953 (N.D. Cal. 2010).
118. 7 U.S.C. §§7701-7772.
119. *Id.* §7712(a).
120. *Id.* §7701(1).
121. *Id.* §7702(14); 7 C.F.R. §340.1.
122. 7 C.F.R. §340.2.
123. *See* 7 U.S.C. §7701(3).
124. *Id.* §7702(10).
125. *Id.* §7714(a)(1)(A).
126. *Id.* §7733.
127. *Id.* §7731(b).

cost-recovery measures[128]; establish quality assurance programs[129]; and develop a classification of status and action levels for risks as well as develop integrated management plans.[130]

The PPA also substantially increased USDA's enforcement authority by increasing the amount in penalties that USDA could impose.[131] Finally, the PPA mandates that all USDA decisions "be based on sound science."[132]

1. Applying the PPA's Plant Pest Authority

As an initial matter, USDA's plant pest authority, by itself, is broad enough to support regulation of the above-discussed harms of GE crops, such as transgenic contamination and the creation of Roundup-resistant weeds. The definition includes plants or organisms that can "directly or indirectly injure or cause disease or damage in any plants or parts thereof, or any processed, manufactured, or other products of plants."[133] A transgenic plant's contamination of conventional or organic crops can damage or injure harvests under the law's definition because it would rob those farmers of their choice and ability to grow a non-GE variety and, as a consequence, cost them their market value, reputation, and customers.[134] The risk of contamination "indirect[ly] injur[es]" these farmers since they must take up onerous burdens to prevent such contamination, such as a regimen of testing and the installment of buffer zones.[135]

The subsequent risk of secondary contamination of the food production chain or supply chain after harvest would also fall within the agency's authority, since it includes damage not only to plants but also to "any processed, manufactured, or other products of plants."[136] Thus, the intertwined economic harm directly stemming from contamination should be covered since the definition includes both "direct and indirect" damage.[137] The environmental harm of transgenic pollution of native ecosystems and the loss of genetic diversity would also logically be covered, since the definition covers harm that "can directly or indirectly injure, cause damage to, or cause disease in *any plant*," not just agricultural crops.[138]

Similarly, an HR crop's contribution to the epidemic of resistant weeds would be a cognizable harm covered by this authority. The unrestricted approval of a Roundup Ready cropping system by USDA, leading to the rapid evolution of these weeds, would directly or indirectly injure and cause damage to farmers and the environment. For the farmer, it would increase costs and labor in the form of additional herbicides, and eventually more toxic herbicide-pesticide cocktails, as well as manual labor.[139] For the environment, it would create resistant invasive weeds, as well as cause an increase in the herbicidal runoff in the environment, both potentially damaging native plants as well as organisms beneficial to them.[140] In short, the risk of both of these harms would seem to increase the potential that a particular transgenic crop would meet the definition of "plant pest" harm, as it is broadly defined by the PPA, and require USDA analysis and potential restrictions.

2. Transgenic Contamination and Economic Impacts Under the Noxious Weed Authority

Under its noxious weed mandate, USDA has even broader authority to address the full range of adverse agricultural, public health, and environmental impacts associated with GE crops,[141] in order to fulfill the

---

128. *Id.* §7735(2)-(3).
129. *Id.* §7721.
130. *Id.* §7714(c)(2).
131. *Id.* §§7732, 7734.
132. *Id.* §7701(4); *see also id.* §7712(b).
133. *Id.* §7702(14); 7 C.F.R. §340.1.
134. *See supra* notes 36-37, 95, 103 & 105.
135. *Monsanto*, 130 S. Ct. at 2755 (finding standing for alfalfa farmers challenging deregulation based on risk of transgenic contamination and burden of preventative measures to avoid it); Center for Food Safety v. Vilsack, 636 F.3d 1166, 1172 (9th Cir. 2011) (same, for sugar beet farmers).
136. 7 C.F.R. §340.1.
137. 7 C.F.R. §340.1.
138. 7 U.S.C. §7702(14) (emphasis added); *see supra* notes 38-39.
139. *See supra* notes 41, 46-47, 51.
140. *See supra* notes 47-59 and *infra* note 154.
141. 7 U.S.C. §7702(10).

PPA's purpose to "protect[ ] the agriculture, environment, and economy of the United States."[142] Indeed, USDA has an affirmative obligation to apply that authority, although it has yet to integrate it into its transgenic crop regulations or otherwise employ it.[143]

As noted above, the term "noxious weed" is defined very broadly, to cover "any plant or plant product that can directly or indirectly injure or cause damage" not only to "crops (including nursery stock or plant products)," but also "livestock, poultry, or other interests of agriculture, irrigation, navigation, the natural resources of the United States, the public health, or the environment."[144] The only court to address its meaning, in *International Center for Technology Assessment (ICTA) v. Johanns*, held in the context of GE, Roundup Ready grasses: "Under basic principles of statutory construction, the term 'noxious weed' as used here should be read according to its articulated statutory definition, as 'any' plant that can injure crops, livestock, or other agricultural or environmental interests."[145] USDA itself has noted that the PPA "grants significant new regulatory authority for noxious weeds"[146] and provides "a much wider and more flexible set of criteria for identifying and regulating noxious weeds."[147]

This expansive authority is broad enough to cover transgenic contamination. It expressly requires that USDA account for the economic impacts of its decisions on transgenic crops, by acting "for the protection of agriculture, environment and *economy* of the United States."[148] Indeed, seven of nine introductory findings of the PPA focus on preventing burdens on commerce and the economy.[149]

Nor does it matter if the potential negative economic impact of transgenic crops is classified as indirect, as the plain language of the noxious weed authority covers "[a]ny plant or plant product that can directly or indirectly injure or cause damage" to, inter alia, "crops . . . or other interests of agriculture. . . ."[150] As discussed above, a number of contamination events have already caused significant damage to the U.S. agricultural economy.[151] Thus, analysis of this potential impact should be completed as part of the agency's decision to permit field trials or deregulate a transgenic crop.

3. The Noxious Weed Authority and HR Weeds

Many transgenic plants could pose broad environmental and agronomic risks that would trigger oversight under the noxious weed authority and should require analysis and restrictions. For example, hypothetical engineered plants with stress and drought-tolerance genes could increase the fitness of transgenic plants or wild relatives, thereby allowing them to spread in natural areas.[152] Increased geographic range of stress-tolerant plants could cause harm to the environment by displacing other species or exposing nontarget organisms to transgene products that could be harmful.

However, one of the most important current uses USDA should make of its noxious weed authority is to assess, and regulate as needed, transgenic HR crop systems[153] as noxious weed risks. The glyphosate-resistant weed epidemic is a symptom of regulatory breakdown, a perfect example of how thoroughly uncoordinated the "coordinated" Framework actually is. USDA regulates the transgenic herbicide-dependent crop,

---

142. 7 U.S.C. §7701(1).
143. 7 U.S.C. §7701(3).
144. 7 U.S.C. §7702(10).
145. *ICTA*, 473 F. Supp. 2d at 25. As that court noted, the PPA "deliberately expanded" the meaning of noxious weed from that of the former Federal Noxious Weed Act, which had limited the term to plants only "of foreign origin" that were "new to or not widely prevalent in the United States"; the PPA removed that limiting language. *See* 7 U.S.C. §2802 (1999) (repealed); *ICTA*, 472 F. Supp. 2d at 25 & n.15; H.R. Rep. No. 106-639 (2000) (Conf. Rep.), *reprinted in* 146 Cong. Rec. H3763, H3803 (daily ed. May 24, 2000).
146. APHIS, USDA, *Plant Protection Act*, http://www.aphis.usda.gov/publications/plant_health/content/printable_version/fs_phproact.pdf.
147. *ICTA*, 473 F. Supp. at n.15 (citing and quoting USDA administrative record documents); 73 Fed. Reg. 60008, 60011 (Oct. 9, 2008) (proposed new rules stating that "[T]echnological advances have led to the possibility of developing GE organisms that do not fit within the plant pest definition, but may cause environmental or other types of physical harm or damage covered by the definition of noxious weed in the PPA. Therefore, we consider that it is appropriate to align the regulations with both the plant pest and noxious weed authorities of the PPA"); *see also* 73 Fed. Reg. 60013 (proposed new regulations stating that "evaluation of noxious weed risk expands what we can consider").
148. 7 U.S.C. §7701(1) (emphasis added).
149. *See* 7 U.S.C. §7701(1), (3)-(6), (8)-(9); Am. Paper Inst., Inc. v. Am. Elec. Power Serv. Corp., 461 U.S. 402, 413 (1983) (factors an agency must consider are determined by purpose of the underlying statute).
150. 7 U.S.C. §7702(10).
151. *See supra* notes 36-37, 95, 103 & 105.
152. Nat'l Research Council, Nat'l Acad. of Sciences, Biological Confinement of Genetically Engineered Organisms 49 (2004).
153. The term "HT crop system" comes from Monsanto, which markets and sells its herbicide and engineered seed as a "crop system." *See, e.g.*, http://www.monsanto.com/products/Documents/safety-summaries/alfalfa_benefits.pdf (describing the "[RRA] weed control system"); http://www.bic.searca.org/seminar_proceedings/4th_corn/day2/Samson-MonsantoRRC.pdf ("Roundup Ready Corn Works as a System").

and EPA separately regulates the associated herbicide. But no agency regulates the *combination*, the HR crop-herbicide system. It is that system—the use of glyphosate made possible and furthered by glyphosate-resistant seeds, for instance—that is responsible for the growing epidemic of glyphosate-resistant weeds. Absent for the first 20-plus years of glyphosate's use, the explosion of weed resistance has occurred in the decade since the widespread adoption of Roundup Ready crop systems. As one federal court noted, "one would expect that *some* federal agency is considering whether there is some risk to engineering all of America's crops to include the gene that confers resistance to glyphosate."[154]

The herbicide-resistance trait increases the potential of the crop to create a noxious weed harm in either of two ways. First, the trait can enter related weeds and make them still "weedier"—that is, more difficult and expensive to control; second, vastly increased use of the HR crop-associated herbicide exerts enormous selection pressure to foster rapid propagation of resistant weeds—in both cases harming the "interests of agriculture." Such resistant weeds can also harm the environment, as seen with the example of Roundup Ready bentgrass, which the U.S. Fish and Wildlife Service (FWS) concluded would, if commercialized, likely cause the extinction of two endangered plants by spreading transgenic resistance to wild relatives, which would then take over the species' critical habitat and be impossible to eradicate.[155] And as use of more toxic herbicides increases to control such a weed, there may also be further harms to the environment and public health. Finally, as with contamination harm, the fact that these harms might be termed "indirect" does not exclude them from the agency's statutory mandate.[156]

4. Following EPA's Example

From the very start, regulation of GE organisms in accordance with the guidance of the Coordinated Framework has been based upon creative application of preexisting statutes that were never intended for that purpose. GE organisms could never have been regulated in the first place without considerable regulatory adaptation of existing laws, as demanded by the Coordinated Framework. EPA has taken active steps to regulate GE pesticide-producing plants, even though, logically, a plant with a transgene is very different than a spray-applied chemical pesticide, the traditional sphere of EPA regulation. This adjustment allowed EPA to place reasonable restrictions on the use of pesticide-producing plants, benefitting both farmers and the environment.

EPA's main role in the Framework is to regulate GE pesticide-producing plants pursuant to FIFRA,[157] the federal statute covering the manufacture, labeling, sale, and application of pesticides.[158] Pursuant to its broad authority over pesticides and pests under FIFRA[159]—an authority not unlike the PPA's broad authority over noxious weed and plant pest harms—EPA established tailored regulatory requirements (under the same FIFRA statutory definition and standard)[160] for several distinct classes or types of genetically engineered organisms, including genetically modified microbial pesticides[161] and plant-incorporated protectants (PIPs, formerly known as plant pesticides).[162] The vast majority of these crops are engineered to produce an insecticidal protein derived from the bacterium *Bacillus thuringiensis* (Bt), with Bt corn and cotton available commercially. EPA tailored its basic regulatory pesticide framework to fit the distinctive characteristics of genetically engineered biological pesticides.[163] The PIPs rules state: "The characteristics

---

154. Geertson Seed Farms, 2007 WL 518624, at *11 (emphasis added).
155. U.S. Fish and Wildlife Serv., *Draft Biological Opinion*, February 2010, at 2 (finding the proposed action likely to jeopardize the continued existence of the Willamette daisy (*Erigeron decumbens var. decumbens*) and Bradshaw's lomatium *(Lomatium bradshawi)* and would likely adversely modify designated critical habitat of the Fender's blue butterfly (*Icaricia icarioides fenderi*) and Willamette daisy (*Erigeron decumbens var. decumbens*)).
156. 7 U.S.C. §7702(10).
157. Plant-Pesticides Subject to the Federal Insecticide, Fungicide, and Rodenticide Act, 59 Fed. Reg. 60519-35 (proposed Nov. 23, 1994) (codified at 40 C.F.R. §§152.1, 152.3, 152.20, 174 et seq.), *available at* http://www.epa.gov/fedrgstr/EPA-PEST/1994/November/Day-23/pr-53.html.
158. 7 U.S.C. §§136-136y et seq.
159. 7 U.S.C. §136(u)(1); 40 C.F.R. §152.3 (pesticide); 7 U.S.C. §136(t); 40 C.F.R. §152.5 (pest).
160. 7 U.S.C. §136(bb)(1); *see, e.g.,* Regulations Under the Federal Insecticide, Fungicide, and Rodenticide Act for Plant-Incorporated Protectants (Formerly Plant-Pesticides), 66 Fed. Reg. 37772-73 (July 19, 2001).
161. 40 C.F.R. §§172.43-.59.
162. 40 C.F.R. Parts 152 and 174, 66 Fed. Reg. 37771-817 (July 19, 2001); *see, e.g.,* 40 C.F.R. §174.71 (submission of information regarding adverse effects for PIPs); U.S. EPA, *Plant Incorporated Protectants*, http://www.epa.gov/pesticides/biopesticides/pips/index.htm.
163. 40 C.F.R. §174.71. EPA also regulates GE microorganisms under its general chemicals authority, the Toxic Substances Control Act (TSCA). *See* 59 Fed. Reg. 45526 (Sept. 1, 1994).

of plant-incorporated protectants such as their production and use in plants, their biological properties, and their ability to spread and increase in quantity in the environment distinguish them from traditional chemical pesticides. Therefore, plant-incorporated protectants are subject to some different regulatory requirements and procedures than traditional chemical pesticides."[164] For example, when assessing the potential risks of genetically engineered PIPs, EPA requires data addressing numerous factors, such as nontarget organisms and the environment, potential for gene flow, and the need for insect resistance management plans.[165]

In the case of USDA, the broad noxious weed authority conferred by the PPA provides an opportunity and additional impetus to regulate similarly. For instance, USDA could adapt its traditional noxious weed regulations to forestall and mitigate the emergence of noxious HR weeds. Like traditional noxious weeds, the HR types damage the interests of farmers and agriculture as a whole. Likewise, GE contamination harms the economic interests of farmers, and when detected often hampers interstate and foreign commerce. Scientifically sound, pro-farmer regulation beyond the prevention and control of "traditional" plant pests and noxious weeds would seem not only good policy but in keeping with the spirit of regulatory adaptation that is a cornerstone of the Coordinated Framework. It would represent good governance, supported by sound scientific, policy, and legal moorings. Regulation of GE plants provides a useful counterpole because, unfortunately, USDA thus far has taken the opposite approach. The agency has declined to adapt (or even apply) its expansive noxious weed authority to address large and growing adverse impacts on farmers as well as the environment. As noted above, USDA implemented regulations for GE crops in 1987 and 1993 and has not revised them since the enactment of the PPA. USDA should modify its regulations to explicitly include criteria regarding certain transgenic crops and their concomitant risks, as EPA has with PIPs.

5. Integrated Resistance Management

The most important component of EPA's regulation of insect-resistant Bt crops is mandatory insect-resistance management (IRM), through which EPA largely forestalled evolution of insect resistance.[166] EPA recognized that Bt crops present a high risk of inducing resistance in insect pests because pests are continually exposed to Bt crop toxins, and thus the agency requires that all farmers who grow Bt crops plant a specified portion of their field with a so-called "refuge" of non-Bt varieties.[167] The aim of this strategy is to maintain an ample population of insect pests that remain susceptible to the Bt toxin. These susceptible insects are then available to mate with any that have evolved resistance in the Bt portion of the field, thus diluting the pool of resistance genes and thereby prolonging the life of this valuable, less-toxic insecticide.[168] EPA adopted mandatory IRM to preserve the efficacy of Bt toxins for the sake of organic growers, many of whom rely on Bt in spray form and would be robbed of this important tool in the event of insect resistance[169]; as well as for the long-term interests of biotech growers, who would likewise lose Bt crops if insect pests evolved resistance; and finally to avert future harm to the environment in the form of more environmentally damaging chemical insecticides that would be employed if Bt toxins were rendered inefficacious. IRM thus represents regulation that protects the environment and serves the long-term interests of both organic and biotech growers, at the small cost of restraints on use of the technology in the short term.

USDA can and should follow EPA's lead in the analogous case of GE HR crops, which foster rapid evolution of resistant and potentially noxious weeds due to continual use of the HR crop-associated herbicide. Since the PPA's noxious weed mandate includes both direct and indirect harms, the statute autho-

---

164. 40 C.F.R. §171.4; 66 Fed. Reg. at 37774 (July 19, 2001).
165. U.S. EPA, *Plant Incorporated Protectants*, http://www.epa.gov/pesticides/biopesticides/pips/index.htm.
166. Bruce E. Tabashnik et al., *Field-Evolved Insect Resistance to Bt Crops: Definition, Theory and Data*, 102 J. Econ. Entomology 2011-25 (2009).
167. U.S. EPA, *EPA's Regulation of Bacillus Thuringiensis (Bt) Crops*, http://www.epa.gov/oppbppd1/biopesticides/pips/regofbtcrops.htm; *see also* U.S. EPA, *Biopesticides Registration Action Document—Bacilllus Thuringiensis Plant-Incorporated Protectants* (Oct. 16, 2001), http://www.epa.gov/pesticides/biopesticides/pips/bt_brad.htm.
168. U.S. EPA, *EPA's Regulation of Bacillus Thuringiensis (Bt) Crops*, http://www.epa.gov/oppbppd1/biopesticides/pips/regofbtcrops.htm.
169. Jim Jones, *Testimony Before the Domestic Policy Subcommittee of the House Oversight and Government Reform Committee Regarding USDA Regulation of Herbicide-Resistant Crops and the Threat Posed by Glyphosate-Resistant Weeds, Part II* (Sept. 30, 2010), *available at* http://oversight.house.gov/wp-content/uploads/2012/01/20100930Jones.pdf.

rizes USDA to regulate the weeds' agricultural pathway, as well as the weeds themselves.[170] As explained above, glyphosate-resistant weeds are directly produced by Roundup Ready crop systems. HR weeds are noxious in that they often trigger substantial yield losses, dramatically higher weed control costs, and large increases in toxic herbicide use.[171]

The PPA further provides USDA the authority to develop integrated management plans to address noxious weed impacts in specific regions or ecological ranges which could also be used to address the HR weed epidemic by restricting Roundup Ready cropping systems.[172] Regulatory actions would be determined on a case-by-case basis, guided by sound science, but might include a prohibition on planting HR crops every year at farm or field scale, or denial of a petition to deregulate an HR crop. Integrated management plans for HR crops could help improve assessment and start to limit the HR weeds epidemic. As with Bt crops, regulation would serve the long-term interests of farmers and the environment.

6. Updating USDA's Scope of Authority

Under USDA's current regulations only transgenic plants that are developed with the use of, or contain genetic material from, one or more of the "plant pest" organisms listed in the regulations are "regulated articles" that must undergo deregulation before commercialization.[173] However, genetic engineering can now be accomplished without the use of listed plant pest organisms, a problematic loophole to USDA's oversight that would allow a transgenic plant to circumvent review.[174] The integration of the full PPA provides USDA the opportunity—and duty—to go beyond the taxonomic trigger, and include all engineered crops with potential harms pursuant to USDA's broader noxious weed mandate. The sound science-based alternative that would close this loophole, recommended by the National Academy of Sciences, is for USDA to replace the current taxonomic trigger and instead use the genetic engineering process itself as USDA's "trigger" for initial regulation.[175]

7. Including Public Health Assessment

The PPA's goals include the protection of human health by including, under its noxious weed mandate, "any plant or plant product that can directly or indirectly injure or cause damage to . . . the public health."[176] Accordingly, USDA has the authority to affirmatively investigate potential adverse health effects of newly proposed GE crops. This is particularly important as applied to transgenic, biopharmaceutical crops over which USDA has purview, since they are extremely potent compounds that can be biologically active at low levels.[177]

The need for health review by USDA is underscored by the very limited review given transgenic foods by FDA.[178] Under FDA's policy, the manufacturer, not FDA, determines whether a transgenic ingredient warrants premarket review as a food additive; any consultation with FDA on that decision is voluntary.[179] Accordingly, FDA does not "approve" transgenic foods, nor undertake any independent analysis of their health safety.[180] The agency merely reviews summaries of the data the industry presents, and issues a "no

---

170. *See, e.g.*, 7 C.F.R. §360.400 (restricting the import and requiring the pre-import treatment of *Guizotia abyssinica* (niger seed), not because that plant is itself a noxious weed, but because it commonly harbors noxious weed seeds).
171. *See supra* notes 41-43 and 46-47.
172. 7 U.S.C. §7714(c)(2).
173. 7 C.F.R. §§340.1, 340.2.
174. Andrew Pollack, *U.S.D.A. Ruling on Bluegrass Stirs Cries of Lax Regulation*, N.Y. Times, July 6, 2011, at http://www.nytimes.com/2011/07/07/business/energy-environment/cries-of-lax-regulation-after-usda-ruling-on-bluegrass.html.
175. *2002 NRC Report, supra* note 66, at 79, 83 (2002).
176. 7 U.S.C. §7702(10).
177. *2002 NRC Report, supra* note 66, at 68.
178. Statement of Policy: Foods Derived From New Plant Varieties, 57 Fed. Reg. 22984-23005 (FDA May 29, 1992); Gregory N. Mandel, *Gaps, Inexperience, Inconsistencies, and Overlaps: Crisis in the Regulation of Genetically Modified Plants and Animals*, 45 Wm. & Mary L. Rev. 2216, 2218, 2243 (2004).
179. FDA, *Consultation Procedures Under FDA's 1992 Statement of Policy—Foods Derived From New Plant Varieties: Guidance on Consultation Procedures*, available at http://www.fda.gov/food/guidancecomplianceregulatoryinformation/guidancedocuments/biotechnology/ucm096126.htm (rev. Oct. 1997) (last visited Aug. 22, 2012); Gregory N. Mandel, *Toward a Rational Regulation of Genetically Modified Food*, 4 Santa Clara L. Rev. 21, 24 (2006).
180. *See* William Freese & David Schubert, *Safety Testing and Regulation of Genetically Engineered Foods*, 21 Biotechnology & Genetic Eng'g. Revs. 299, 304-05 (2004).

questions" letter conveying the developer's assurances.[181] While FDA would have more expertise in assessing and implementing such a system, the agency has declined to do so and only provides limited voluntary safety oversight. Given this situation, USDA should include in its transgenic crop regulations a mandatory human health safety assessment and improve interagency cooperation on health assessments.

8. Implementing USDA's Authority to Promulgate Partial Deregulations and Continuing Post-Market Oversight

Finally, integrating and applying the full PPA provides USDA the opportunity to clarify the agency's post-market authority over transgenic crops, including its ability to place restrictions on such crops after commercialization under its partial deregulation authority. These restrictions can and should include monitoring, geographic restrictions, and isolation distances from nontransgenic varieties to protect against contamination, as well as integrated management plans to limit the proliferation of HR weeds.

a. Post-Market Monitoring

To date, USDA oversight of transgenic crops has largely focused on assessing and preventing risks posed by GE crops in the testing phase. However, many stakeholders have warned that widespread use of transgenic crops can have unintended consequences that should be monitored. Many observers have noted that field trials are inadequate for resolving all environmental issues.[182] Risk assessment during field trials is hampered by the small scale of field trials, which are often incapable of detecting risks that emerge only with commercial-scale plantings. Such risks can be several orders of magnitude larger. It may take several years for such adverse impacts to become evident, according to the National Academy of Sciences, which recommended post-commercial monitoring and testing.[183] The 2008 GAO report similarly recommended post-market monitoring.[184] Other unanticipated risk issues may emerge based on scientific or agronomic findings not available at the time of the initial deregulation. For example, there is now over a decade of experience with GE, HR crop systems, primarily with Roundup Ready crops planted on roughly 150 million acres; nature has proven wrong industry experts who predicted slow development of weed resistance. Hence, it would be prudent of USDA to use its broad authority to impose restrictions, as needed, on all transgenic crops the agency deregulates.

As discussed above, the PPA provided USDA broad post-commercialization remedial enforcement authority[185] to hold, seize, quarantine, or apply other measures to destroy or dispose[186]; to issue subpoenas[187] and conduct inspections[188]; and to issue monetary penalties.[189] Monitoring and continuing regulatory oversight is also in line with the framework of EPA, which grants time-limited registrations to PIPs, enabling it to consider new data that emerges after the initial registration. If new scientific research indicates that a PIP has a previously unnoticed risk, EPA can re-register the PIP to ameliorate the harm. For example, in 2011, EPA tightened its Bt crop refuge requirements based on reports showing lack of compliance.[190]

USDA can make use of its PPA authority to serve similar ends. Conditional or partial deregulation should be accompanied by mandatory monitoring to assess potential risk issues, with USDA retaining authority to impose restrictions, as needed, to address any problems that develop after commercialization. Such monitoring could be conducted by USDA or independent third parties. The results of monitoring

---

181. *Id.*
182. Peter Kareiva et al., *Can We Use Experiments and Models in Predicting the Invasiveness of Genetically Engineered Organisms?*, 77 ECOLOGY 1670-75; *2002 NRC Report*, *supra* note 66, at Sec. 6; ALLISON A. SNOW ET AL., ECOLOGICAL SOC'Y OF AMERICA, GENETICALLY ENGINEERED ORGANISMS AND THE ENVIRONMENT: CURRENT STATUS AND RECOMMENDATIONS (2004), *available at* http://www.esa.org/pao/policyStatements/pdfDocuments/geo-positionPaper0405.pdf.
183. *2002 NRC Report*, *supra* note 66, at 192.
184. *2008 GAO Report*, *supra* note 79, at 45-48.
185. 7 U.S.C. §7732.
186. 7 U.S.C. §7714(a)(1)(A).
187. 7 U.S.C. §7733.
188. 7 U.S.C. §7731(b).
189. 7 U.S.C. §7734.
190. David Bennett, *Bt Corn Refuge Requirements Tightened*, DELTA FARM PRESS, Mar. 1, 2011, at http://deltafarmpress.com/corn/bt-corn-refuge-requirements-tightened; U.S. EPA, OFFICE OF PESTICIDE PROGRAMS, TERMS AND CONDITIONS FOR BT CORN REGISTRATIONS 87-88 164, 171-72, 205 (2010), *available at* http://www.epa.gov/opp00001/biopesticides/pips/bt-corn-terms-conditions.pdf-.

activity would provide the basis for any needed restrictions on cultivation of the pertinent GE crop. Post-commercialization monitoring would perform a function similar to that of post-market drug surveillance for adverse effects not detected in clinical trials, where size is necessarily limited.

b. Partial Deregulations With Restrictions

Even under its current, pre-PPA regulations, one way that USDA could impose requirements—such as post-market monitoring, or other restrictions on planting such as use limitations, geographic restrictions, or isolation distances—is through partial deregulation.[191] However, the regulations do not elaborate on when the agency can deregulate "in part," and, until 2011, USDA never utilized the authority.[192]

Monsanto v. Geertson Seed Farms, the first Supreme Court case on genetically engineered crop impacts, provided important clarity on USDA's power to partially deregulate. In that case, which challenged USDA's approval of Monsanto's Roundup Ready alfalfa (RRA), the district court ordered the first EIS required of any transgenic crop approval, analyzing, among other harms, transgenic contamination and superweeds stemming from the deregulation.[193] The court then halted its planting pursuant to both an injunction and vacatur (i.e., vacating the deregulation and thereby reverting the crop to its previous status of a regulated article that could not be grown commercially).[194] After the U.S. Court of Appeals for the Ninth Circuit affirmed,[195] the Supreme Court set aside the district court's injunction for two reasons. It held, first, that the vacatur independently halted the planting, and so the injunction was unnecessary.[196] Second, and as important, the Court held that the injunction was overbroad because it improperly barred USDA from trying to prevent transgenic contamination and RRA-created HR weeds with a partial deregulation. The Court posited that, absent the overbroad injunction, USDA could have enacted a "limited deregulation" that "would not pose a significant risk of gene flow or harmful weed development."[197] That is to say, the Court predicated its analysis and holding on USDA's ability to enact partial deregulations with restrictions over transgenic crops, to mitigate the exact types of harms discussed here and in that case.

The Supreme Court's decision was in accord with the two earlier district court opinions requiring EISs on transgenic contamination. Under NEPA doctrine, agencies are not required to analyze impacts that they do not have statutory control over, since NEPA's procedural purpose is meaningless if the agency has no authority to act on that information.[198] Those courts rejected arguments that the agency need not analyze transgenic contamination impacts based on lack of authority, holding instead that, because USDA could have partially deregulated to address that harm, it thus did necessarily have to analyze those impacts in an EIS.[199]

Following the decision in *Monsanto*, USDA promulgated a partial deregulation for the first time, allowing commercialization of Roundup Ready sugar beets under certain mandatory conditions during the pendency of a court-ordered EIS. The deregulation conditions were put in place specifically to limit the risk of transgenic contamination of related conventional and organic species, and USDA retained the authority to re-regulate the crop if the conditions were violated.[200] Moreover, the future implementation and integration of the PPA noxious weed authority should further expand the types of harms for which USDA can

---

191. 7 C.F.R. §340.6(d)(3).
192. *See Monsanto*, 130 S. Ct. 2743, at 2768 n.6 (Stevens, J., dissenting).
193. Geertson Seed Farms, 2007 WL 518624 (N.D. Cal. 2007).
194. Geertson Seed Farms, 2007 WL 776146, at *3 (N.D. Cal. Mar. 12, 2007); Geertson Seed Farms, 2007 WL 1302981, *8-9 (N.D. Cal. May 3, 2007).
195. Geertson Seed Farms, 541 F.3d 938 (9th Cir. 2008), *amending opinion and denying petition for rehearing and rehearing en banc*, 570 F.3d 1130 (9th Cir. 2009).
196. *Monsanto*, 130 S. Ct at 2761 (holding that "the vacatur of APHIS' deregulation decision means that virtually no [genetically engineered alfalfa] can be grown or sold until such time as a new deregulation decision is in place. . . .").
197. *Id.* at 2759-60.
198. *See* Department of Transportation v. Public Citizen, 541 U.S. 752 (2004).
199. In *Geertson Seed Farms*, the court noted that USDA could have deregulated with geographic restrictions to prevent contamination, and held that USDA failed to analyze measures that could potentially contain or limit transgenic contamination, because such data were critical, the "very information it needs to determine if such an [partial deregulation] option is warranted." *Geertson Seed Farms*, 2007 WL 518624, at *1 & *6. Similarly in *Center for Food Safety*, USDA argued that it did not have to analyze the socioeconomic contamination impacts of the deregulation of Roundup Ready sugar beets because such impacts were beyond the scope of the agency's authority; the court rejected the argument, holding that the agency could have partially deregulated to address these contamination harms, and because of that authority, thus did necessarily have to analyze the impacts in an EIS. *Center for Food Safety*, 2009 WL 3047227, at *8.
200. Roundup Ready Sugar Beets Partial Deregulation, 78 Fed. Reg. 6759, 6760 (Feb. 8, 2011).

partially deregulate. In the interim, however, in a case challenging USDA's second unrestricted approval of RRA, a district court agreed with USDA's position that contrary to the Supreme Court's decision in *Monsanto* and the other earlier precedent, the agency could not prevent the crop's harms through a partial deregulation, at least not pursuant to its current regulations.[201] That decision, issued in January 2012, is currently on appeal.[202]

Conclusion

The need for improved oversight of transgenic crops is manifest. One option is for USDA to utilize its authority to analyze, regulate, and, where appropriate, restrict transgenic crops from continuing to cause harm to agriculture and the environment. Clear and unequivocal statutory authority to do so exists in the Plant Protection Act of 2000. To be sure, transgenic crops are not the same as traditional noxious weeds or traditional plant pests. However, that need not stop USDA from applying its authority over these crops and addressing their environmental and economic risks. The gravamen of that implementation should be how to apply the statutory PPA standards—the same standards it applies to conventional crops—to transgenic crops in its transgenic crop regulation, with the goal of best fulfilling the purpose of the statute. This is precisely what EPA has done regarding GE plant-pesticides pursuant to its FIFRA authority.

Although it has yet to take any final action, for a number of years, USDA has been considering amending its GE crop regulations, issuing a scoping notice regarding the needed rule amendment in 2004,[203] a draft programmatic EIS (PEIS) on potential regulatory changes in 2007,[204] and finally, a proposed new rule in 2008 and 2009.[205] However, as of this writing, there have been no further developments. The draft PEIS of 2007 provided some hope that USDA would amend its regulations to provide needed improvements in oversight. In the draft PEIS, the agency noted that the PPA had "redefined authorities and responsibilities for the agency" beyond those it had used "historically," which it concluded "may not be of sufficient breadth to cover . . . the full range of potential agricultural and environmental risks posed by these organisms."[206] The draft PEIS made preliminary decisions that USDA would implement its new PPA noxious weed authority beyond plant pest risks, which would "increase oversight of GE plants by increasing the scope of what is regulated and by allowing a broader consideration of risks,"[207] including public health risks,[208] socioeconomic impacts,[209] potentially the inclusion of post-commercialization monitoring for some transgenic crops,[210] and partial deregulations.[211] The draft PEIS also proposed replacing the current taxonomic trigger with a genetic engineering trigger, as recommended by the National Academy of Sciences.[212]

However, USDA's subsequent 2008 proposed rule was a reversal: instead of strengthening oversight based on USDA's newer, more robust statutory authority, the rule proposed lessening oversight. USDA would cabin the application of its PPA authority,[213] including limiting it to only direct types of harm[214]—for example, not assessing impacts from herbicides that are part and parcel of the engineered HR crop system that the agency was assessing. A decision limiting application of the PPA's noxious weed mandate to only "traditional" noxious weeds would make transgenic crop application perfunctory and likely leave significant impacts unregulated. It would also arguably be contrary to sound science, given that transgenic crops present new risks and often trigger substantial adverse impacts that closely resemble those of tradi-

---

201. Center for Food Safety v. Vilsack, 2012 WL 27787 (N.D. Cal. 2012).
202. Center for Food Safety v. Vilsack, No. 12-15052 (9th Cir. 2012).
203. 69 Fed. Reg. 3271-72 (Jan. 23, 2004); 69 Fed. Reg. 16181-01 (Mar. 29, 2004).
204. *Introduction of Organisms and Products Altered or Produced Through Genetic Engineering*, 72 Fed. Reg. 39021 (July 17, 2007).
205. 74 Fed. Reg. 16797 (Apr. 13, 2009).
206. U.S. Dep't. Agric., APHIS, Draft Programmatic Environmental Impact Statement: Introduction of Genetically Engineered Organisms 131 (July 2007).
207. *Id.* at ix, 21.
208. *Id.*
209. *Id.* at 49.
210. *Id.* at vi.
211. *Id.* at 143.
212. *Id.* at 20, 168.
213. 73 Fed. Reg. 60008, 60012-14 (Oct. 9, 2008).
214. *Id.* at 60013.

tional noxious weeds, as well as contrary to the PPA's express language, given that the above harms fit the broad statutory definition of noxious weed and plant pest harms.

Regardless of when and how USDA eventually finalizes its new regulations and implements the PPA, improving USDA's oversight via its existing authority is no panacea for long-standing structural problems of the Coordinated Framework. Eventually, the Framework must be reframed to account for the realities and risks now apparent and worsening, and properly balance these impacts with potential future development. U.S. biotech policy remains an outlier worldwide and must eventually be set in line with the more transparent and precautionary approach of the majority of the world's nations.[215]

In fact, the entire issue of agricultural biotechnology is a microcosm of a larger U.S. policy failing: over 40 years have passed since the enactment of our major environmental laws. The necessity to rethink agricultural biotechnology oversight is an opportunity to ameliorate long-festering problems in U.S. oversight structures. Agricultural biotechnology developments highlight the outdated nature of our current regulatory systems and how ill-equipped they are to deal with the issues of the 21st century. Without a new generation of laws more in line with the ecological and technological realities of this century, those entrusted with protecting public health and the environment are forced to continue to try and squeeze blood from the existing statutory stones. The agricultural biotechnology dialogue provides the challenge and the opportunity to rethink these existing paradigms.

However, until and unless the American populace convinces its policymakers to implement needed legislative changes, urgently needed improvements in environmental protection, health safety assessments, and economic stability are there for the taking, just lying in the (likely HR) weeds.

---

215. *See, e.g.*, Cinnamon Carlarne, *From the USA With Love: Sharing Home-Grown Hormones, GMOs, and Clones With a Reluctant Europe*, 37 ENVTL. L. 301, 318 (2007); Debra M. Strauss, *The International Regulation of Genetically Modified Organisms: Importing Caution Into the U.S. Food Supply*, 61 FOOD & DRUG L.J. 167, 186 (2006); Debra M. Strauss, *Feast or Famine: The Impact of the WTO Decision Favoring the U.S. Biotechnology Industry in the EU Ban of Genetically Modified Foods*, 45 AM. BUS. L.J. 775, 784, 822-23 (2008).

Chapter 17
The Future of Food Eco-Labeling: A Comparative Analysis
Jason J. Czarnezki

The environmental costs of modern industrial and large-scale food production and consumption have begun to enter public consciousness. The true costs of the modern food system are not adequately reflected by the low prices most consumers pay. Food choices shape our waistlines, the natural landscape, and ecological health. Society has become increasingly aware that choices about food contribute to the climate crisis, cause species loss, impair water and air quality, and accelerate land use degradation. The causes of these environmental costs are many—the livestock industry, diet, agricultural practices like the use of pesticides and fertilization, and large-scale food transportation, processing, packaging, and distribution systems.

Recent legal scholarship suggests both that environmental policy will focus more on changing individual behavior,[1] and that consumer informational labeling can be an effective regulatory tool in encouraging eco-friendly choices. Individuals in the United States contribute 30-35% of greenhouse gas emissions nationwide, which accounts for 8% of the world's total.[2] A European Union study showed that products from only three areas—food and drink, private transportation, and housing—are responsible for 70-80% of the environmental impacts of personal consumption.[3]

Can better information and dissemination lead to consumer-driven environmental improvement leading to fewer toxins in the environment, decreased greenhouse gas emissions, and more sustainable use of natural resources? What labeling schemes and legal policies best support environmentally friendly food consumption? Previous scholarship argued for the creation of more ambitious informational labeling regimes such as "eco-labeling," product labels evaluating the ecological and carbon footprint of products including foods, and for promoting a more local and organic food system.[4] This chapter builds on this work and considers the role and implementation of eco-labeling in promoting a sustainable food system. While the entire American food system needs to be reevaluated, the farm bill requires modification,[5] and local, organic, nonindustrially processed food systems should be promoted, the incremental step of better food labeling is necessary, given the dominant industrial food system and emerging industrial organic market. The objective of an eco-label would be to provide consumers with information about the environmental costs of food choices, resulting in changes in consumer preferences and buying practices.

This chapter is, with permission, a significantly revised, updated, and adapted version of an article previously published as Jason J. Czarnezki, *The Future of Food Eco-Labeling: Organic, Carbon Footprint, and Environmental Life-Cycle Analysis*, 30 Stan. Envtl. L.J. 3 (2011). All mistakes and errors, of course, are the author's.

1. *See, e.g.*, Hope M. Babcock, *Assuming Personal Responsibility for Improving the Environment: Moving Toward a New Environmental Norm*, 33 Harv. Envtl. L. Rev. 117 (2009); Jason J. Czarnezki, Everyday Environmentalism: Law, Nature, and Individual Behavior (2011); Michael P. Vandenbergh, *The Individual as Polluter*, 35 ELR 10723 (Nov. 2005); Michael P. Vandenbergh, *Order Without Social Norms: How Personal Norm Activation Can Protect the Environment*, 99 Nw. U. L. Rev. 1101 (2005).
2. Anne E. Carlson et al., *The Forum: Creating the Carbon-Neutral Citizen*, 24 Envtl. F. 46, 46 (2007); Michael P. Vandenbergh & Anne Steinemann, *The Carbon-Neutral Individual*, 82 N.Y.U. L. Rev. 1673 (2007). In defining individual behavior, MIchael Vandenbergh and Anne Steinemann include emissions from personal motor vehicle use, personal air travel, mass transport, and emissions attributable to household electricity use.
3. Bo P. Weidema et al., Environmental Improvement Potentials of Meat and Dairy Products 5, 17 (Peter Eder & Luis Delgado eds., 2008).
4. Czarnezki, *supra* note 1, at Introduction and Chapter Four (Food).
5. *See, e.g.*, William S. Eubanks II, *Paying the Farm Bill: How One Statute Has Radically Degraded the Natural Environment and How a Newfound Emphasis on Sustainability Is the Key to Reviving the Ecosystem*, 27 Envtl. F. 56 (2010).

Labeling already exists for some purposes. For example, organic labeling is primarily concerned with prohibiting the use of synthetic chemicals, which may result in less risk to consumers from chemicals in their food and may have some environmental benefits such as less risk to wildlife and soil from pesticides. But such labeling does not explicitly say anything about other environmental concerns such as water usage and greenhouse gas emissions. Likewise, carbon footprint labeling does not address ecological concerns beyond greenhouse gas emissions.

The objective of any new food eco-label program should be to achieve a broader objective based on a definition of "sustainable food" that combines many objectives—lowering the carbon footprint of food at all stages (agriculture, distribution, and packaging), reducing consumption, supplying healthier food, promoting sustainable agriculture (less resource-intensive and less polluting), and encouraging water and land use efficiency. Food would have to be environmentally evaluated at all stages of its production and processing. "Evaluated" here could mean quantitative environmental life-cycle analysis or qualitative best-practices standards.

Organic labeling programs exist, carbon labeling programs and environmental best practices for food production and processing are under development, and environmental life-cycle assessments for foods are under consideration. Both the United States and European Union have developed organic food certification and labeling programs. The U.S. Organic Foods Production Act (OFPA) establishes a national organic certification program in which agricultural products may be labeled as organic if produced and handled without the use of synthetic substances. European Union regulations on organic production and labeling, at least on paper, exhibit a broader and more ambitious model than their U.S. counterpart. The European Union organic model attempts to offer a holistic paradigm reflecting animal welfare, environmental pollution, and biological diversity, in addition to chemical and synthetic inputs. Carbon footprint labeling is now occurring in the United Kingdom through the Carbon Trust, and many private companies around the world are engaging in environmental labeling. Food might also be labeled based on qualitative best practices standards or quantitative life-cycle analyses that would include consideration of natural resource and chemical inputs starting at the production process or raw extraction stage, and emissions and pollution outputs during the production, distribution and use, and disposal stages.

This chapter discusses public and private efforts to inform consumers about environmentally preferable food choices, comparing such efforts in the United States and Europe. The chapter first briefly describes the environmental consequences of the modern food system. It then describes existing public and private eco-labeling regimes, including organic labeling, carbon footprint labeling, and country of origin labeling. The chapter then describes the "Swedish experiment," in which that country, a leader in reducing greenhouse gases, has embarked upon an ambitious carbon labeling and dietary information program. The Swedish National Food Administration developed new dietary guidelines, formally proposed to the European Commission, which give equal weight to climate and health, though unfortunately these guidelines were not adopted by Sweden. Additionally, Sweden's largest organic certification organizations have embarked upon a program called "Climate Labelling for Food" that requires food to be both produced organically and to emit low levels of greenhouse gases to meet certification requirements. The Swedish efforts offer a potential model for more ambitious eco-labeling in the United States.

Against the backdrop of those European labeling efforts, the chapter then considers the future of food eco-labeling in the United States and abroad, addressing the merits of creating a U.S. national eco-labeling program similar to the Swedish program or other European Union programs, replacing current federal organic food legislation. A key issue is the extent to which an American state could engage in environmental federalism and develop a stringent eco-labeling program that does not run afoul of the existing federal regulations about organic labeling under the OFPA. That section draws an analogy between member states in the EU and American states in our federal system in evaluating the ability to have stricter eco-labeling for food than suggested by American and European Community law. The chapter then considers the difficulties in developing an eco-label that considers a wider range of environmental assessments than existing organic and climate labeling programs, focusing in particular on the goals of the European Food Sustainable Consumption and Production Round Table, as well as discussing the merits of a best practices versus life-cycle approach. The chapter closes by addressing the challenges to creating effective food eco-labeling.

A. Food and the Environment

Food choices contribute to the climate crisis, cause species loss, impair water and air quality, and accelerate land use degradation. For example, "[a]n estimated 25 percent of the emissions produced by people in industrialized nations can be traced to the food they eat."[6] The causes of these environmental costs are many—practices of the livestock industry, a processed and meat-heavy diet, agricultural practices such as the use of pesticides and fertilization, fossil-fuel intensive food transportation, factory processing, packaging, and large-scale distribution systems. These are traits of the dominant industrial food model. Given the ecological costs of the industrial food system, as well as the growing industrial organic market that also relies on processed and packaged foods and significant transportation costs, eco-labeling is warranted. Several major characteristics of the modern food system contribute to environmental degradation and are relevant to the development of a food eco-labeling program.

1. Agricultural Practices

Many growers of fruits, vegetables, beans, nuts, and grains engage in high-input and nonorganic production, employing synthetic pesticides and fertilizers. Chemical use has perhaps the most direct environmental impact of any agricultural practice. A 2006 study by the U.S. Geological Survey released the following findings:

> At least one pesticide was detected in water from all streams studied and . . . pesticide compounds were detected throughout most of the year in water from streams with agricultural (97 percent of the time), urban (97 percent), or mixed-land-use watersheds (94 percent). In addition, organochlorine pesticides (such as DDT) and their degradates and by-products were found in fish and bed-sediment samples from most streams in agricultural, urban, and mixed-land-use watersheds—and in more than half the fish from streams with predominantly undeveloped watersheds. Most of the organochlorine pesticides have not been used in the United States since before the [National Water-Quality Assessment] studies began, but their continued presence demonstrates their persistence in the environment.[7]

In 2004, nearly 500 million pounds of pesticides were used in the United States.[8] In 2007, over 22 million tons of inorganic fertilizer (nitrogen, phosphate, and potash) were used in the United States.[9] While this amount of fertilizer has remained fairly steady since the mid-1970s (around 20 million tons), this is over triple the amount used in 1960.[10]

Mary Jane Angelo has chronicled industrial agriculture's impact on the environment.[11] Chemical inputs, in the form of fertilizers and pesticides, have the potential, through runoff, to pollute groundwater and streams, induce algae blooms and oxygen depletion in waterways, contribute to soil acidification, kill beneficial insects, and potentially poison wildlife and their reproductive systems. Industrial farming techniques such as over-tilling, a lack of crop rotation, use of inorganic fertilizers and pesticides, and monoculture mine the soil of its natural nutrients, destroy soil biota and its habitat, and increase erosion. In addition, petroleum remains the single most important ingredient in the modern food system, not only used as fuel for transportation and production of food, but also to produce fertilizers and pesticides.

Water resources, in terms of both quantity and quality, are particularly endangered by the industrial food model.[12] For example, corn (perhaps the iconic example of modern commodity-driven agriculture[13])

---

6. Elisabeth Rosenthal, *To Cut Global Warming, Swedes Study Their Plates*, N.Y. Times, Oct. 23, 2009, at A6.
7. U.S. Geological Survey, Pesticides in the Nation's Streams and Ground Water, 1992-2001—A Summary (2006), *available at* http://pubs.usgs.gov/fs/2006/3028/.
8. Craig Osteen & Michael Livingston, *Pest Management Practices*, *in* Agricultural Resources and Environmental Indicators 107, 108 (Keith Wiebe & Noel Gollehon eds., 2006).
9. U.S. Dep't of Agric., ERS, *U.S. Fertilizer Use and Price, Table 1—U.S. Consumption of Nitrogen, Phosphate, and Potash, 1960-2007* (June 30, 2010), http://www.ers.usda.gov/data-products/fertilizer-use-and-price.aspx.
10. *Id.*
11. Mary Jane Angelo, *Corn, Carbon, and Conservation: Rethinking U.S. Agricultural Policy in a Changing Global Environment*, 17 Geo. Mason L. Rev. 593 (2010).
12. *Id.* at 603 (citing William S. Eubanks II, *A Rotten System: Subsidizing Environmental Degradation and Poor Public Health With Our Nation's Tax Dollars*, 28 Stan. Envtl. L.J. 213, 269-70 (2009)).
13. *See, e.g.*, King Corn (Mosaic Films 2007); Michael Pollan, The Omnivore's Dilemma: A Natural History of Four Meals (2006).

has a very large "water footprint,"[14] and is a pesticide-intensive crop.[15] As weeds become more resistant and more-toxic pesticides are used,[16] the ecological costs of runoff increase, as described by Angelo:

> When rain or irrigation water comes into contact with farm fields, certain agricultural chemicals, including water soluble pesticides [such as atrazine] and nutrients, such as nitrites found in fertilizers, easily leach into groundwater. This contamination can render groundwater sources of water unacceptable for drinking. Where ground water naturally flows into surface water, such as is the case with artesian springs, surfacewaters become contaminated as well . . . . Scientific studies demonstrate that agricultural intensification via increased chemical fertilizer and other inputs is directly linked to increased environmental damage. Large quantities of these compounds are carried in rain run-off into waterbodies where they exert their plant growth enhancing effect resulting in overgrowth of algae.[17]

In addition to water-quality concerns, "[t]he fossil-fuel-intensive inputs required in industrialized agriculture exacerbate the daunting challenge of reducing carbon emissions to stem climate change."[18] Industrial agriculture remains fossil fuel-intensive as pesticides, fertilizers, harvesting and tilling machinery, food processing factories, and transportation all use fossil fuels. Agriculture accounts for about 20% of U.S. fossil fuel consumption, as well as 37% of U.S. and 15% of worldwide greenhouse gas emissions.[19]

2. Livestock and Fishing Industries

Americans raise and kill nearly 10 billion animals a year for food, more than 15% of the world's total (despite accounting for less than 5% of the world's population).[20] The livestock industry has a substantial carbon footprint, contributes to waste runoff and water pollution, and creates potentially harmful ecosystem effects. According to the United Nations Food and Agriculture Organization, meat production accounts for 18% of world's total greenhouse gas emissions.[21] By contrast, all transportation forms combined represent 13%.[22] In terms of negative environmental impacts caused by personal choice, meat and dairy products are the most greenhouse gas-intensive products purchased, and they contribute to the most environmental impacts.[23]

Pasture-fed cattle eat grass, their growth fueled by the sun's energy. Concentrated animal feed operations (CAFOs), however, are now the norm in the modern livestock industry as they produce meat quickly and cheaply. Due to government subsidies, cattle now are fed cheap grain instead of their natural preference of grass.[24] Corn feed production uses fertilizer and requires delivery, making animal growth fossil-fueled. Natural inefficiency is coupled with practical inefficiency, compounding the emissions problems, as it takes 20 pounds of grain to produce one pound of beef, four pounds of chicken, or seven pounds of

---

14. Angelo, *supra* note 11, at 604 (citing Adell Amos, *Freshwater Conservation in the Context of Energy and Climate Policy: Assessing Progress and Identifying Challenges in Oregon and the Western United States*, 12 U. Denv. Water L. Rev. 1, 6 (2008)).
15. *See, e.g.*, William Neuman & Andrew Pollack, *Rise of the Superweeds: Herbicide's Wide Use Fosters the Spread of Resistant Pests*, N.Y. Times, May 4, 2010, at B1.
16. *Id.*
17. Angelo, *supra* note 11, at 605-06 (internal citations omitted).
18. *Id.* at 602.
19. *Id.* at 612 (citing William S. Eubanks II, *The Sustainable Farm Bill: A Proposal for Permanent Environmental Change*, 39 ELR 10493, 10504, (June 2009).
20. Mark Bittman, *Rethinking the Meat-Guzzler*, N.Y. Times, Jan. 27, 2008, at WK1.
21. Food & Agric. Org. of the U.N., *Livestock a Major Threat to Environment*, FAO Newsroom, Nov. 29, 2009, at http://www.fao.org/newsroom/en/news/2006/1000448/index.html.
22. Richard Black, *Shun Meat, Says UN Climate Chief*, BBC News, Sept. 7, 2008, at http://news.bbc.co.uk/2/hi/science/nature/7600005.stm.
23. Weidema et al., *supra* note 3, at 6 ("The study finds that the consumption of meat and dairy products contributes on average 24% of the environmental impacts from the total final consumption in EU-27, while constituting only 6% of the economic value. . . . The four main product groups (dairy, beef, pork and poultry products) contribute respectively 33-41%, 16-39%, 19-44%, and 5-10% to the impact of meat and dairy products consumption in EU-27 on the different environmental impact categories."); Øresund Food Network & Øresund Env't Acad., Climate Change and the Food Industry: Climate Labelling for Food Products: Potential and Limitations (Maria Olofsdotter & Jacob Juul eds., 2008) (citing Arnold Tukker et al., Environmental Impact of Products (EIPRO): Analysis of the Life Cycle Environmental Impacts Related to the Final Consumption of the EU-25 (2006), *available at* http://ec.europa.eu/environment/ipp/pdf/eipro_report.pdf ("Out of 25 top GHG intensive product categories, 52% are related to food production. It was shown that meat, dairy, fats and oils are the most GHG intensive products within the food category. The authors estimated that meat's and meat products' contribution to GWP ranges from about 4 to 12% of all products studied across the EU.")).
24. *See, e.g.*, Daniel Imhoff, Food Fight: The Citizen's Guide to a Food and Farm Bill 59-60 (2007); Pollan, *supra* note 13, at 39-40.

pork.[25] Processing plants, making everything from hamburger patties to chicken nuggets, emit greenhouse gases as do the cattle themselves in the form of methane from manure.

Christopher Weber and Scott Matthews have described meat's high carbon footprint as it relates to food consumption.[26] Shifting one day per week of protein from meat or dairy to vegetables, or even another protein source (fish, chicken, eggs), has the same carbon footprint effect as buying all household food from local providers. A one-day-per-week protein shift from red meat to chicken, fish, or eggs saves the equivalent of 760 miles per year driven, and a one day per week shift to vegetables saves 1,160 miles per year.[27]

In addition to the problem of carbon output, the waste stream from CAFOs and processing plants contribute to damaging runoff and water pollution. Hogs, chickens, and cows produce mounds of manure, requiring the creation of "poop lagoons" and, like fertilizer, leading to a high nutrient load in runoff. Land simply cannot absorb enough nitrogen and phosphorus, so precipitation washes them into streams, rivers, and underground aquifers. The nutrients foster the growth of algae, which sucks oxygen from the water and endangers aquatic species.

Fishing practices can also lead to ecological destruction. Monterey Bay Aquarium's Seafood Watch® is well known for developing its pocket guide and rating system for picking sustainable seafood.[28] The nonprofit aquarium devoted to ocean conservation notes that industrial-scale fishing has led to the decline of fish populations.

> By the 1980s, these fishing practices had made it impossible for natural fish stocks to keep up. Seventy percent of the world's fisheries are now exploited, overexploited or have collapsed. Meanwhile, demand has continued to rise, to about 110 tons in 2006—over eight times what it was in 1950. It's estimated that by 2030, the world will need an additional 37 million tons of farmed fish per year to maintain current levels of consumption.[29]

Seafood Watch® considers the major environmental concerns for wild seafood to be overfishing, illegal fishing, habitat destruction, bycatch (i.e., unintended harm to other marine populations), and poor regulation and enforcement.[30]

3. Food Processing and Distribution Systems

The large production and distribution systems of modern agriculture and commercial processing are powered by fossil fuels. Thus, the life cycles of food products have significant carbon footprints. In the United States, food production from the farm to the store accounts for 20% of fossil fuel consumption.[31] According to a study identifying the environmental impact of European consumption, the "food and drink" category causes 20-30% of the various environmental impacts of private consumption.[32] In a study of Iowa food production, researchers found that the "conventional system used four to 17 times more fuel than the Iowa-based regional and local systems, depending on the system and truck type. The same conventional system released five to 17 times more CO_2 from the burning of this fuel than the Iowa-based regional and local systems."[33] But while there is limited evidence that the carbon footprint of food production is too high

---

25. Paul Roberts, The End of Food 210 (2008).
26. Christopher L. Weber & H. Scott Matthews, *Food-Miles and the Relative Climate Impacts of Food Choices in the United States*, 42 Envtl. Sci. & Tech. 3508 (2008).
27. *Id.* at 3512-13.
28. Monterey Bay Aquarium Seafood Watch®, *Seafood Recommendations*, http://www.montereybayaquarium.org/cr/cr_seafoodwatch/sfw_recommendations.aspx (last visited Nov. 24, 2012).
29. Monterey Bay Aquarium Seafood Watch®, *Wild Seafood*, http://www.montereybayaquarium.org/cr/cr_seafoodwatch/issues/wildseafood.aspx (last visited Nov. 24, 2012).
30. *Id.*
31. Dan Imhoff, Paper or Plastic: Searching for Solutions to an Overpackaged World 102 (2002).
32. Øresund Food Network & Øresund Env't Acad., *supra* note 23, at 2 (discussing an analysis that included the full food production and distribution chain "from farm to fork').
33. Leopold Ctr. for Sustainable Agric., Food, Fuel and Freeways: An Iowa Perspective on How Far Food Travels, Fuel Usage, and Greenhouse Gas Emissions 1-2 (2001), *available at* http://www.leopold.iastate.edu/pubs/staff/ppp/food_mil.pdf ("The *conventional system* represented an integrated retail/wholesale buying system where national sources supply Iowa with produce using large semitrailer trucks. The *Iowa-based regional system* involved a scenario modeled after an existing Iowa-based distribution infrastructure. In this scenario, a cooperating network of Iowa farmers would supply produce to Iowa retailers and wholesalers using large semitrailer and midsize trucks. The *local system* represented farmers who market directly to consumers through community supported agriculture (CSA) enterprises and farmers markets, or through institutional markets such as restaurants, hospitals, and conference centers. This system used small light trucks.").

and food miles are too many, no systematic measurement methodology exists to fully quantify the carbon footprint along food supply chains.[34]

B. Environmental Labeling Regimes for Food in the United States and Europe

Food labeling has expanded from being just a marketing and sales effort to include public and environmental health. Nutritional labeling began in the United States in 1990 under the Nutrition Labeling and Education Act.[35] Under the OFPA of 1990 and the National Organic Program (NOP), the U.S. government sets labeling standards for organic agricultural products.[36] In January 2009, similar organic product regulations went into effect in the European Union.[37] In addition, public and private carbon footprint labeling is being promoted on both sides of the Atlantic. Finally, the United States has enacted "country of origin" labeling legislation that requires retailers to inform consumers about the source of certain foods.[38]

1. Organic Labeling

The organic food market is flourishing. People want chemical-free foods for personal health and environmental reasons. In light of the economic benefits of organic production—organic products sell for much more than conventional ones[39]—the modern organic production and distribution system is now dominated by large-scale "industrial organic" or "big organic" producers.[40] Along with large-scale production, even if organic, comes increased greenhouse gas emissions and questionable agricultural methods. Yet, organic production also meets demand for food produced and processed in a chemical-free environment. Organic food has almost quadrupled its market share in the last decade,[41] and organic food sales have grown from $1 billion in 1990 to over $20 billion today.[42] But all of this might not have happened without a regulatory model creating a value-added food label like "organic."

a. The U.S. OFPA and the NOP

Under the OFPA and the NOP, the U.S. government creates production, handling, and labeling standards for organic agricultural products. Individuals buy organic products to promote sustainable and chemical-free agriculture, keep their bodies healthy and free of synthetics and pesticides, and protect animal welfare.[43] The OFPA establishes a national organic certification program under which agricultural products may be labeled as organic if produced and handled without the use of synthetic substances.[44] The program prohibits using synthetic fertilizers, growth hormones, and antibiotics in livestock, or adding synthetic ingredients during processing.[45]

---

34. Gareth Edwards-Jones et al., *Testing the Assertion That "Local Food Is Best": The Challenges of an Evidence-Based Approach*, 19 TRENDS FOOD SCI. & TECH. 265 (2008) ("We conclude that food miles are a poor indicator of the environmental and ethical impacts of food production. Only through combining spatially explicit life cycle assessment with analysis of social issues can the benefits of local food be assessed. This type of analysis is currently lacking for nearly all food chains.").
35. Nutrition Labeling and Education Act of 1990, 21 U.S.C. §343.
36. Organic Foods Production Act of 1990, 7 U.S.C. §§6501-6523.
37. Council Regulation 834/2007, The New EU Regulation for Organic Food and Farming, 2007 O.J. (L 189) 1 (EC).
38. U.S. DEP'T OF AGRIC., AGRICULTURAL MARKETING SERVICE, *Country of Origin Labeling*, http://www.ams.usda.gov/AMSv1.0/Cool (last visited Nov. 24, 2012).
39. Kate L. Harrison, *Organic Plus: Regulating Beyond the Current Organic Standards*, 25 PACE ENVTL. L. REV. 211, 211 (2008) (citing Tom Philpott, *Up Against the Wal-Mart: Big Buyers Make Organic Farmers Feel Smaller Than Ever*, GRIST: ENVTL. NEWS & COMMENT. (Aug. 23, 2006), http://grist.org/article/buyers/, for the proposition that "[t]oday, organic foods command a 20-30% price premium over their conventionally produced counterparts").
40. Harrison, *supra* note 39, at 212 (citing James Temple, *The "O" Word: Some Organic Farmers Opt Out of Federal System*, CONTRA COSTA TIMES, Oct. 29, 2006, at 6B; HighJump Software, Earthbound Farm Gains Efficiencies With Supply Chain Execution Solutions From HighJump Software (June 14, 2005), http://news.3m.com/press-release/company/earthbound-farm-gains-efficiencies-supply-chain-execution-solutions-highjump-s).
41. ORGANIC TRADE ASS'N, EXECUTIVE SUMMARY, ORGANIC TRADE ASSOCIATION'S 2007 MANUFACTURER SURVEY (2007), *available at* http://www.ota.com/pics/documents/2007ExecutiveSummary.pdf (stating that organic food sales accounted for 0.8% of total food sales in 1997, and 2.8% in 2006).
42. Organic Trade Ass'n, *The Organic Industry*, http://www.ota.com/pics/documents/Mini%20fact%201-08%20confirming.pdf.
43. Gemma C. Harper & Aikaterini Makatouni, *Consumer Perception of Organic Food Production and Farm Animal Welfare*, 104 BRITISH FOOD J. 287 (2002).
44. Organic Foods Production Act, 7 U.S.C. §6504(1)-(3).
45. *Id.* §§6508(b)(1), 6509(c)(3), 6510.

Agricultural practices must follow an organic plan approved by an accredited certifying agent and agreed to by the producer and handler of the product.[46] The OFPA creates process-based standards but does not implement standards or require tests for actual chemical content in food, nor assessment of overall land use practices. Thus, "certified organic" labeling informs consumers about the food production process, but does not directly describe food quality or environmental considerations, though organic food is still likely to have fewer chemicals than conventional counterparts.[47]

The NOP, created under the OFPA, establishes a four-tiered labeling system for organic foods.[48] First, a product can be labeled "100 percent organic" and carry the seals of the U.S. Department of Agriculture (USDA) and certifying agent if it is entirely composed of organically produced ingredients that are defined by the OFPA as free from synthetic substances.[49] Second, a product must contain at least 95% organic ingredients to be labeled simply "organic" and use USDA and private certifying agent seals.[50] Third, a product with at least 70% organically produced ingredients can be labeled "made with organic ingredients" and carry the seal of a private certifying agent.[51] For products containing less than 70% organic ingredients, organic ingredients may be listed on the label, but neither the word "organic" nor any seal can be used. Small farmers who gross less than $5,000 annually and only sell directly to consumers via farmers markets and family farm stands can avoid the certification process by simply signing a declaration of compliance that their practices meet organic standards.[52] However, if these farmers sell any of their products through conventional distribution channels, they may not use the term "certified organic" on products without also obtaining official certification, a process that can be expensive and time-consuming.[53]

b. The European Union Organic Program

Like the United States, the European Union has established organic product legislation detailing production rules for plants, livestock, and processed products, their labeling and control, and import rules.[54] Compliance with European Union organic farming regulations permits display of the European Union's organic food logo. The European Union regulations on organic production and labeling, at least on their face, are broader and more ambitious than their U.S. counterpart.[55] The European Union organic model attempts to offer a more holistic paradigm that considers animal welfare, environmental pollution, biologi-

---

46. *Id.* §§6504-6505.
47. Michelle T. Friedland, *You Call That Organic?—The USDA's Misleading Food Regulations*, 13 N.Y.U. Envtl. L.J. 379, 398-99 (2005). However, Michelle Friedland notes:
    > Because food produced in accordance with the NOP regulations will not be intentionally sprayed with pesticides or intentionally grown or raised using genetically engineered seed or other inputs, the likelihood of the presence of pesticide residue or genetically engineered content will clearly be lower than in foods intentionally produced with pesticides and genetic engineering techniques. But organic food will not be free of such contamination. Evidence clearly indicates that both pesticides and genetically engineered plant materials often drift beyond their intended applications, and organic food, like any food, may be accidentally contaminated.

    Id.
48. 7 C.F.R. §205.301. In addition to looking for "organic" labeled foods, consumers can look at five-digit PLU codes. Organic foods all start with 9.
49. *Id.* §§205.301(a), 205.303. The OFPA defines "synthetic" as "a substance that is formulated or manufactured by a chemical process or by a process that chemically changes a substance extracted from naturally occurring plant, animal, or mineral sources, except that such term shall not apply to substances created by natural occurring biological processes." 7 U.S.C. §6502(12).
50. 7 C.F.R. §§205.301(b), 205.303.
51. *Id.* §205.301(c).
52. Harrison, *supra* note 39, at 219 (citing Andrew J. Nicholas, *As the Organic Industry Gets Its House in Order, the Time Has Come for National Standards on Genetically Modified Foods*, 15 Loy. Consumer L. Rev. 277, 285 (2003)).
53. *Id.*
54. Commission Regulation 889/2008, 2008 O.J. (L 250) 1 (EC), http://ec.europa.eu/agriculture/organic/eu-policy/legislation_en (last visited Nov. 24, 2012) (laying down detailed rules for the implementation of Council Regulation 834/2007, 2007 O.J. (L 189) 1 (EC), on organic production and labeling of organic products with regard to organic production, labeling and control); Commission Regulation 1254/2008, 2008 O.J. (L 337) 1 (EC) (amending Commission Regulation 889/2008, 2008 O.J. (L 250) 1 (EC), and laying down detailed rules for implementation of Council Regulation 834/2007, 2007 O.J. (L 189) 1 (EC), on organic production and labeling of organic products with regard to organic production, labeling and control); Int'l Fed'n of Organic Agric. Movements EU Group, The New EU Regulation for Organic Food and Farming 14 (2007), http://www.ifoam-eu.org (last visited Nov. 24, 2012) ("The major part of the revision of the EU regulatory framework for organic farming is now finished. The agreement reached in the Council in 2007 led to the publication of Council Regulation (EC) No. 834/2007 in the Official Journal of July 20, 2007. Since then, it has been completed with two sets of implementing rules in 2008: Commission Regulation (EC) No. 889/2008 on detailed production rules for plants, livestock and processed products including yeast, and their labeling and control, and Commission Regulation (EC) No. 1235/2008 on detailed rules for imports."); European Commission, Organic Farming, Legislation, http://ec.europa.eu/agriculture/organic/eu-policy/legislation_en (last visited Nov. 24, 2012).
55. Council Regulation 834/2007, On Organic Production and Labelling of Organic Products and Repealing Regulation (EC) No. 2092/91, 2007 O.J. (L 189) 1 (EC).

cal diversity, and renewable energy, in addition to chemical and synthetic inputs. According to the European Commission website:

> The goal of this new legal framework is to set a new course for the continued development of organic farming. Sustainable cultivation systems and a variety of high-quality products are the aim. In this process, even greater emphasis is to be placed in future on environmental protection, biodiversity and high standards of animal protection. Organic production must respect natural systems and cycles. Sustainable production should be achieved insofar as possible with the help of biological and mechanical production processes, through land-related production and without the use [sic] genetically modified organisms.[56]

As a result, the Europeans are ahead of the curve worldwide in terms of food labeling and production. For example, egg labeling is compulsory. All eggs produced in the European Union must be stamped with a code to show whether they arrive from a free-range environment, a barn, or a caged battery, and egg packaging must indicate the method of production.[57]

Like the American counterpart, European foods can only be labeled "organic" and carry the European Union organic logo if at least 95% of their agricultural ingredients are organic.[58] Similarly, nonorganic food may be listed as organic in the list of ingredients, and genetically modified organisms are prohibited in organic production.[59]

2. Carbon Footprint Labeling

Food author Michael Pollan wrote: "The way we feed ourselves contributes more greenhouse gases to the atmosphere than anything else we do—as much as 37 percent, according to one study."[60] When buying a food product, in addition to wanting to know if the production process was chemically intensive, a consumer may wish to know the carbon footprint of the product. Food can be energy intensive given factory processing and distribution methods powered by fossil fuels (also known as "food miles").

> Therefore, the impetus for carbon labels is that by providing consumers with information about the carbon content of a product, they will be able to make informed decisions about the goods they purchase and ultimately choose products with a smaller carbon footprint, and therefore less carbon emissions. This will decrease carbon footprints from individuals and consequently lead to a reduction in carbon emissions worldwide.[61]

Carbon input data into production is generated through life-cycle analyses based on individual data, national averages, or a hybrid of the two.[62] The goal of such generating such information, and ultimately putting it on a label, is to allow consumers to pick environmentally-friendly purchases and to encourage manufacturers to reduce the environmental impact resulting from the goods' production.[63]

Carbon labeling is gaining traction both through governmental implementation and through private industry seeking to tap into new consumer markets. Carbon footprint information is not only being conveyed on food, but on other goods as well. In the United States, for example, California and New York have developed carbon emission labeling for new motor vehicles.[64] California has now gone further in proposing a voluntary carbon labeling program aimed at standardizing the labeling of life-cycle carbon footprints for consumer products sold in the state.[65]

Compared to relatively slow movement in the United States, European countries, and Great Britain in particular, are leading the way on carbon labeling. The Carbon Trust was created by the British gov-

---

56. EUROPEAN COMMISSION, ORGANIC FARMING, LEGISLATION, *supra* note 54.
57. Council Regulation 2052/2003, Amending Regulation (EEC) No. 1907/90 on Certain Marketing Standards for Eggs, 2003 O.J. (L 305) 1 (EC).
58. EUROPEAN COMMISSION, ORGANIC FARMING, LEGISLATION, *supra* note 54.; Council Regulation 834/2007, The New EU Regulation for Organic Food and Farming, 2007 O.J. (L 189) 1 (EC).
59. EUROPEAN COMMISSION, ORGANIC FARMING, LEGISLATION, *supra* note 54.
60. Michael Pollan, *Farmer in Chief*, N.Y. TIMES MAG., Oct. 12, 2008, at MM62.
61. Stacey R. O'Neill, *Consuming for the Environment: A Proposal for Carbon Labels in the United States*, 39 CAL. W. INT'L L.J. 393, 396 (2009).
62. *Id.* at 403-04.
63. *Id.* at 401.
64. *See* CAL. HEALTH & SAFETY CODE §43200.1; N.Y. ENVTL. CONSERV. LAW §19-1103.
65. O'Neill, *supra* note 61, at 397, 431 (citing Carbon Labeling Act of 2009, CAL. HEALTH & SAFETY CODE §44570). *See also* CAL. AB 19, http://www.leginfo.ca.gov/pub/09-10/bill/asm/ab_0001-0050/ab_19_bill_20081201_introduced.pdf.

ernment in 2001 as an independent entity tasked with accelerating the move to a low-carbon economy,[66] and its standard for measuring carbon labeling (known as PAS 2050) has since developed into the industry norm.[67]

The Carbon Trust helps individuals and companies, including food and drink producers, use and communicate carbon emissions information. The company's most recognizable achievement is the creation of the "Carbon Reduction Label"—a black footprint with "CO_2" written on it along with the number of grams or kilograms of total greenhouse gas emissions emitted at every stage of the product's life cycle, including production, transportation, preparation, use, and disposal.[68]

In addition to publicly supported carbon information programs, private companies, both for-profits and nonprofits, are exploring environmental carbon labeling options.[69] For example, Home Depot uses its "Eco Options" label,[70] and Timberland footwear displays a "nutritional label" that lists climate impact, chemicals used, and resource consumption.[71] In the food industry, British supermarkets have taken the lead. Tesco announced that it will begin labeling all 70,000 of its products with the quantity, in grams of carbon dioxide equivalent, emitted into the atmosphere through their manufacture and distribution.[72] Sweden's businesses are at the forefront of advocating ecological awareness.[73]

In the nonprofit sector, U.S. organizations like the Carbon Fund have created the "Certified Carbon Free" label, and California's Climate Conservancy has developed the "Carbon Conscious" label.[74] Additionally, the Carbon Trust has set up a U.S. office.

3. Country of Origin Labeling and Other Food Labels

Country of origin labeling (COOL) requires that a food product inform consumers of its source.[75] While the underlying rationale for COOL in the United States is assurance of the safety of foreign goods and economic protectionism for domestic products,[76] COOL also allows consumers to choose food products originating closer to their own market, and thus with a lower carbon footprint (i.e., lower food miles). Also, COOL may implicitly provide information to buyers because educated consumers may know, for example, whether produce was grown out of season in a greenhouse or came from an unsustainable fishery based on its source country. COOL requirements were enacted in under the Farm Security and Rural Investment Act of 2002 (better known as the 2002 farm bill)[77] and its implementing regulations,[78] and reauthorized in the Food, Conservation and Energy Act of 2008 (the 2008 farm bill).[79]

Despite objections to COOL by many producers and retailers, the idea had support from consumer and product safety organizations.[80] Under COOL, retailers, such as grocery stores, supermarkets, and club

---

66. The Carbon Trust, http://www.carbontrust.com/home (last visited Nov. 24, 2012).
67. Using a cradle-to-grave life-cycle analysis, under the Carbon Trust methodology all GHG emissions should be measured and then converted in carbon dioxide equivalent emissions using 100 year global warming potential (GWP) coefficients. *See also* British Standards Institute, *PAS 2050—Assessing the Life Cycle Greenhouse Gas Emissions of Goods and Services* (2009), http://www.bsigroup.com/upload/Standards%20&%20 Publications/Energy/PAS2050.pdf.
68. The Carbon Trust, http://www.carbontrust.com/home.
69. Michael P. Vandenbergh, *Climate Change: The China Problem*, 81 S. Cal. L. Rev. 905, 949 (2008) ("[L]abeling may occur through inclusion of greenhouse gas emissions-disclosure requirements in existing labels or new labels managed by private standard-setting organizations, or through unilateral action by firms.").
70. The Home Depot, *The Eco Options Program*, http://www.ecooptions.homedepot.com/ (last visited Nov. 24, 2012).
71. Timberland, Grading Our Products: Timberland's Green Index® Program (2009), *available at* http://responsibility.timberland.com/wp-content/uploads/2011/05/Timberlands_Green_Index_Program_2009_report.pdf. *See also* Amy Cortese, *Friend of Nature? Let's See Those Shoes*, N.Y. Times, Mar. 6, 2007, at http://www.nytimes.com/2007/03/06/business/businessspecial2/07label-sub.html?r=2.
72. Julia Finch, *Tesco Labels Will Show Products' Carbon Footprints*, The Guardian, Apr. 16, 2008, at http://www.guardian.co.uk/environment/2008/apr/16/carbonfootprints.tesco.
73. *See, e.g.*, Rosenthal, *supra* note 6 ("Max, Sweden's largest homegrown chain of burger restaurants, now puts emissions calculations next to each item on its menu boards. Lantmannen, Sweden's largest farming group, has begun placing precise labels on some categories of foods in grocery stores, including chicken, oatmeal, barley and pasta.").
74. The Carbon Fund, http://www.carbonfund.org/ (last visited Nov. 24, 2012); The Climate Conservancy, http://www.climateconservancy.org/index.php (last visited Nov. 24, 2012) (for an English translation of this website, type the URL into translate.google.com).
75. For discussions of COOL, see Peter Chang, *Country of Origin Labeling: History and Public Choice Theory*, 64 Food & Drug L.J. 693, 702 (2009); Anastasia Lewandoski, *Legislative Update: Country-of-Origin Labeling*, 9 Sustainable Dev. L. & Pol'y 62 (2008).
76. Chang, *supra* note 75, at 693.
77. *See* Farm Security and Rural Investment Act of 2002, Pub. L. No. 107-71, 116 Stat. 134, 533 (codified as amended at 7 U.S.C. §1638).
78. *See* 7 C.F.R. §§60, 65.
79. Food, Conservation and Energy Act of 2008, Pub. L. No. 110-234, §11002, 122 Stat. 923, 1352-1354 (2008).
80. Chang, *supra* note 75, at 702.

warehouse stores must provide customers with information regarding the source of certain foods.[81] Food products subject to the legislation include cut and ground meats (beef, veal, pork, lamb, goat, and chicken), wild and farm-raised fish and shellfish, fresh and frozen fruits and vegetables, nuts (peanuts, pecans, and macadamia nuts), and ginseng.[82] In addition, future expansion of COOL legislation may seek to require all food labels to identify the country of final processing or mandate that manufacturers' websites identify each ingredient's country (or countries) of origin.[83]

In addition to organic, carbon footprint, and country of origin labeling, other private food labeling schemes inform consumers about the ecological consequences of their purchases. Food-sector businesses often seek to show such labels on their products since consumers may be willing to pay more for "green" food items. For example, the "Bird Friendly" label created by the Smithsonian Migratory Bird Center identifies shade grown and organic coffee that protects bird habitat.[84] The blue Marine Stewardship Council label certifies sustainable fisheries and seafood businesses.[85] Similarly, co-op grocers often display and sell food with labels such as "localvore region" and "locally grown" (grown or produced within the same geographic area of the store, often within 100 miles). There is no shortage of environmental labels for food. *Consumer Reports*, on its Greener Choices website, evaluated 78 labels for food alone, mostly state organic certification logos.[86]

C. The Swedish Experiment

In Sweden, new labels listing the carbon dioxide emissions associated with food production are appearing on grocery items and restaurant menus around the country. The *New York Times* has described the Swedish efforts as a "new experiment."[87] The Swedish experiment is driven by (1) the creation of proposed new national dietary guidelines, and (2) an initiative called "Klimatmärkning för Mat," or, in English, "Climate Labelling for Food."[88]

Sweden's proposed national recommended dietary guidelines give equal weight to climate impacts and public health.[89] Scandinavia's organic certification programs will begin requiring farmers to convert to low-emissions techniques if they want to display the organic seal.[90] The Swedish effort grew out of a study by Sweden's national environmental agency on how personal consumption generates emissions, finding that 25% of national per capita emissions were attributable to eating.[91] Sweden has proved to be a leader in greenhouse gas emission reductions, and the United States may learn from its efforts.

1. Swedish Dietary Guidelines

The Sweden National Food Administration (SNFA), in collaboration with the Swedish Environmental Protection Agency and based on a scientific assessment published by the Swedish University of Agricultural Sciences, developed dietary food guidelines that account for and balance health and environmental impacts.[92] This is a unique approach. Sweden notified the European Union of its new dietary guidelines,[93] and similar concerns about the relationship between diet and the environment have been

---

81. U.S. Dep't. of Agric., Agric. Mktg. Serv., *Country of Origin Labeling*, supra note 38.
82. *Id. See also* 7 C.F.R. §§60, 65.
83. *See* Chang, *supra* note 75, at 713 (citing Draft Bill, Food and Drug Globalization Act of 2009, H.R. 759, 111th Cong. (2009)).
84. *See* Migratory Bird Ctr., http://nationalzoo.si.edu/scbi/MigratoryBirds/default.cfm (last visited Nov. 24, 2012).
85. *See* Marine Stewardship Council, http://www.msc.org/get-certified (last visited Nov. 14, 2010).
86. *See* Consumer Reports, *Greener Choices: Products for a Better Planet*, http://www.greenerchoices.org/ (last visited Nov. 24, 2012).
87. Rosenthal, *supra* note 6.
88. Note that the word "labeling" is spelled "labelling" in Europe.
89. Rosenthal, *supra* note 6, ¶ 5.
90. *Id.* ¶ 20.
91. *Id.* ¶ 11. *See also* Naturvårdsverket, The Climate Impact of Swedish Consumption, Report 5992 (Jan. 2010); Adrian Leip et al., Evaluation of the Livestock Sector's Contribution to the EU Greenhouse Gas Emissions (GGELS)—Final Report, European Commission, Joint Research Centre (2010).
92. *See* Swed. Nat'l Food Admin., The National Food Administration's Environmentally Effective Food Choices (2009), *available at* http://www.slv.se/upload/dokument/miljo/environmentally_effective_food_choices_proposal_eu_2009.pdf.
93. The European Union was notified pursuant to and in accordance with procedure established by Directive 98/34/ (EC) of the European Parliament and the Council of 22 June 1998. *See* Swed. Nat'l Food Admin., *supra* note 92.

raised by the European Commission.[94] However, the new Swedish dietary guidelines no longer exist in this form, and, after numerous rounds of revisions, the Swedish Agricultural Ministry decided not to proceed with the guidelines.[95]

Despite their lack of adoption, this is potentially an attractive model to pursue. Ulf Bohman, Head of the Nutrition Department at the SNFA, said: "We're used to thinking about safety and nutrition as one thing and environmental as another."[96] If the new food guidelines were heeded, Sweden could cut its emissions from food production by as much as 20-50%.[97]

As seen in Table 1, the new Swedish guidelines would have divided dietary choices into six categories and suggest ways to change eating behaviors within each category to promote personal health and a sustainable environment.[98]

Table 17.1
Swedish Dietary Guidelines Food Categories and Recommendations[99]

Food Categories	Recommendations
Meat, including beef, lamb, pork and chicken	Eat less and reduce portion size. Eat locally produced and grass-fed meat.
Fish and shellfish	Choose from stable stocks. Pick seafood with existing eco-labels.
Fruits and berries, vegetables, and leguminous plants	Choose seasonal and locally grown fruits and vegetables. Pick organic and pesticide-free fruits, berries, and vegetables. Choose fiber-rich vegetables. Eat beans, lentils, and peas. Store fruits and vegetables properly.
Potatoes, cereal products, and rice	Choose locally grown potatoes and cereal products. Reduce rice consumption in favor of other grains.
Cooking fat	Choose rapeseed oil or olive oil. Reduce use of palm oil.
Water	Choose tap water whenever possible. Choose locally produced packaged water if necessary.

These recommendations were suggested for a whole host of environmental reasons in addition to acknowledged health benefits. For example, the proposed guidelines account for the high climate impact of beef due to methane released in cattle digestion, the depletion of many fish stocks, the energy-heavy refrigerated transport required for delicate fruits and vegetables, the fact that fiber-rich root vegetables are more likely to be grown outdoors than in greenhouses requiring fossil fuels, that water-soaked rice fields produce more greenhouse gases than potato farms, that oil palms are often cultivated on former rainforest lands, and even the high carbon footprint of plastic water bottles.[100]

2. Klimatmärkning för Mat *(Climate Labeling for Food)*

Sweden is also pursuing labeling programs that would provide consumers with information about the environmental costs, and carbon footprints in particular, of their food purchases. Swedish food certification organizations KRAV and Swedish Seal (*Svenskt Sigill* in Swedish) are developing a label for climate-friendly products under a program called "*Klimatmärkning för Mat*" (KFM) or "Climate Labelling For Food."[101]

---

94. *See* Arnold Tukker et al., Environmental Impacts of Diet Changes in the EU: Annex Report (JRC Eur. Comm'n 2009), *available at* http://ftp.jrc.es/EURdoc/EURdoc/ EURdoc/JRC50544_Annex1.pdf.
95. Interview with Monika Pearson, Swedish National Food Administration (Livmedels Verket) on Mar. 4, 2011.
96. Rosenthal, *supra* note 6, ¶ 5.
97. *Id.* ¶ 11.
98. Swed. Nat'l Food Admin., *supra* note 92.
99. *Id.*
100. *See* Swed. Nat'l Food Admin. *supra* note 92.
101. Øresund Food Network & Øresund Env't Acad., *supra* note 23.

KRAV in particular has long been the key player in the Swedish organic market.[102] Until recently, it was difficult to market organic products in Sweden without the KRAV label, and KRAV remains the dominant domestic organic label.[103] Both KRAV and Swedish Seal had for some time considered incorporating climate labeling, also known as carbon footprint labeling, into their certification systems.[104] In 2007, the two parties decided to cooperate in producing standards for climate labeling, and were joined in the project by key industry groups and food producers. The Swedish Board of Agriculture also took part as a co-opted member by contributing its expertise.[105] Fortunately, large Swedish retail chains have come to an agreement not to use climate marking as a competitive weapon among themselves, but rather to cooperate to avoid multiple labels, minimize customer confusion, and pool label development and research resources.[106] What is appearing to develop is a comprehensive, country-wide labeling scheme that incorporates information on public health, chemical-free foods, environmental health, and lower greenhouse gas emissions.

What does this mean in practice? It means that Sweden's primary organic certification programs, like KRAV, will require farmers and producers to convert to low-emission production and processing in order to display their seal. It means that farmers seeking to stay certified may need to change their agricultural practices. For example, most foods transported by air will not be eligible for the more stringent label.[107] An organic tomato produced in a hot house may no longer receive the Swedish label since, while it would be produced without synthetic or chemical inputs, the growing environment (i.e., the greenhouse and its lights) likely would be powered completely by fossil fuels.[108] The same could be said for an "organic" apple flown in from New Zealand.[109]

According to the KFM initiative, there are two approaches for creating standards for climate labeling for food.[110] The first option is to calculate exact emissions per product based on life-cycle analysis measuring total carbon emissions from production to distribution to use to disposal. The KFM project has concluded that such exact calculations, because they would require extensive knowledge about individual product history and continuous updates due to changes in the production methods and modified emission factors, would be prohibitively time consuming and expensive to develop at this time.[111] (Despite these difficulties, the ecological and consumer choice advantages of a life-cycle analysis for food are strong and are currently being evaluated by the European Union through its European Food Sustainable Consumption and Production Roundtable, as discussed below.)

The second approach is to use existing knowledge to create general standards.[112] For example, certain processes have large carbon impact: use of concentrates based on soy protein, high consumption of fossil fuels, and production of nitrous oxides for artificial fertilizer.[113] Using assumptions based on the known environmental harm caused by such factors, the hope is that standards can be produced more quickly and simply in a way that can be developed and refined over time.[114] The KFM process can be thought of as creating first and second generations of climate standards, that is, "[a] first one that is simpler and includes

---

102. U.S. Dep't of Agric. Foreign Agric. Serv., Food and Agricultural Import Regulations and Standards—Narrative, FAIRS Country Report: Sweden 4 (2009), *available at* http://gain.fas.usda.gov/Recent%20GAIN%20Publications/Food%20and%20Agricultural%20Import%20Regulations%20and%20Standards%20-%20Narrative_Brussels%20USEU_EU-27_8-7-2009.pdf.
103. *Id.*
104. Climate Labelling for Food, Project Description for the Project: Standards for Climate Marking of Foods Version No. 2.3 1 (2009) [hereinafter Climate Labelling 1].
105. *Id.*
106. *Id.*
107. *See* Øresund Food Network & Øresund Env't Acad., *supra* note 23, at 12. ("According to Johan Cejie, Head of Standards Development (regelutvecklingschef) at KRAV, products transported by air will not be eligible to be awarded with this label (Local Tidningen, 2007).").
108. Rosenthal, *supra* note 7, at ¶ 20.
109. *See* Øresund Food Network & Øresund Env't Acad., *supra* note 23, at 12.
110. For discussion of standard-setting techniques, see Climate Labelling for Food, Project Description: Standards for Climate Label for Food Version No. 2009:1 1 (2009), *available at* http://www.klimatmarkningen.se/wp-content/uploads/2009/02/project-description-english.pdf [hereinafter Climate Labelling 2]. *See also* Climate Labelling 1, *supra* note 104, at 1.
111. Climate Labelling 1, *supra* note 104, at 3 ("Another difficulty is that the climate impact of the product varies throughout the season. The message to the consumers is also unclear because different products are not directly comparable from a nutritional point of view. However, it shall be stressed that the heaviest workload is at the initial stages, for when an LCA analysis is made it is easy to change individual factors and produce analyses for several farms.").
112. *Id.* at 3.
113. *Id.*
114. *Id.*

identified activities that have a large climate impact, to be followed by a second version where we can specify the climate impact of every product."[115]

KFM has already created detailed labeling rules (i.e., standards) for some foods. In addition to organic certification, it provides a labeling system that incorporates greenhouse gas emissions. In an effort to merge with other eco-labeling interests, KFM rules require that users of the new labeling system hold another quality certificate related to environmental protection, animal welfare, or social welfare.[116] According to KFM standards, the criteria for achieving the label "shall cover production from the manufacturing of production means to shop's loading bay, including distribution from the farm gates to the shop's loading bay."[117]

As seen in Table 17.2, the KFM rules tap into a number of areas for which standards have been created in an effort to lessen the carbon footprint.

Table 17.2
Current "Climate Labelling for Food" (KFM) Rule: Categories and Areas of Interest[118]

Categories	Areas
General farm activities	Energy consumption on the farm Storage of food and use of refrigerants Transport operations and use of machines within the farm and when selling products Cultivation of organogenic soils
Crop production	Nitrogen flows Use of manure Fertilizers Feed production Crop rotation
Milk production	Animal health Fodder Handling manure
Greenhouses	Energy consumption Use of refrigerants Transport operations and use of machines within the business and when selling products
Fisheries	Fish stocks Fuel demand

In order to make them eligible to receive the KFM label, future rules are planned for pork, beef and chicken production, egg production, fish production, processed products, packaging, rules for imported products, and general transportation rules.[119]

With such a diverse array of environmental standards and with agreement among third-party certification labelers like KRAV and Swedish Seal, as well as the food industry and government, the hope is that the label can be used by several standard owners and certification bodies. KFM's stated objective in creating its label is to "reduce climate impact by creating a marking system for food where the consumers make a conscious climate choice and businesses can strengthen their competitiveness best practices approach,[120] with the latter winning, certifiers also had to choose between using a separate carbon label or integrating the best practices standards into their existing certifying."[121]

---

115. *Id.*
116. Climate Labelling for Food, Standard for Reducing Climate Impact in the Production and Distribution of Food Version 2009:1 (2009), *available at* http://www.klimatmarkningen.se/wp-content/uploads/2009/02/climate-labelling-for-food-2009-1.pdf [hereinafter Climate Labelling 3].
117. *Id.* at 2.
118. *Id.* at 1.
119. *Id.* at 2.
120. This is a decision to use verifiable production process standards, or standards based on quantitative data/statistics about environment costs.
121. Climate Labelling 1, *supra* note 104, at 4.

In addition to deciding between a life cycle or label.[122] KRAV chose the latter approach, thus adding climate standards to their existing label.[123] The label does not provide quantifiable emissions numbers, but it does ensure that measures have been taken to reduce climate impact. KRAV labeling takes organic and climate factors into account, as well as standards for animal welfare, social responsibility, and public health. Standards must be created in a variety of product areas, and are being done so in the following order: greenhouses; fisheries; imports; animal husbandry; and crops.[124]

As is often the case, the response to the Swedish dietary guidelines and KFM labeling by certifiers like KRAV has not been universally positive, eliciting especially harsh criticism from some types of food producers. For example, the dietary guidelines have been attacked by Europe's meat industry, Norwegian salmon farmers, and Malaysian palm oil growers.[125] Similarly, many farmers are not happy since greenhouse produce will lose its organic label, and, for example, farmers with high concentrations of peat soil on their property may no longer be able to grow carrots, since plowing peat releases huge amounts of carbon dioxide.[126] As with many other labeling systems, there are criticisms that labels will be ignored or improperly used by consumers, and that product comparison remains difficult. On the other hand, the KRAV initiative provides a consistent labeling regime, and it should make buyers more aware of a product's general ecological footprint.

D.   Environmental Federalism in the United States and Europe

The term "environmental federalism" refers to the ability of states to establish more rigorous or creative environmental protection legislation than that of the national government.[127] This idea is not new. In his dissenting opinion in *New State Ice Co. v. Liebmann*, Justice Louis Brandeis stated, "[i]t is one of the happy incidents of the federal system that a single courageous State may, if its citizens choose, serve as a laboratory; and try novel social and economic experiments without risk to the rest of the country."[128] An individual American state should similarly take on such an experiment to establish a comprehensive, creative, and rigorous eco-labeling scheme for food. One can also draw an analogy between member states in the EU and American states in our federal system in evaluating the ability to have stricter eco-labeling for food than suggested by American and European Community law.

The goal behind a state-sponsored voluntary eco-labeling program in the United States would be to test an eco-labeling model that merges and moves beyond existing organic, carbon footprint, and country-of-origin labeling,[129] and perhaps the same can be done in European countries through private labeling regimes. Such an eco-label could engage in environmental life-cycle analysis or create best-practices standards, in an effort to provide consumers with greater information about the production's overall ecological footprint. This information is necessary given the existing industrial food market, where large-scale production, chemical usage, and significant transportation miles are the norm. Consumers then can choose products that in the aggregate are more likely to have been produced closer to point-of-purchase, have fewer chemical and synthetic additives, require fewer greenhouse gases in production and processing, and were produced more sustainably in terms of water usage and land degradation.

---

122. Karl Johan Bonnedahl & Jessica Erikkson, *The Role of Discourse in the Quest for Low-Carbon Economic Practices: A Case of Standard Development in the Food Sector*, European Mgmt. J. (2010) ("At the time, the question of whether climate criteria would be integrated with KRAV's organics standards or result in freestanding label remained open.").
123. KRAV has noted this in its slogan accompanying its label. The label is a green "KRAV" surrounded by an oval with the slogan, "Du får mer," meaning "you get more." *See also* KRAV-labelled Organic Food Is Worth More, http://www.krav.se/System/Spraklankar/In-English/For-Consumers/KRAV-labelled-Organic-Food-is-Worth-More/.
124. KRAV, KRAV Standards January 2011, http://www.krav.se/KravsRegler/; Interview at KRAV on Mar. 4, 2011 (KRAV standard implementation timeline on file with author).
125. Rosenthal, *supra* note 6, at ¶ 18.
126. *Id.*
127. *See generally* Robert V. Percival, *Environmental Federalism: Historical Roots and Contemporary Models*, 54 Md. L. Rev. 1141 (1995).
128. New State Ice Co. v. Liebmann, 285 U.S. 262, 311 (1932) (Brandeis, J., dissenting).
129. *Contra* O'Neill, *supra* note 61, at 432 ("However, in a nation where interstate commerce is at the core of our economy and a world where the international marketplace is prominent, such a small, state-based policy is far too limited in scope.").

While some scholars have suggested that an "organic plus" model should be pursued, merging organic labeling with a local or carbon label,[130] an even more holistic and wholesale change is needed.[131] The current organic label in the United States is insufficient because it does not give consumers the opportunity to make distinctions between different kinds of organic products (e.g., the carbon footprint of eggs versus beef, or processed organic snacks versus raw organic produce).[132] The objective of any proposed new food eco-label program should have the broader objective of a sustainable food system by combining multiple interests—lowering the carbon footprint of food at all stages (agriculture, distribution, and packaging); reducing consumption; supplying healthier food; promoting sustainable agriculture (less resource-intensive and less polluting agriculture); and encouraging water and land use efficiency.

The U.S. NOP is a good start. Sweden's efforts with its environmental and health-based dietary guidelines and organic plus carbon labeling under the "Climate Labelling for Food" are a significant step forward in the process of developing a more sustainable food system. But a creative approach using best-practices, like KRAV, and then life-cycle analysis, like that used in the European Union Flower Logo labeling program for consumer durables,[133] and ideas currently being studied by the European Food Sustainable Consumption and Production Round Table (Round Table), can be the next significant steps in the future of food eco-labeling. Due to the existing marketing power of the "organic" brand in the United States, there are advantages to pursuing an environmental life-cycle eco-labeling program within the confines of the U.S. OFPA or federally redefining "organic" as something more similar to the Swedish model.

Admittedly, new federal legislation creating an improved eco-label for food would be an ideal model, though perhaps politically unrealistic. A national eco-labeling program could be similar to the Swedish program, essentially integrating carbon emissions concerns into the current USDA organic certification model.[134] Or, more ambitiously, the U.S. Congress could replace current legislation in order to pursue a best-practices approach or a life-cycle model like that of the European Union Flower Logo or the model under discussion by the European Union Round Table discussed below.

Despite the advantages of a national model, an individual state could pursue a voluntary eco-labeling program that goes beyond the current federal organic program and even the emerging carbon labeling programs in Europe. Such a program could be adopted by other states, or serve as a model for future national legislation.

This section considers how a state food eco-label program can be established without running afoul of existing federal labeling laws, specifically considering whether such an eco-label can be consistent with the OFPA. It addresses the challenges in creating food eco-labels, with specific focus on initiatives taking place in Europe. The key legal and policy question is whether an American state can create a voluntary label for products sold within a given state that does not run afoul of the existing federal law and that effectively engages in life-cycle analysis of environmental impacts. Finally, this section discusses the practical challenges of designing any eco-label that will successfully impact consumer choice.

1. The Merits of Federal Legislation

New federal legislation creating an environmental life-cycle eco-label for food would be the ideal model, though politically unrealistic given the historic challenges surrounding COOL legislation, including strong lobbying by special interests that have impacted organic regulation. A national eco-labeling program could be similar to the Swedish program discussed above, essentially integrating carbon emissions concerns into the current USDA organic certification model by redefining what it means to be "organic,"[135] or could

---

130. *See* Harrison, *supra* note 39, at 233.
131. Edwards-Jones et al., *supra* note 34 ("We conclude that food miles are a poor indicator of the environmental and ethical impacts of food production. Only through combining spatially explicit life cycle assessment with analysis of social issues can the benefits of local food be assessed. This type of analysis is currently lacking for nearly all food chains.").
132. *See* Harrison, *supra* note 39, at 227-28. I note that this sort of environmental comparison across product categories does not exist in the Swedish labeling initiatives.
133. The EU Flower Logo, also known as the EU Ecolabel. Is designed to help consumers "identify products and services that have a reduced environmental impact throughout their life cycle, from the extraction of raw material through to production, use and disposal," http://ec.europa.eu/environment/ecolabel/.
134. *Cf.* O'Neill, *supra* note 61 (advocating a federal carbon labeling program implemented by the U.S. Environmental Protection Agency).
135. *See id.* at 408.

develop an "organic plus" label by merging organic labeling with a local or carbon label.[136] Alternatively, it could develop best practices in a number of categories, such as is already done for organic production processes, or, more ambitiously, Congress could replace current legislation and pursue a brand new environmental life-cycle labeling model like that of the European Union Flower Logo or the one under discussion by the European Union Round Table discussed below.

Given concerns about a regulatory patchwork among states, product labeling may be better suited to federal standards. A federal standard would avoid subjecting manufacturers to potentially different sets of state standards and confusing consumers by many different labeling schemes. That said, organic food certifiers exist in many states to implement the same federal substantive standards (with variable procedural requirements as discussed below). Under the current NOP, a single state could create a more environmentally-conscious food labeling model that surpasses federal standards. Doing so would create two sets of substantive food standards: (1) the existing federal standards implemented by federal government and the states, and (2) a better state standard incorporating far more information and with a much broader purpose than the existing federal program.

A national program could then be modeled after a more ambitious state model, or other states could adopt this state model, similar to the way in which states voluntarily adopt California auto emissions standards under the Clean Air Act. In addition, and as discussed below, current federal law asks for, but has not yet received, a creative approach for environmental food labeling by an entrepreneurial state. It is important, therefore, that an individual state pursue a voluntary eco-labeling program that goes beyond current federal organic program in the United States and emerging carbon labeling programs in Europe.

2.   A State-Sponsored Eco-Label in the United States

At first glance, the legal barriers to creating a new environmental life-cycle eco-label seem formidable, but this is not the case.[137] Marketing is actually the greatest barrier to building an effective eco-label that incorporates a wider range of environmental concerns. The OFPA monopolizes the use of the term "organic," requiring all products labeled as "organic" to be certified through the government-approved certifiers that comply with all OFPA regulations under the NOP.[138]

Under state law, a new eco-label could be created that considers a wide array of concerns like the use of synthetic substances in the production process, greenhouse gas emissions, and ecological degradation. This program could use any new word or logo, outside of the term "organic," thus developing and marketing a new label or logo without any of the existing advantages or disadvantages of using the term.[139] However, the word "organic" has developed significant cache and marketing power in the food industry, and sends an important message to consumers (despite the disconnect between consumer understanding of the label's meaning and the label's function).[140] In fact, a criticism of the organic label is that it does not mean what consumers may think it means. For example, consumers may identify "organic" with small farms, local production, healthy foods, and environmental consciousness. These characteristics are often not true of many organic foods. Many consumers are not aware that products labeled "organic" are not 100% organic. Products that are only 95% organic can be labeled "organic," and if made with as little as 70% synthetic-free ingredients, can still be labeled "made with organic ingredients."[141] A preferable scenario would be to create a more rigorous eco-label that encompasses organic production but also considers a whole host of other environmental concerns at every stage of a food product's life. In other words, organic foods with a new label might have the ecologically friendly characteristics that many consumers already think they have.

---

136. *See* Harrison, *supra* note 39, at 213.
137. This is not to discount potential World Trade Organization issues that might arise due to eco-labeling. *See* Erich Vranes, *Climate Labelling and the WTO: The 2010 EU Ecolabelling Programme as a Test Case Under WTO Law* (Vienna Univ. of Econ. & Bus. Admin., Research Inst. for European Affairs, Working Papers Series, Mar. 10, 2010), *available at* http://papers.ssrn.com/sol3/papers.cfm?abstract_id=1567432.
138. 7 U.S.C. §6505(a)(1)(A) (2010).
139. Renate Gertz, *Eco-Labelling—A Case for Deregulation?*, 4 Law, Probability & Risk 127, 136 (2005) (pointing out that European Union flower label is meeting only limited success, and it will take a while for the new label to gather traction).
140. *See* David Conner & Ralph Christy, *The Organic Label: How to Reconcile Its Meaning With Consumer Preferences*, 35 J. Food Distribution Res. 40 (2004).
141. 7 C.F.R. §205.301 (2010).

Fortunately, under §6507(a) of the OFPA, individual U.S. states have the right to seek approval for their own organic certification program.[142] The statute reads:

> A State organic certification program . . . may contain more restrictive requirements governing the organic certification of farms and handling operations and the production and handling of agricultural products that are to be sold or labeled as organically produced under this chapter than are contained in the program established by the Secretary.[143]

In other words, states can create more rigorous food standards and use the term "organic."[144] (Interestingly and in contrast to American law, greater organic labeling standards can be created in Europe, but only through private certifiers and not through member states.[145])

The OFPA implementing regulations state that "certifying agents certifying production or handling operations within a State with more restrictive requirements, approved by the Secretary, shall require compliance with such requirements as a condition of use of their identifying mark by such operations."[146] Thus, a more rigorous state certification program can have its own label, mark, or logo with which it could describe the more restrictive standards that must be followed.

However, if a state pursues an organic certification program with additional requirements, the standards (1) must further the goals of the OFPA, (2) may not be inconsistent with the statute, and (3) cannot be discriminatory towards agricultural commodities organically produced in other states.[147] The stated purposes of the OFPA are to establish national standards governing the marketing of certain agricultural products as organically produced, to assure consumers that organically produced products meet a consistent standard, and to facilitate interstate commerce in fresh and processed food that is organically produced.[148]

Obviously, a state program would be at least as rigorous as the federal rules and provide a consistent standard for its program. And if the state eco-label is successful, it should increase commerce in organically produced goods and create a market for food products that go beyond the federal standards.[149] The eco-label would not directly discriminate against food products from other states. A state could not limit the eco-label to local food or food from the label's home state in a form of intrastate economic protectionism, though presumably the label would be attractive to intrastate farmers and producers.[150] The label would likely neutrally and indirectly discriminate on the basis of distance in terms of average food miles required to get the product to market. In fact, producers from all states might seek the label of a single state certification since it would meet federal standards, be more rigorous, and potentially become preferred by environmentally-minded consumers. Consumers often prefer cheese "made in Wisconsin," Florida oranges, Vermont maple syrup, Maine lobster, and Washington apples—why, then, not prefer "Organically and Environmentally Certified in the State of X?" According to the U.S. Court of Appeals for the First Circuit

---

142. 7 U.S.C. §6507(a) (2010) ("The governing State official may prepare and submit a plan for the establishment of a State organic certification program to the Secretary for approval. A State organic certification program must meet the requirements of this chapter to be approved by the Secretary.").
143. *Id.* §6507(b).
144. *See also* Int'l Dairy Foods Ass'n v. Boggs, 2:08-CV-628-29, 2009 WL 937045, at *15 (S.D. Ohio Apr. 2, 2009), *aff'd in part, rev'd in part*, Nos. 09-3515, 09-3526, 2010 WL 3782193 (6th Cir. Sept. 30, 2010) ("The OFPA allows states to create a plan for the establishment of a State organic program only if the plan is submitted to and approved by the Secretary of Agriculture. Such a plan may contain more restrictive requirements governing the organic certification of farms and handling operations and the production and handling of agricultural products that are to be sold or labeled as organically produced and must further the purposes of the Act, not be inconsistent with the Act, and not be discriminatory towards agricultural commodities organically produced in other States." (internal citations omitted)).
145. Council Reg (EC) 834/2007 of 28 June 2007 at Art. 25 ("National and private logos may be used in the labelling, presentation and advertising of products which satisfy the requirements set out under this Regulation."); Art. 34 ("Competent authorities, control authorities and control bodies may not, on grounds relating to the method of production, to the labelling or to the presentation of that method, prohibit or restrict the marketing of organic products controlled by another control authority or control body located in another Member State, in so far as those products meet the requirements of this Regulation. In particular, no additional controls or financial burdens in addition to those foreseen in Title V of this Regulation may be imposed. . . . Member States may apply stricter rules within their territory to organic plant and livestock production, where these rules are also applicable to non-organic production and provided that they are in conformity with Community law and do not prohibit or restrict the marketing of organic products produced outside the territory of the Member State concerned.").
146. 7 C.F.R. §205.501(21)(b)(2).
147. 7 U.S.C. §6507(b)(2).
148. *Id.* §6501.
149. An analogy can be drawn to California's ability to set different emissions standards under the Clean Air Act, and the interest of other states, in order to better serve the public health and environment of its citizens, in using these more rigorous standards than federal ones.
150. Similarly, a state eco-label program would need to be a voluntary labeling program so as to avoid any dormant commerce clause concerns putting undue impediments on interstate commerce, and concerns that a mandatory eco-label would make food producers reluctant to supply a given state.

in the case *Harvey v. Veneman*, the OFPA allows and encourages competition in developing more stringent organic standards.[151]

A disadvantage of a new state organic certification program is that the certification process cannot be relaxed, since the statute only authorizes "more restrictive" organic certification rules. This poses a problem for small farmers and producers that would wish to meet the organic criteria but may lack the resources to go through the certification process. The OFPA only provides an exemption for "persons who sell no more than $5,000 annually in value of agricultural products."[152] But small and local farmers could gain an advantage through an environmental life-cycle eco-label since it will consider greenhouse gas emissions in both production and food miles.

While state entities serve as USDA organic certifiers, none have exercised their authority to develop more rigorous standards than those required by federal law. At present, there are 19 states that are state organic certifiers,[153] but none have applied under the OFPA's §6507 to certify organic food under more restrictive substantive state standards under a state program. Most states simply manage organic certification programs by adopting the NOP standards.[154] While states have sought more restrictive programs from USDA, these standards are not substantive and merely create additional procedural requirements such as registration for organic food producers and private certifiers and applicable fee tables.

For example, California applied to have a state organic program with additional procedural requirements not identified in the OFPA.[155] California requires registration by organic producers through the submission of public information sheets, maintenance of detailed records (about livestock history, substances applied to fields, and agricultural practices), and following labeling rules that prohibit the use of the terms "transitional organic" and "organic when available." It also defines how organic producers can describe the percentage of organic ingredients.[156] While no substantive standards under the California Organic Products Act of 2003 are more rigorous than those of the OFPA and NOP, the Act does create room for rules about organic products not subject to federal organic certification rules.[157] The state of Washington established similar procedural requirements regarding registration[158] but also developed its own separate standard for mushrooms.[159] No specific federal organic standard yet exists explicitly for mushrooms, and mushroom farmers generally use the standard organic crop regulations, which are less applicable to mushroom harvesting.

---

151. Harvey v. Veneman, 396 F.3d 28, 45 (1st Cir. 2005) ("OFPA further provides that state certification programs may be more restrictive than the federal program. This provision, incidentally, allows for the type of competition developing more stringent organic standards sought by Harvey. . . ." Furthermore, "nothing in the challenged regulation prevents private certifiers from making truthful claims about the products they certify; it only bars such certifiers from applying more stringent requirements as a condition of use of their USDA accredited certifying mark" (internal citations omitted)). However, the OFPA likely bars the use of the word organic by private certifiers that have more restrictive standards, as opposed to state certifiers that apply for more restrictive standards under 7 U.S.C. §6507.

152. 7 U.S.C. §6505(d) (2010).

153. The states that have organic certification programs through state agencies are California, Colorado, Iowa, Idaho, Kentucky, Louisiana, Maryland, Mississippi, Montana, Nevada, New Hampshire, New Jersey, New Mexico, Oklahoma, Oregon, Rhode Island, Texas, Utah, and Washington. U.S. Dep't of Agric., Agric. Mktg. Serv., *National Organic Program* (Nov. 12, 2010), http://www.ams.usda.gov/AMSv1.0/ams.fetchTemplateData.do?template=TemplateJ&navID=NationalOrganicProgram&leftNav=NationalOrganicProgram&page=NOPACAs&description=USDA%20Accredited%20Certifying%20Agents&acct=nopgeninfo.

154. *See, e.g.*, Cal. Dep't of Food and Agric., *California Organic Program*, http://www.cdfa.ca.gov/is/i_&_c/organic.html (last visited Nov. 14, 2010) (incorporating the NOP regulations by reference in Title 3 California Code of Regulations Article 6.1).

155. After passing the California Organic Products Act in 2003, the state applied to have a state organic program and was approved in 2004, despite not meeting all the criteria, because the secretary of agriculture thought compliance would eventually occur. A March 2010 audit report, however, found that California is still not in compliance with the NOP. Thus, although they applied and have operated some version of an organic certification program, California's program is not fully operational and cannot be considered more stringent than the federal program from a substantive or procedural standpoint. *See* Office of Inspector Gen., U.S. Dep't of Agric., Oversight of the National Organic Program, Audit Report 01601-03-Hy 2, 4, 14-16, 20 (March 2010); Miguel A. Caceres, U.S. Dep't of Agric., Livestock and Seed Program Audit, Review, and Compliance Branch Quality System Audit Report, NP3140MA NC Annual Update Report MCCO Salinas CA 1-2 (June 17, 2003).

156. Cal. Food & Agric. Code §46013 (2010) (requiring public information sheet); *id.* §46028 (records requirement); *id.* §46024(h) (prohibiting use of term "transitional organic"); *id.* §46027 (prohibiting use of term "organic when available"); Cal. Health & Safety Code §110838 (2010) (defining how to describe the percentage of organic ingredients).

157. Cal. Health & Safety Code §110835 ("The director may adopt regulations allowing or prohibiting the use of substances in the processing of products that are exempt or excluded from certification under the NOP, and animal food and cosmetics sold as organic."). To this end, California has adopted rules for cosmetic products.

158. Wash. State Dep't of Agric., Organic Food Program, Organic Rules and Regulations 58-81, (2008), *available at* http://agr.wa.gov/FoodAnimal/Organic/Certificate/2008/ProcessorHandlerRetailerBroker/NewApplicant/OFPOrganicRulesandRegs1_08.pdf.

159. Wash. Admin. Code §16-157-120 (2010).

In any case, no state so far has created more restrictive substantive organic standards.[160] While federal organic certification has too narrow a definition, the word "organic" has built up marketing power that cannot be ignored.[161] An improved eco-label could do more to inform consumers about environmentally friendly foods, and could incorporate the value of the term "organic" if properly conforming to the OFPA. Such an eco-label might accomplish precisely what people already think the organic label does, and if producers could not meet the more restrictive standards, they could still simply use the existing USDA organic label.

3. Environmental Life-Cycle Analysis

In food labeling, the choice between environmental life-cycle analysis or a best-practices approach must be considered. Environmental life-cycle analysis is more difficult and costly to generate,[162] but would provide greater information than a best-practices approach.[163] Given that progress has already been made in creating best-practices standards (see the discussion of KFM and KRAV above), how might environmental life-cycle analysis for food be pursued?

Food eco-labels can be based on an assessment of the food's life cycle: its raw materials, production process, distribution, use, and disposal, including consideration of pollution, waste, and carbon footprint. "The main objective of eco-labeling programs is to harness market forces and channel them towards promoting more environmentally friendly patterns of production."[164] However, effective eco-labeling of food requires accurate and verifiable information, and it must provide life-cycle information on production, processing, and distribution. Consumers must have access to aggregated information that considers the chemical additives, land stewardship practices, and fossil fuel consumption required to bring any food to market.

An effective environmental life-cycle eco-labeling system for food would inform consumers about the environmental costs of their food purchases and provide a baseline comparison for food in different categories. An eco-label seal should be available for products within a food category meeting defined environmental criteria. While eco-labels would be based on a technical assessment of a product's life cycle providing consumers with a visual seal, products also could list descriptive information of interest such as location of production or carbon footprint.

Outside the food industry, many life-cycle labeling schemes already exist. For example, the European Union's voluntary flower logo program indicates products that are more environmentally friendly than conventional products based on a life-cycle ecological assessment.[165] But the European Union flower logo eco-label is not used for food, and the European Union is only beginning to consider what life-cycle analysis for food would look like, as discussed below. The European Union uses five administrative layers to implement its eco-label scheme, and has developed product groups and ecological criteria to harmonize environmental labeling in its member countries.[166] The eco-label can be affixed to those products that meet established product group criteria for the entire life cycle of the product.

---

160. State implementation and requests for additional procedural elements to implement national organic standards shed little light on the potential success or failure in developing a more expansive eco-labeling system, except that some states have struggled to implement and comply with the NOP.

161. *See* Marvin T. Batte, *Putting Money Where Their Mouths Are: Consumer Willingness to Pay for Multi-Ingredient, Processed Organic Food Products*, 32 Food Pol'y 145 (2007).

162. Effective Approaches to Environmental Labelling of Food Products (Univ. of Hertfordshire), http://randd.defra.gov.uk/Document.aspx?Document=FO0419_9996_FRP.pdf ("Our principal conclusion from the work that has been undertaken in this project is that we do not believe that the science is sufficiently robust to develop an outcome-based, environmentally broad, omni-label at this time. Additionally, the costs that such a scheme may incur could be unacceptably high in relation to the potential benefits that could be realised.").

163. *Id.* ("The lack of Type I, II and III food eco-labels can be explained by a number of factors including the diversity of products and production systems; the complexities of determining environmental impacts; issues involved in communicating environmental information to consumers via product labels and the use of labels by consumers, including matters of trust, preferences and motivations; and the lack of evidence showing that labels can help deliver environmental benefits (as it is difficult to differentiate the influence of labels amongst many other drivers, such as regulatory or other market influences).") (" As such, the labels that are placed on food products from farms achieving certification under these schemes do not directly imply that environmental benefits have been achieved. This approach is generally less costly and more practical than an outcome-based approach, and can be appropriate where the objectives of a scheme do not rely on measuring product-specific impacts or communicating these to consumers. Indeed, in communicating environmental performance to consumers, it is important that the limitations of any scheme are respected and perhaps more importantly not exaggerated.").

164. Surya P. Subedi, *Balancing International Trade With Environmental Protection: International Legal Aspects of Eco-Labels*, 25 Brook. J. Int'l L. 373, 375 (1999).

165. *See* European Commission Environment, *EU Eco-label*, http://ec.europa.eu/ environment/ecolabel/index_en.htm (last visited Nov. 14, 2010).

166. Julian Morris, Green Goods?: Consumers, Product Labels and the Environment 42 (1997).

The flower logo, however, has met only limited success since the label is still widely unknown and only beginning to gather traction with consumers.[167] This lends additional support for the argument to continue using the already established cache of "organic" in pursuing more rigorous food labeling options. The European Union flower logo program is an ambitious project since its goal is to introduce one eco-label that would eventually replace all national labels within the European Union, including those on food.[168] In July 2008, the European Commission presented a proposal to widen the scope of European Union eco-labeling efforts, "taking in the particularly complex food and drink market."[169] There is no doubt that life-cycle eco-labeling for food is ambitious. According to the Environmental Audit Committee in the United Kingdom House of Commons:

> Attempts to reach lifecycle footprints even for basic products can result in complex calculations based on a highly hypothetical average usage.... [A] carrot could be eaten raw, cooked in a microwave, or boiled in a pan of water. It is difficult to see how any in use measurement for food and drink products could ever be of genuine use to a consumer, whereas labels allowing them to select locally-produced or organically grown carrots could engage their interest and have a significant impact in at least one environmental dimension.[170]

In addition to continuing movement in the European Union, individual European countries have led in the creation of eco-labels outside the food sector with the Nordic Council Program (of Norway, Sweden, and Finland) and Germany's Blue Angel Program.[171] In the latter, an environmental label jury—comprised of representatives from environmental groups, science organizations, consumer associations, industry, trade unions, and the media—reviews life-cycle reports to determine if the "Unweltzeichen" (environmental label) is appropriate.[172]

Germany's program, the oldest eco-labeling program in Europe, is perhaps the most successful as studies have shown that German consumers make frequent and continuous use of the eco-label as a means of obtaining product information and shopping accordingly.[173] Given the success of eco-labeling in Germany and Scandinavia, one concern about any state-sponsored eco-label in the United States is whether it could achieve a degree of success only in a geographic location with a relatively high environmental consciousness among its population.[174] That might prove beneficial if the state-sponsored label is developed by a state like Vermont or Oregon, where there is a high level of ecological awareness and a reputation for environmentalism.[175] Like building on the "organic" label, the state could build on its own "green" reputation, perhaps even generating state revenue by certifying the most environmentally friendly food products in the country.

The European Commission, with the support of the United Nations Environmental Programme, European Environment Agency, and experts of several member states, is currently laying the groundwork for a life-cycle eco-labeling scheme for food with the creation, in 2009, of the European Food Sustainable Consumption and Production Round Table. The agenda of the Round Table is to use environmental assessment methods to "examine key sustainability challenges along the food value chain (e.g., climate change, water conservation, resource efficiency and waste reduction) and develop adequate strategies to address them."[176] The Round Table seeks to establish reliable life-cycle environmental assessment methodologies for foods, and determine the best way to supply information to consumers to enable them to make informed choices.[177] More specifically stated, the key objectives of the Round Table are to:

---

167. Gertz, *supra* note 139, at 128.
168. *Id.*
169. U.K. HOUSE OF COMMONS, ENVIRONMENTAL AUDIT COMMITTEE—SECOND REPORT OF SESSION 2008-09: ENVIRONMENTAL LABELLING, ¶ 21 (Mar. 3, 2009), *available at* http://www.publications.parliament.uk/pa/cm200809/cmselect/cmenvaud/243/24302.htm (last visited Nov. 14, 2010).
170. *Id.* ¶ 62.
171. Other public and private eco-labels include Green Seal, Sweden's Bra Miljöval (Good Environmental Choice), Canada's EcoLogo; Japan's Eco-Mark. Also see the ISO 14024 standards for eco-labelling, http://www.iso.org/iso/catalogue_detail.htm?csnumber=23145.
172. Subedi, *supra* note 164, at 378.
173. Gertz, *supra* note 139, at 136.
174. *See id.*
175. Brian Wingfield & Miriam Marcus, *America's Greenest States*, FORBES (Oct. 17, 2007), http://www.forbes.com/2007/10/16/environment-energy-vermont-biz-beltway-cxbw mm1017greenstates.html (referring to the top three "greenest" states of Vermont, Oregon, and Washington as being "synonymous with environmentalism.").
176. Joint Press Release, Key Food Chain Partners to Launch Sustainability Roundtable 1 (Feb. 26, 2009), http://www.ciaa.eu/documents/press-releases/ PREFSCPRT final260209.pdf.
177. Press Release, European Food Sustainable Consumption and Production Round Table, European Food SCP Round Table Welcomes 14 New Member Organisations 1 (Dec. 9, 2009), http://www.ciaa.eu/documents/press_releases/Food%20SCP%20RT%20Press%20 Release%20

(1) Identify scientifically reliable and uniform environmental assessment methodologies for food and drink products, including product category specifications where relevant, considering their significant impacts across the entire product life cycle;

(2) Identify suitable communication tools to consumers and other stakeholders and develop guidance on their use, looking at all channels and means of communication;

(3) Promote and report on continuous environmental improvement along the entire food supply chain and engage in an open dialogue with its stakeholders.[178]

In early 2010, the Round Table drafted a document laying down guiding principles to develop "a harmonised framework methodology for the environmental assessment specifically of food and drink products."[179] The document lists seven key questions to be researched in order to create its methodology.

(1) How to measure, verify, collect and consolidate environmental information along the entire food chain in an efficient way?

(2) How to consider the various environmental aspects and/or impacts of the production and consumption of different categories of food and drink products in a consistent framework methodology?

(3) How to consider specificities of highly diverse food and drink products with different beneficial and adverse environmental impacts at different stages of their life cycle?

(4) What costs and benefits are involved as well as what challenges are the various food chain operators, including SMEs [small and medium enterprises], facing or going to face in this respect?

(5) How should a uniform environmental assessment methodology be designed in order to support the identification of continuous environmental improvement potentials at all stages of the food chain?

(6) How effective are existing and emerging environmental information tools along the food chain and vis-à-vis the consumer? What kind of information is relevant for consumers? What type of questions could we and should we expect consumers and food chain partners to have now and in the near future? How can consumer confusion be avoided?

(7) What is already available at the European and international level to help assess and communicate the potential environmental impacts associated with the production and consumption of food and drink products?[180]

These questions are important because, at present, no commonly applied methodology exists to assess and communicate environmental information along the food chain, including to consumers.[181] The Round Table's goal structure and future methodology may eventually provide a replicable avenue for creating a comprehensive eco-labeling in the United States.

4. Implementing an Eco-Labeling Program

In practicality, implementing a state-sponsored (or federally legislated) organic certification program and eco-label based on best-practices standards or environmental life-cycle analysis is no small task. An eco-label informational and certification scheme can provide engaged consumers with a measurable analysis created by experts, and provide a single point of product comparison for the less engaged consumer. How might an eco-labeling scheme be implemented?[182]

FINAL%20091209.pdf.
178. *Id.* at 2.
179. European Food Sustainable Consumption and Production Round Table, Voluntary Environmental Assessment and Communication of Environmental Information Along the Food Chain, Including to Consumers: Guiding Principles 2-3 (Mar. 15, 2010), *available at* http://www.foodscp.eu/files/consultation/FoodSCPRTGuidingPrinciplesforConsultation.pdf.
180. *Id.*
181. *Id.* at 1.
182. For a discussion of a potential eco-label model, see Morris, *supra* note 166, at 30-34. *See also* Czarnezki, *supra* note 1.

First, a group of experts, under the direction of a state agency, must pick food categories that have significant adverse environmental impacts and where, therefore, eco-labels would make a significant improvement to the environment.[183] These categories might include meats and seafood, pesticide-intensive produce like berries, spinach and potatoes, and heavily processed foods. For example, research on carbon footprinting shows that there are product categories that have high variability in footprints within a single category, so it makes sense to inform consumers about these differences, as it "will give them genuine options that make a difference" since "consumers need options, not just information."[184]

Second, an environmental life-cycle analysis methodology and/or best-practices standards must be developed and used, including consideration of natural resource and chemical inputs (starting at the production process or raw extraction stage), and emissions and pollution output during the production, distribution, use, and disposal stages. The key is to inventory materials that make up food and allow for food production, but equally important—and more difficult—to determine is how to inventory their environmental impact. As stated earlier in regard to the Carbon Trust, British PAS 2050 is a publicly available specification for assessing product life-cycle greenhouse gas emissions for goods and services, and perhaps it could be modified to apply to food as well. Food miles, the distance food travels from farm to table, should also be considered. To calculate how far a food product travelled, the most commonly used tool is a weighted average source distance (WASD), a single figure that combines information on the distances from production to point of sale and the amount of food product transported.[185]

No widely accepted environmental life-cycle assessment methodology for food exists. To determine the key food categories and the environmental footprint of food products, a state could use the Round Table's process as a model to determine the appropriate environmental life-cycle methodology, asking the same questions and using the same principles, but giving particular consideration to the American domestic market and analyzing existing informational regulatory tools in the United States.

Third, products must be evaluated according to scientific criteria and a seal awarded to those products meeting or surpassing a designated benchmark. It is important to determine what factors influence the success of any eco-labeling program. In other words, what labels work? It is hard to overemphasize the importance of first identifying what food categories would help the environment if their carbon, chemical, and waste footprints were reduced. What is also known is that centralized government eco-labels are more effective than numerous private ones, and that simple, clear, obvious and transparent seal-of-approval logos and labels have generally shaped consumer behavior more than the complex information-disclosure labels.[186]

Rather than simply requiring products to meet certain criteria to be eligible for a particular seal or logo, it might be possible to require "environmental product declarations" (EPD) similar to nutritional facts currently required under the Nutrition Labeling and Education Act of 1990. EPDs are "industry-created statements containing a variety of information about the composition and environmental characteristics of a product based on life-cycle assessment."[187] This approach would inform consumers about a wide range of life-cycle environmental concerns associated with the product, such as water usage, chemicals usage, pollution and carbon emissions, and waste disposal. Presumably, the environmental characteristics listed would be those categories of most significance as part of the environmental life-cycle analysis methodology.

---

183. Karl Johan Bonnedahl & Jessica Erikkson, *The Role of Discourse in the Quest for Low-Carbon Economic Practices: A Case of Standard Development in the Food Sector*, EUROPEAN MGMT. J. (2010) ("KRAV felt that "[t]he label should build in climate impact within broad categories but not distinguish the categories themselves. . . . Somewhat inconsistent however, under the heading 'what to do awaiting the [climate] label on its homepage, KRAV did advice consumers to eat less meat, giving the example that meat causes CO_2 emissions that may be 67 times higher than beans do.").
184. Tom Berry et al., *Check-Out Carbon: The Role of Carbon Labelling in Delivering a Low-carbon Shopping Basket*, F. FUTURE 7, 12 (June 2008), *available at* http://www.forumforthefuture.org/files/Checkout%20carbon%20 FINAL300608.pdf.
185. ØRESUND FOOD NETWORK & ØRESUND ENV'T ACAD., *supra* note 23, at 8-9.
186. *See* Abhijit Banerjee & Barry D. Solomon, *Eco-Labeling for Energy Efficiency and Sustainability: A Meta-Evaluation of US Programs*, 31 ENERGY POL'Y 109 (2003); Berry et al., *supra* note 184, at 5.
187. Nancy J. King & Brian J. King, *Creating Incentives for Sustainable Buildings: A Comparative Law Approach Featuring the United States and the European Union*, 23 VA. ENVTL. L.J. 397, n.232 (2005) (citing EUROPEAN COMMISSION, SUMMARY OF DISCUSSIONS AT THE 2ND INTEGRATED PRODUCT POLICY EXPERT WORKSHOP: ENVIRONMENTAL PRODUCT DECLARATIONS (ISO 14025 Technical Report) 2 (2001), *available at* http://ec.europa.eu/environment/ipp/pdf/epd.pdf (last visited Nov. 14, 2010)).

Unlike an eco-label seal, an EPD alone would disclose information "in a neutral way that enables consumer evaluation but that does not seek to judge the environmental characteristics of a product."[188]

Part of developing an eco-label for food is determining how to best convey information to consumers in a manner that will effectively shift buying preferences. For example, would a logo or seal for products that meet a particular environmental standard in addition to an EPD label overwhelm consumers with information?[189] In addition, eco-labels require a good quality assurance scheme, which would benefit from governmental ownership of the label, and a successful marketing program.[190] Absent unlikely federal legislation, a state-sponsored label—"Organically and Environmentally Certified in the State of X"—can embody these characteristics.

Conclusion

The ecological consequences of the modern diet are simply too high. Produce is farmed with inorganic fertilizer and pesticides, processed foods are made in factories that burn fossil fuel, and food miles have increased as consumers can buy anything in any season from anywhere on the planet. This path is simply unsustainable, and it is unhealthy for people and the environment in which we live.

Development of the organic food market in the United States and worldwide has been, for the most part, a positive development. The marketing and economic success of the NOP and the Organic Foods Product Act will likely continue to expand the industrial organic food market in the United States until it becomes a dominant market. Other nations, like Sweden, have moved a step ahead, taking environmental protection into consideration when establishing national dietary guidelines, and attempting to incorporate greenhouse gas emissions into their organic labeling scheme.

But we must go even further. The European Commission is already assembling experts to design an effective environmental life-cycle assessment methodology and label scheme for food. Absent federal legislation overhauling the national organic certification program or scrapping that program entirely in favor of a more sustainable food eco-label, an American state with a strong reputation for environmental awareness should, within the boundaries of the national organic certification program, develop a new food eco-label approach that conveys a wider array of environmental information to consumers.

Improved eco-labeling is only a start. In addition to improving labeling schemes and legal policies to support environmentally-friendly food consumption, the market for available food products must be improved. Legal policies and marketing should better support local, low-input, and nonindustrial unprocessed food markets through streamlined organic certification for small farmers, low-carbon diets, community-supported agriculture, farmers' markets, and increased consumer access to sustainable food products. The industrial food sector will continue to shift to organic production (to the point where perhaps organic food rivals the conventional food market). Demand for value-added products (i.e., those with the organic label, or with an environmental life-cycle label in the future) will increase. With these trends, improved eco-labeling regimes will enhance consumer awareness by revealing the environmental costs of consumer purchases, and will create shifts in consumer choice and, consequently, the norms of food production and distribution by farmers and corporations.

---

188. *Id.*
189. I note that a mandatory EPD labeling requirement, unlike a voluntary eco-label seal, would most certainly require federal legislation.
190. *See* Helen Nilsson et al., *The Use of Eco-Labeling Like Initiatives on Food Products to Promote Quality Assurance—Is There Enough Credibility?*, 12 J. Cleaner Production 517 (2004).

Chapter 18
Into the Future:
Building a Sustainable and Resilient Agricultural System for a Changing Global Environment
Mary Jane Angelo

During the time this book was being written, the world's population crossed the threshold of seven billion inhabitants.[1] While many of the more affluent countries, including the United States and most of the European Union, were muddling through severe economic turmoil, parts of the developing world continued to experience unprecedented improvements in their standard of living. As the global population grows, demand for food and fiber will continue to increase. As populations in some countries, particularly many in Asia, become more affluent, their diets will become more varied and they will consume more animal products. These changes will inevitably add increasing stressors to an already precarious global food system.

To complicate matters, all of these changes are occurring against the backdrop of immense uncertainty surrounding not only worldwide financial stability but also global climate change. One recent sign that agriculture may be faced with changing climactic conditions is the recent revision of the U.S. Department of Agriculture's (USDA's) Plant Hardiness Zone Map,[2] which shows a shift in most areas to warmer zone designations.[3] As we move forward into an uncertain future, our global and national food systems must adapt to changing conditions.

As described in this book, our modern industrial agricultural system has a number of significant shortcomings. The current form of centralized industrial agriculture is a major contributor to many problems, including air and water pollution, inefficient energy use, climate change, loss of biodiversity, and human health effects. It is not the type of sustainable and resilient system that will be needed to ensure food security in a world with a growing population and complex and unpredictable modifications likely to occur as a result of global climate change.

A. The Link Between Agriculture and Climate Change

The academic and popular literature is filled with discussions of the link between carbon emissions and climate change, and the potential global harms that are likely to occur as a result.[4] According to most scientists, no environmental problem in human history is as potentially harmful as the climate change crisis.[5] The vast majority of scientists predict that without dramatic and timely reductions in releases of carbon into the atmosphere, various global climatic changes will occur that will make all other environmental

---
1. *See* Sam Roberts, *U.N. Says 7 Billion Now Share the World*, N.Y. Times, Oct. 31, 2011, at http://www.nytimes.com/2011/11/01/world/united-nations-reports-7-billion-humans-but-others-dont-count-on-it.html?_r=1.
2. USDA Plant Hardiness Zone Map, http://planthardiness.ars.usda.gov/PHZMWeb/ (last visited Nov. 26, 2012).
3. USDA is careful to point out that the new map should not be used as evidence of climate change. *See* http://planthardiness.ars.usda.gov/PHZMWeb/AboutWhatsNew.aspx.
4. *See, e.g.*, Intergovernmental Panel on Climate Change, Climate Change 2007: The Physical Science Basis, Summary for Policymakers 10 (2007), *available at* http://www.ipcc.ch/pdf/assessment-report/ar4/wg1/ar4-wg1-spm.pdf (stating that most of the increase in global temperatures is very likely attributable to greenhouse gas (GHG) concentrations).
5. *See* Raymond B. Ludwiszewski & Charles H. Haake, *Climate Change: A Heat Wave of New Federal Regulation and Legislation*, Fed. Law., June 2009, at 32 (explaining that global climate change is currently the top environmental concern).

crises pale in comparison.⁶ Anticipated consequences of climate change include future warming,⁷ increased frequency of heat waves,⁸ increased heavy precipitation in some areas,⁹ increased droughts,¹⁰ more intense tropical storms,¹¹ and rises in sea level.¹²

Climate change and agriculture are closely linked in several ways. The likely changes in temperature and rainfall patterns as a result of climate change have the potential to dramatically impact worldwide food production.¹³ Conversely, current agricultural practices are themselves significant contributors to greenhouse gas (GHG) emissions that are fueling climate change.¹⁴ Agriculture is therefore in the unique position of both contributing to climate change and having the potential to mitigate some of climate change's impacts.

Strategies for managing the impacts of climate change can generally be grouped into two broad approaches: mitigation strategies and adaptation strategies. Climate change mitigation is often referred to as "avoiding the unmanageable" through policies that seek to reduce net GHG emissions. Climate change adaptation, on the other hand, is referred to "managing the unavoidable" impacts that will result as the globe warms. The majority of strategies that have been employed to address the impacts of climate change have primarily fallen under the category of mitigation. Nevertheless, research suggests that even if the atmospheric concentrations of GHGs could be stabilized through mitigation measures, climate change's impacts on agricultural production will continue without stabilizing for some time after GHG emissions reach equilibrium.¹⁵ It is therefore critical that policymakers focus not only on mitigation but also on adaptation in order for agriculture to respond to the impacts of global climate change. Meeting the global population's future food supply demands within the context of climate change will require both mitigation in the form of policies that reduce fossil fuel inputs and GHG emissions from agriculture and policies that encourage the development of agricultural systems that are resilient enough to be able to adapt to the likely changes that will occur.

B. Agriculture's Contribution to Climate Change

As described throughout this book, high-intensity industrial agriculture has a large "carbon footprint." Inputs such as pesticides and fertilizers that are relied on in industrial agriculture are derived from fossil fuels.¹⁶ Nitrogen fertilizers are made from natural gas,¹⁷ and most synthetic pesticides are made from fossil fuels.¹⁸ Fossil fuels, especially diesel and gasoline, are used for heavy machinery such as tractors and combines, as well as for transportation of agricultural products to processing facilities and ultimately to retail grocery stores.¹⁹ Agriculture accounts for approximately 20% of U.S. fossil fuel consumption as well as 15% of worldwide GHG emissions.²⁰ It is estimated that it takes "[10] calories of petroleum to yield just one calorie of industrial food" and about two-thirds of a gallon of gasoline to produce one bushel of corn.²¹

---

6. *See* Linda R. Larson & Jessica K. Ferrell, *Precautionary Resource Management and Climate Change*, NAT. RESOURCES & ENV'T, Summer 2009, at 51, 52.
7. According to the Intergovernmental Panel on Climate Change (IPCC) Report, it is "[v]irtually certain" (>99% probability of occurrence) that future warming will occur. INTERGOVERNMENTAL PANEL ON CLIMATE CHANGE, CLIMATE CHANGE 2007: SYNTHESIS REPORT 53 (2007), *available at* http://www.ipcc.ch/pdf/assessment-report/ar4/syr/ar4_syr.pdf. For explanation of the probability terminology, see *id*. at 27.
8. According to the IPCC Report, it is "[v]ery likely" (>90% probability of occurrence) that there will be an increased number of heat waves. *Id*.
9. According to the IPCC Report, it is "[v]ery likely" (>90% probability of occurrence) that there will be increased heavy precipitation in some areas of the globe. *Id*.
10. According to the IPCC Report, it is "[l]ikely" (>66% probability of occurrence) that there will be an increased number of droughts. *Id*.
11. According to the IPCC Report, it is "[l]ikely" (>66% probability of occurrence) that there will be more intense tropical storms. *Id*.
12. According to the IPCC Report, it is "[l]ikely" (>66% probability of occurrence) that there will be increased incidents of high sea level. *Id*.
13. *See* Christina Ross et al., *Limiting Liability in the Greenhouse: Insurance Risk-Management Strategies in the Context of Global Climate Change*, 43A STAN. J. INT'L L. 251, 297-98 (2007).
14. William S. Eubanks II, *A Rotten System: Subsidizing Environmental Degradation and Poor Public Health With Our Nation's Tax Dollars*, 28 STAN. ENVTL. L.J. 213, 269-70 (2009).
15. *See* Steven K. Rose & Bruce A. McCarl, *Greenhouse Gas Emissions, Stabilization and the Inevitability of Adaption: Challenges for U.S. Agriculture*, 23 CHOICES, 1st Quarter 2008, *available at* http://www.choicesmagazine.org/2008-1/theme/05.pdf.
16. *See id.; see also* Peter Warshall, *Tilth and Technology: The Industrial Redesign of Our Nation's Soils*, *in* FATAL HARVEST: THE TRAGEDY OF INDUSTRIAL AGRICULTURE 221, 225 (Andrew Kimbrell ed., 2002).
17. *See id.*
18. Warshall, *supra* note 16, at 225.
19. Eubanks, *supra* note 14, at 10504; Warshall, *supra* note 16, at 225.
20. Eubanks, *supra* note 14, at 10504.
21. *Id*. (citing DANIEL IMHOFF, FOOD FIGHT: THE CITIZEN'S GUIDE TO A FOOD AND FARM BILL 102 (2007)) (internal quotation marks omitted).

Another significant agricultural GHG contributor is methane production.[22] Animals, particularly cows that are kept in confined feeding operations and fed large quantities of corn and other grains, emit substantial amounts of methane gas.[23] Methane is a GHG that has been demonstrated to be approximately 20 times more powerful than carbon dioxide in exerting a greenhouse effect.[24] While methane gas is obviously a natural waste product produced by animals, the enormous quantities of methane gas produced in modern agriculture are directly attributable to the sheer numbers of animals in confined feeding operations, which would not exist if not for cheap corn and soy production.[25]

A complicating factor arises from the search for alternative renewable fuels, which has resulted in substantial increases in corn ethanol production. With the intense focus on both climate change and the desire for domestic energy independence in recent years, scientists and policymakers have searched for alternative energy sources that could be produced domestically and that would not contribute to climate change to the extent that fossil fuels do. One of the major alternative energy supplies heavily subsidized by the federal government is corn ethanol.[26] Production of corn ethanol has increased from approximately 175 million gallons in the early 1980s to almost 6.5 billion gallons in 2007.[27] In 2010, approximately 44% of all U.S. corn production was used for ethanol production, up from approximately 20% just five years prior.[28] Often touted as a "renewable" or "alternative" energy,[29] the use of ethanol as a major source of fuel is not without controversy.[30] As described in Chapter 2, U.S. policy continues to promote corn ethanol as an alternative fuel source despite the fact that scientific studies consistently demonstrate that reliance on corn ethanol will not help to solve the climate change crisis and poses additional environmental and social problems.[31] The rapid acceleration in corn ethanol production is at least in part attributable to the heavy subsidies that have been provided since the 1970s.[32] Although the U.S. Congress allowed corn ethanol subsidies to lapse at the end of 2011, other programs still in effect continue to encourage its production. Specifically, the Energy Policy Act of 2005 and the Energy Independence and Security Act of 2007 created a renewable fuel standard which provides a ready market for corn ethanol.[33] These policies contribute to the ongoing practice of growing large-scale monocultures of industrialized corn.

C. Climate Change Impacts on Agriculture

In addition to being a major contributor to climate change, industrial agriculture is also vulnerable to the likely impacts of climate change. Climate change has the potential to greatly impact global food security as its effects become more prevalent.[34] The effect of changing weather patterns on the volume and quality of global and regional food production will greatly impact food availability and food accessibility. Research suggests that the most significant changes in precipitation and temperature will be in the world's poorest

---

22. *See id.*; Joshua A. Utt et al., *Carbon Emissions, Carbon Sinks, and Global Warming*, in Agricultural Policy and the Environment 151, 156 (Rodger E. Meiners & Bruce Yandle eds., 2003).
23. William S. Eubanks II, *The Sustainable Farm Bill: A Proposal for Permanent Environmental Change*, 39 ELR 10493, 10504 (June 2009).
24. *Id.*
25. *See id.*
26. *See* Gary D. Libecap, *Agricultural Programs With Dubious Environmental Benefits: The Political Economy of Ethanol*, in Agricultural Policy and the Environment, *supra* note 22, at 89 (explaining that ethanol has received over $10 billion in subsidies).
27. James A. Duffield et al., *Ethanol Policy: Past, Present, and Future*, 53 S.D. L. Rev. 425, 425 (2008); *see also* Karl R. Rabago, *A Review of Barriers to Biofuel Market Development in the United States*, 2 Envtl. & Energy L. & Pol'y J. 211, 212 (2008) (describing the remaining barriers to full commercial success for biofuels in the United States).
28. USDA Economic Research Service, *Corn*, http://www.ers.usda.gov/topics/crops/corn/background.aspx (last visited Jan. 29, 2013).
29. *See, e.g.*, Growth Energy, *About Growth Energy*, http://www.growthenergy.org/about-growth-energy/ (last visited Jan. 29, 2013).
30. *See, e.g.*, Christopher Jensen, *Caution Flags Raised Over Ethanol Industry's 15% Solution*, N.Y. Times, May 10, 2009.
31. *Id.*
32. *See* Wallace E. Tyner, *The U.S. Ethanol and Biofuels Boom: Its Origins, Current Status, and Future Prospects*, 58 Bioscience 646, 646 (2008); *see also* Robert W. Hahn, *Ethanol: Law, Economics, and Politics*, 19 Stan. L. & Pol'y Rev. 434, 437-45 (2008) (describing how federal subsidies have driven the development of the ethanol fuel industry in the United States); Libecap, *supra* note 26, at 89.
33. Energy Policy Act of 2005, Pub. L. No. 109-58, 119 Stat. 594 (2005); Energy Independence and Security Act of 2007, Pub. L. No. 110-140, 121 Stat. 1492 (2007). *See also* Mark Holt & Carol Glover, Cong. Research Serv., Energy Policy Act of 2005: Summary and Analysis of Enacted Provisions 100 (2006), *available at* http://lugar.senate.gov/energy/links/pdf/Energy_Policy_Act.pdf; Fred Sissine, Cong. Research Serv., Energy Independence and Security Act of 2007: A Summary of Major Provisions 6 (2007), *available at* http://www.seco.noaa.gov/Energy/2007_Dec_21_Summary_Security_Act_2007.pdf.
34. The United Nations Food and Agriculture Organization (FAO) defines "food security" as existing when "all people at all times have physical or economic access to sufficient safe and nutritious food to meet their dietary needs and food preferences for an active and healthy life." The FAO definition encompasses four dimensions: food availability, food accessibility, food utilization, and food systems stability. FAO CC & Food Security, 3.

and most vulnerable regions.[35] Climate change's disproportionate impacts on the livelihoods and food security of the poor will present significant challenges as we struggle to meet the ever-increasing global population's demands for food and resources.

Closer to home, climate change has the potential to alter growing seasons and change the kinds of crops and crop varieties that can be grown in particular regions of the United States. Changes in rainfall patterns are likely to result in increased droughts in some areas and increased flooding in others, and creating the need for new or different water management practices in many agricultural regions.[36] Moreover, probable increases in insect pest damage, weeds, and crop disease have the potential to alter agricultural production in the United States in ways not yet fully understood. Sea level rise and saltwater intrusion will cause problems for agriculture in some coastal regions. All of these impacts of climate change on agriculture will indirectly impact human health. For example, if crop yields are reduced, food prices and child malnutrition are likely to rise.[37]

There is considerable uncertainty about the extent to which crop yields are likely to increase or decrease in a warming climate. Changes in crop yield will vary dramatically by geographical region depending on whether a particular locale experiences changes in rainfall and other conditions. However, general trends are very difficult to predict. While it may seem intuitive that warmer temperatures will lead to increased yields due to longer growing seasons and lower vulnerability to frost, the reality is much more complex. The U.S. Climate Change Science Program has reported that for numerous reasons decreased yields are likely for many crops including corn, rice, and sorghum. For example, as changes in rainfall result in less water availability in many areas, longer growing seasons will require increased water for irrigation. Many studies suggest that weeds, pest insects, and diseases are likely to increase, all of which could adversely affect crop yield.[38] Studies also show that increased carbon dioxide levels promote weed growth. Warmer temperatures promote increased pests and diseases and can create hospitable conditions that result in new pests and diseases moving into areas previously inhospitable due to cold temperatures. Moreover, there are likely to be many unanticipated effects. For example, one study shows that growing soybeans in increased carbon dioxide environments results in dramatic increases in pest damage to soybean plants. Other studies indicate that some weed control chemicals that are widely relied upon, such as glyphosate, will lose their efficacy in an environment with elevated carbon dioxide.

Most studies predict that the likely impacts of climate change on agriculture over the coming century will be both positive and negative. The type and intensity of the impacts will vary by location. The production practices and crop types grown in a particular region will determine how well-suited that region is to adapt to the likely changes.[39] Many experts believe that the likely negative impacts on developing regions of the world could be slightly offset by some limited positive impacts in developed regions.[40] Thus, while the aggregate overall impact to global food production may not be large, many regions may suffer significant impacts.[41]

D. Adapting to Climate Change

Strategies for adapting agriculture to climate change include a mix of technological and institutional policy changes, and they typically distinguish between changes that could be made at the individual and institutional levels. Technology-based proposals include increased development of crop varieties; innovations in resource management including water conservation measures; development of forecasting systems; development of improved irrigation systems; and changes in land use and timing of planting.

Institutional and policy proposals typically involve increased government support for agriculture, including subsidies and incentives. Although technological fixes and increased government support may be

---

35. Susan Charles, *Climate Change: Impacts on Food Safety*, 26 SUM NAT. RESOURCES & ENV'T 44 (2011).
36. *Id.*
37. INTERNATIONAL FOOD POLICY RESEARCH INSTITUTE, FOOD POLICY REPORT—CLIMATE CHANGE: IMPACT ON AGRICULTURE AND COSTS OF ADAPTATION (2009).
38. Simon N. Gosling, *A Review of Recent Developments in Climate Change Science. Part II: The Global-Scale Impacts of Climate Change*, 35(4) PROGRESS PHYSICAL GEOGRAPHY 451-53 2011.
39. Charles, *supra* note 35, at 44.
40. *Id.*
41. *Id.*

necessary components of any climate change adaptation plan, these types of proposals for piecemeal fixes ignore the need to look at the entire farming system to ensure it will be able to adapt to the inevitable and unpredictable changes that will occur. Of particular concern is the fact that our current industrial agriculture system has become so highly dependent on large fossil fuel inputs that, if such inputs become scarce or too costly, farm production will plummet. Equally as significant, through this type of large-scale farming, we have eliminated the natural ecosystem functions that make ecosystems more resilient and better able to adapt to changing conditions. For farming systems to be able to readily adapt, they will need to become more ecologically resilient.

1. Ecological Resilience

Ecological resilience has been described as "a measure of the amount of change or disruption that is required to transform a system from being maintained by one set of mutually reinforcing processes and structures to a different set of processes and structures."[42] The concept of ecological resilience is based on the understanding that ecosystems can exist in multiple stable states.[43] Ecological resilience should not be confused with "engineering resilience," which is a measure of the time it takes for a system to return to a steady state after experiencing a perturbation."[44] In contrast, ecological resilience, a measure of the magnitude of a perturbation that a system can absorb before the disturbance, causes the system to shift into a different regime of behavior with different controlling processes.[45] As such, ecological resilience captures the strength of redundancies in an ecosystem stemming from reinforcing processes and compensating functions provided by more than one species. These redundancies enable the system to absorb disturbances and persist despite the disruption.[46] When applied to an agricultural system, ecological resilience is a measure of an agricultural system's ability to continue to function and provide yield despite changes or perturbations such as increased pest populations disease or changed rainfall patterns. By ensuring that the ecological resilience of an ecosystem, including an agricultural ecosystem, is maintained or reintroduced, it is more likely that the ecosystem will be able to withstand a greater range of perturbations without undergoing a shift to, for example, a nonproductive agricultural system.

To ensure that any ecological system, including an agricultural ecosystem, is resilient, a number of factors must be present. Research suggests that one of the most significant factors in increasing a system's ecological resilience is to increase its species richness. Because individual species are only able to perform limited ecosystem functions, the greater the species richness the greater functional diversity in the ecosystem.[47] The ability of a system to dampen the effects of perturbations depends in part on the extent to which one species can compensate for the loss of a function previously provided by another species.[48] Thus, to create a resilient agricultural system it will be necessary to ensure a sufficient amount of redundancy in ecosystem controlling processes, such that unexpected disturbances, whether anthropogenic or natural, can be absorbed without causing the system to shift states. This can only be accomplished by introducing biodiversity back into the agricultural system. The planting of large-scale monocultures strips farms of the diversity needed for ecological resilience. Monocultures by definition and design are comprised of only one crop variety, often spanning hundreds or thousands of acres. In contrast, alternative farming systems—where diverse numbers of crop types and varieties are planted in close proximity, where crops are rotated, or where natural refugia are provided on the farm—contain the biological diversity that reduces vulnerability to change. Not only does a variety of crops provide a safety net in case one crop is lost due to an outbreak of pests or disease, but by reducing the chemical inputs and providing natural refugia on site, natural populations of beneficial species such as predators and parasites of pest species, pollinators, soil microbes, and a

---

42. Garry Peterson, *Contagious Disturbance and Ecological Resilience*, at 216 (Ph.D. dissertation, Univ. of Florida, 1999).
43. *Id.* at 217.
44. C.S. Holling & Lance H. Gunderson, *Resilience and Adaptive Cycles, in* Panarchy: Understanding Transormations in Human and Natural Systems 28 (Lance H. Gunderson & C.S. Holling eds., 2002).
45. Lance H. Gunderson et al., *Resilience of Large-Scale Resource Systems, in* Resilience and the Behavior of Large-Scale Systems 4 (Lance H. Gunderson & Lowell Pritchard Jr. eds., Island Press 2002).
46. *Id.* at 6.
47. Peterson, *supra* note 42, at 209.
48. Gunderson et al., *supra* note 45, at 9.

diverse array of other organisms ensure that natural ecosystem functions are maintained and redundancies are built into the system to provide resilience.[49]

2. Building a Sustainable and Resilient Agro-Ecosystem

To ensure food security for a growing global population in a time of significant change, it will be necessary to build a more ecologically resilient agricultural system that contains the biodiversity, redundancies, and ecosystem functions that enhance its ability to adapt to new conditions. It will be necessary not only to have farms that are run in a more environmentally friendly, sustainable, and resilient way, but also to have food distribution systems that ensure food availability. Changes are needed both to environmental laws—which, as demonstrated in Chapters 8-12, are not adequate to address the environmental risks posed by agriculture, and to current agricultural policy—which, as shown in Chapters 1 and 2, encourage unsustainable, nonresilient, industrialized practices. Many of the proposals for change set forth in Chapters 14-17 of this book are the very types of changes that will help our agricultural system to be more sustainable and resilient.

Over the past few decades, scientists, policymakers, farm organizations, environmentalists, and others have called for new sustainable agricultural approaches to replace industrialized agriculture. One such approach, eco-agriculture, seeks to limit the harm to wildlife, biodiversity, and ecosystem services resulting from industrial agriculture by advancing farming practices that view the farm as a healthy sustainable living system rather than an industrial facility. The premise of eco-agriculture is that the farm is a kind of ecosystem—an "agro-ecosystem"—made up of soil, plants, insects, and animals. A healthy farm ecosystem will have healthy and fertile soils, and healthy populations of natural predators and parasites, pollinators, and other beneficial species. The farm maintains many ecosystem services that provide benefits beyond the farm, and it reduces the negative externalities of farming. Farming practices are geared toward maintaining and enhancing ecosystem health and function, while at the same time maximizing yields within the constraints of maintaining a healthy ecosystem. The agro-ecosystem mindset represents a shift in thinking away from the idea that human activity and functioning ecosystems are mutually exclusive.

In a recent report, the National Research Council of the National Academies described a "farming systems continuum," with conventional farming at one end and ecologically-based farming at the other and a potentially infinite set of combinations of farming practices falling somewhere in between.[50] The concept of eco-agriculture relies on modern knowledge about the interactions within natural systems, as well as cutting-edge technologies, to achieve its results. When done properly, it can produce high yields and profits for farmers while protecting human health, animal health, and the environment. Eco-agriculture draws on the key techniques of sustainable agriculture including crop rotation; cover crops; reduced tillage and no-till practices; soil enrichment; maintenance and enhancement of natural pest enemies; integrated pest management; precision farming that utilizes detailed spatial information about soil conditions and crop performance to target crop management practices to the specific place they are needed; diversification of farm enterprises, which helps increase biodiversity; other agricultural best-management practices such as buffer or filter strips and wildlife habitat enhancement; and enhanced genetic resistance to climatic extremes, pests, and other threats.[51] These techniques may be used in a variety of combinations depending on the specific circumstances of the farm at issue. These techniques not only help to build a sustainable and ecologically sound system, but more specifically help to increase resilience within the agricultural system by building diversity, thereby making the system more adaptable to change. Working farmlands can reap the benefits provided by nature in the form of pest control through predators and parasites by promoting biodiversity on the farmlands themselves. Biodiversity on the farm can be enhanced through reduced pesticide use, intercropping, crop rotation, and the creation and maintenance of refugia on or near the farm field. Crops on working farmlands can become more resistant to disease and pests through practices that

---

49. *See generally* Robin Kundis Craig, *"Stationarity Is Dead"—Long Live Transformation: Five Principles for Climate Change Adaptation Law*, 34 Harv. Envtl. L. Rev. 9 (2010) (describing some activities that promote resilience in agricultural systems).
50. National Research Council, Toward Sustainable Agriculultural Systems in the 21st Century 20-23 (2010) (NRC Report).
51. *Id.* at 21.

maintain the health and fertility of soils, such as rotating crops and planting legumes and other cover crops that improve soil fertility.[52]

In recent years, attempts to adopt eco-agricultural approaches by using various combinations of the above-described techniques have been made throughout the world. A number of examples have been cited as eco-agriculture "success stories." For example, in parts of Central America, farmers have integrated trees into livestock pastures to provide habitat for forest birds. Also in Central America, farmers have had success with planting biodiversity-friendly coffee plantations. In California, a type of eco-agriculture has been implemented by flooding rice fields for birds during fallow seasons. In Indonesia, there are several successful examples of "agro-forests" that integrate agricultural and forestry, as has the integration of rice terraces with fish and vegetables. Each of these examples has been successful in some measure by introducing a level of biodiversity into the farming system. A number of domestic sustainable farming success stories are outlined in the 2010 NRC Report. Most of the farms in those stories share certain commonalities. As described by the NRC, one such commonality is that "[m]any farmers emphasized the importance of maintaining or building up their natural resources base and maximizing the use of internal resources as key parts of their farming strategies."[53]

The report emphasized the use of various combinations of farming approaches in each of the successful farms and also described how successful farms tended to readily adapt to new information.[54] In fact, many of the successful farms either carried out their own trials and experiments or participated in research conducted by universities "because they recognize the importance of adapting their farming approaches to local conditions."

The reliance on specifically tailored combinations of ecologically based practices combined with the active participation in experimenting with and adapting farming practices to local conditions have led to the types of ecologically based sustainable farming that will be more resilient and better able to adapt to climate change.

E. Policy Solutions

The chapters in Part IV of this book set forth a number of proposals geared toward reforming U.S. law and policy in order to build the type of environmentally responsible, sustainable, and resilient agricultural and food system that will be able to adapt to the inevitable global climate changes and likely political and economic challenges to food security. No one proposal will solve all of our problems, but each of these chapters contains proposals that independently and in the aggregate can help to move us in the direction of sustainability and resilience. In Chapter 14, J.B. Ruhl proposes paying farmers to do the "new right thing." His proposal would provide financial incentives to farmers to encourage the preservation and maintenance of ecosystem services on working farmlands. This proposal recognizes that farms can serve as houses of "natural capital capable of providing a diverse stream of goods and services, including ecosystem services such as increased biodiversity, carbon sequestration, pollination, groundwater recharge, and improvement of water quality." By paying farmers to preserve the ecosystem services that are critical to the maintenance of a healthy environment, we are also encouraging farmers to engage in practices that produce more sustainable and resilient eco-agricultural systems.

In Chapter 15, William Eubanks proposes a number of revisions to the farm bill that would eliminate or shift subsidies that currently promote large-scale industrialized farming in favor of more sustainable and ecologically sound agricultural practices. The reforms offered by Eubanks not only would achieve many of the same agricultural benefits as those that would result from paying farmers to preserve ecosystem services, but they would also promote sustainability, resilience, and food security in a broader sense by enhancing the ability of smaller farms to compete, and by strengthening local food systems and rural communities. Removing subsidies to large industrial growers that allow them to undercut smaller growers and providing more flexibility in what farmers can grow while receiving subsidies will have the added benefit of promoting local and regional food systems capable of producing a range of healthy foods close to home.

---

52. *Id.* at 94-110.
53. *Id.* at 355.
54. *Id.* at 397.

This, in turn, will provide greater food security and infrastructure development for rural communities, which have generally been devastated by decades of agricultural policies encouraging farm aggregation that has resulted in dwindling rural populations.

George Kimbrell, in Chapter 16, focuses on one increasingly important aspect of industrialized farming—the growing of genetically modified crops, which has become a predominant force in global agriculture in recent years—and offers suggestions for better regulatory oversight to address the potential risks associated with planting genetically modified organisms (GMO) in the environment. The chapter outlines the major regulatory gaps in GMO law and describes the need for more precautionary approaches to ensure environmental protection, human health and safety, and economic stability. The fact that many scientists have identified new GMO crop varieties that are resistant to disease, pest, drought, or salinity as important tools in adapting to climate change underscores the need for an effective, transparent, and precautionary regulatory system to be in place before climate change impacts press us to introduce even more GMO crops into the ecosystem.

Finally, in Chapter 17, Jason Czarneski explores the use of food labeling as a means of achieving a more sustainable and resilient agricultural system. He stresses the importance of individual behavior and the ability of consumer choice to influence the way in which food is produced, processed, and distributed. Czarneski compares a number of food labeling regimes and proposes a new "eco-labeling" program that would employ environmental life-cycle analysis and best-practices standards to ensure consumers are aware of the full range of environmental and health implications of the foods they purchase. Well-informed consumers who demand that their food be produced, processed, and distributed in ecologically sustainable ways can provide market incentives that will encourage more ecologically sound agricultural practices, which will be more resilient to the challenges likely to accompany climate change.

Conclusion

Each of the proposals set forth in Part IV of this book provides important ideas for shifting our current industrialized agricultural system to a more sustainable and resilient system. While none of these proposals alone can solve all of the environmental, health, social, and economic issues related to agriculture, they each offer significant paths to improvement. Moreover, when viewed in combination, these proposals provide a comprehensive roadmap for developing the type of sustainable and resilient food system that will be needed to adapt to the impacts of climate change and to provide food security for a growing and changing global population.

Index

Agricultural Exemption, 159, 177, 178, 179, 205, 218

Air quality, 23, 82-86, 88-89, 119, 121, 124, 164-66, 174, 176-77, 179-81, 183, 270, 301, 303

Ammonia, 69, 72, 75, 79, 81-85, 87-89, 157, 163, 165-74, 176, 179, 181, 195

Antibiotic resistance, 77, 85, 235

Antibiotics, 74-75, 77-78, 106, 109, 115, 203-204, 228, 231, 285, 306

Bioconcentration/Bioaccumulation, 37, 42-43, 104, 204

Biodiversity, 23, 25, 28-29, 39, 41, 47, 65, 73, 75, 79, 83, 90, 96-98, 111, 120, 122, 185, 193, 195, 228-29, 242, 267, 289, 308, 325, 329, 330-31

Biofuel(s), 13, 25-30, 33, 213, 327

Biotechnology, 27, 38, 43-44, 86, 93-94, 99, 100-01, 109, 111, 220, 281-84, 286-89, 296, 299

Budget, 1, 7, 9, 14, 25, 31-33, 83-84, 126, 226, 236-37, 240, 271

Cancellation, 40, 43, 135, 136, 137, 143

Child nutrition, 223, 233, 234, 235, 237, 238, 239

Clean Air Act (CAA), 163-69, 171-72, 174-79, 181-83, 316-17

Clean Water Act (CWA), 44-45, 58, 69-70, 76, 141-42, 147-61, 171-72, 188, 195, 204-05, 218-19, 241

Climate change, 25-26, 29-30, 33, 48, 61-62, 65, 73, 82, 86, 88-90, 98, 113-14, 121, 124-28, 183-85, 198, 201, 240-41, 270, 304, 309, 320, 325-28, 330-32

Commodity crops, 6, 16, 18, 20, 29-30, 33, 35, 114-15, 117, 119, 123-24, 128, 234, 236-37, 266, 268-70, 274, 283

Comprehensive Environmental Response, Compensation, and Liabilty Act (CERCLA), 169-75, 177

Concentrated animal feeding operation (CAFO), 38, 45, 67-69, 74, 80-82, 85, 87-89, 91, 150, 155-58, 163, 167, 169, 173-74, 181-82, 203-04, 219, 241, 270, 304-05

Congress, 3-9, 11, 13-14, 25, 29, 31-32, 53, 61, 74, 82, 84, 108, 115, 131, 134-35, 142, 144, 149-51, 153-54, 164-65, 168-69, 177, 179, 185-88, 191, 194, 196, 204-05, 207, 210-14, 217-19, 222, 226-27, 229, 232, 234-36, 238, 241, 243, 248, 259, 264-66, 268-78, 281, 285, 287, 289-90, 315-16, 327

Conservation, 3, 6-15, 17, 19, 21-25, 27-33, 35-36, 39-41, 47, 49-50, 54-55, 62, 67, 74, 76, 81, 90, 96, 115, 118, 121, 123, 157, 160-61, 179, 185-88, 202, 206, 212, 215-17, 228, 232, 238-39, 241-43, 247-49, 251, 253, 256-58, 260, 264-67, 269, 271-78, 289, 303-05, 309, 320, 328

Conservation compliance, 32, 249, 274

Conservation Reserve Program (CRP), 7, 21-22, 25, 90, 215-17, 242-46

Conservation Stewardship Program (CSP), 23-24, 234, 242, 264-67, 275-77

Coordinated Framework, 99, 105, 107, 111, 281, 286-87, 293-94, 299

Cosmetic standards, 225

Cost/Benefit Balancing, 131, 138, 140, 142, 145

Countercyclical payment(s), 9, 11, 16, 18-21, 274

Crop insurance, 3, 11-12, 14, 19, 24, 29, 216-17, 271-74

Cultural Control, 50, 129

Dead zone(s), 46, 121, 185, 195, 197

Decoupling, 3, 8-9, 15

Direct payment(s), 9, 11, 16-21, 32, 216, 270-72, 274

Ecosystem services, 41, 241-48, 250-53, 255-61, 330-31

Endangered Species Act (ESA), 39, 41, 138, 142-45, 185-206, 209, 220, 241

Energy, 13-15, 17, 20, 23-33, 48-49, 72, 78-79, 81, 109, 113-14, 117, 120-21, 124, 127-28, 182, 212-13, 232, 238, 241-43, 245-46, 260, 264-65, 267, 270-72, 276-78, 284, 289, 295, 304, 308-09, 311, 313, 320, 322, 325, 327

Environmental justice, 65, 68, 73, 90, 217

Environmental Quality Incentives Program (EQIP), 9, 23, 29, 216, 242

Emergency Planning and Community Right-to-Know Act (EPCRA), 169-77

Eutrophication, 44-48, 75-76, 81-83, 148, 195-97, 204

Factory farm, 163, 165, 167-79, 181-84, 204

Family farm, 2, 3, 5, 10-11, 30, 66, 198, 229, 269, 272, 283, 307

Farm Bill, 1-16, 18-27, 29-33, 115, 119, 123-24, 185, 192, 201, 204, 206, 211-17, 223, 232, 235, 237, 239, 241-43, 245, 248-49, 259-60, 263-79, 289, 301, 304, 309, 326-27, 331

Federal Food, Drug, and Cosmetic Act (FFDCA), 100, 105-07, 110-11, 135, 224-25, 227

Federal Insecticide Fungicide and Rodenticide Act (FIFRA), 44, 99-100, 130-38, 140-45, 151-52, 195, 201-03, 205, 217-18, 224, 287, 293, 298

Fertilizer(s), 5, 22, 25, 30, 35-36, 38, 42, 44-50, 58, 73, 75, 79, 81, 96, 115, 117, 119-23, 128, 148-49, 167-68, 176, 191-92, 194-98, 204, 228, 231, 264, 267, 270, 303-06, 312-13, 323, 326

Food and Drug Administration (FDA), 77, 99-100, 103, 105-11, 204-05, 221-22, 224-27, 286-87, 295-96

Food miles, 114, 127, 228, 232, 306, 308-09, 315, 317-18, 322-23

Food safety, 76, 93, 100, 108, 111, 183, 194, 219-21, 223-24, 226-27, 237, 240, 266-67, 269, 281, 283-86, 288-91, 297-98, 328

Genetically engineered (GE), 43-44, 93-95, 98, 103, 105-10, 192, 205, 219, 229, 281-82, 284-90, 292-94, 296-98, 307

Greenhouse gas, 23, 48, 82, 85-87, 96, 113-17, 121, 123-26, 128, 163, 166-67, 177, 181-83, 228, 301-02, 304-06, 308-14, 316, 318, 322-23, 325-26

Herbicides, 5, 37, 40, 48-49, 93-98, 101-02, 120, 129, 192-93, 201, 218, 281, 283-86, 291, 293-95, 299, 304

Hydrogen sulfide, 69, 82, 84, 88-89, 163, 165, 166-74, 176, 179-81

Income Supports, 4, 14-15, 242

Insecticide(s), 36-38, 41, 48-49, 95-96, 98, 100, 120, 129-30, 137, 141, 151, 195, 201, 217, 224-26, 287, 293-94

Irrigation, 5, 24, 44-45, 49, 51-55, 57-62, 72, 80, 88, 96, 116-19, 121, 123, 128, 141, 147-48, 150-51, 198-201, 292, 304, 328

Labeling, 97, 100, 102-03, 105, 108-12, 130, 132-33, 135-36, 139, 141, 206, 223-24, 228-33, 240, 267, 281, 293, 301-03, 306-23, 332

Local food systems, 227, 236, 273, 331

Marketing loan(s), 7, 9, 16, 19, 20-21, 216, 272

National Environmental Policy Act (NEPA), 100-01, 106-07, 194, 207-08, 210-22, 289-90, 297

National Organic Program, (NOP) 29, 228-31, 267, 306-07, 315-16, 318-19, 323

Natural Resources Conservation Service (NRCS), 22, 24, 157, 160-61, 264, 267, 274-75

National School Lunch Program (NSLP), 4, 223-40, 278

Nitrogen, 5, 38, 42, 44-48, 73, 75-76, 81-84, 86-88, 119-24, 139, 148-49, 157, 164, 194-95, 264, 267, 303, 305, 313, 326

Nonrecourse loans, 19, 21

Nuisance, 46, 73, 88, 164, 180, 193

Nutrient(s), 1, 29, 36, 38, 44-47, 50, 54, 58, 67, 70-76, 81-83, 88-89, 97, 104, 118, 120-23, 148-49, 157-58, 194-97, 203, 235-37, 242, 245-46, 249, 258, 267-68, 276, 303-05

Nutrition, 3, 6, 8, 29, 103-04, 106, 110-11, 179, 217, 223, 233-40, 243, 264-65, 267-68, 270, 273, 276, 278-79, 283, 306, 309, 311-12, 322, 328

Organic, 6, 13- 14, 21, 24, 27, 29-31, 33, 37-39, 41, 45-48, 74-75, 77, 79, 82-85, 88, 97-98, 102, 114-18, 120-21, 123-24, 127-28, 157, 166-68, 183, 206, 217, 223, 227-31, 236, 245, 264-65, 267-69, 273, 276-78, 281, 284-85, 289, 291, 294-95, 298, 301-03, 306-08, 310-14, 315-21, 323

Organochlorine(s), 36-37, 39, 41-43, 121, 303

Organophosphate(s), 36-37, 40, 43, 78

Particulate matter, 82, 84-85, 163-66

Pesticide(s), 5, 25, 29-30, 35-45, 47-50, 54, 58, 74-75, 78, 88, 95-96, 98-100, 102, 115, 117, 119, 120-24, 128-45, 148-49, 151-53, 167-68, 191-95, 201-03, 205, 217-18, 223-26, 228-29, 231, 235, 246, 266, 268, 270, 283-87, 291, 293-94, 297-98, 301-04, 306-07, 311, 322-23, 326, 330

Phosphorus, 44, 47-48, 73, 75-76, 78, 81, 109, 149, 157, 194, 305

Plant Protection Act (PPA), 100, 193-95, 219-21, 281-82, 287, 290-99

Planting flexibility, 9, 15, 17, 271-73

Price Supports, 4, 6, 8, 10, 15

Prior Converted Cropland, 160-61

Registration, 37, 40, 130-38, 141-44, 201-03, 217-18, 224, 227, 286, 294, 296-97, 318

Runoff, 22, 43-47, 50, 54-58, 60, 62, 68, 74-75, 78-81, 96, 118, 121-22, 148-50, 156-57, 160, 192, 194, 196-97, 246, 266, 291, 303-05

Species, 22-23, 35, 37, 39-47, 52, 61, 68, 73-75, 79, 82, 84, 86, 90, 94-95, 97-99, 103-07, 109, 129, 132, 138-40, 142-45, 148, 185-205, 209, 220-21, 241, 243, 250, 258, 260, 281, 284, 292-93, 298, 301, 303, 305, 329-30

Subsidies, 4-5, 7-9, 12-17, 25-26, 30, 32-33, 35, 53, 62, 214, 223, 229, 236-37, 241-43, 245-47, 250, 253, 259-60, 264-67, 269-73, 327

Suspension, 134-37

Tolerances, 135, 224, 286

Total maximum daily loads (TMDLs), 156

Trade, 1, 5, 6, 8-9, 12, 14-15, 17-18, 20, 29-33, 66, 74, 107, 115, 141, 213-14, 217, 227, 230, 233, 238, 243, 247-48, 253, 256-58, 261, 268, 276, 282, 306, 316, 319-20

Transportation, 28, 48, 78, 82-83, 113-14, 127, 164, 226, 264, 266-67, 270, 277, 297, 301, 303-04, 309, 313-14, 326

U.S. Department of Agriculture (USDA), 2-3, 6, 9-10, 12-16, 18-24, 26, 29, 36, 38, 43, 65, 67, 69, 72-73, 75, 77, 81, 93-94, 99, 100-02, 105, 118, 147, 157, 160-61, 170-71, 183, 192-93, 201, 212-17, 219-22, 224-26, 228-39, 248-49, 253, 260, 264, 266-68, 270, 272-78, 281-84, 286-99, 303, 306-07, 312, 315, 318-19, 325, 327

U.S. Environmental Protection Agency (EPA), 37, 69, 99, 130, 147, 163, 195, 224, 286, 315

Water quality, 8, 23-24, 45, 47, 52, 54, 58, 67, 69-70, 73-76, 78-83, 117, 121, 142, 147-49, 151, 155-56, 161, 197-98, 242-43, 258, 260-61, 331

Water quantity, 23, 147, 198

Wetlands, 8, 21-23, 25, 27, 45, 47, 56-57, 59-60, 148-49, 151, 153-55, 158-61, 209, 219, 242, 252-54, 257, 260, 274

Workers, 37, 83-84, 88-89, 102, 125, 139, 168